Advanced Materials Engineering Fundamentals

Richard Skiba

AFTER MIDNIGHT
PUBLISHING

Skiba, Richard (author)

Advanced Materials Engineering Fundamentals

ISBN 978-1-7638440-5-6 (Paperback) 978-1-7638440-6-3 (eBook) 978-1-7638440-7-0 (Hardcover)

Non-fiction

Contents

Chapter 1: Introduction to Advanced Materials Engineering   1

  What are Advanced Materials?   1

  Historical Evolution of Material Science   7

  Role of Advanced Materials in Modern Industries   9

  The Scope of Advanced Materials Engineering   15

Chapter 2: The Science Behind Advanced Materials   17

  Atomic and Molecular Structure of Materials   17

  Mechanical, Thermal, and Chemical Properties   42

  Crystallography and Material Defects   53

  Material Characterization Techniques   69

Chapter 3: Composite Materials   112

  Definition and Types of Composites   112

  Matrix and Reinforcement Components   129

  Processing Techniques for Composites   133

  Applications of Composites in Aerospace, Automotive, and Construction   158

Chapter 4: Nanomaterials   160

  Introduction to Nanotechnology and Nanomaterials   160

  Synthesis Methods for Nanomaterials   166

  Properties and Behaviour at the Nanoscale   172

  Applications in Electronics, Medicine, and Energy Storage   174

Chapter 5: Bioplastics and Sustainable Materials   176

  Overview of Bioplastics and Their Types   176

  Polymer Chemistry and Biodegradability   181

  Production Methods for Bioplastics   190

  Industrial Applications and Environmental Impact   198

Chapter 6: Material Development, Testing and Characterization   200

  Developing a New Advanced Material   200

Mechanical Testing: Tensile, Compression, and Fatigue 213

Thermal Analysis: DSC, TGA, and DMA 233

Chemical and Spectroscopic Characterization: XRD, SEM, and TEM 241

Non-Destructive Testing Techniques 243

Chapter 7: Sustainability and Environmental Considerations 265

Lifecycle Analysis of Advanced Materials 265

Recycling and Upcycling Strategies 271

Reducing Carbon Footprint in Material Production 275

Future Trends in Sustainable Material Design 285

Chapter 8: Computational Materials Engineering 287

Role of Computational Modelling in Material Design 287

Chapter 9: Challenges in Advanced Materials Engineering 342

Economic Viability and Scalability 342

Overcoming Technical Limitations in Material Properties 343

Ensuring Regulatory and Safety Compliance 346

Balancing Innovation with Sustainability 348

Chapter 10: Emerging Trends in Advanced Materials 351

Self-Healing Materials 351

Smart and Responsive Materials 363

Quantum Materials and Topological Insulators 364

Bio-Inspired and Biomimetic Materials 372

Chapter 11: The Advanced Materials Engineer's Toolkit 377

Essential Equipment for Material Design and Testing 377

Key Software Tools for Analysis and Simulation 394

Skills and Certifications for Advanced Material Engineers 396

Networking and Professional Organizations 398

Chapter 12: Career Pathways and Opportunities 402

Educational Background for Advanced Materials Engineers 402

Industries and Roles for Advanced Materials Professionals 403

Trends in Employment and Salaries 405

Entrepreneurial Opportunities in Material Innovation   408

References   411

Index   466

Chapter 1

Introduction to Advanced Materials Engineering

What are Advanced Materials?

Advanced materials are a class of engineered substances that exhibit superior properties or functionalities compared to traditional materials, allowing for their application in specialized sectors across various industries. These materials are often the result of significant advancements in material science, chemistry, and engineering, leading to innovations that enhance performance, sustainability, and application-specific designs.

The 21st century has witnessed unprecedented progress in material science. Scientists are now able to manipulate substances at an atomic level, creating purpose-built materials that vastly outperform naturally occurring counterparts. These innovations have been successfully integrated into high-tech manufacturing, medical procedures, renewable energy, and even food production.

Despite these advances, we are still at the threshold of a new technological revolution. The coming decades are expected to bring transformative leaps in material science, comparable to or exceeding the innovations of the Industrial Revolution. The primary constraints are the limits of physical laws as we currently understand them and the reach of human imagination— though the latter may soon be expanded by artificial intelligence.

Advanced materials represent a convergence of curiosity and capability. The desire to understand and manipulate the world around us has driven human progress for millennia, from alloying metals in ancient times to today's atomic-scale innovations. As research and development continue to expand, advanced materials will remain at the forefront of technological revolutions, shaping industries, improving lives, and redefining what is possible in the 21st century and beyond.

Advanced materials are characterized by their enhanced performance, which includes superior strength, stiffness, and toughness compared to traditional materials. These materials are engineered to endure extreme conditions such as high temperatures, corrosion, and wear, making them suitable for demanding applications across various industries, particularly aerospace and automotive sectors. For instance, advanced aerospace materials like superalloys and carbon-fibre composites exhibit remarkable physical properties that allow them to withstand the rigorous demands of flight [1, 2]. Additionally, the use of metal matrix composites (MMCs) has gained traction due to their lightweight and enhanced mechanical properties, which are crucial for applications where performance and weight are critical [3].

The specialized properties of advanced materials can be finely tuned to meet specific needs. This includes tailoring their thermal, optical, or electrical characteristics to achieve desired outcomes. For example, the development of advanced polymer composites has been pivotal in various applications, including aerospace and automotive, where specific mechanical properties are required [4]. Moreover, materials such as carbon nanotubes (CNTs) and MXenes are being explored for their multifunctional capabilities, including high conductivity and flexibility, which are essential for applications like electromagnetic interference (EMI) shielding [5, 6]. These materials can also be engineered to respond dynamically to external stimuli, such as light or heat, enhancing their functionality in applications like soft robotics and wearable technology [7, 8].

Sustainability is increasingly becoming a defining feature of advanced materials. Many of these materials are designed to be biodegradable or recyclable, significantly reducing their environmental impact throughout their lifecycle. The aerospace industry, in particular, is under pressure to adopt more sustainable practices, leading to the development of materials that align with eco-friendly solutions [9]. For instance, the use of 3D printing technologies in ceramics and polymers allows for the creation of complex structures while minimizing waste, thus contributing to sustainability efforts [10, 11]. Furthermore, the integration of nanotechnology in material development has opened avenues for creating materials that not only perform well but also adhere to environmental standards [12].

Lastly, advanced materials are often application-specific, tailored to meet the unique demands of various industries. In aerospace, lightweight components made from advanced composites are essential for improving fuel efficiency and performance [1, 2]. In electronics, materials with enhanced conductivity are critical for the development of next-generation devices [13, 14]. Similarly, in healthcare, biocompatible materials are being engineered for implants and medical devices, showcasing the versatility and application-specific nature of advanced materials [15, 16]. This customization ensures that advanced materials deliver optimal performance tailored to the specific requirements of their respective fields.

Advanced materials can be categorized into several distinct groups, each characterized by unique properties and applications. This synthesis will explore the categories of composites, nanomaterials, biomaterials, metamaterials, smart materials, high-performance polymers, and ceramics and glasses, supported by relevant literature.

Composites are materials formed by combining two or more distinct materials to achieve superior properties. For instance, carbon fibre-reinforced polymers (CFRPs), example shown as Figure 1, and glass-reinforced plastics (GRPs) are widely recognized for their high strength-to-weight ratios and durability. These composites are extensively used in aerospace, automotive, and sports equipment industries due to their lightweight nature and enhanced mechanical properties [17-19]. The versatility of composites allows for tailored properties that meet specific application requirements, making them a crucial component in modern engineering [19].

Figure 1: Small piece of Laminated Uni-directional Carbon Fibre. Simon.white.1000, CC BY-SA 3.0, via Wikimedia Commons.

Nanomaterials are engineered at the nanoscale, typically ranging from 1 to 100 nanometres. Examples include graphene, carbon nanotubes (see Figure 2), and quantum dots, which exhibit remarkable electrical, thermal, and mechanical properties. These materials have significant applications in electronics, medical imaging, and energy storage systems due to their unique characteristics that differ from their bulk counterparts [18, 20]. The manipulation of materials at the nanoscale opens up new avenues for innovation in various fields, particularly in enhancing the performance of electronic devices and improving energy efficiency [18].

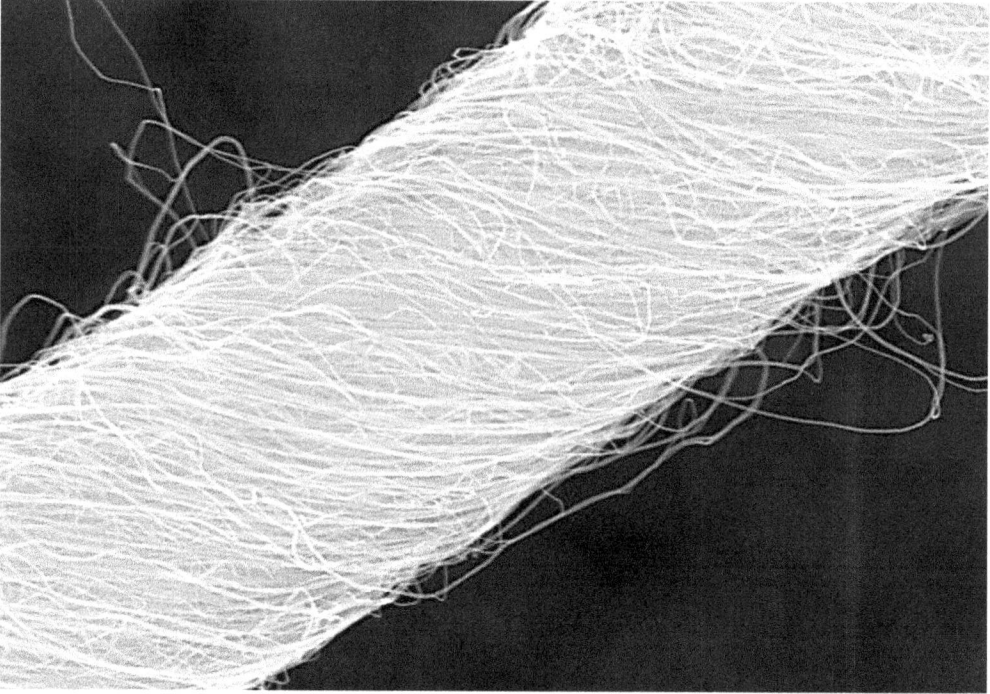

Figure 2: Carbon nanotubes spun to form a yarn. The yarn contains hundreds of thousands of fibres in cross section. Each fibre is one ten-thousandth the diameter of a typical human hair. Carbon nanotube fibres are thermally and electrically conductive , can withstand extremes of temperature and are resistant to radiation-induced degradation. CSIRO, CC BY 3.0, via Wikimedia Commons.

Biomaterials are designed to be compatible with biological systems, making them essential in medical applications. Examples include biodegradable polymers (see Figure 3) and hydrogels, which are used in implants, drug delivery systems, and tissue engineering. The development of biomaterials focuses on ensuring biocompatibility and functionality, which are critical for successful integration into the human body [18, 20]. The ongoing research in this field aims to create materials that not only support biological functions but also promote healing and regeneration [18].

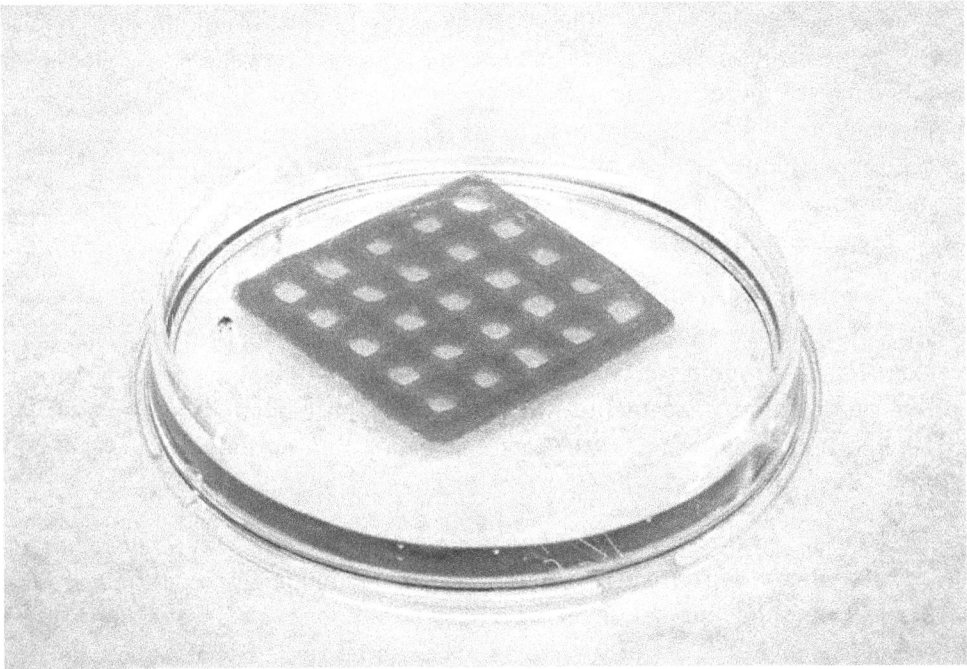

Figure 3: A "living material," made of a natural polymer combined with genetically engineered bacteria. This could offer a sustainable and eco-friendly solution to clean pollutants from water. David Baillot/UC San Diego Jacobs School of Engineering, CC BY-NC-SA 4.0, via Superinnovators.

Metamaterials are artificial materials engineered to have properties not found in nature, achieved through structural design rather than chemical composition. They can manipulate electromagnetic waves in novel ways, leading to applications such as cloaking devices, advanced optics, and antennas. For example, negative refractive index materials have been developed to create invisibility cloaks that can render objects undetectable to certain wavelengths of light [21-24]. The theoretical and experimental advancements in metamaterials have opened up possibilities for innovative applications in telecommunications and sensor technologies [25, 26].

Smart materials respond dynamically to external stimuli, such as temperature, pressure, or electric fields. Examples include shape-memory alloys and piezoelectric materials, which are used in sensors, actuators, and adaptive structures. These materials can change their properties in response to environmental changes, making them ideal for applications in robotics, aerospace, and medical devices [18, 19]. The integration of smart materials into systems enhances functionality and adaptability, paving the way for more responsive technologies [18].

High-performance polymers are characterized by exceptional thermal stability, mechanical strength, and chemical resistance. Notable examples include polyether ether ketone (PEEK) and Kevlar, which are utilized in medical devices, aerospace components, and protective gear. These polymers are engineered to withstand extreme conditions, making them suitable for demanding applications where reliability and performance are critical [18-20]. The development of high-performance polymers continues to evolve, focusing on enhancing their properties for specialized applications [18].

Ceramics and glasses are non-metallic materials known for their exceptional heat resistance and strength. Advanced ceramics and transparent ceramics are increasingly used in turbine engines, armour, and optoelectronics. Their ability to withstand high temperatures and corrosive environments makes them invaluable in various industrial applications [18-20]. Research in this area aims to improve the mechanical properties and processing techniques of ceramics, expanding their utility in advanced engineering applications [18].

The applications of advanced materials span various industries, each leveraging unique properties to enhance performance, efficiency, and sustainability. In the aerospace and automotive sectors, the integration of lightweight composites has become crucial for improving fuel efficiency. The use of materials such as carbon fibre and glass fibre composites significantly reduces vehicle weight, leading to lower fuel consumption and reduced carbon emissions. Studies have shown that lightweight materials can improve fuel efficiency without compromising structural integrity, with weight reductions of 40-60% compared to traditional metals like steel and aluminium [27-29]. Additionally, heat-resistant ceramics are increasingly utilized in engine components due to their ability to withstand high temperatures while maintaining mechanical strength, which is essential for enhancing engine performance and longevity [30].

Advanced materials are revolutionizing healthcare through the development of biodegradable implants and drug-delivery systems. These materials are designed to safely dissolve in the body, reducing the need for surgical removal and minimizing long-term complications [31]. Furthermore, antimicrobial coatings applied to prosthetics and implants are critical in preventing infections, enhancing patient outcomes. Research indicates that these coatings can significantly reduce bacterial adhesion and biofilm formation, which are common challenges in medical device applications [32].

In the energy sector, high-capacity batteries utilizing nanomaterials are at the forefront of technological advancements. Nanomaterials enhance the electrochemical properties of batteries, leading to improved energy density and faster charging times, which are vital for electric vehicles and renewable energy storage systems [33]. Photovoltaic materials for solar panels are also evolving, with new materials being developed to increase efficiency and reduce production costs. The integration of advanced materials in solar technology is essential for meeting global energy demands sustainably [34].

The electronics industry is experiencing a transformation with the introduction of conductive polymers and nanomaterials, which enable the creation of flexible displays and lightweight electronic devices. These materials offer significant advantages in terms of flexibility and durability, allowing for innovative designs in consumer electronics [35]. Additionally, advanced semiconductors are crucial for developing faster processors, enhancing computational power and energy efficiency in electronic devices [36].

In construction, self-healing concrete represents a significant advancement in materials technology. This innovative concrete can autonomously repair cracks, thereby extending the lifespan of structures and reducing maintenance costs. Studies have demonstrated that self-healing mechanisms, such as the incorporation of bacteria or microcapsules, can effectively restore the mechanical properties of concrete after damage [37-39]. Moreover, high-strength composites are used in infrastructure projects to improve durability and reduce the weight of structural components, contributing to more sustainable construction practices [40]. Insulating materials are also being developed to enhance energy efficiency in buildings, addressing the growing need for sustainable construction solutions [41].

Historical Evolution of Material Science

The evolution of materials has been a pivotal factor in defining various eras of human history, illustrating the profound relationship between material innovation and societal development. The Stone Age, Bronze Age, Iron Age, and Steel Age serve as historical markers that reflect how advancements in materials have shaped civilizations, economies, and technological progress. For instance, the transition from stone tools to bronze implements marked significant advancements in craftsmanship and trade, while the Iron Age introduced stronger and more durable materials that revolutionized agriculture and warfare [42, 43]. The understanding of materials science, which has roots in ancient practices such as metallurgy and ceramics, underscores humanity's long-standing quest to manipulate the physical world for practical applications [42, 44].

The late 19th century heralded a significant transformation in materials science, primarily due to the work of Josiah Willard Gibbs, who elucidated the relationship between thermodynamic properties and atomic structures. This foundational insight allowed scientists to understand how atomic behaviour influences material properties, setting the stage for modern materials science [45]. The implications of Gibbs' work were far-reaching, facilitating technological advancements during the 20th century, particularly during the Space Race, where innovations in metallic alloys and carbon-based materials were crucial for the development of spacecraft [44].

The early 20th century saw materials science largely focused on metallurgy and ceramics, reflecting the industrial priorities of the time. However, a paradigm shift occurred in the 1960s, driven by initiatives such as the Advanced Research Projects Agency (ARPA), which promoted interdisciplinary collaboration in materials research. This shift enabled researchers to not only utilize existing materials but also to design new materials tailored to specific properties, thereby expanding the scope and application of materials science [45]. The emergence of the Materials Genome Initiative further emphasized the importance of accelerating the discovery and deployment of new materials, which is essential for addressing contemporary challenges such as sustainable energy and advanced manufacturing [46].

As materials science evolved, it diversified to include a wide array of materials, such as polymers, semiconductors, and biomaterials. This diversification allowed the field to tackle a broader range of applications, from high-performance plastics to advanced electronic components [42, 43]. The integration of computational tools has further revolutionized the field, enabling researchers to simulate material properties and predict behaviours, which has led to breakthroughs in various sectors, including renewable energy and medicine [47].

The Cold War era marked the formal establishment of materials science as a distinct discipline, characterized by top-down institutional efforts that fostered interdisciplinary research. ARPA's Materials Research Laboratories exemplified this approach, bringing together experts from various fields to develop advanced materials for defence and space exploration [45]. This model of collaboration has inspired similar initiatives globally, leading to the integration of materials science into academic and industrial frameworks in regions such as Europe, Japan, and China, each adapting the principles of materials science to meet local needs [42, 43].

In the latter part of the 20th century, the focus of materials science expanded to include advanced materials such as nanomaterials and biomaterials, which have become central to modern applications. The manipulation of materials at the nanoscale has opened new avenues for innovation in fields ranging from medicine to [47]. The establishment of professional organizations, such as the Materials Research Society, has further solidified the discipline's status, providing platforms for collaboration and dissemination of research findings [42, 43].

Looking forward, materials science continues to evolve in response to global challenges and technological advancements. Current research emphasizes sustainable materials, smart materials that respond to environmental stimuli, and advanced biomaterials for medical applications. The reliance on computational tools and interdisciplinary collaboration ensures that materials science remains adaptable and relevant in addressing emerging societal needs [45]. As humanity continues to explore new frontiers in materials, the field stands as a testament to our enduring quest for innovation and improvement in the physical world.

Role of Advanced Materials in Modern Industries

Advanced materials are at the forefront of modern industrial innovation, offering superior performance, sustainability, and functionality compared to traditional materials. These purpose-built substances are designed to meet specific requirements in various industries, making them indispensable in driving technological progress and addressing global challenges. Their ability to deliver enhanced mechanical, thermal, electrical, and chemical properties has transformed sectors ranging from aerospace to healthcare, ensuring efficiency, safety, and sustainability.

In the aerospace sector, advanced materials are vital for creating lightweight, durable, and heat-resistant components. Materials such as carbon-fibre-reinforced polymers and titanium alloys are extensively used in aircraft and spacecraft, reducing weight and enhancing fuel efficiency without compromising structural integrity. High-temperature ceramics and advanced composites enable the construction of jet engines and thermal shields capable of withstanding extreme conditions during space exploration. These materials have redefined aerospace engineering, allowing for more efficient and safer travel while pushing the boundaries of exploration.

The automotive industry relies heavily on advanced materials to improve vehicle performance, safety, and environmental impact. Lightweight materials such as aluminium alloys, magnesium composites, and high-strength steels contribute to improved fuel efficiency and reduced emissions. In electric vehicles, advanced materials like lithium-ion battery components and thermally conductive polymers enhance energy storage and heat management. Additionally, advanced safety materials, including impact-resistant polymers and structural composites, play a crucial role in improving crash protection.

In healthcare, advanced materials have revolutionized medical devices, diagnostics, and treatments. Biomaterials, such as biodegradable polymers and titanium alloys, are used for implants, prosthetics, and surgical tools. Nanomaterials enable targeted drug delivery systems, enhancing the effectiveness of treatments while minimizing side effects. Advanced materials like hydrogels and 3D-printable biopolymers support tissue engineering and regenerative medicine, providing new solutions for previously untreatable conditions. These innovations have improved patient outcomes and expanded the possibilities of modern medicine.

The energy sector benefits immensely from advanced materials in its transition to sustainable and renewable sources. High-performance materials like silicon for solar panels and advanced polymers for wind turbine blades optimize energy capture and efficiency. In energy storage, nanostructured electrodes and solid-state electrolytes improve the capacity and lifespan of batteries, enabling the widespread adoption of renewable energy systems. Advanced materials also play a critical role in developing hydrogen fuel cells and thermal management systems, paving the way for a cleaner energy future.

Advanced materials are the backbone of modern electronics, powering devices from smartphones to supercomputers. Semiconductors, such as silicon and gallium arsenide, are foundational to microchips and integrated circuits, enabling the miniaturization and efficiency of electronic devices. Conductive polymers and flexible substrates are crucial for developing wearable technology and flexible displays. Advanced dielectrics and superconductors enhance energy efficiency in data centres and telecommunications infrastructure, driving progress in the digital age.

The construction industry leverages advanced materials to create stronger, more sustainable, and energy-efficient structures. Self-healing concrete and high-performance steel composites increase the durability of buildings and infrastructure, reducing maintenance costs. Insulating materials, such as aerogels and phase-change materials, improve energy efficiency in residential and commercial buildings. Advanced materials also enable innovative designs, such as lightweight yet robust bridges and earthquake-resistant structures, ensuring safety and resilience in construction.

Advanced materials have significantly impacted the consumer goods sector, improving product quality, functionality, and sustainability. High-performance plastics and composites are used in sports equipment, packaging, and household items, enhancing durability and reducing environmental impact. Smart materials, which respond to changes in temperature, pressure, or light, are increasingly incorporated into wearable devices and interactive home technologies, offering personalized and adaptive solutions.

In environmental engineering, advanced materials are integral to addressing pressing global issues such as pollution and resource scarcity. Nano-enabled filters and membranes enhance water purification and desalination processes, ensuring access to clean water. Advanced sorbents and catalysts are used in air purification and carbon capture technologies, mitigating the effects of climate change. Biodegradable and recyclable materials reduce waste and promote circular economy practices, contributing to a more sustainable future.

Advanced materials have transformed manufacturing processes by enabling additive manufacturing (3D printing), automation, and precision engineering. Materials such as high-performance alloys and thermoplastics facilitate the creation of complex geometries and customized components. These innovations reduce production costs, waste, and lead times while maintaining high quality. In industries like aerospace, healthcare, and automotive, advanced materials allow for rapid prototyping and on-demand manufacturing, fostering innovation and adaptability.

The role of advanced materials in modern industries is both transformative and essential. By enabling groundbreaking innovations and addressing critical challenges, these materials drive progress across diverse sectors. Their impact extends beyond improving efficiency and performance; they are key to creating sustainable solutions, enhancing human health, and shaping the future of technology and society. As research and development continue to push the boundaries of material science, the potential for advanced materials to redefine industries remains limitless.

Social Implications of Advanced Materials

As these materials become increasingly integrated into modern life, their societal impact expands, presenting both opportunities and challenges. Advanced materials play a crucial role in promoting sustainability by enabling more efficient resource use and minimizing environmental impacts. Innovations such as biodegradable plastics, self-healing concrete, and lightweight composites are designed to reduce waste and foster eco-friendly practices [48, 49]. However, the production of certain advanced materials, particularly rare-earth elements used in electronics, raises concerns regarding resource extraction, environmental degradation, and geopolitical tensions [50]. Addressing these challenges is essential to ensure that the benefits of advanced materials do not come at an unsustainable cost, necessitating a balance between innovation and environmental stewardship [51].

The healthcare sector has been revolutionized by advanced materials, which provide new treatments, enhanced diagnostic tools, and improved medical devices. For instance, biomaterials used in implants and regenerative medicine significantly enhance patient outcomes and longevity [52, 53]. Nanomaterials facilitate targeted drug delivery systems, thereby reducing side effects and increasing treatment efficacy [52, 53]. However, the high costs associated with these innovations can exacerbate healthcare inequalities, leaving underserved populations without access to these advancements [52]. Thus, while advanced materials improve quality of life, they also highlight the need for equitable access to healthcare innovations.

The integration of advanced materials into manufacturing and technology sectors generates new job opportunities and necessitates specialized skills [54, 55]. Industries such as aerospace, automotive, and renewable energy benefit from these innovations, yet they also create disparities between regions with access to education and those without [54]. This shift underscores the importance of widespread training programs to ensure inclusive job growth, as the demand for skilled labour in advanced materials processing continues to rise [48, 55]. The challenge lies in bridging the skills gap to foster equitable economic development across diverse regions.

Advanced materials are pivotal in driving economic growth by fostering new industries and expanding existing markets. Their applications in sectors such as renewable energy, healthcare, and electronics enhance global competitiveness [48, 50]. However, this rapid progress can lead to unequal access to the benefits of advanced materials, with wealthier nations or corporations often monopolizing the advantages while developing regions struggle to participate [50]. Thus, while advanced materials can stimulate economic growth, they also necessitate policies that promote equitable access and participation in these advancements.

The production and use of advanced materials raise significant ethical questions, particularly regarding their accessibility and the conditions under which they are produced. For instance, the extraction of materials like cobalt and lithium often involves exploitative labour practices and environmental harm in resource-rich developing countries [50]. Additionally, the application of advanced materials in military technologies presents ethical dilemmas concerning their role in global conflict and power dynamics [50]. Ensuring equitable access and ethical practices in the development and deployment of advanced materials is critical to addressing these concerns.

The introduction of advanced materials into everyday products and infrastructure presents both opportunities and risks. While materials such as fire-resistant composites and antimicrobial coatings enhance safety, others, particularly nanomaterials, may pose unknown health risks due to their small size and reactivity [50]. Comprehensive research and regulation are necessary to ensure that these materials are safe for users and the environment, highlighting the need for ongoing scrutiny and oversight in their application [50].

Advanced materials significantly contribute to the transformation of urban infrastructure, enabling the development of smart cities and energy-efficient buildings. Innovations such as self-healing concrete and advanced insulation materials enhance structural resilience and sustainability [49]. However, these advancements may also exacerbate disparities between well-funded urban centres and under-resourced rural or developing areas, necessitating a focus on inclusive urban development strategies [50].

The role of advanced materials in renewable energy technologies, such as solar panels and wind turbines, is critical for supporting the transition to sustainable energy [50]. This transition is essential for mitigating climate change and reducing reliance on fossil fuels. However, the extraction and processing of materials required for renewable technologies can create social and environmental challenges that must be carefully managed [50]. Thus, while advanced materials facilitate the renewable energy transition, they also require a balanced approach to resource management.

Advanced materials are integral to the electronics that power modern communication and information systems, potentially bridging the digital divide by making technology more accessible [50]. However, the complex supply chains involved in their production can reinforce existing inequalities, emphasizing the need for policies that promote equitable access to technology [50]. Addressing these disparities is crucial for ensuring that the benefits of advanced materials are widely shared.

The global nature of advanced materials research, production, and distribution fosters international collaboration but also creates geopolitical tensions, particularly regarding access to critical raw materials [50]. These dynamics can lead to economic dependency and resource conflicts, necessitating cooperative policies to ensure fair distribution and sustainable practices [50]. As advanced materials continue to shape global collaboration, addressing these geopolitical challenges will be essential for fostering a sustainable and equitable future.

Employment Opportunities in Advanced Materials

The field of advanced materials encompasses a wide range of employment opportunities across industries such as aerospace, healthcare, energy, electronics, and construction. These opportunities are driven by the increasing demand for innovative materials that offer enhanced performance, sustainability, and cost-effectiveness. As advanced materials continue to revolutionize industries, the job market in this field is rapidly expanding, with roles available in research, manufacturing, product development, and more.

Research and development are foundational to the advanced materials sector. Materials scientists focus on creating new materials and improving existing ones, often in specialized labs exploring applications for composites, nanomaterials, or biomaterials. Nanotechnologists work at the atomic and molecular levels, designing materials for use in electronics, renewable energy, and medicine. Biomaterials researchers develop materials for medical applications like implants and tissue engineering, while metallurgists study metals to enhance their performance in aerospace and automotive manufacturing.

Manufacturing and production roles are integral to transforming advanced materials into usable products. Process engineers design and optimize production processes to ensure efficiency and scalability, while quality control specialists ensure materials meet industry standards through rigorous testing. Composite technicians fabricate and assemble advanced composites for aerospace and automotive use, and ceramic engineers develop high-performance ceramics for applications in electronics and medical devices.

Design and product development professionals play a key role in integrating advanced materials into new products. Product development engineers focus on improving performance and cost-effectiveness, while application engineers collaborate with clients to tailor materials for specific uses. Electronics materials engineers develop materials for semiconductors and sensors, and textile engineers design high-performance fabrics for healthcare, sports, and defence.

The renewable energy and environmental sectors benefit immensely from advanced materials. Energy materials specialists develop solutions for solar panels, wind turbines, and energy storage systems, while sustainability analysts evaluate the environmental impact of materials and propose eco-friendly alternatives. Carbon capture materials developers work on innovative solutions to combat climate change, and water purification specialists design nanomembranes for clean water technologies.

Healthcare and biomedical fields are transforming with the use of advanced materials. Biomedical engineers create medical devices, implants, and drug delivery systems, while tissue engineering specialists develop scaffolds for regenerative medicine. Pharmaceutical materials scientists improve drug formulations, and orthopaedic device engineers design advanced prosthetics and implants to enhance patient outcomes.

In aerospace and automotive industries, advanced materials play a critical role in improving efficiency and performance. Aerospace engineers use lightweight, high-strength materials for aircraft and spacecraft designs, while automotive materials engineers focus on energy storage and safety in electric vehicles. Structural engineers ensure the reliability of load-bearing components in these industries.

The construction and infrastructure sectors utilize advanced materials to create sustainable and resilient structures. Structural materials engineers design solutions like self-healing concrete and high-performance steel, while insulation materials developers focus on energy-efficient building materials. Urban planners with materials expertise incorporate advanced materials into infrastructure projects to promote durability and sustainability.

Academic and training roles are essential for advancing the field of materials science. University professors and researchers mentor the next generation of professionals and conduct cutting-edge research. Technical trainers provide industry professionals with specialized knowledge, and curriculum developers design educational programs to prepare students for careers in materials science.

Business and policy roles are vital for integrating advanced materials into industries and society. Materials consultants advise on selecting and implementing innovative materials, while intellectual property specialists handle patents and legal aspects. Policy analysts focus on safety and ethical considerations, and entrepreneurs establish startups to commercialize breakthroughs in advanced materials.

Emerging roles in advanced materials highlight the field's dynamic nature. Specialists in artificial intelligence for materials discovery use machine learning to predict properties and find new materials. Quantum materials scientists focus on innovations for quantum computing, while 3D printing materials specialists develop advanced materials for additive manufacturing. Circular economy specialists design systems for recycling and repurposing materials to reduce environmental impact.

The field of advanced materials is vast and interdisciplinary, offering diverse career opportunities in research, manufacturing, design, and policy. As industries increasingly adopt advanced materials to achieve technical and sustainability goals, the demand for skilled professionals is set to grow. With ongoing innovations and applications, careers in advanced materials provide fertile ground for growth and the chance to contribute to transformative changes in technology and society.

The Scope of Advanced Materials Engineering

Advanced Materials Engineering is a multidisciplinary field that focuses on the discovery, design, processing, and application of materials with superior properties tailored to meet specific industrial and technological demands. It bridges multiple scientific disciplines, including physics, chemistry, biology, and engineering, to create innovative solutions that address complex challenges in modern society.

Materials Design and Development: One of the primary scopes of Advanced Materials Engineering is the design and development of new materials. Engineers in this field work on creating materials with enhanced mechanical, thermal, electrical, and optical properties. These include nanomaterials, biomaterials, smart materials, and composites. The process often involves manipulating the atomic and molecular structures of materials to achieve specific characteristics, enabling their use in high-performance applications such as aerospace, healthcare, and renewable energy systems.

Characterization and Analysis: Understanding the microstructure and properties of materials is a crucial aspect of this field. Advanced Materials Engineering employs sophisticated techniques such as electron microscopy, spectroscopy, and X-ray diffraction to analyse materials at the nanoscale. These tools provide insights into the behaviour of materials under different conditions, guiding their improvement and customization for specific applications.

Manufacturing and Processing: This field also encompasses the development and optimization of manufacturing processes for advanced materials. Engineers focus on techniques that enhance efficiency, reduce costs, and improve the scalability of production. Innovations such as additive manufacturing (3D printing) and advanced casting methods are integral to producing complex geometries and integrating novel materials into industrial applications.

Application and Integration: Advanced materials play a transformative role in diverse industries. Aerospace relies on lightweight composites and heat-resistant alloys; healthcare uses biomaterials for implants and prosthetics; electronics depend on semiconductors and superconductors; and renewable energy systems benefit from high-efficiency photovoltaic materials and energy storage solutions. Engineers work on integrating these materials into practical applications, ensuring they meet performance and safety requirements.

Sustainability and Environmental Focus: A growing area within Advanced Materials Engineering is the focus on sustainability. This includes designing eco-friendly materials, such as biodegradable plastics and recyclable composites, and improving the lifecycle sustainability of products. Materials engineers also contribute to developing solutions for renewable energy, carbon capture, and water purification, addressing critical environmental challenges.

Computational Materials Science: The use of computational tools and simulations is expanding the scope of Advanced Materials Engineering. By modelling material behaviour and properties in silico, engineers can predict performance, optimize compositions, and discover new materials faster and more cost-effectively than traditional experimental methods. This computational approach is particularly useful in exploring materials for emerging technologies such as quantum computing and nanotechnology.

Interdisciplinary Collaboration: Advanced Materials Engineering thrives on interdisciplinary collaboration. It integrates expertise from fields such as nanotechnology, biotechnology, and environmental science to tackle challenges that require a holistic approach. This collaborative nature broadens its scope and ensures its relevance in addressing diverse scientific, industrial, and societal needs.

Future-Oriented Innovations: The field is inherently forward-looking, with a focus on developing materials for next-generation technologies. These include materials for wearable electronics, self-healing structures, advanced batteries, and smart textiles. Advanced Materials Engineering is pivotal in pushing the boundaries of what is technologically possible, driving progress in areas like artificial intelligence, robotics, and space exploration.

The scope of Advanced Materials Engineering is vast and continually evolving. It encompasses the creation, analysis, processing, and application of materials that define and shape modern technologies and industries. By addressing critical challenges in sustainability, healthcare, energy, and beyond, this field plays a central role in fostering innovation and ensuring a sustainable and advanced future.

Chapter 2

The Science Behind Advanced Materials

Atomic and Molecular Structure of Materials

Materials are fundamentally defined as substances, primarily in solid form but also including other condensed phases, that are designed for specific applications. This definition encompasses a broad spectrum of materials, from everyday objects to advanced technologies. Advanced materials, such as nanomaterials, biomaterials, and energy materials, represent a significant frontier in scientific innovation, showcasing the versatility and critical importance of materials science in various industries, including healthcare, energy, and electronics [56-58]. These advanced substances are meticulously engineered to meet the specific demands of their applications, illustrating the dynamic interplay between material properties and their intended uses.

At the core of materials science lies the understanding of the relationship between a material's structure, the processing methods employed, and the resultant properties. This relationship is crucial for predicting and enhancing a material's performance in various applications. The structure of a material can be analysed at multiple scales, from its atomic arrangement and chemical composition to its microstructure and macroscopic features shaped by processing techniques [59-61]. These interconnected characteristics dictate how a material behaves under different conditions, such as stress, temperature, or chemical exposure, highlighting the necessity for a comprehensive understanding of these factors in materials design

Processing methods significantly influence the final properties of materials. Techniques such as casting, forging, additive manufacturing, and nanofabrication play a pivotal role in determining the microstructure of materials, which directly affects their mechanical strength, electrical conductivity, thermal resistance, and other essential attributes [56, 61]. The complexity of these interactions indicates that even minor alterations in processing can lead to substantial variations in material performance. Consequently, materials scientists engage in detailed analyses of these effects to optimize materials for their intended functionalities, ensuring that they meet the rigorous demands of modern applications [56, 58].

Furthermore, understanding material behaviour involves principles of thermodynamics and kinetics. Thermodynamics provides insights into the stability and energy states of materials, influencing phenomena such as phase transitions and chemical compatibility [60, 61]. Kinetics, conversely, examines the rates of processes like diffusion and phase transformations, which are critical for understanding a material's evolution during processing and use. By integrating these principles, materials scientists can predict and control the development of a material's properties, facilitating the design of materials with tailored characteristics for specific applications [60, 61].

The integration of structure, processing, properties, and performance forms the foundation of materials science. This holistic approach enables researchers to innovate and design materials with enhanced properties, which are crucial for advancements across diverse fields. Whether developing nanomaterials with superior strength and flexibility, creating biomaterials that replicate natural tissues for medical implants, or designing energy materials for improved batteries and solar cells, the principles of materials science are instrumental in driving transformative advancements [56-58].

The Materials Paradigm

The materials paradigm, represented as a tetrahedron in Figure 4, is a foundational concept in materials science and engineering, describing the interdependent relationship between four critical components: structure, processing, properties, and performance. This framework guides the understanding, design, and optimization of materials for specific applications. By systematically exploring these elements and their interactions, scientists and engineers can predict how a material will behave and tailor it to meet the demands of various industries.

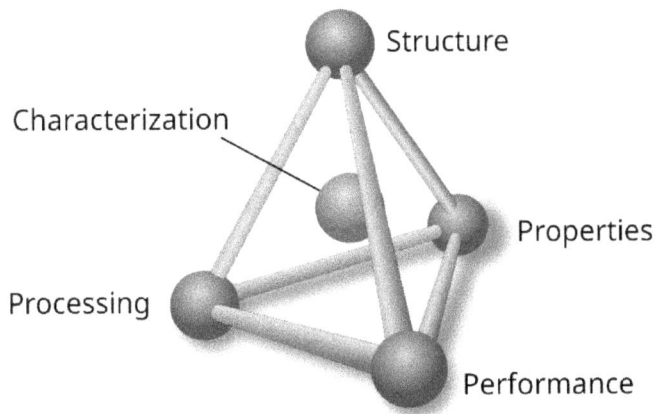

Figure 4: The materials paradigm represented in the form of a tetrahedron. Dhatfield, Public domain, via Wikimedia Commons.

Structure refers to the arrangement of a material's internal components across different length scales, encompassing atomic, nanoscopic, microscopic, and macroscopic features. At the atomic level, it involves the arrangement of atoms, bonding types, and crystal lattices. Nanostructure focuses on features such as grain boundaries, defects, and nanoparticles, while microstructure includes aspects like grain size, phase distributions, and inclusions observable under a microscope. Macrostructure, on the other hand, pertains to large-scale features such as the overall shape and arrangement of components within the material. These levels collectively define the material's structure, which directly impacts properties like strength, ductility, thermal conductivity, and optical behaviour. For example, in metals, smaller grain sizes can enhance hardness and strength, a phenomenon explained by the Hall-Petch relationship.

Processing involves the techniques used to manufacture or modify materials, shaping their structure and influencing their properties. It includes physical, chemical, and thermal treatments such as casting, forging, annealing, welding, and modern techniques like additive manufacturing (3D printing). Advanced methods, such as nanofabrication and chemical vapor deposition (CVD), are used to create high-performance materials. Processing affects the material's microstructure by influencing factors like grain orientation, phase distribution, and defect density. For instance, the cooling rate during metal casting determines whether the material develops a coarse-grained or fine-grained structure, which subsequently affects its mechanical properties.

Properties are the measurable attributes of a material, reflecting how it responds to external forces or environmental conditions. These properties are broadly classified into mechanical, thermal, electrical, optical, and chemical categories. Mechanical properties include strength, hardness, ductility, toughness, and elasticity, while thermal properties cover heat capacity, thermal conductivity, and expansion. Electrical properties include conductivity and dielectric strength, optical properties involve refractive index and transparency, and chemical properties focus on corrosion resistance and reactivity. The material's properties are intricately linked to its structure and processing history. For example, alloying metals can modify the atomic structure, improving strength and resistance to corrosion. These interdependencies highlight the importance of understanding the interplay between structure, processing, and properties in materials science.

The materials paradigm is fundamentally characterized by the interconnections among structure, processing, properties, and performance. This interconnectedness implies that modifications in one aspect can lead to significant changes in the others, creating a cascade effect that is critical for material design and application. For instance, altering a processing technique can lead to changes in microstructure, which subsequently affects the material's properties and ultimately its performance in practical applications.

One illustrative example of this paradigm is the development of high-strength steel for automotive applications. Engineers often select specific processing methods, such as quenching and tempering, to refine the microstructure of the steel. This refinement is crucial as it enhances mechanical properties like tensile strength and toughness, ensuring that the material meets stringent performance requirements in automotive contexts [62]. The relationship between processing and microstructure is emphasized in the literature, where machine learning and data-driven approaches are increasingly utilized to analyse these interconnections, thereby accelerating materials discovery and optimization [62, 63].

Moreover, the choice of materials and processing techniques is frequently driven by desired performance metrics. The Materials Genome Initiative (MGI) exemplifies this approach by advocating for a systematic integration of theory, computation, and experimentation to expedite the discovery and deployment of new materials [64]. This initiative highlights the importance of understanding the processing-structure-property-performance relationships, which are essential for achieving specific functional outcomes in material applications [62, 65].

Structure

Structure is one of the most critical components in the field of materials science. The discipline itself revolves around investigating the intricate relationships between the structure and properties of materials. Structure in materials science is studied at multiple scales, ranging from the atomic level to the macroscopic level. The interplay of these structural levels influences a material's behaviour, performance, and suitability for various applications. Materials scientists employ advanced characterization methods such as X-ray diffraction, spectroscopy, and electron microscopy to examine these structures and gain insights into their properties and potential applications.

Advanced Material Engineering Fundamentals

Atomic structure focuses on the arrangement of atoms within a material. This level of structure defines the fundamental chemical, electrical, and magnetic properties of the material. The scales involved in atomic structures are in the range of angstroms (Å), and the study often requires an understanding of chemical bonding and crystallography. Chemical bonding, whether ionic, covalent, or metallic, determines how atoms interact and combine to form molecules or larger structures. Solid-state physics and quantum chemistry are essential disciplines in understanding these interactions. Crystallography, a specialized field within atomic structure studies, examines how atoms are arranged in crystalline solids. A key concept is the unit cell, the smallest repeating unit in a crystal lattice, which builds the larger macroscopic structure. Crystal defects, such as dislocations, vacancies, and interstitial atoms, significantly affect a material's physical properties, such as its strength, ductility, and conductivity. While many materials occur in polycrystalline forms, some exhibit amorphous structures, like glass and certain polymers, lacking long-range atomic order.

Nanostructures encompass materials with features in the 1 to 100 nm range. At this scale, atoms or molecules aggregate to form nanoscale objects, leading to unique and often extraordinary properties. These properties include enhanced electrical conductivity, magnetic behaviour, and mechanical strength, which are not present in their bulk counterparts. Nanostructures are categorized based on their dimensions. For instance, nanotextured surfaces have one nanoscale dimension, nanotubes have two nanoscale dimensions, and spherical nanoparticles have three. These distinctions are critical in applications ranging from magnetic technologies to biological systems, where nanoscale structures are referred to as ultrastructures. Nanomaterials, such as buckyballs and carbon nanotubes, are intensively researched for their potential in fields like medicine, electronics, and energy.

Microstructure refers to the structure of a material visible under a microscope, typically magnified 25× or more, and includes features ranging from 100 nm to a few centimetres. It strongly influences a material's mechanical and physical properties, such as strength, toughness, ductility, and corrosion resistance. Traditional materials like metals and ceramics are often defined by their microstructures, which include grains, phases, and defects. Imperfections such as grain boundaries, dislocations, and precipitates are unavoidable in crystalline materials and play a significant role in determining their properties. Advances in simulation technologies have provided deeper insights into how these defects can be engineered to enhance material performance. For example, the Hall-Petch relationship describes how grain size affects the strength of a material, with smaller grains generally increasing strength.

Macrostructure deals with the material as it appears to the naked eye, spanning scales from millimetres to metres. This level of structure often relates to the overall geometry, surface features, and visible defects of a material. Macrostructural analysis is essential in industries like construction and manufacturing, where the large-scale properties of materials determine their usability and performance. For example, a steel beam's macrostructure may reveal surface imperfections or inclusions that could affect its load-bearing capacity.

Understanding the structure of materials across all scales is fundamental to materials science. From atomic arrangements that dictate intrinsic properties to macrostructures that influence practical applications, each level offers insights that contribute to the design and development of new and improved materials. Advanced characterization techniques and theoretical modelling enable scientists to explore these structures in unprecedented detail, unlocking innovations that drive technological progress in industries ranging from aerospace to healthcare. As materials science evolves, the study of structure remains central to addressing the challenges of modern engineering and industrial design.

The atomic and molecular structure of advanced materials serves as the foundation for their unique properties and performance. Understanding the arrangement of atoms and the nature of their interactions is essential for tailoring materials to specific applications, from electronics and energy storage to healthcare and environmental solutions. Advanced materials, including nanomaterials, biomaterials, and energy materials, derive their remarkable characteristics from precise control and manipulation at the atomic and molecular levels.

Atomic Structure

Atomic structure refers to the precise arrangement of atoms within a material, a foundational aspect that defines its intrinsic properties and behaviours. The atomic structure encompasses bonding types, atomic configurations, and crystal structures, all of which collectively determine how a material interacts with its environment and performs in specific applications.

Bonding types—covalent, ionic, metallic, and van der Waals—play a critical role in shaping a material's properties. Covalent bonds, where atoms share electrons, are found in materials like diamond, giving it exceptional hardness and high thermal conductivity [66]. This bond type results in tightly bound atomic networks, making materials highly durable and resistant to deformation.

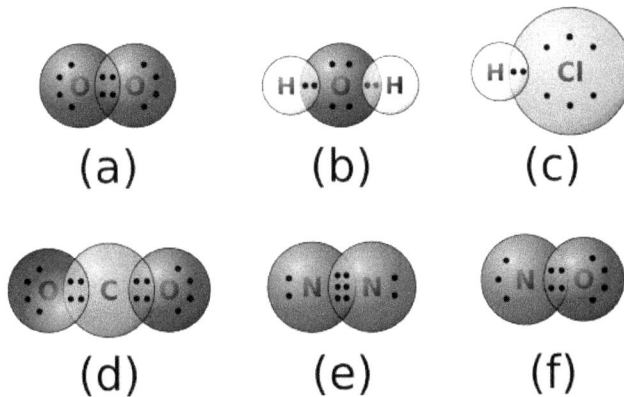

Figure 5: Covalent bond examples: (a) Oxygen, (b) Water, (c) Hydrogen chloride, (d) Carbon dioxide, (e) Dinitrogen, (f) Nitric oxide. MikeRun, CC BY-SA 4.0, via Wikimedia Commons.

Covalent bonds are one of the primary types of chemical bonds, formed when two atoms share electrons to achieve a stable electron configuration. This type of bonding typically occurs between nonmetal atoms with similar electronegativities, as neither atom fully gains or loses electrons. Instead, they share electron pairs, which allows each atom to satisfy the octet rule (having eight electrons in its outer shell) or, in the case of hydrogen, a duet (two electrons). Covalent bonds are foundational to the structure and properties of many materials, from everyday substances like water to advanced materials like diamond and graphene.

Covalent bonds form when the atomic orbitals of two atoms overlap, allowing them to share electrons. The shared electrons occupy the overlapping orbitals, which are regions of space where the probability of finding these electrons is highest. This sharing creates a strong electrostatic attraction between the positively charged nuclei and the shared electron pairs, holding the atoms together.

For example, in a simple molecule like methane (CH_4), each hydrogen atom shares an electron with the carbon atom, resulting in four covalent bonds. In this arrangement, both the carbon atom and the hydrogen atoms achieve stable electron configurations.

The strength and characteristics of covalent bonds depend on the nature of the atoms involved and the number of shared electron pairs. A single covalent bond involves one pair of shared electrons, a double bond involves two pairs, and a triple bond involves three pairs. As the number of shared pairs increases, the bond strength and bond energy increase, while the bond length decreases.

In materials science, covalent bonds play a pivotal role in determining the properties of a wide range of materials. The strength and nature of these bonds directly influence mechanical, thermal, and optical characteristics, making them essential for designing materials with specific functionalities. Materials dominated by covalent bonding often exhibit exceptional performance in various applications, highlighting the importance of understanding these bonds.

Diamond serves as a quintessential example of a material defined by covalent bonding. Each carbon atom in diamond forms four strong covalent bonds with neighbouring carbon atoms in a tetrahedral configuration. This three-dimensional bonding network gives diamond unparalleled hardness, making it one of the hardest known materials. The robust covalent structure also contributes to diamond's exceptional thermal conductivity, allowing it to dissipate heat efficiently. Additionally, the rigidity of its atomic arrangement makes diamond optically transparent, as the absence of free electrons prevents the absorption of visible light, enabling its widespread use in optics and high-precision tools.

Graphene and carbon nanotubes are other remarkable materials that derive their properties from covalent bonding. In graphene, carbon atoms are covalently bonded in a two-dimensional hexagonal lattice, resulting in a material with extraordinary mechanical strength, electrical conductivity, and flexibility. This unique combination of properties has made graphene a cornerstone for innovations in electronics, energy storage, and composite materials. Similarly, carbon nanotubes, which are essentially rolled-up sheets of graphene, exhibit exceptional properties due to the covalent bonding of carbon atoms in their cylindrical structures. These nanotubes are incredibly strong, lightweight, and conductive, enabling applications ranging from nanotechnology to aerospace.

Polymers also rely on covalent bonds for their structural integrity and versatility. In polymers, long chains of repeating molecular units, or monomers, are held together by covalent bonds. The properties of polymers are highly dependent on the type of covalent bonds and the degree of cross-linking between chains. For instance, in vulcanized rubber, covalent cross-links between polymer chains impart elasticity and durability, making it suitable for applications like tires and seals. The ability to tailor covalent bonding in polymers allows for the creation of materials with diverse mechanical, thermal, and chemical properties, from flexible plastics to high-strength composites.

Covalent ceramics, such as silicon carbide (SiC) and boron nitride (BN), further illustrate the significance of covalent bonding in materials science. These ceramics are characterized by their exceptional hardness, thermal stability, and resistance to chemical corrosion, all of which stem from the strength of the covalent bonds in their structures. Silicon carbide, for instance, is used in high-temperature applications and abrasive tools due to its ability to maintain integrity under extreme conditions. Boron nitride, often referred to as "white graphite," offers excellent thermal conductivity and electrical insulation, making it valuable in electronics and heat management systems.

Advanced Material Engineering Fundamentals

Covalent bonds are fundamental to the design and functionality of advanced materials. Whether in diamonds, graphene, polymers, or ceramics, the properties imparted by covalent bonding enable materials to meet the demanding requirements of modern technologies. By manipulating these bonds, materials scientists can create innovative solutions for industries ranging from electronics to aerospace, highlighting the transformative power of covalent bonding in materials science.

Covalent bonding significantly influences the properties of materials, making it a cornerstone in materials science. One of the defining characteristics of covalently bonded materials is their high melting and boiling points. The strength of covalent bonds requires substantial energy to break, resulting in exceptional thermal stability. This property is especially prominent in materials like diamond, where the robust three-dimensional network of covalent bonds ensures stability even at extremely high temperatures. This thermal resilience makes such materials ideal for applications in high-temperature environments and industries.

Another hallmark of covalent materials is their hardness. The rigid bonding networks in covalent solids create structures that resist deformation and wear. Diamond, one of the hardest substances known, exemplifies this property. Its covalent bonding network not only imparts extraordinary hardness but also contributes to its durability, making it suitable for cutting tools, abrasives, and industrial applications where resistance to mechanical stress is essential.

Covalently bonded materials typically exhibit low electrical conductivity because their electrons are localized within the bonds rather than being free to move. This contrasts with metallic bonding, where delocalized electrons contribute to electrical and thermal conductivity. However, there are notable exceptions, such as graphene, where delocalized π-electrons allow for exceptional electrical conductivity. This makes graphene a standout material for applications in electronics, energy storage, and advanced computing technologies.

Transparency is another distinctive feature of some covalent materials, such as diamond. The absence of free electrons in its structure prevents the absorption of visible light, allowing it to transmit light efficiently. This optical clarity is not only aesthetically valuable in gemstones but also functional in high-precision optical instruments and lasers.

Understanding the properties imparted by covalent bonds is essential for designing materials tailored to specific applications. In materials science, these principles are leveraged to develop high-performance composites and coatings that combine strength, durability, and resistance to environmental degradation. Covalent bonding is also fundamental in the design of semiconductors, where the bonds between elements like silicon and germanium form the structural basis of electronic devices. These semiconductors are integral to the functionality of computers, smartphones, and photovoltaic cells.

In the realm of biomaterials, covalent bonding enables the creation of structures that mimic natural systems. Organic molecules connected by covalent bonds form materials used in medical implants, drug delivery systems, and tissue engineering. These materials are designed to interact harmoniously with biological systems, promoting compatibility and functionality.

Covalent bonding also drives innovation in nanotechnology. Carbon-based nanostructures, such as nanotubes and graphene, exhibit unique properties derived from their covalent bonds. These materials are revolutionizing industries by enabling advancements in lightweight composites, energy storage solutions, and next-generation electronics.

The properties and applications of covalently bonded materials underscore their importance in advancing modern technology. From thermal stability and mechanical strength to breakthroughs in nanotechnology and biomaterials, the study and manipulation of covalent bonds continue to shape the future of materials science and engineering.

Ionic bonds, on the other hand, involve the transfer of electrons between atoms, creating a strong electrostatic attraction. This bonding is typical in ceramics, which exhibit high melting points and brittleness due to the rigid nature of their atomic arrangements.

Ionic bonds are a fundamental type of chemical bond formed through the electrostatic attraction between oppositely charged ions. These bonds are central to the structure and properties of many materials, particularly ceramics and salts, and play a critical role in determining their mechanical, thermal, and electrical behaviours. In materials science, understanding ionic bonding is essential for designing and engineering materials with specific functional characteristics.

Ionic bonds are formed when electrons are transferred from one atom to another, creating charged particles called ions. This process typically occurs between atoms with significantly different electronegativities. Metals, which have low electronegativity, tend to lose electrons and form positively charged ions (cations), while nonmetals, with high electronegativity, gain those electrons to become negatively charged ions (anions). The resulting electrostatic attraction between these oppositely charged ions forms the ionic bond.

For example, in sodium chloride ($NaCl$), a classic ionic compound, a sodium atom donates one electron to a chlorine atom. This electron transfer results in a positively charged sodium ion (Na^+) and a negatively charged chloride ion (Cl^-). The strong electrostatic force between these ions holds them together in a lattice structure, forming the crystalline material we know as table salt.

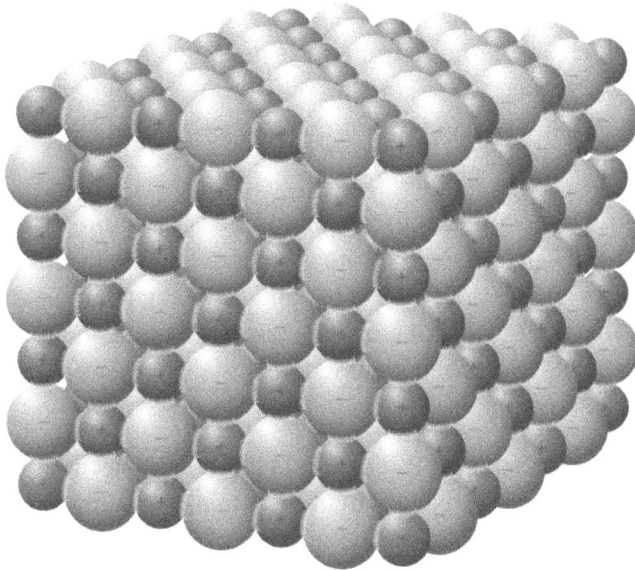

Figure 6: The crystal structure of sodium chloride, NaCl, a typical ionic compound. Goran tek-en, CC BY-SA 4.0, via Wikimedia Commons.

Ionic bonds are characterized by their exceptional strength, which arises from the significant electrostatic forces of attraction between positively charged cations and negatively charged anions. This strength contributes to the formation of rigid and stable crystalline structures, which are a hallmark of ionic materials. The magnitude of an ionic bond's strength is determined by several factors, including the charges on the ions and their respective radii.

The charge on the ions plays a critical role in the bond's strength. Higher charges on the cations and anions result in stronger electrostatic forces and, consequently, stronger ionic bonds. For example, in magnesium oxide (MgO), the ions involved are doubly charged (Mg^{2+} and O^{2-}), leading to a stronger bond compared to sodium chloride (NaCl), where the ions (Na^+ and Cl^-) carry only single charges. This increased bond strength in magnesium oxide manifests in its higher melting point and greater stability under extreme conditions.

The radii of the ions also influence the strength of ionic bonds. Smaller ions can pack closer together, minimizing the distance between their charges and enhancing the electrostatic attraction. This closer packing results in stronger bonds and contributes to the material's structural integrity. For instance, the small size of lithium ions (Li^+) in lithium fluoride (LiF) allows for a tightly bound lattice, giving the material high thermal stability and a robust crystalline structure.

Ionic compounds generally form repeating lattice structures where ions are arranged to maximize attraction and minimize repulsion. This ordered arrangement ensures that every positive ion is surrounded by negative ions and vice versa, creating a stable and energetically favourable configuration. The lattice structure is responsible for many of the distinctive properties of ionic materials. For example, the strong ionic bonds and regular lattice arrangement result in high melting and boiling points, as significant energy is required to break these bonds and disrupt the lattice.

The rigid crystalline structures formed by ionic bonds also contribute to the brittleness of ionic materials. While the bonds are strong, they are directional, meaning any significant force that displaces the lattice can cause ions of the same charge to align and repel each other, leading to material fracture. Despite this brittleness, the strength and stability of ionic bonds make these materials invaluable in applications requiring high thermal resistance, chemical stability, and electrical insulation.

Understanding these characteristics of ionic bonds is crucial in materials science, as they underpin the behaviour and functionality of many essential materials, from everyday table salt to advanced ceramics and energy storage components. The interplay between ionic charge, radii, and lattice structure defines the performance of these materials in diverse applications.

In materials science, ionic bonding is a fundamental concept, particularly significant in ceramics, minerals, and various crystalline materials. These bonds, formed by the electrostatic attraction between positively and negatively charged ions, impart unique properties to materials, making them suitable for a wide range of applications.

Ceramics are one of the most prominent classes of materials dominated by ionic bonding. In compounds such as alumina (Al_2O_3) and zirconia (ZrO_2), the ionic bonds between metal cations and oxygen anions create extremely stable and rigid structures. These bonds confer exceptional hardness and resistance to high temperatures, as well as remarkable chemical inertness. These properties make ceramics indispensable in demanding applications, such as thermal barrier coatings for jet engines, cutting tools for machining, and biomedical implants used in dentistry and orthopaedics. Despite their strength and stability, the brittleness inherent to ionic bonding in ceramics poses challenges, as their rigid lattice structures can fracture under stress.

Minerals, commonly found in nature, also owe much of their properties to ionic bonding. Halite (sodium chloride, NaCl) and fluorite (calcium fluoride, CaF_2) are prime examples. The ionic bonds in these minerals determine critical characteristics such as hardness, cleavage, and solubility. Halite, for instance, dissolves readily in water due to the dissociation of its ionic bonds, a property that underpins its widespread use in food and de-icing applications. Fluorite, with its characteristic cleavage and transparency, is used in optics and as a source of fluorine for industrial processes.

Ionic crystals, such as sodium chloride and calcium carbonate, further exemplify the effects of ionic bonding. The regular, repeating arrangement of ions in their crystalline lattice results in transparency and brittleness. This ordered structure minimizes energy within the crystal but makes the material susceptible to fracture under force. These materials find applications in construction and optics, where their transparency and rigidity are advantageous.

Ionic bonding plays a pivotal role in the development of advanced materials for modern applications. In the field of energy storage, ionic materials are critical components of batteries and fuel cells. Lithium-ion batteries, for example, rely on the movement of lithium ions through ionic conductors, enabling efficient energy storage and release. To enhance safety and efficiency, researchers are developing solid-state electrolytes, often composed of ionic materials, to replace traditional liquid electrolytes in batteries. These advancements aim to improve battery longevity and reduce risks such as leakage and flammability.

In electronics, ionic materials like lead zirconate titanate (PZT) are utilized in piezoelectric devices. These materials exploit the displacement of ions within their lattice under applied mechanical forces, converting mechanical stress into electrical signals and vice versa. This property is crucial in sensors, actuators, and ultrasound equipment.

Environmental technologies also benefit significantly from ionic materials. Zeolites and other ion-exchange materials, which consist of intricate ionic frameworks, are widely used to purify water and remove contaminants. These materials facilitate the exchange of ions to trap impurities, making them invaluable in water treatment processes and air purification systems.

The role of ionic bonding in materials science is vast and varied. From the stability and hardness of ceramics to the functionality of piezoelectric devices and the environmental utility of ion-exchange materials, ionic bonds are a cornerstone of modern technology. Their study and application continue to drive innovations across industries, addressing challenges in energy, electronics, and environmental sustainability.

Metallic bonds, formed by the pooling of electrons among atoms, impart properties such as ductility and electrical conductivity, as seen in metals like aluminium and steel. Van der Waals forces, although weaker, play a significant role in materials like graphite, where these interactions allow for layers of carbon atoms to slide over each other, contributing to its lubricative properties.

Metallic bonds are a fundamental type of chemical bond that define the structure and properties of metals and alloys. These bonds form the basis for the strength, conductivity, and malleability of metallic materials, making them essential to countless industrial and technological applications. In materials science, understanding metallic bonding is crucial for designing and manipulating metals to achieve desired performance characteristics.

Metallic bonds are unique in that they arise from the sharing of free electrons among a lattice of metal atoms. Unlike covalent or ionic bonds, where electrons are localized between specific atoms or ions, metallic bonds involve a "sea of electrons" that is delocalized throughout the structure. This electron sea forms as metal atoms lose their valence electrons, creating positively charged metal ions (cations). These ions are then held together by the electrostatic attraction between themselves and the delocalized electrons.

The delocalized electrons in metallic bonds are not confined to any particular atom but move freely throughout the lattice. This freedom of movement is responsible for many of the characteristic properties of metals. The strength of metallic bonds depends on several factors, including the number of delocalized electrons and the charge and size of the metal cations. For example, transition metals, which have more delocalized electrons, typically form stronger metallic bonds than alkali metals.

Metallic bonds impart a range of distinctive properties to metals and alloys, making them crucial in materials science and engineering. These properties are a direct result of the unique "sea of electrons" that defines metallic bonding, where delocalized electrons move freely among a lattice of positively charged ions. This bonding mechanism not only determines the behaviour of metals under various conditions but also enables their diverse applications across industries.

One of the most prominent properties of metallic bonds is their ability to facilitate electrical and thermal conductivity. The delocalized electrons in metallic structures act as efficient carriers of charge and energy, allowing metals to conduct electricity and heat with remarkable efficiency. This property is particularly vital in applications such as electrical wiring, where the rapid transmission of electricity is essential, and in heat exchangers, where metals like copper and aluminium dissipate heat effectively.

Another defining characteristic of metallic bonding is its contribution to malleability and ductility. The flexibility of metallic bonds allows metal atoms to slide past one another without breaking the bond. This capability enables metals to be deformed into thin sheets, a property known as malleability, or drawn into wires, referred to as ductility. These properties make metals ideal for creating complex shapes, from intricate jewellery to structural components in construction and manufacturing.

The interaction of delocalized electrons with light also gives metals their characteristic lustre and reflectivity. When light strikes a metal surface, the free electrons reflect it, producing the shiny appearance commonly associated with metals. This optical property is widely utilized in decorative applications, mirrors, and reflective coatings, enhancing both the functionality and aesthetic appeal of metals.

The strength and toughness of metals are also directly tied to metallic bonding. The electrostatic attraction between the positively charged ions and the electron sea creates a strong, cohesive lattice that can withstand significant mechanical stress. This durability makes metals and alloys ideal for structural applications, such as in buildings, bridges, and vehicles. The strength of metallic bonds can be further enhanced in alloys, where the combination of different metal elements optimizes performance. For instance, adding carbon to iron creates steel, an alloy with enhanced strength and toughness compared to pure iron.

Finally, the high energy required to disrupt metallic bonds results in metals having high melting and boiling points. This thermal stability is crucial in applications where materials must perform under extreme conditions, such as in engines, turbines, and industrial furnaces. The strength of metallic bonds, and consequently the melting and boiling points, are influenced by the number of delocalized electrons and the charge density of the metal ions. Metals with a higher density of delocalized electrons, such as tungsten, exhibit exceptionally high melting points, making them suitable for demanding applications.

The properties conferred by metallic bonding underscore the versatility and indispensability of metals in modern technology and industry. By leveraging these characteristics, materials scientists and engineers continue to innovate, creating advanced alloys and materials that meet the evolving needs of society.

In materials science, metallic bonding is the foundation for understanding and manipulating metals and alloys to achieve specific performance criteria. Metals like iron, aluminium, and copper owe their versatility and widespread use to the characteristics conferred by metallic bonds.

Alloy Design: Metallic bonding allows for the formation of alloys, which are mixtures of two or more metallic elements. By altering the composition of the alloy, materials scientists can fine-tune properties such as strength, corrosion resistance, and thermal expansion. For instance, the addition of carbon to iron creates steel, an alloy with enhanced strength and toughness compared to pure iron.

Superalloys: Metallic bonds are integral to the development of superalloys, which are engineered to perform under extreme conditions, such as high temperatures and pressures. These materials, often used in aerospace and power generation, combine strong metallic bonding with precise microstructural control to deliver exceptional performance.

Lightweight Metals: Materials such as aluminium and titanium rely on metallic bonding to achieve high strength-to-weight ratios. These properties are crucial in industries like automotive and aerospace, where reducing weight while maintaining structural integrity is critical.

Metallic Glasses: Metallic bonds are also key to the development of amorphous metals, or metallic glasses, which lack the crystalline structure of traditional metals. These materials exhibit unique properties such as high strength and corrosion resistance, making them valuable for advanced applications in medical devices and electronics.

Figure 7 provides a visual representation of metallic bonding, a fundamental concept in materials science that explains the unique properties of metals. Metallic bonding occurs in a lattice structure where positively charged metal ions, known as cations, are surrounded by a "sea" of delocalized electrons. These metal atoms lose their valence electrons, becoming positively charged, and arrange themselves in a regular, repeating pattern characteristic of crystalline structures in metals. This ordered arrangement of cations is central to the stability and organization of metallic materials.

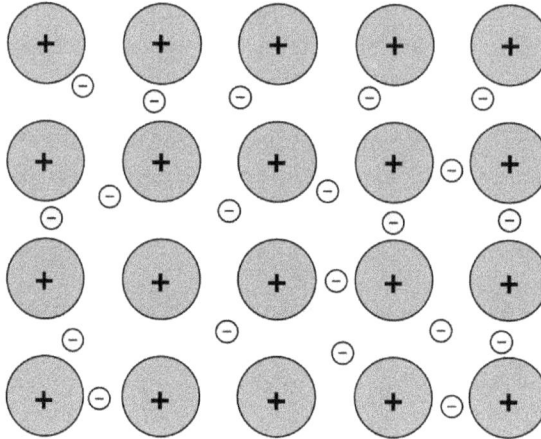

Figure 7: Scheme for metallic bonding. Muskid, CC BY-SA 3.0, via Wikimedia Commons.

The delocalized electrons, which are no longer bound to any specific atom, move freely throughout the metal lattice. This mobility is a defining feature of metallic bonds and is often referred to as the "electron sea." These electrons play a crucial role in maintaining the cohesion of the lattice, as they are attracted to the positively charged cations. The result is a strong electrostatic force that holds the entire structure together, forming the metallic bond. This type of bonding is unique compared to covalent or ionic bonds, as the electrons are not confined to discrete bonds or pairs but instead flow throughout the structure, creating a highly conductive and flexible material.

Several key features can be inferred from the diagram. The uniform arrangement of cations reflects the orderly and symmetrical nature of metal crystals, which contributes to the strength and stability of metallic materials. The presence of delocalized electrons highlights their role in imparting essential properties to metals. These free electrons enable metals to conduct electricity and heat efficiently, as they can move freely in response to an electric field or thermal gradient. Additionally, the flexibility of the electron sea allows metal ions to slide past one another without breaking bonds, giving metals their malleability and ductility. This explains why metals can be hammered into thin sheets or stretched into wires without losing their structural integrity.

The cohesive strength of metallic bonds is another notable characteristic. The uniform distribution of the electron sea around the cations ensures that the lattice remains stable under various mechanical stresses. This property makes metals durable and able to withstand significant loads, which is why they are widely used in construction and engineering applications. The diagram simplifies the complex interactions within metallic bonds to help visualize the mechanism that gives metals their distinct mechanical, thermal, and electrical properties, making it easier to understand their behaviour and applications.

In crystalline materials, atoms are organized in regular, repeating patterns known as crystal lattices. This arrangement provides structural stability and symmetry to the material, influencing its macroscopic properties. The unit cell, the smallest repeating unit of a crystal lattice, defines the symmetry and geometry of the overall structure. Common crystal structures include face-centred cubic (FCC), body-centred cubic (BCC), and hexagonal close-packed (HCP) arrangements. Each of these configurations offers distinct mechanical and physical characteristics. For instance, FCC structures, found in materials like aluminium and copper, have closely packed atoms that allow for excellent ductility and high strength. BCC structures, typical of metals like tungsten and chromium, offer greater strength but less ductility due to the relatively lower atomic packing density. HCP structures, as seen in titanium and magnesium, combine strength with lightweight properties, making them ideal for aerospace and biomedical applications.

The study of atomic structure extends beyond understanding how atoms are arranged; it also involves exploring how this arrangement affects a material's performance. For example, the precise alignment of carbon atoms in a diamond lattice results in its unparalleled hardness, while the delocalized electrons in metallic bonds enable efficient electrical and thermal conductivity in metals. These relationships are fundamental to materials science and form the basis for designing new materials with tailored properties.

The ability to manipulate atomic structures enables the creation of advanced materials with enhanced functionalities. By understanding and controlling how atoms are arranged and bonded, scientists can develop materials for cutting-edge applications in electronics, energy storage, healthcare, and more. The study of atomic structure is thus central to advancing technology and addressing the challenges of modern industry.

Crystallographic Defects and Their Role

In crystalline materials, atomic perfection is exceedingly rare. The lattice structures of these materials are typically disrupted by various crystallographic defects, which play a crucial role in determining their physical, chemical, and mechanical properties. These imperfections, while often viewed as flaws, are essential for tuning the behaviour and performance of materials in both traditional and advanced applications.

One common type of crystallographic defect is a vacancy, where an atom is missing from its expected position in the lattice. Vacancies can influence diffusion processes within a material, which are essential for applications such as heat treatment in metals. Another type of defect is the interstitial atom, where an additional atom is situated in the small spaces between the regular lattice points. These interstitial atoms can introduce stress into the material, altering its mechanical properties. Dislocations, which are irregularities in the lattice arrangement, are perhaps the most impactful defects when it comes to mechanical behaviour. They allow layers of atoms to slip past one another under stress, facilitating plastic deformation. This mechanism explains why metals are often malleable and ductile, allowing them to be shaped and formed without breaking.

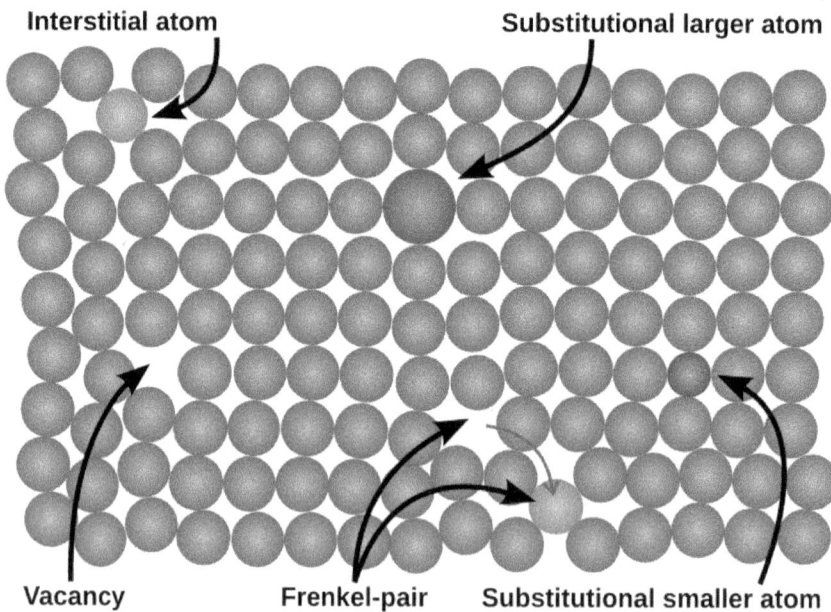

Figure 8: Point defects in crystal structures: Interstitial and substitutional atom, Vacancy, Frenkel defect. DaniFeri, CC BY-SA 3.0, via Wikimedia Commons.

Defects also exist on a larger scale in the form of grain boundaries, which are the interfaces where crystals of different orientations meet in polycrystalline materials. Grain boundaries significantly affect a material's properties. For example, they can enhance a material's strength by impeding the motion of dislocations, a principle known as grain boundary strengthening. However, this improvement in strength often comes at the cost of reduced electrical conductivity because the boundaries disrupt the free flow of electrons. This trade-off is utilized in materials like polycrystalline silicon, which is widely used in solar cells and other semiconductor devices.

In the field of advanced materials, defect engineering has emerged as a critical technique for tailoring material properties to meet specific needs. By intentionally introducing or manipulating defects, scientists can enhance or create desired characteristics. For instance, controlled dislocations in metals can improve ductility, making them more suitable for applications such as aerospace components, where both strength and flexibility are essential. In ionic conductors, introducing vacancies can enhance the movement of ions, improving conductivity and performance in devices like fuel cells and batteries. Similarly, in optical materials, defects can be engineered to modify light absorption and emission properties, enabling the development of materials for lasers and photonics.

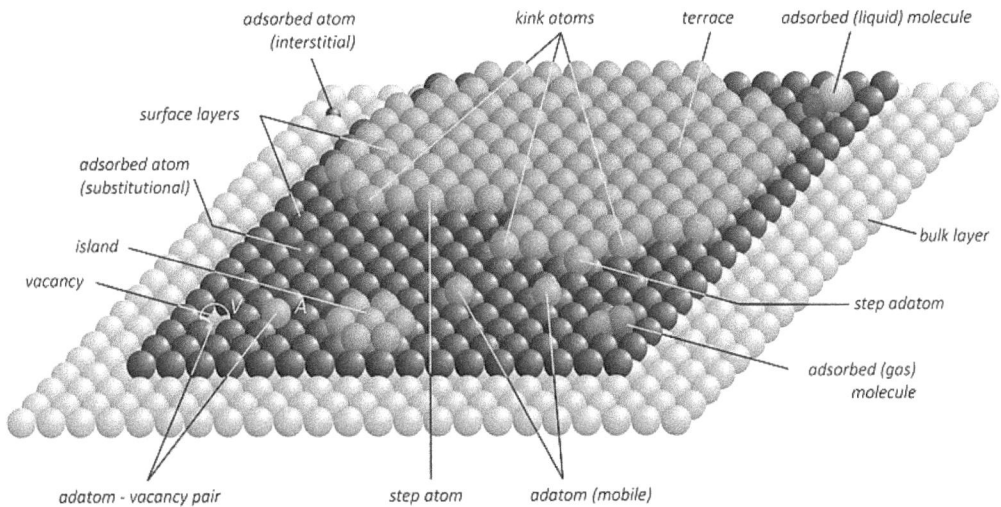

Figure 9: Ball model representation of a real (atomically rough) crystal surface with steps, kinks, adatoms, and vacancies in a closely-packed crystalline material. Adsorbed molecules, substitutional and interstitial atoms are also illustrated. ShutterWaves, CC BY-SA 4.0, via Wikimedia Commons.

The ability to manipulate defects at the atomic and microscopic levels has opened new avenues for material design and innovation. By understanding the nature and behaviour of crystallographic defects, researchers can transform what might traditionally be seen as imperfections into tools for creating materials with superior performance, tailored to the challenges of modern technology. This approach underscores the importance of defects as both a fundamental aspect of materials science and a gateway to new possibilities in advanced material applications.

Molecular Structure

Molecular structure plays a critical role in defining the properties and functionality of advanced materials, especially in polymers and biomaterials. While atomic arrangements are crucial in crystalline and metallic materials, the organization and interactions of molecules become the dominant factor in these complex materials. Molecular structure encompasses the way molecules are arranged and interact within a material, profoundly influencing its mechanical, thermal, and chemical behaviour.

In polymers, the molecular structure is composed of long chains of repeating molecular units known as monomers. These chains can be tailored to exhibit a range of properties depending on their length, degree of cross-linking, and alignment. For example, a polymer with long, flexible chains may exhibit high elasticity, as seen in materials like rubber, whereas one with shorter or more rigid chains may form a brittle plastic. The degree of cross-linking, or the number of chemical bonds linking one polymer chain to another, also plays a significant role. Highly cross-linked polymers, such as epoxy resins, are rigid and durable, while lightly cross-linked ones, like hydrogels, are soft and absorbent. Furthermore, molecular alignment can determine whether a polymer is amorphous (lacking long-range order) or semi-crystalline, affecting properties like transparency, strength, and thermal resistance.

Biomaterials, which are inspired by or derived from biological systems, rely heavily on hierarchical molecular structures to mimic the complexity of natural tissues. Proteins, for instance, are composed of amino acid chains folded into specific three-dimensional shapes, enabling applications such as tissue scaffolds and drug delivery systems. Polysaccharides, another class of biomaterials, consist of sugar molecules arranged in intricate molecular networks that replicate the properties of natural tissues like cartilage or skin. These hierarchical structures provide the versatility needed for biomaterials to perform in highly specialized roles, such as medical implants or wound dressings.

Molecular interactions, such as hydrogen bonding and van der Waals forces, are equally crucial in determining the performance of advanced materials. Hydrogen bonding, which occurs when a hydrogen atom is shared between two electronegative atoms (like oxygen or nitrogen), plays a pivotal role in materials like Kevlar. In Kevlar, hydrogen bonds form between adjacent polymer chains, creating a tightly bonded network. This structure gives Kevlar its extraordinary tensile strength, making it ideal for high-performance applications like bulletproof vests and aerospace components. Similarly, van der Waals forces, which are weaker interactions arising from the polarization of molecules, can contribute to the cohesion and stability of molecular assemblies, especially in soft materials like gels and thin films.

In advanced materials, the ability to design and control molecular structure enables a wide range of applications across industries. By understanding how molecules interact and organize themselves, materials scientists can develop polymers with specific mechanical properties, biomaterials that integrate seamlessly with human tissue, or high-strength composites that withstand extreme conditions. The molecular structure is not merely a foundational characteristic but a versatile tool for engineering materials that meet the diverse demands of modern technology. Through meticulous control of molecular design, scientists continue to unlock new possibilities in materials science, from sustainable packaging to cutting-edge medical devices.

Modern technologies have significantly advanced our ability to manipulate atomic and molecular structures, leading to the development of materials with unique properties. One of the most pivotal advancements in this field is nanotechnology, which allows for the precise placement of atoms to create nanostructures with tailored electronic and mechanical behaviours. For instance, carbon nanotubes and graphene are prime examples of nanostructures that exhibit exceptional strength, electrical conductivity, and thermal properties due to their unique atomic arrangements [67, 68]. The ability to synthesize these materials with atomic precision has been demonstrated through various methods, including surface-assisted coupling of molecular precursors, which enables the production of graphene nanoribbons with specific topologies and widths [67].

In addition to nanotechnology, molecular self-assembly plays a crucial role in the design of advanced materials. This process involves the spontaneous organization of molecules into functional structures, which is particularly beneficial in applications such as organic electronics and drug delivery systems. The self-assembly of inorganic nanostructures has been extensively reviewed, highlighting the forces that guide this process and its implications for device applications [69]. The ability to control molecular architectures through self-assembly techniques is essential for creating materials that can respond dynamically to environmental stimuli, enhancing their functionality in various applications [69].

High-resolution spectroscopy and microscopy techniques, such as scanning tunnelling microscopy (STM) and atomic force microscopy (AFM), have revolutionized our ability to visualize and manipulate atomic and molecular arrangements directly.

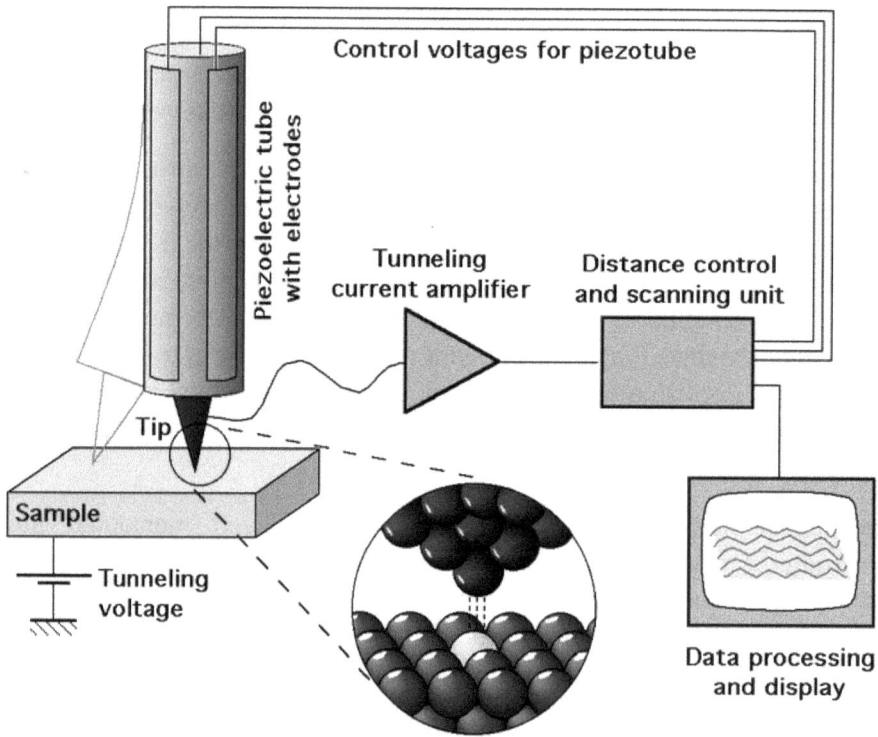

Figure 10: Schematic diagram of a scanning tunnelling microscope. Michael Schmid, CC BY-SA 2.0 AT, via Wikimedia Commons.

STM, in particular, allows for the imaging of surfaces at atomic resolution, enabling researchers to observe phenomena such as quantum tunnelling and charge transfer at the molecular level [70, 71]. AFM complements this by providing insights into surface properties through force interactions between the tip and the sample, allowing for detailed characterization of nanostructures [72, 73].

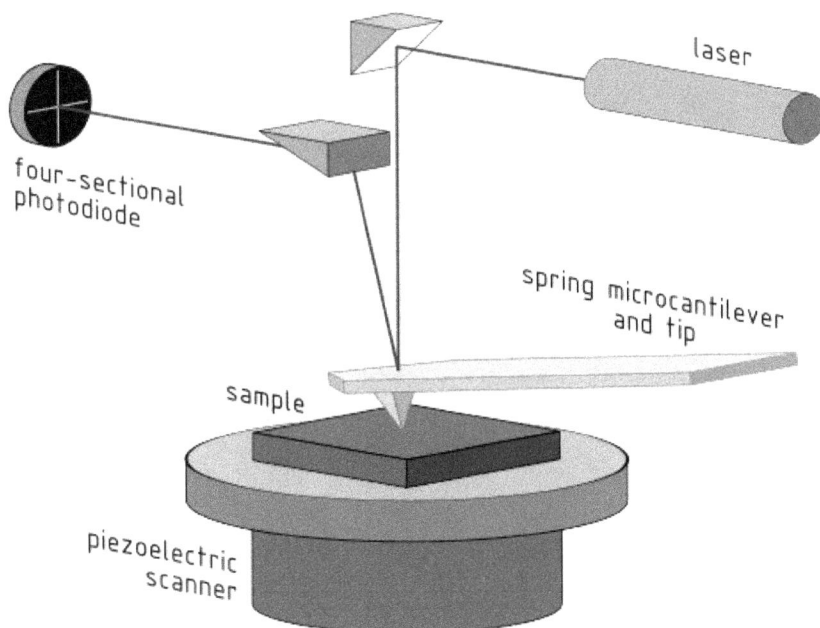

Figure 11: Schematic of an atomic force microscope with optical detection of the deflection of the microcantilever. GregorioW, CC BY-SA 3.0, via Wikimedia Commons.

These techniques not only facilitate the exploration of fundamental physical properties but also enable the development of novel materials with tailored functionalities [74-76]. As an example of such development, Hexabenzocoronene (HBC), see Figure 12, is a prominent member of the polycyclic aromatic hydrocarbon (PAH) family, characterized by its disc-like molecular structure and high symmetry. Its molecular formula, $C_{42}H_{18}$, indicates a fully conjugated π-electron system, which facilitates extensive electron delocalization across the molecule. This unique structural arrangement contributes to HBC's remarkable electronic and optical properties, including high thermal stability, efficient electrical conductivity, and distinctive photophysical characteristics [77]. The planar geometry of HBC, classified under the $D6h$ point group, enhances its potential applications in materials science and nanotechnology [78].

The synthesis of HBC typically employs cyclization reactions, with the Scholl reaction being a prevalent method. This process involves the oxidative cyclodehydrogenation of precursors such as hexaphenylbenzene, allowing for the formation of the hexagonal core of HBC [77, 79]. Recent advancements in synthetic methodologies have enabled precise functionalization of HBC, allowing researchers to tailor its properties for specific applications. For instance, the introduction of various functional groups or heteroatoms can modify HBC's solubility, electronic conductivity, and optical absorption [80, 81].

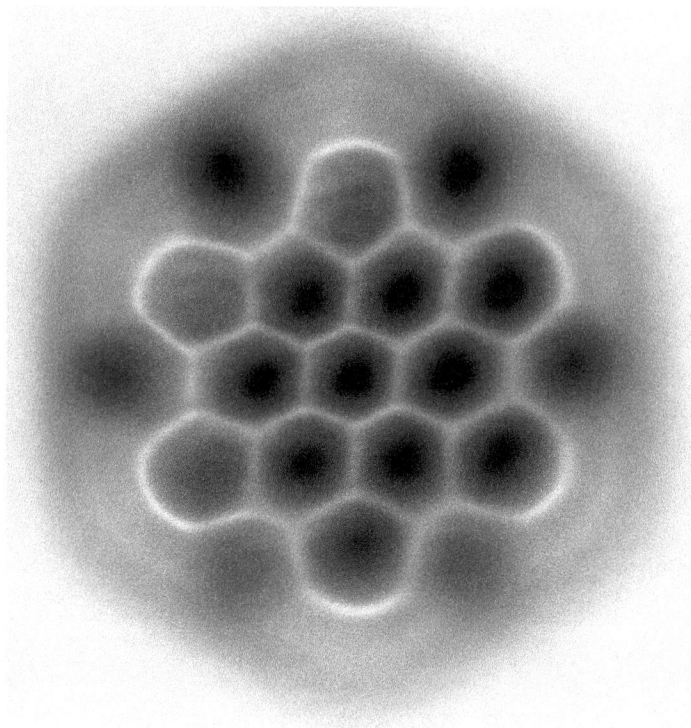

Figure 12: Hexabenzocoronene Atomic Force Microscope image, black and white. This molecule was synthesized at the Centre National de la Recherche Scientifique (CNRS) in Toulouse. Patrik Tschudin, CC BY 2.0, via Wikimedia Commons.

HBC exhibits exceptional thermal stability, attributed to its rigid, conjugated π-system, making it suitable for high-temperature applications [75, 82]. Its delocalized π-electrons facilitate efficient charge transport, positioning HBC as a valuable material in organic electronics, including organic light-emitting diodes (OLEDs), organic field-effect transistors (OFETs), and organic photovoltaics (OPVs) [76, 81]. Additionally, HBC's ability to self-assemble into columnar liquid-crystalline structures, driven by π-π stacking interactions, enhances its utility in nanotechnology [83].

The versatility of HBC extends to its applications in sensors and energy storage systems. Functionalized HBC derivatives are employed in chemical and biological sensors due to their fluorescence and electronic responsiveness to environmental changes [75, 84]. Furthermore, HBC-based materials are being explored for use in supercapacitors and batteries, leveraging their efficient charge conduction and energy storage capabilities [85].

Current research endeavours are focused on expanding the functionalization of HBC to enhance its applicability. By attaching side chains or introducing heteroatoms, researchers aim to improve solubility and create novel electronic and optical properties [86, 87]. The exploration of HBC in the development of two-dimensional materials, such as graphene nanoribbons, is also a significant area of interest, highlighting its potential in advanced electronic and energy-related technologies [78, 81].

Applications of Atomic and Molecular Design

In the realm of semiconductors, atomic doping in silicon has emerged as a critical technique for controlling electrical conductivity, which is essential for the functionality of microchips and solar cells. Doping allows for the precise manipulation of the electronic properties of silicon, enabling the development of high-performance devices. For instance, the introduction of specific dopants can enhance the charge carrier concentration, thereby improving the efficiency of solar cells and the speed of microchips [88]. Furthermore, the integration of nanostructures, such as silicon nanoparticles, has been shown to enhance luminescent properties and catalytic activities, further broadening the applications of silicon-based materials in electronics [88].

Nanomaterials, particularly quantum dots and nanowires, exhibit unique optical and electronic properties due to their precise atomic arrangements. Quantum dots, which are semiconductor nanocrystals, have gained attention for their tuneable photoluminescence and potential applications in sensors and displays. Their size-dependent properties allow for customization in various applications, including bioimaging and optoelectronics [89, 90]. The development of carbon quantum dots has also been significant, as they offer advantages such as low toxicity and excellent solubility, making them suitable for a wide range of applications in biomedicine and sensor technology [91, 92]. Additionally, the synthesis of nanowires has enabled the creation of high-performance electronic devices, leveraging their unique conductive properties [93].

In the field of biomaterials, the molecular design of hydrogels and biodegradable polymers has opened new avenues for tissue regeneration and targeted drug delivery. These materials can be engineered at the molecular level to mimic natural tissues, providing scaffolds that support cell growth and differentiation [89]. For example, hydrogels can be tailored to respond to specific biological stimuli, enhancing their functionality in drug delivery systems [91]. The incorporation of nanomaterials into biomaterials has further improved their properties, allowing for enhanced bioactivity and controlled release mechanisms [94].

Energy materials have also benefited from atomic-level engineering, particularly in the development of lithium-ion battery materials and catalysts for hydrogen production. The optimization of electrode materials at the atomic scale has been shown to enhance energy density and cycle life, which are critical for the performance of batteries [95]. Moreover, advanced catalysts designed for hydrogen evolution reactions have demonstrated improved efficiency, contributing to the development of sustainable energy solutions [96, 97]. The integration of nanostructured materials in these applications has been pivotal, as they often exhibit superior catalytic properties due to their high surface area and unique electronic characteristics [98].

Mechanical, Thermal, and Chemical Properties

Mechanical Properties

The mechanical properties of advanced materials describe their behaviour under applied forces, encompassing attributes such as strength, elasticity, toughness, hardness, and flexibility. These properties are crucial for determining the suitability of materials in demanding applications, including aerospace, construction, and biomedical implants.

Advanced materials often exhibit exceptional strength, enabling them to endure high levels of stress without undergoing permanent deformation. This characteristic is particularly valuable in industries like aerospace, where materials such as carbon fibre-reinforced composites are widely used. These composites boast a high tensile strength-to-weight ratio, making them ideal for lightweight and robust structural components.

Toughness is another critical mechanical property, reflecting a material's ability to absorb energy and resist fracture. Many advanced materials strike a delicate balance between strength and toughness. For example, titanium alloys, often employed in medical implants, are valued for their ability to combine high strength with slight deformation, preventing breakage under mechanical stress.

Toughness describes a material's capacity to absorb energy and deform without fracturing. This property is essential for materials subjected to dynamic or impact forces. It is quantified as energy per unit volume, typically expressed in Joules per cubic meter (J/m^3). Mild steel demonstrates toughness values around 5–20 MJ/m^3, while advanced materials like titanium alloys range from 30 to 50 MJ/m^3, and carbon fibre composites achieve 10–20 MJ/m^3. Titanium alloys are particularly valued in medical implants for their combination of toughness and strength, ensuring durability and safety in critical applications.

Elasticity and plasticity also play significant roles in advanced materials. Elasticity refers to a material's capacity to return to its original shape after being deformed, while plasticity involves the ability to undergo permanent deformation without failure. Shape-memory alloys are a prime example, as they exhibit tailored elastic and plastic behaviours, allowing them to return to predefined shapes after being bent or stretched. These properties are especially useful in robotics and medical devices where precise, controlled deformation is required.

Figure 13: Three-point flexural test on a composite beam at speed=10 mm/min. Instron universal testing machine with a 300 kN dynamometer. Cjp24, CC BY-SA 3.0, via Wikimedia Commons.

Elasticity and plasticity are measures of a material's deformation behaviour. Elasticity, determined by Young's modulus, represents the material's stiffness and is expressed in GigaPascals (GPa). Steel has an elasticity range of about 200 GPa, while rubber exhibits values between 0.01 and 0.1 GPa. Advanced materials like graphene achieve an extraordinary elasticity of ~1 TPa, and shape-memory alloys range from 30 to 80 GPa. Plasticity, evaluated through strain measurements such as percent elongation, indicates a material's ability to undergo permanent deformation. Aluminium typically exhibits 10–25% elongation, whereas shape-memory alloys demonstrate 8–15% elongation. These properties make shape-memory alloys particularly useful in robotics and medical devices, where tailored elastic and plastic responses are necessary.

Hardness, defined as a material's resistance to deformation or scratching, is another vital property of advanced materials. Advanced ceramics like silicon carbide and boron nitride are renowned for their exceptional hardness, making them indispensable for applications such as cutting tools, protective coatings, and wear-resistant surfaces.

Hardness measures a material's resistance to deformation or scratching and is expressed using units like the Vickers Hardness Number (VHN), Brinell Hardness Number (BHN), or GigaPascals (GPa) for microhardness evaluations. Mild steel typically exhibits hardness values of 100–300 VHN. Advanced ceramics like silicon carbide and boron nitride demonstrate exceptional hardness, with values ranging from 25 to 30 GPa and 30 to 50 GPa, respectively. This makes silicon carbide highly suitable for cutting tools and protective coatings due to its superior wear resistance.

In addition to these properties, lightweight and flexibility are increasingly important in the development of advanced materials. Lightweight materials, including aerogels and composite polymers, are essential in industries where weight reduction is critical, such as automotive and aerospace engineering. These materials offer not only reduced weight but also the flexibility needed to adapt to complex geometries and dynamic loads.

Lightweight and flexibility are critical for materials used in industries such as automotive and aerospace, where reducing weight is a priority. Lightweight materials are characterized by their density, measured in kilograms per cubic meter (kg/m^3). Common materials like steel have a density of ~7850 kg/m^3, while aluminium is ~2700 kg/m^3. Advanced materials like aerogels exhibit densities as low as 1–100 kg/m^3, and carbon fibre composites average around 1600 kg/m^3. Flexibility, often evaluated through the flexural modulus, is also expressed in GPa. Polymers generally range between 0.5 and 5 GPa, while carbon fibre composites achieve 100–150 GPa. Aerogels are widely used for thermal insulation due to their extremely low density, while carbon fibre composites combine lightweight properties with high flexibility, making them indispensable in aerospace and automotive engineering.

These properties are determined and measured using standardized tests and methodologies, which provide insights into how materials behave under specific conditions.

Strength is measured by subjecting a material to external forces and determining the maximum stress it can withstand before failure. This is typically evaluated through tensile, compressive, or flexural tests, where samples are pulled, compressed, or bent until they break. For example, carbon fibre-reinforced composites are subjected to tensile tests to measure their tensile strength, which is a critical parameter in aerospace applications. The strength-to-weight ratio of these materials makes them ideal for lightweight and robust structures.

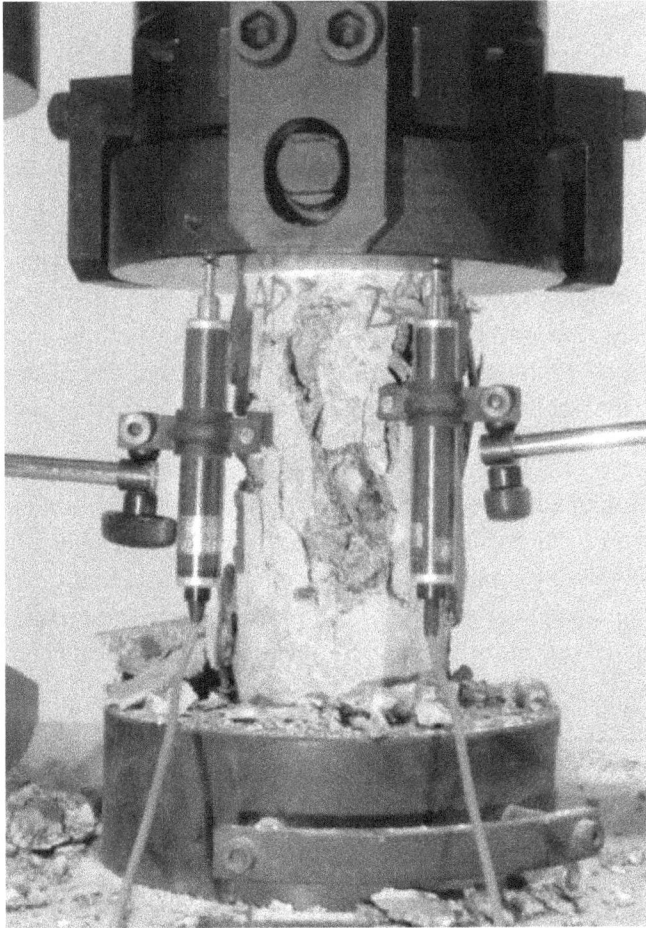

Figure 14: Compressive strength test of concrete in UTM. Nirmaljoshi, CC BY-SA 4.0, via Wikimedia Commons.

Strength refers to a material's ability to endure applied stress without failing, a property fundamental to structural and mechanical applications. It is typically quantified in units of Pascals (Pa) or MegaPascals (MPa) in engineering contexts. Tensile strength, a key measure, evaluates resistance to pulling forces. Common materials like steel exhibit tensile strengths ranging from 250 to 1500 MPa, while aluminium ranges from 70 to 500 MPa. Advanced materials such as carbon fibre-reinforced composites offer tensile strengths between 500 and 6000 MPa, and Kevlar achieves approximately 3600 MPa. These properties make materials like carbon fibre composites ideal for aerospace applications, where strength-to-weight ratios are critical. Compressive strength, another measure of material performance under compressive forces, is typically observed in materials like concrete (~20–40 MPa) and advanced ceramics such as silicon carbide (~2500 MPa).

Toughness is evaluated by measuring the energy a material can absorb before fracturing. This is often done using impact tests like the Charpy or Izod test, where a pendulum strikes a notched specimen to determine its ability to resist fracture. Titanium alloys, known for their toughness, are tested in this way to ensure they can endure impacts and deformation without breaking, which is vital for medical implants and other high-stress applications.

Elasticity and plasticity are assessed using stress-strain curves obtained from tensile tests. Elasticity is determined by measuring the material's ability to return to its original shape after being deformed, represented by the elastic modulus (Young's modulus). Plasticity, on the other hand, is observed beyond the elastic limit, where the material undergoes permanent deformation. Advanced materials like shape-memory alloys are tested for their unique behaviour, including their ability to recover their shape when heated, making them suitable for applications in robotics and medical devices.

Hardness is a measure of a material's resistance to localized deformation or scratching. It is commonly determined using techniques such as the Vickers, Brinell, or Rockwell hardness tests, which involve indenting the material with a specific force and measuring the depth or size of the indentation. Advanced ceramics like silicon carbide and boron nitride demonstrate high hardness levels, making them ideal for cutting tools and protective coatings.

Lightweight and flexibility are important parameters for advanced materials used in automotive and aerospace industries, where reducing weight is essential without compromising performance. The lightweight nature of materials like aerogels and composite polymers is determined by their density, measured through volume and mass calculations. Flexibility is assessed by testing the material's ability to bend or stretch without breaking, often using bending or tensile tests to quantify its ductility and resilience.

Thermal Properties

Thermal properties describe how advanced materials respond to changes in temperature, including their capacity to conduct heat, resist thermal stress, and maintain stability under extreme environmental conditions. These properties are critical for applications across industries such as electronics, aerospace, energy, and construction, where materials must reliably perform under varying temperature conditions.

Thermal conductivity is one of the key properties of advanced materials. Materials like graphene and diamond are highly conductive, allowing them to effectively dissipate heat in electronic devices, where managing thermal loads is critical for performance and durability. On the other hand, advanced insulators such as aerogels and phase-change materials are specifically designed to minimize heat transfer. These materials play a vital role in energy-efficient building designs, helping to reduce energy consumption by maintaining controlled temperatures.

Advanced Material Engineering Fundamentals

Thermal conductivity measures a material's ability to conduct heat and is expressed in watts per meter per Kelvin (W/m·K). This property determines how effectively a material can transfer heat under a temperature gradient. Materials with high thermal conductivity, such as diamond (up to 2000 W/m·K) and graphene (~5000 W/m·K), are ideal for applications like heat dissipation in electronics. In contrast, low-conductivity materials like aerogels (0.013–0.03 W/m·K) and phase-change materials are used as insulators in energy-efficient building designs. Common metals such as copper and aluminium have thermal conductivities of ~385 W/m·K and ~205 W/m·K, respectively, making them suitable for heat exchangers and electrical components.

Thermal expansion is another essential property that engineers carefully consider when designing advanced materials. Materials are often engineered to exhibit controlled thermal expansion to prevent deformation during temperature fluctuations. This property is especially crucial in applications such as spacecraft, where materials with low coefficients of thermal expansion ensure structural integrity despite the extreme temperature changes experienced in space.

Thermal expansion is quantified by the coefficient of thermal expansion (CTE), expressed in parts per million per Kelvin (ppm/K). This measures how much a material expands or contracts with temperature changes. Materials with controlled or low CTE, such as Invar (~1 ppm/K) or advanced ceramics (~0.1–10 ppm/K), are essential in applications requiring dimensional stability under temperature fluctuations, such as spacecraft components or precision instruments. Common materials like steel (~11–13 ppm/K) and aluminium (~23 ppm/K) exhibit higher CTE values, making them less suitable for such applications without modification.

Heat resistance is a hallmark of many advanced materials, enabling them to withstand very high temperatures without degrading. Refractory metals and ceramics, for example, are used in high-temperature applications like jet engines, turbines, and nuclear reactors. These materials are specifically designed to endure prolonged exposure to extreme heat while maintaining their structural and functional properties.

Heat resistance is defined by a material's ability to withstand high temperatures without degradation, often measured as the maximum service temperature in degrees Celsius (°C). Refractory metals like tungsten (3422°C) and molybdenum (2623°C) exhibit excellent heat resistance and are commonly used in jet engines and turbines. Advanced ceramics such as silicon carbide (~2700°C) and zirconia (~2400°C) are also valued for their high-temperature performance. Common materials like steel (~500–800°C) degrade at significantly lower temperatures, limiting their use in extreme environments.

Thermal stability is another critical property, ensuring that materials retain their mechanical characteristics at elevated temperatures. Superalloys, for instance, are specifically designed for use in high-stress, high-temperature environments. Their ability to maintain strength, toughness, and resistance to thermal creep under such conditions makes them indispensable in demanding applications such as gas turbines and power plants.

Thermal stability refers to a material's ability to retain its mechanical and chemical properties under prolonged exposure to high temperatures. It is typically evaluated through stress-strain testing at elevated temperatures or by measuring creep resistance. Superalloys like Inconel 718 (service temperature up to ~700–1000°C) and titanium alloys are widely used in aerospace and power generation due to their exceptional thermal stability. In contrast, common metals like aluminium (service temperature ~150–200°C) lose strength and deform more readily under heat.

Thermal insulation is equally important for applications requiring the minimization of heat transfer. Materials like aerogels and multilayer polymer composites are widely used as thermal insulators in various industries. These materials are essential for cryogenic storage, where maintaining extremely low temperatures is critical, and in thermal protection systems for spacecraft re-entry, where managing heat buildup is crucial for safety and performance.

Thermal insulation is characterized by a material's resistance to heat flow, commonly measured by its thermal resistivity or R-value. The unit of R-value depends on the system (e.g., $m^2 \cdot K/W$ in the SI system). Advanced materials like aerogels exhibit extremely high thermal resistivity, making them ideal for cryogenic storage and thermal protection systems, with R-values exceeding $10 \ m^2 \cdot K/W$. Common insulating materials like fiberglass (~2–3 $m^2 \cdot K/W$) are less effective but more cost-efficient for general use in building construction.

Thermal Conductivity is measured by determining a material's ability to transfer heat. This is typically evaluated using techniques like the steady-state method or transient techniques such as the laser flash method. In the steady-state method, a constant heat flow is applied to one side of a material, and the temperature difference across it is measured. Highly conductive materials like graphene and diamond exhibit exceptional thermal conductivity, making them suitable for dissipating heat in electronic devices. Conversely, materials with low thermal conductivity, such as aerogels and phase-change materials, are measured for their insulating properties in applications like energy-efficient construction.

Thermal Expansion is determined by measuring the dimensional changes of a material as it is heated or cooled. This is done using a dilatometer, which precisely measures the change in length or volume of a specimen as a function of temperature. Materials with low coefficients of thermal expansion, such as those used in spacecraft, are essential to prevent deformation during extreme temperature fluctuations. The linear coefficient of thermal expansion (CTE) is calculated from the change in dimensions over a specific temperature range.

Heat Resistance is evaluated by exposing materials to high temperatures and observing their structural and chemical stability. Refractory metals and ceramics, for instance, are tested in controlled environments to determine their melting points and resistance to thermal degradation. Techniques such as thermogravimetric analysis (TGA) and differential thermal analysis (DTA) are commonly used to measure how materials respond to heat over time, ensuring their suitability for high-temperature applications like jet engines and nuclear reactors.

Thermal Stability refers to a material's ability to maintain its mechanical properties at elevated temperatures. This property is assessed using creep testing, where a material is subjected to constant stress at high temperatures for extended periods. Superalloys, which are widely used in high-stress environments, are tested for their ability to resist deformation and maintain strength under thermal stress. Dynamic mechanical analysis (DMA) can also be employed to measure changes in stiffness and damping as a function of temperature.

Thermal Insulation is determined by evaluating a material's ability to resist heat transfer. This is often measured using guarded hot plate apparatus or heat flow meters, which assess the thermal resistance (R-value) of the material. Advanced insulators like aerogels and multilayer polymer composites are characterized by their low thermal conductivity, making them effective for applications in cryogenic storage and thermal protection systems. Thermal imaging and infrared thermography are also used to study heat distribution and insulation performance in real-world conditions.

Chemical Properties

The chemical properties of advanced materials define their behaviour in the presence of other substances, including their resistance to chemical reactions, environmental degradation, and their ability to interact with various elements or compounds. These properties are critical for ensuring the durability, functionality, and performance of materials in specific applications.

Corrosion resistance is a fundamental chemical property in advanced materials. Materials such as stainless steel, titanium alloys, and advanced coatings are engineered to withstand corrosive environments, including chemical exposure, seawater, and extreme weather conditions. This resistance is particularly vital in marine, automotive, and industrial applications, where prolonged exposure to harsh elements can compromise structural integrity and functionality.

Corrosion resistance is a material's ability to withstand chemical degradation, particularly in harsh environments such as exposure to seawater, acids, or oxidizing agents. It is typically measured in terms of the corrosion rate, expressed in millimetres per year (mm/year) or mils per year (mpy). Low corrosion rates (e.g., less than 0.1 mm/year) indicate high resistance. Stainless steel, commonly used in marine and automotive applications, exhibits corrosion rates of less than 0.01 mm/year in moderate environments, while titanium alloys are even more resistant, particularly in seawater. Advanced coatings like ceramic-based or polymer-based coatings further enhance resistance by acting as physical and chemical barriers.

Chemical stability is another key characteristic, allowing materials to endure contact with aggressive substances such as acids, bases, and oxidizing agents without degrading. Advanced ceramics and inert materials like Teflon are examples of highly stable materials used in chemical processing equipment, laboratory tools, and medical devices, where durability and resistance to chemical attacks are paramount.

Chemical stability refers to a material's resistance to chemical reactions that could degrade its structure. This property is assessed using techniques like chemical exposure tests, where materials are subjected to acids, bases, or oxidizing agents for specified durations, and the degradation is quantified. Stability is often expressed qualitatively or through weight loss percentages or reaction rates. Advanced ceramics like alumina and zirconia remain inert in highly acidic or basic environments, making them ideal for chemical processing. Teflon (PTFE), with its exceptional resistance to nearly all chemicals, finds applications in chemical reactors and medical devices.

Some advanced materials are engineered for specific chemical reactivity to enable targeted applications. Catalytic materials, such as zeolites and platinum-based nanoparticles, are designed to facilitate chemical reactions efficiently. These materials play a crucial role in industries like petrochemical refining and environmental remediation, where they help accelerate processes or break down harmful substances.

Reactivity is evaluated based on the material's tendency to participate in chemical reactions. This property is often quantified through reaction rates (measured in moles per second or specific catalytic activity), typically determined under controlled conditions. Catalytic materials like zeolites and platinum nanoparticles are assessed for their turnover frequency (TOF), a measure of their efficiency in facilitating reactions. For instance, zeolites exhibit high reactivity in petrochemical refining, while platinum catalysts are critical in environmental applications, such as vehicle emission control systems.

Surface properties of advanced materials often determine their interactions with the environment. Surface modifications, such as coatings or the functionalization of nanoparticles, can enhance chemical resistance or provide catalytic activity. For example, superhydrophobic coatings are engineered to repel water and resist contamination, making them ideal for self-cleaning surfaces, protective coatings, and anti-fouling applications.

Surface properties include chemical resistance, wettability, and catalytic activity. These are measured using techniques such as contact angle measurements (for wettability), X-ray photoelectron spectroscopy (XPS) for surface composition, and activity tests for catalytic surfaces. Wettability, for instance, is expressed in degrees of contact angle: superhydrophobic surfaces exhibit angles greater than 150°, as seen in self-cleaning coatings. Nanostructured materials with functionalized surfaces, such as titanium dioxide nanoparticles, exhibit enhanced catalytic properties and are used in photocatalytic applications.

Advanced Material Engineering Fundamentals

Biocompatibility is a critical property for materials used in healthcare and biomedical fields. Advanced biomaterials, such as hydroxyapatite and biodegradable polymers, are designed to interact safely with biological environments. These materials are commonly used in medical implants, tissue engineering, and drug delivery systems, where compatibility with human tissues and processes is essential for success.

Biocompatibility refers to a material's ability to interact safely with biological systems without causing adverse reactions. It is evaluated through biological assays, cytotoxicity tests, and in vivo studies. Units of measure include cell viability percentages or inflammatory response markers. Biomaterials like hydroxyapatite and polylactic acid (PLA) demonstrate excellent biocompatibility, making them suitable for bone implants and biodegradable drug delivery systems. Titanium alloys, widely used in orthopaedic implants, exhibit negligible reactivity with body tissues, ensuring compatibility.

Environmental interactions highlight the role of advanced materials in addressing global challenges. Materials like carbon capture compounds and photocatalysts are specifically designed to mitigate environmental issues. These materials exhibit unique chemical properties, such as the ability to absorb carbon dioxide or degrade pollutants under sunlight, contributing to sustainability and environmental protection efforts.

Environmental interactions describe a material's response to external environmental factors, such as pollutants or greenhouse gases. Metrics include absorption capacity (e.g., millimoles of CO_2 per gram for carbon capture materials) and degradation rates under specific conditions. Photocatalysts like titanium dioxide are assessed for their ability to degrade pollutants under UV light, measured in terms of the reduction in pollutant concentration over time. Advanced carbon capture materials, such as metal-organic frameworks (MOFs), can absorb CO_2 at rates exceeding 10 mmol/g, offering significant potential for addressing climate change.

Through the careful design and control of their chemical properties, advanced materials enable innovations across a wide range of industries, from healthcare and energy to environmental management and beyond. Their resistance to degradation, tailored reactivity, and environmentally friendly characteristics make them indispensable in modern technological and industrial advancements.

Corrosion Resistance is assessed by exposing materials to corrosive environments and evaluating their behaviour over time. Standard tests include salt spray testing (ASTM B117), where materials are subjected to a saline mist to simulate marine environments, and immersion tests in acidic or alkaline solutions. Electrochemical techniques, such as potentiodynamic polarization and electrochemical impedance spectroscopy (EIS), are also used to measure the rate of corrosion and the material's resistance to it. Materials like stainless steel and titanium alloys are engineered to resist degradation, making them ideal for marine, automotive, and industrial applications.

Chemical Stability is determined by exposing materials to harsh chemicals, including acids, bases, and oxidizing agents, and evaluating their ability to maintain structural integrity. Tests such as acid immersion and chemical soak tests assess material degradation over specific timeframes. Advanced ceramics and inert materials like Teflon are often subjected to high-temperature chemical environments to ensure their stability in chemical processing equipment and medical devices.

Reactivity is measured through chemical reaction rate analysis and catalytic performance tests. Materials like zeolites and platinum-based nanoparticles are tested in controlled reactors to evaluate their efficiency in catalysing specific chemical reactions. Techniques such as gas chromatography and mass spectrometry are used to monitor reaction byproducts and determine catalytic effectiveness in applications like petrochemical refining and pollution control.

Surface Properties are characterized using advanced microscopy and spectroscopy techniques. Tools like scanning electron microscopy (SEM) and atomic force microscopy (AFM) analyse surface topography and modifications. Wettability tests, such as contact angle measurements, assess the effectiveness of superhydrophobic coatings in repelling water and contamination. Surface functionality is also evaluated using X-ray photoelectron spectroscopy (XPS) and energy-dispersive X-ray spectroscopy (EDS) to analyse chemical composition and bonding states.

Biocompatibility is crucial for materials used in medical applications and is assessed through biological assays and in vitro tests. Cytotoxicity tests measure the material's effect on living cells, while protein adsorption and blood compatibility tests ensure safe interactions with biological systems. Advanced biomaterials like hydroxyapatite and biodegradable polymers undergo rigorous preclinical and clinical evaluations to confirm their safety and efficacy for implants, tissue engineering, and drug delivery.

Environmental Interactions are evaluated by testing materials in conditions that mimic environmental challenges. For example, carbon capture compounds are analysed in CO_2-rich environments using adsorption isotherms to measure their capacity and efficiency. Photocatalysts are tested under simulated sunlight or UV light to determine their ability to degrade pollutants or facilitate chemical reactions. These tests are crucial for developing sustainable solutions to environmental challenges.

Integration of Properties in Advanced Materials

Advanced materials often integrate mechanical, thermal, and chemical properties to meet the specific demands of high-performance applications. For example, carbon-carbon composites combine exceptional mechanical strength, thermal resistance, and chemical inertness, making them suitable for aerospace applications like rocket nozzles. Similarly, advanced polymers like Kevlar are lightweight, chemically resistant, and mechanically strong, enabling their use in bulletproof vests and protective gear.

The ability to tailor these properties through advanced processing techniques, such as nanostructuring, additive manufacturing, and defect engineering, has expanded the scope of materials science. Researchers continue to develop materials with optimized combinations of properties, addressing challenges in energy, healthcare, transportation, and beyond. This holistic understanding of mechanical, thermal, and chemical properties is crucial for designing materials that meet the complex needs of modern technology.

Crystallography and Material Defects

Crystallographic defects are fundamental interruptions in the orderly arrangement of atoms or molecules within crystalline solids, leading to deviations from the ideal periodic structure defined by the unit cell parameters. These defects can significantly influence the physical and mechanical properties of materials, making their characterization essential in materials science. The primary categories of crystallographic defects include point defects, line defects, planar defects, and bulk defects, each exhibiting unique characteristics and implications for material behaviour.

Kröger–Vink Notation

Kröger–Vink notation is a standardized method used in materials science and solid-state chemistry to describe point defects in crystals. It conveys information about the type, charge, and site of defects, helping to analyse defect chemistry and predict the behaviour of materials under different conditions.

This notation is particularly useful for ionic solids, where charge balance and defect interactions are critical. Defects are categorized based on the location in the lattice, their type (e.g., vacancy, interstitial, or substitutional), and their effective charge relative to the surrounding lattice.

Key Components of Kröger–Vink Notation:

1. Species of the Defect: The symbol for the atom, ion, or vacancy involved in the defect. For example:

- *V* represents a vacancy (missing atom or ion).

- *O* represents an oxygen atom or ion.

- *Zn* represents a zinc atom or ion.

2. Subscript Denoting the Site: Specifies where the defect is located. Common examples include:

- s for substitutional lattice sites.

- ii for interstitial sites.

- O, Zn, or other specific lattice sites.

3. Superscript Denoting the Charge: Indicates the effective charge of the defect relative to the perfect lattice:

- ' (prime) represents a negative charge.

- · (dot) represents a positive charge.

- x (cross) indicates no net charge (neutral defect).

The general format is:

$$\text{Defect}_{\text{site}}^{\text{charge}}$$

Examples of Kröger–Vink Notation:

1. Vacancy in an Oxygen Lattice Site: A missing oxygen ion in a lattice site is represented as:

$$V_O^{\cdot\cdot}$$

This indicates a doubly positively charged vacancy (missing O^{2-} at an oxygen site.

2. Interstitial Cation: A zinc ion (Zn^{2+}) occupying an interstitial site is represented as:

$$Zn_i^{\cdot\cdot}$$

This shows a doubly positively charged ion at an interstitial position.

3. Substitutional Defect: A calcium ion (Ca^{2+}) replacing a sodium ion (Na^+) in a lattice is represented as:

$$Ca_{Na}^{\cdot}$$

This indicates a singly positively charged substitutional defect.

4. Frenkel Defect in a Crystal: A Frenkel defect involves a cation vacancy and an interstitial cation. For silver (Ag) in silver chloride ($AgCl$):

- The vacancy is:

$$V'_{Ag}$$

(negatively charged due to the missing Ag^+).

- The interstitial is:

$$Ag_i^{\cdot}$$

(positively charged due to the Ag^+ ion in an interstitial site).

5. Aliovalent Substitution: A Al^{3+} ion substituting for a Mg^{2+} ion in a lattice creates a charge imbalance:

$$Al'_{Mg}$$

This indicates a singly negatively charged substitutional defect.

Applications of Kröger–Vink Notation:

- **Predicting Ionic Conductivity**: In materials like yttria-stabilized zirconia (ZrO_2), oxygen vacancies are described as $V_O^{\cdot\cdot}$, which are critical for oxygen ion transport in fuel cells.

- **Describing Defect Chemistry**: In semiconductors, defects such as dopants or vacancies are described using this notation to understand electrical behaviour. For example, a phosphorus dopant in silicon might be represented as $P_{Si}^{\cdot}$

- **Analysing Solid-State Reactions**: In reactions involving defect formation, such as oxidation or reduction, Kröger–Vink notation provides a concise way to represent changes in defect concentrations.

Point Defects

Point defects are critical imperfections in the crystal structure of materials, occurring at or around a single lattice point and influencing various material properties. These defects can manifest as vacancies, interstitials, or substitutions, each playing a significant role in determining the physical, chemical, and electrical characteristics of materials. For instance, vacancies arise when a lattice site that should be occupied by an atom is vacant, disrupting the ordered arrangement of atoms in a perfect crystal. This disruption can lead to neighbouring atoms moving to fill the vacancy, facilitating ionic transport essential for electrochemical reactions [99, 100]. The stability of the surrounding structure is crucial, as it prevents atoms from collapsing into the vacancy, a phenomenon particularly relevant in ionic solids where vacancy pairs, known as Schottky defects, maintain charge neutrality [101].

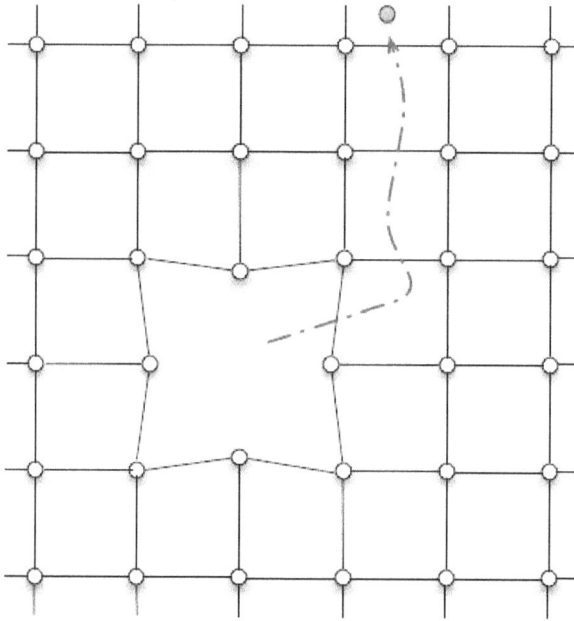

Figure 15: Point defect in a crystal lattice. Safe cracker, CC BY 3.0, via Wikimedia Commons.

Interstitial defects occur when additional atoms occupy positions within the crystal lattice that are typically unoccupied. These configurations can arise from impurities or naturally occurring processes. For example, small atoms such as hydrogen can occupy interstitial sites in materials like palladium, significantly affecting their properties [102]. The presence of interstitial defects can lead to high-energy configurations that alter the electronic and mechanical properties of the material [102]. Furthermore, Frenkel defects, which consist of a vacancy and an interstitial defect in close proximity, are common in ionic solids and have notable implications for ionic conductivity and material behaviour under stress [103].

Substitutional defects arise when impurity atoms replace regular atoms in the lattice. The impact of these defects is contingent upon the size and charge of the substituting atom. For instance, if the impurity atom is significantly smaller than the atom it replaces, this can lead to off-centre ions and affect the material's charge balance [102]. Substitutional defects can be classified as isovalent or aliovalent, with aliovalent substitutions necessitating additional mechanisms, such as the creation of vacancies, to maintain charge neutrality [104]. Antisite defects, which occur when atoms exchange positions in ordered alloys, also disrupt the regularity of the crystal structure without necessarily creating vacancies or interstitials [102].

Topological defects represent another category of imperfections, where the chemical bonding environment deviates from the surrounding structure. In materials like graphene, variations in the number of atoms in the rings can lead to defects such as the Stone-Wales defect, which consists of adjacent five- and seven-membered rings [105]. These defects can significantly influence the electronic properties and stability of the material.

In addition to crystalline materials, point defects also exist in amorphous solids. For example, in amorphous silica, dangling bonds—oxygen atoms with only one silicon bond—are considered defects that can affect optical and mechanical properties [102]. The interaction of different types of defects can lead to complex defect structures, such as vacancy-impurity complexes or split interstitials, which can further modify material behaviour [106].

Point defects play a significant role in determining the properties of advanced materials. The following table (Table 1) presents some examples of advanced materials that benefit from point defects and their applications:

Table 1: Advanced materials, their associated point defects, and the roles these defects play in enabling innovative applications across various industries.

Material	Defect Type	Application	Role of Point Defects
Yttria-Stabilized Zirconia (YSZ)	Oxygen vacancies	Solid oxide fuel cells (SOFCs)	Facilitate oxygen ion transport, enhancing ionic conductivity for high-temperature electrochemical devices.
Doped Silicon	Substitutional dopants (e.g., P, B)	Semiconductors and microelectronics	Dopant atoms replace silicon atoms, introducing free carriers (electrons/holes) for electrical conductivity.
Gallium Arsenide (GaAs)	Antisite defects and vacancies	Optoelectronic devices (lasers, LEDs)	Influence electronic and optical properties for efficient light-to-electric signal conversion.
Titanium Dioxide (TiO₂)	Oxygen vacancies	Photocatalysts and solar cells	Introduce energy levels within the bandgap, improving

Material	Defect Type	Application	Role of Point Defects
			photocatalytic activity and light absorption.
Perovskite Oxides (e.g., $SrTiO_3$)	Oxygen vacancies and substitutional dopants	Fuel cells, memristors, dielectric materials	Enhance ionic conductivity, dielectric properties, or memory storage capabilities.
Graphene and Carbon Nanotubes	Vacancies, Stone-Wales defects, substitutional impurities	Electronics, sensors, composite materials	Modify electronic properties, enhance chemical reactivity, or improve interaction with other components.
Cerium Oxide (CeO_2)	Oxygen vacancies	Catalysts for automotive emissions and fuel cells	Improve redox reactions, enhancing catalytic efficiency.
Zinc Oxide (ZnO)	Oxygen vacancies and zinc interstitials	Transparent conductive oxides, UV light detectors	Tune electrical conductivity and optical transparency for optoelectronic applications.
Magnesium Aluminate Spinel ($MgAl_2O_4$)	Cation vacancies	Transparent ceramics for armour and aerospace	Enhance optical clarity and mechanical strength for high-performance applications.
Nitride-Based Semiconductors (GaN, InGaN)	Vacancies and substitutional impurities	LEDs, laser diodes, high-electron-mobility transistors	Control optical emission efficiency and electrical properties for optoelectronics.
Lithium-Ion Battery Materials (e.g., $LiFePO_4$, $LiCoO_2$)	Lithium vacancies and interstitials	Rechargeable batteries	Enable lithium-ion transport, influencing battery capacity and cycling stability.
Transition Metal Dichalcogenides (e.g., MoS_2, WS_2)	Vacancies and substitutional doping	2D materials for electronics and energy storage	Tailor electronic band structure and catalytic properties for advanced applications.

Advanced materials, their associated point defects, and the roles these defects play in enabling innovative applications across various industries.

Line Defects

Line defects, commonly referred to as dislocations, are critical one-dimensional imperfections in crystal lattices that significantly influence the mechanical and electrical properties of materials. These defects disrupt the orderly arrangement of atoms, leading to variations in material behaviour under stress and thermal conditions. The understanding of line defects is essential in materials science, particularly in the context of dislocation theory, which describes their behaviour and interactions within the lattice structure [107, 108].

Dislocations can be primarily categorized into two types: edge dislocations and screw dislocations. Edge dislocations occur when an extra half-plane of atoms is introduced into the crystal structure, causing a distortion that can be visualized as an additional layer of paper inserted into a stack. The surrounding atomic planes bend to accommodate this defect, maintaining the overall structure on either side of the dislocation line. The Burgers vector, which is perpendicular to the dislocation line, quantifies the magnitude and direction of this distortion [107, 109]. In contrast, screw dislocations involve a helical arrangement of atomic planes around the dislocation line, with the Burgers vector aligned parallel to the dislocation line. This creates a spiral distortion that can be likened to a ramp extending through the material [108].

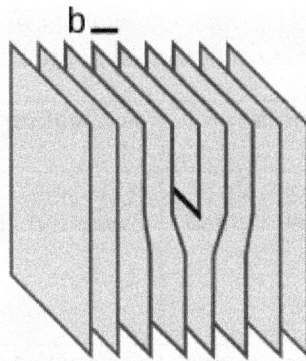

Figure 16: Simplified atomic plane diagram of an edge dislocation. Magasjukur2, CC BY-SA 2.5, via Wikimedia Commons.

The Burgers vector is a pivotal parameter in characterizing dislocations, as it directly correlates with the lattice strain induced by the dislocation. In metallic materials, the mobility of dislocations under applied stress is crucial for plastic deformation, allowing materials to exhibit ductility and malleability without fracturing [107, 110]. The ability of dislocations to move and interact under stress is fundamental to understanding the mechanical properties of materials, particularly in the context of their strength and toughness [109].

Advanced techniques such as transmission electron microscopy (TEM) and atom probe tomography are employed to observe and characterize dislocations at the atomic level. These methods provide insights into the structural and electronic properties of materials, revealing how dislocations affect carrier mobility and recombination rates in semiconductors [107, 111]. Additionally, disclinations, which are another type of line defect characterized by rotational distortions, have been shown to play a role in the mechanical properties of materials, potentially enabling self-healing mechanisms by redistributing stress [108].

The significance of line defects extends beyond mechanical properties; they also influence the electronic characteristics of materials. For instance, in two-dimensional materials like graphene and boron nitride, line defects can modulate electronic band structures and enhance magnetic properties [107]. Research indicates that the presence of line defects can lead to the emergence of localized states that alter the electronic behaviour of materials, which is particularly relevant in the development of advanced electronic devices [107].

By influencing mechanical strength, ductility, electrical conductivity, and other characteristics, these defects can be engineered or controlled to optimize material performance. Below (Table 2) are examples of advanced materials that benefit from line defects and their corresponding applications:

Table 2: Engineered line defects in advanced materials, enabling exceptional performance in a variety of cutting-edge applications.

Advanced Material	Line Defects	Application	Role of Line Defects
Single-Crystal Turbine Blades (Nickel-Based Superalloys)	Dislocations are deliberately limited.	Jet engines and power turbines.	Enhance creep resistance and maintain structural integrity under extreme stress and high temperatures.
High-Strength Steels	Dislocations pinned or impeded by alloying elements.	Automotive and construction industries.	Strengthening through strain and precipitation hardening, restricting dislocation movement.
Titanium Alloys	Mixed dislocations due to alloying and processing.	Aerospace, biomedical implants, marine uses.	Dislocation interactions provide a balance of strength, toughness, and corrosion resistance.
Graphene and Carbon Nanotubes	Stone-Wales defects and edge dislocations.	Nanoelectronics, sensors, composites.	Modify electronic properties, improve chemical reactivity for

Advanced Material	Line Defects	Application	Role of Line Defects
			functionalization, and enhance mechanical strength.
Silicon Wafers in Semiconductors	Slip bands and threading dislocations.	Integrated circuits and solar cells.	Minimize dislocations to ensure high electrical performance and carrier mobility.
Shape-Memory Alloys (e.g., NiTi)	Dislocation networks interacting with martensitic transformations.	Actuators, medical stents, robotics.	Facilitate reversible deformation, enabling materials to "remember" their original shape.
Perovskite Oxides (e.g., $SrTiO_3$)	Dislocations coupled with oxygen vacancies.	Solid oxide fuel cells, memristors.	Enhance ionic conductivity and create pathways for electron transport, improving device performance.
High-Entropy Alloys (HEAs)	Complex dislocation interactions from multi-element distribution.	Aerospace, nuclear reactors.	Improve strength and toughness by increasing lattice friction and impeding dislocation motion.
Gallium Nitride (GaN) for LEDs	Threading dislocations.	LEDs and high-power electronics.	Minimize non-radiative recombination, enhancing optical and electrical efficiency.
Ceramic Matrix Composites (CMCs)	Grain boundary dislocations and interphase strains.	Heat shields, turbine components.	Control dislocation density to improve thermal shock resistance and mechanical durability.
Transition Metal Dichalcogenides (TMDs, e.g., MoS_2)	Screw and grain boundary dislocations.	Lubricants, 2D electronics, catalysis.	Enhance catalytic activity and mechanical strength.
Nuclear Reactor Materials (e.g., Zr Alloys)	Dislocations formed under radiation damage.	Nuclear fuel cladding.	Serve as sinks for radiation-induced defects, reducing material degradation.
Ultra-Fine-Grained Metals	High-density dislocations at grain boundaries.	Lightweight structural materials.	Increase yield strength through grain boundary strengthening.

Advanced Material	Line Defects	Application	Role of Line Defects
Topological Insulators	Screw dislocations altering surface states.	Quantum computing, spintronics.	Modify electronic band structures to enable unique edge conduction properties.
Amorphous Metals (Metallic Glasses)	Dislocation-like shear bands.	High-strength gears, medical devices.	Enable localized deformation, improving toughness and resistance to wear.

Planar defects

Planar defects in crystalline materials are critical to understanding various phenomena in materials science, particularly in the context of stacking faults, grain boundaries, and twin boundaries. These defects can significantly influence the physical properties of materials, including their mechanical strength, electrical conductivity, and adsorption characteristics.

Grain boundaries are planar defects that occur when two crystals with different orientations meet. This abrupt change in the crystallographic direction can lead to a variety of properties that differ from those of the bulk material. For instance, the presence of grain boundaries can enhance the diffusion of atoms and affect the mechanical properties of the material [112, 113]. The interaction between the grains and the surrounding environment can also influence the adsorption of molecules, as defects can create localized energy states that affect how substances interact with the surface [114].

Antiphase boundaries, which are particularly relevant in ordered alloys, maintain the same crystallographic direction but exhibit opposite phases on either side of the boundary. This phenomenon can be observed in structures such as hexagonal close-packed crystals, where the stacking sequence might alternate in a manner that disrupts the periodicity of the lattice. The presence of such boundaries can lead to unique electronic and mechanical properties, as they can act as sites for localized strain and stress concentration.

Stacking faults are another type of planar defect commonly found in close-packed structures. They arise from a local deviation in the stacking sequence of atomic layers, which can be illustrated by sequences such as ABABCABAB. These faults can significantly alter the material's properties, including its mechanical strength and electrical conductivity [112, 113, 115]. The formation of stacking faults is influenced by various factors, including temperature and the rate of crystal growth, and they can serve as nucleation sites for further defects or phase transformations [113, 115].

Twin boundaries introduce a plane of mirror symmetry within the crystal structure, which can be particularly important in cubic close-packed crystals. The stacking sequence across a twin boundary may resemble ABCABCBACBA, which can lead to unique mechanical properties such as increased ductility and toughness. The presence of twin boundaries can also affect the material's response to external stresses, making them a focal point for research into improving material performance.

Additionally, steps between atomically flat terraces on single crystal planes can be considered planar defects. These steps can significantly influence the adsorption of organic molecules, as the geometry of the defects can create preferential sites for adsorption, thereby affecting the overall interaction between the surface and adsorbates [116, 117]. The presence of such defects can enhance the reactivity of the surface, leading to increased adsorption rates and altered molecular arrangements on the surface.

Table 3 below demonstrates the significant impact of planar defects, including grain boundaries, twin boundaries, and stacking faults, on the properties and applications of advanced materials across diverse industries.

Table 3: How planar defects, such as grain boundaries, twin boundaries, and stacking faults, critically influence the properties and applications of advanced materials across various industries.

Advanced Material	Planar Defects	Application	Role of Planar Defects
Polycrystalline Silicon	Grain boundaries.	Solar cells, microelectronics.	Grain boundaries serve as barriers to charge carrier flow, influencing electrical conductivity and efficiency.
Graphene	Grain boundaries and twin boundaries.	Electronics, sensors, composites.	Grain boundaries influence electrical and thermal conductivity and mechanical strength; defects can enhance reactivity for functionalization.
High-Performance Ceramics	Grain boundaries and phase interfaces.	Cutting tools, thermal barriers.	Planar defects at grain boundaries improve toughness and thermal resistance but may reduce electrical conductivity.
Superalloys (e.g., Ni-based)	Grain boundaries and stacking faults.	Jet engines, power turbines.	Controlled planar defects increase creep resistance and thermal stability at high temperatures.

Advanced Material	Planar Defects	Application	Role of Planar Defects
Perovskite Oxides (e.g., $SrTiO_3$)	Twin boundaries and grain boundaries.	Fuel cells, capacitors, memristors.	Twin boundaries enhance dielectric properties and influence ionic conductivity critical for energy storage and conversion devices.
Carbon Nanotube (CNT) Composites	Stacking faults and interlayer boundaries.	Lightweight structural materials.	Planar defects between CNT layers improve load transfer and mechanical reinforcement in composites.
Magnesium Alloys	Grain boundaries and twin boundaries.	Automotive, aerospace, lightweight designs.	Planar defects like twin boundaries improve ductility and toughness in lightweight applications.
Metal Matrix Composites (MMCs)	Grain boundaries and interphase boundaries.	Aerospace, automotive.	Planar defects strengthen the material by impeding dislocation motion and improving stress transfer between matrix and reinforcement.
Gallium Nitride (GaN)	Grain boundaries and stacking faults.	LEDs, high-power electronics.	Planar defects influence optical and electrical properties, affecting device efficiency and performance.
Nanocrystalline Metals	High-density grain boundaries.	Wear-resistant coatings, structural components.	Grain boundaries strengthen materials by impeding dislocation motion (grain boundary strengthening).
Layered Transition Metal Dichalcogenides (TMDs)	Stacking faults and grain boundaries.	Catalysis, electronics, energy storage.	Defects influence electronic band structures, enhancing catalytic activity and tuning electrical properties.
Titanium Alloys	Grain boundaries and twin boundaries.	Aerospace, medical implants.	Controlled planar defects balance mechanical strength and toughness for high-performance applications.

Advanced Material	Planar Defects	Application	Role of Planar Defects
Amorphous Metals (Metallic Glasses)	Interfacial boundaries and shear bands.	Medical devices, structural components.	Planar defects enable localized deformation, improving wear resistance and toughness.
Ferritic-Martensitic Steels	Grain boundaries and lath boundaries.	Nuclear reactors, thermal power plants.	Grain and lath boundaries improve creep resistance and thermal stability under extreme operating conditions.
Boron Nitride Nanotubes (BNNTs)	Stacking faults and interlayer boundaries.	Thermal insulation, composite materials.	Planar defects affect mechanical reinforcement and thermal management properties, making BNNTs ideal for high-temperature environments.
Zinc Oxide (ZnO)	Grain boundaries and twin boundaries.	Transparent conductive films, UV sensors.	Grain boundaries influence carrier transport and optical properties critical for optoelectronic applications.
Transparent Ceramics (e.g., AlON)	Grain boundaries and twin boundaries.	Armor, aerospace optics.	Planar defects are controlled to optimize transparency and mechanical strength in high-performance optical applications.
Cobalt-Based Alloys	Stacking faults and twin boundaries.	Biomedical implants, turbines.	Planar defects improve wear resistance and mechanical durability for medical and industrial applications.
Layered Perovskites (e.g., Ruddlesden-Popper phases)	Stacking faults and grain boundaries.	Catalysts, batteries, fuel cells.	Planar defects enhance ionic and electronic conductivity, improving catalytic and energy storage properties.
Quartz and Silicon Carbide (SiC)	Grain boundaries and planar stacking faults.	High-temperature electronics.	Defects impact piezoelectric properties and thermal conductivity, making them crucial in high-performance

Advanced Material	Planar Defects	Application	Role of Planar Defects
			electronic and sensor applications.

Bulk Defects

Bulk defects, which encompass three-dimensional irregularities in materials, significantly impact their mechanical properties and overall performance. These defects include pores, cracks, voids, and inclusions, each of which can alter the material's strength, toughness, and durability under various conditions.

Pores are a prevalent type of bulk defect, often formed during manufacturing processes such as casting, sintering, or welding. They act as stress concentrators, leading to a reduction in mechanical strength and an increased likelihood of crack initiation and propagation under load. For instance, studies have shown that the compressive strength of porous ceramics is closely related to their pore structure, with smaller pore sizes generally correlating with higher strength [118]. Additionally, the distribution of pores within a material can significantly influence its mechanical properties, as demonstrated by research indicating that a uniform pore size distribution enhances the overall strength of aluminium foams [119]. Furthermore, the relationship between pore structure and mechanical performance has been extensively documented, highlighting the critical need to control pore formation in applications requiring high structural integrity, such as in aerospace and automotive components [120, 121].

Cracks represent another significant form of bulk defect, often resulting from mechanical stress, thermal cycling, or processing flaws. These defects serve as weak points within materials, drastically reducing their ability to withstand tensile forces. The propagation of cracks can be influenced by the presence of microcracks and pores, which may either shield the crack tip or exacerbate crack growth under repeated loading [122]. Advanced non-destructive testing techniques, such as ultrasonic testing and X-ray computed tomography, are essential for detecting and monitoring cracks in critical applications, ensuring the reliability and safety of structural components [122, 123].

Voids, which are regions devoid of atoms, can also compromise material integrity, particularly in high-temperature or radiation-exposed environments. The growth of voids over time can lead to significant degradation of materials, as observed in graphite used in nuclear reactors [122]. The understanding of void dynamics is crucial for predicting material behaviour under extreme conditions, as voids can coalesce and lead to catastrophic failure if not properly managed [124].

Inclusions, or foreign materials embedded within a host material, can either enhance or detract from mechanical properties depending on their size, shape, and distribution. While small, uniformly distributed inclusions may reinforce certain composites, larger or irregular inclusions often act as stress concentrators, reducing toughness [125, 126]. The interaction between inclusions and the matrix material is complex and can lead to varied outcomes in terms of mechanical performance, necessitating careful consideration during material design and processing [126].

Precipitates, which are clusters of impurities or alloying elements that form distinct phases within a material, can also influence mechanical properties. While controlled precipitate formation can enhance strength, uncontrolled growth may lead to embrittlement [126]. The balance between beneficial and detrimental effects of precipitates is critical in the development of high-performance alloys, particularly in aerospace applications where material reliability is paramount [126].

Bulk defects, such as pores, cracks, voids, and inclusions, can significantly influence the properties of advanced materials. While often associated with weaknesses, these defects can also be controlled or utilized to enhance material performance in various applications as outlined below (Table 4) with examples of advanced materials where bulk defects play a critical role:

Table 4: Examples of advanced materials where bulk defects play a critical role.

Advanced Material	Type of Bulk Defect	Application	Role of Bulk Defects
Aerogels	Controlled porosity	Thermal insulation, energy storage, catalysts	Pores provide ultralow density, high surface area, and excellent thermal insulation properties.
Metallic Foams	Porosity	Lightweight structures, impact absorption	Pores reduce weight and improve energy absorption, making them ideal for automotive and aerospace.
Porous Silicon	Controlled voids and pores	Biosensors, drug delivery, and photovoltaics	Void structures increase surface area, enhancing chemical reactivity and optical properties.
Ceramic Matrix Composites (CMCs)	Microcracks and pores	High-temperature turbines, aerospace components	Microcracks improve thermal shock resistance; controlled pores reduce brittleness.
Sintered Metals	Residual pores from sintering	Filters, structural components	Controlled porosity aids in filtration and fluid flow while maintaining mechanical strength.

Advanced Material	Type of Bulk Defect	Application	Role of Bulk Defects
High-Entropy Alloys (HEAs)	Voids and inclusions	Nuclear reactors, extreme environments	Voids and inclusions are minimized to enhance strength, toughness, and radiation resistance.
Metal-Organic Frameworks (MOFs)	Structural porosity	Gas storage, separation, and catalysis	Large, well-defined pores allow for gas adsorption and molecular sieving.
Lightweight Concrete	Entrained air voids	Construction, earthquake-resistant structures	Voids reduce density and improve thermal insulation and energy absorption properties.
Transparent Ceramics	Inclusions and voids	Optical devices, armour	Controlled inclusions improve transparency while minimizing scattering and strength loss.
Graphene Aerogels	Porous structure	Energy storage, sensors, and environmental remediation	Bulk porosity provides ultralight properties and high electrical conductivity.
Nuclear Reactor Materials (e.g., Zr Alloys)	Voids from radiation damage	Fuel cladding, reactor components	Voids trap radiation-induced defects, reducing swelling and maintaining dimensional stability.
Additively Manufactured Metals	Residual pores and inclusions	Aerospace, automotive	Optimized bulk defects improve weight-to-strength ratio and impact resistance.
Catalyst Supports (e.g., Alumina)	Structural porosity	Catalysis, petrochemical refining	Pores enhance surface area, improving catalyst dispersion and reaction efficiency.
Foamed Polymers	Porosity	Packaging, thermal insulation, buoyancy aids	Bulk porosity reduces weight while providing cushioning and insulation properties.
Hydroxyapatite Scaffolds	Porous network	Bone regeneration, implants	Pores facilitate cell growth and nutrient transport, critical for biomedical applications.

Advanced Material	Type of Bulk Defect	Application	Role of Bulk Defects
Advanced Composites	Voids and fibre/matrix inclusions	Aerospace, defence	Controlled voids reduce weight, while inclusions reinforce strength and toughness.
Ultra-High-Performance Concrete (UHPC)	Microcracks and inclusions	Bridges, high-load construction	Microcracks improve durability and energy absorption in dynamic loading conditions.
Metallic Glasses (Amorphous Metals)	Shear bands and voids	Medical devices, structural applications	Bulk defects enhance toughness and enable energy absorption under impact.
Porous Bio-ceramics	Controlled porosity	Dental implants, bone scaffolds	Pores mimic natural bone structures, aiding integration and healing.
Polymeric Foams	Gas-filled voids	Insulation, flotation devices	Controlled bulk defects provide low density and thermal insulation.

These examples illustrate the dual nature of bulk defects in advanced materials. While often a source of weakness, bulk defects can also be engineered to optimize functionality, such as enhancing thermal, mechanical, and catalytic properties. By carefully designing and controlling these defects, material scientists enable novel applications across industries, from construction and aerospace to medicine and energy storage.

Material Characterization Techniques

Material characterization techniques are essential tools used in materials science to understand and analyse the structure, composition, properties, and performance of materials. These techniques are critical for developing advanced materials, optimizing their performance, and ensuring their suitability for specific applications. The techniques can be broadly categorized into structural analysis, compositional analysis, and property measurement methods.

Structural Characterization Techniques

Structural characterization techniques are essential in materials science for investigating the arrangement of atoms, molecules, and microstructures within a material. These techniques enable scientists to understand the internal and external structures of materials, which directly influence their properties and performance.

X-Ray Diffraction (XRD): XRD is a powerful tool for determining the crystal structure, phase composition, and lattice parameters of crystalline materials. When X-rays are directed at a material, they are scattered by the atomic planes within the crystal lattice. The resulting diffraction pattern reveals information about the material's internal structure, including symmetry, unit cell dimensions, and the presence of defects. XRD is widely used in analysing polycrystalline materials such as metals, ceramics, and minerals, offering insights critical for applications in construction, electronics, and aerospace. For instance, XRD can confirm the phase composition of a ceramic material used in thermal barrier coatings.

Figure 17: Dr. Ching-Hua Su, Materials Researcher, At In-Situ X-Ray Diffraction Imaging and Scattering Facility. NASA, Public Domain, via Picryl.

Using X-Ray Diffraction (XRD):

X-Ray Diffraction (XRD) is a powerful analytical technique used to study the crystal structure, phase composition, and lattice parameters of crystalline materials. Here's how to use XRD effectively:

1. Preparation of the Sample

- **Material State:** Ensure the sample is in a solid state. Powdered samples are often preferred for bulk materials, but thin films or single crystals may also be analysed.

- **Sample Size:** Grind the material into a fine powder to create uniformity if analysing a bulk sample. This ensures that the sample contains randomly oriented crystallites.

- **Mounting:** Place the sample on a flat sample holder or substrate to ensure even exposure to the X-ray beam.

- **Thickness and Surface:** Ensure the sample is not too thick, as excessive thickness can cause X-ray absorption and reduce accuracy. The surface should be smooth and free of contaminants.

2. Setting Up the XRD Instrument

- **Select the Target Material for the X-ray Tube:** Commonly used materials are copper (Cu) or molybdenum (Mo), as they generate characteristic X-rays suitable for diffraction.

- **Calibrate the Instrument:** Perform a calibration using a standard material with a known diffraction pattern (e.g., silicon or quartz).

- **Detector Setup:** Ensure the detector is positioned to capture diffracted X-rays accurately. Modern XRD systems often use 2θ detection, where 2θ is the angle of diffraction.

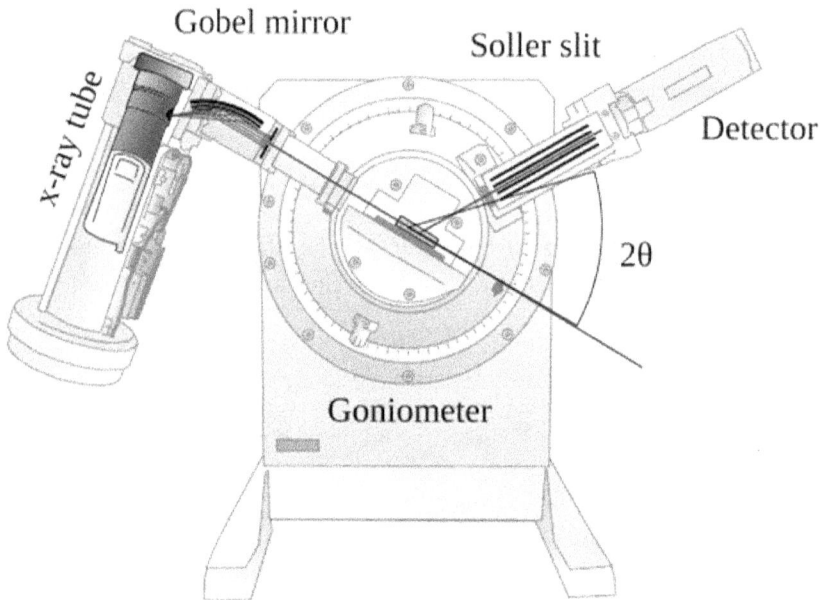

Figure 18: This is a drawing of a XRD diffractometer (modelled after an image of Siemens D5000) in grazing incidence x-ray diffraction configuration. DrBoStefanov, CC BY-SA 4.0, via Wikimedia Commons.

3. Choosing the Measurement Mode

- **Bragg-Brentano Mode:** This is the most common mode for powder XRD. The sample is stationary while the X-ray tube and detector move to record diffraction patterns.

- **Grazing Incidence XRD:** For thin films and coatings, this mode uses a low-angle X-ray beam to maximize surface sensitivity.

- **High-Resolution XRD:** Used for single-crystal samples or to analyse epitaxial layers in semiconductors.

Bragg-Brentano Mode: The Bragg-Brentano mode is one of the most widely used configurations for powder X-ray diffraction (XRD). It is specifically designed for analysing polycrystalline samples, such as powders, ceramics, and bulk materials. In this mode, the sample remains stationary while the X-ray tube and the detector move in a coordinated manner to maintain a constant angle between the X-ray beam and the detector as diffraction patterns are recorded. This alignment ensures that the diffraction condition is met consistently across various crystallographic planes within the sample. The resulting diffraction pattern provides information about the crystal structure, phase composition, and lattice parameters of the material. The Bragg-Brentano mode is particularly valued for its simplicity and effectiveness in identifying and quantifying phases in materials research and industrial quality control.

Grazing Incidence XRD: Grazing Incidence XRD (GIXRD) is a specialized technique tailored for studying thin films, coatings, and surface layers. In GIXRD, the X-ray beam is directed at the sample at a very low angle, typically just above the critical angle for total reflection. This configuration enhances the surface sensitivity of the measurement, allowing for detailed analysis of thin films and surface modifications without significant penetration into the underlying substrate. GIXRD is instrumental in determining the crystallographic orientation, phase composition, and thickness of thin films. It is widely used in fields like microelectronics, photovoltaics, and advanced coatings, where understanding surface and interface properties is crucial for optimizing material performance.

High-Resolution XRD: High-Resolution X-ray Diffraction (HRXRD) is a technique designed for the precise analysis of single-crystal materials and epitaxial layers, particularly in semiconductors. HRXRD employs a highly collimated and monochromatic X-ray beam to achieve fine angular resolution, making it capable of detecting subtle differences in lattice parameters, strain, and layer thickness. This method is essential for characterizing epitaxial layers, which are thin crystalline films grown on a substrate with a matching lattice structure. By analysing rocking curves and reciprocal space maps, HRXRD provides critical insights into the structural quality, strain distribution, and defect density in advanced materials. HRXRD is indispensable in the development of semiconductor devices, optoelectronics, and other technologies where material precision is paramount.

4. Running the Measurement

- **Adjust Parameters:**

  - **X-ray Beam Intensity:** Set the appropriate beam intensity for the sample material and desired resolution.

  - **Scan Range:** Define the 2θ range (e.g., 10° to 90°) to capture relevant diffraction peaks.

  - **Step Size:** Smaller step sizes provide higher resolution but increase scan time.

  - **Scan Speed:** Choose a speed that balances resolution and time efficiency.

- **Begin the Scan:** Start the XRD measurement. The instrument will emit X-rays that interact with the sample, producing a diffraction pattern captured by the detector.

5. Analysing the Diffraction Pattern

- **Identify Peaks:** The diffraction pattern consists of peaks at specific 2θ angles corresponding to the crystal planes in the material.

- **Compare with Standards:** Match the obtained pattern with standard reference databases (e.g., the International Centre for Diffraction Data (ICDD)) to identify phases and crystal structures.

- **Calculate Parameters:**

  - Use Bragg's Law: $n\lambda = 2d\sin\theta$, where:

    - n: Order of diffraction (usually 1).

    - λ: Wavelength of the X-rays.

    - d: Interplanar spacing of the crystal planes.

    - θ: Diffraction angle.

  - Determine lattice parameters, crystallite size, and strain.

6. Interpreting Results

- **Phase Identification:** Identify the crystalline phases present and their proportions.

- **Crystal Structure:** Confirm the lattice type (e.g., cubic, hexagonal) and calculate lattice constants.

- **Texture and Orientation:** Analyse preferred grain orientation in polycrystalline samples.

- **Defect Analysis:** Study peak broadening for insights into defects, dislocations, and macrostrain.

7. Reporting and Documenting

- **Graphical Representation:** Plot the diffraction pattern (intensity vs. 2θ).

- **Quantitative Data:** Provide tables of identified phases, lattice parameters, and other calculated results.

- **Error Analysis:** Include any uncertainties or potential sources of error in the results.

X-Ray Diffraction (XRD) has a wide range of applications that make it indispensable in materials science and engineering. One of its primary uses is in material identification, where it helps determine the crystalline structure of unknown materials by comparing their diffraction patterns to standard databases. This capability is crucial in fields like geology, chemistry, and materials research.

Another important application of XRD is phase analysis. This involves quantifying the proportion of different crystalline phases present in alloys, composites, or multiphase materials. Such analysis is essential for optimizing material properties and understanding their behaviour in specific applications.

XRD is also extensively used for strain and stress measurement. By analysing shifts and broadening in diffraction peaks, researchers can evaluate residual stresses within a material, which are critical for assessing mechanical performance and durability in engineering components.

For thin film analysis, XRD is invaluable in determining film thickness, texture, and composition. These properties are vital in the development of coatings, semiconductors, and other advanced materials. By providing detailed insights into the structural characteristics of thin films, XRD supports innovation in electronics, optics, and surface engineering.

As a versatile and powerful technique, XRD requires a solid understanding of both theoretical principles and practical operation to fully harness its potential. Proper sample preparation, meticulous execution of measurements, and accurate interpretation of data ensure that XRD delivers meaningful results, making it an essential tool in materials research and development.

Transmission Electron Microscopy (TEM): TEM provides high-resolution imaging that reveals the internal structure of materials down to the atomic scale. This technique uses a focused beam of electrons that passes through an ultra-thin specimen, creating a detailed image of its internal features. TEM is invaluable for studying defects, grain boundaries, nanostructures, and dislocations, which are critical in determining material behaviour. For example, TEM is used to investigate graphene and quantum dots, where the analysis of individual atomic layers is necessary for understanding their exceptional electronic and mechanical properties.

Using Transmission Electron Microscopy (TEM):

Transmission Electron Microscopy (TEM) is a powerful technique that allows for high-resolution imaging of materials at the atomic scale. It works by transmitting a focused beam of electrons through a thin specimen. The interaction of electrons with the sample produces images or diffraction patterns that reveal detailed information about the material's structure, composition, and properties. TEM is widely used in materials science, biology, and nanotechnology.

Preparing the Specimen

Sample preparation is critical for successful TEM analysis. The specimen must be extremely thin, typically less than 100 nanometres, to allow electrons to pass through it.

1. **Cutting and Thinning**: For bulk materials, techniques such as mechanical grinding, polishing, and ion milling are used to achieve the required thickness. For biological samples, ultramicrotomy is used to slice specimens into ultra-thin sections.

2. **Deposition**: Nanoparticles, thin films, or similar materials are often deposited directly onto a TEM grid.

3. **Staining and Contrast Agents**: In biological applications, samples may be stained with heavy metals like uranium or lead to enhance contrast.

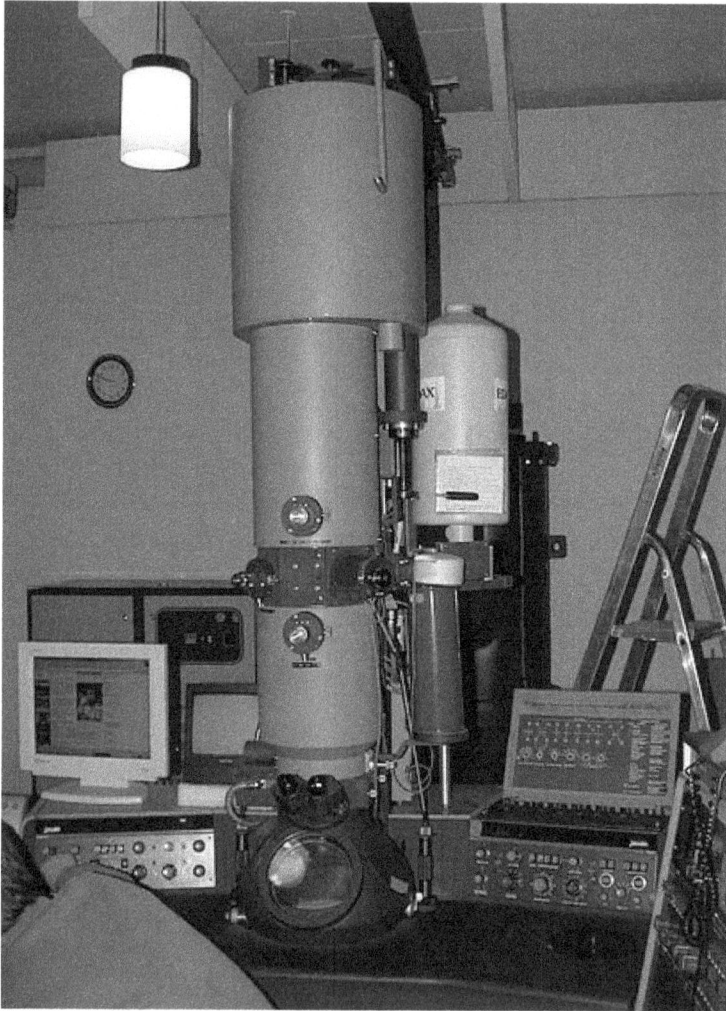

Figure 19: A Philips EM 430 Transmission Electron Microscope. Opensource Handbook of Nanoscience and Nanotechnology, CC BY 2.5, via Wikimedia Commons.

Operating the TEM

1. **Specimen Loading**: Place the prepared sample onto a TEM grid, which is then inserted into the microscope's sample holder. The holder is carefully positioned inside the TEM chamber.

2. **Vacuum Environment**: TEM requires a high vacuum to prevent electron scattering by air molecules. Ensure the vacuum system is operational before proceeding.

3. **Electron Beam Generation**: Electrons are emitted from an electron source (e.g., a tungsten filament or a field emission gun) and accelerated to high energies, typically 100–300 keV.

4. **Beam Focusing**: Electromagnetic lenses focus the electron beam into a fine spot. Adjustments to the condenser lenses ensure optimal beam intensity and diameter for imaging.

5. **Imaging Mode**: The transmitted electrons interact with the sample, producing an image. Adjust the objective lens to focus on areas of interest. Contrast is formed based on differences in electron density and thickness within the sample.

6. **Diffraction Mode**: Switch to diffraction mode to analyse the crystalline structure. This mode provides patterns based on the interaction of the electron beam with the atomic lattice, revealing information about crystallinity, orientation, and defects.

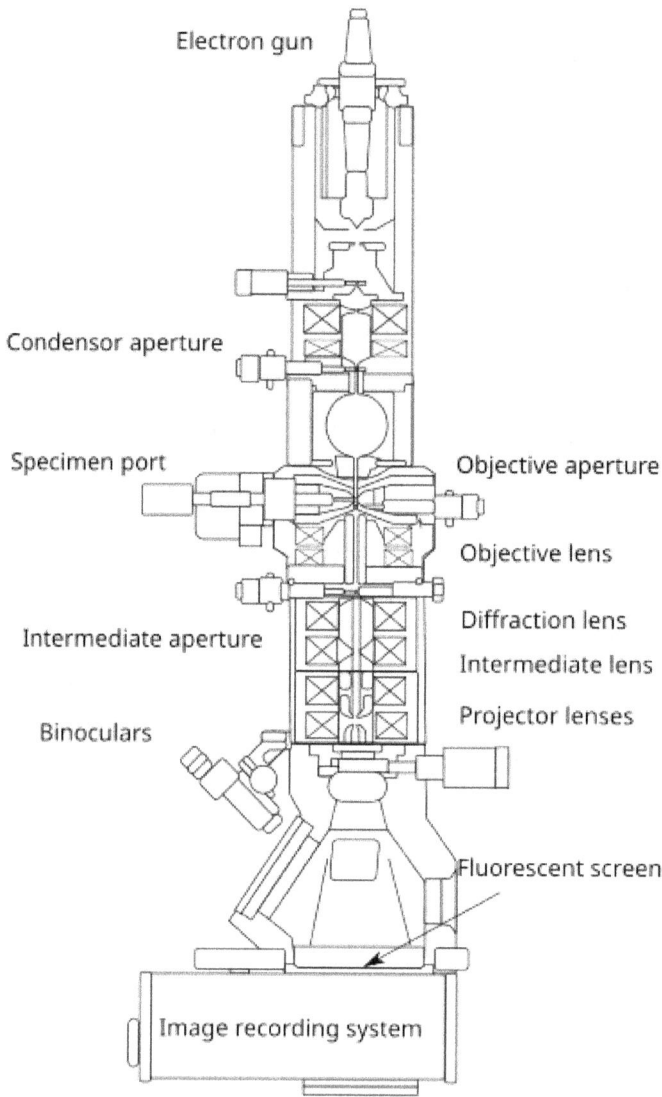

Figure 20: Diagram outlining the internal components of a basic TEM system. Gringer, CC BY-SA 3.0, via Wikimedia Commons.

Adjusting Parameters

1. **Magnification**: TEM can achieve magnifications up to several million times. Adjust magnification settings to focus on specific features of the sample.

2. **Resolution**: Fine-tune the focus and lens alignments to maximize resolution, achieving atomic-level detail.

3. **Brightness and Contrast**: Modify these settings for optimal visualization of the sample features.

Data Acquisition

1. **Image Capture**: Use specialized detectors or cameras to capture high-resolution images.

2. **Diffraction Patterns**: Save diffraction patterns for structural and crystallographic analysis.

3. **Spectroscopy**: Advanced TEMs equipped with energy-dispersive X-ray spectroscopy (EDS) or electron energy loss spectroscopy (EELS) allow for chemical composition analysis.

Post-Analysis

Analyse the collected images and data to extract relevant information about the sample. This may include identifying defects, grain boundaries, or nanostructures, determining chemical composition, or studying electron diffraction patterns.

Transmission Electron Microscopy (TEM) finds extensive applications across a wide range of fields due to its ability to provide high-resolution imaging and detailed structural analysis. In nanotechnology, TEM is crucial for imaging nanoparticles, nanowires, and graphene. These applications enable researchers to explore the morphology, crystallinity, and defects of nanostructures, driving innovations in fields such as electronics, energy storage, and catalysis.

In materials science, TEM is a powerful tool for investigating crystal defects, grain boundaries, and phase compositions. It allows for an in-depth understanding of the microstructural properties of advanced materials, such as alloys, ceramics, and composites, and supports the development of materials with enhanced performance.

In biology, TEM is invaluable for examining cellular structures, viruses, and macromolecules. Its high resolution enables visualization of intricate details at the sub-cellular level, making it a cornerstone technique in structural biology, virology, and biomedical research.

The semiconductor industry relies on TEM for characterizing thin films and device layers. TEM provides insights into the structural integrity, composition, and defects of semiconductor materials, which are critical for optimizing the performance of integrated circuits, LEDs, and other electronic components.

Scanning Electron Microscopy (SEM): SEM is used for detailed imaging of material surfaces by scanning them with a focused electron beam. The interaction between the beam and the material generates signals that provide information about the surface topography and composition. SEM is commonly employed to analyse surface morphology, grain size, fracture surfaces, and microstructural features. This technique is particularly useful for evaluating coatings, composites, and failure analysis in structural materials. For example, SEM can be used to examine the surface texture of a thin-film solar cell to optimize light absorption.

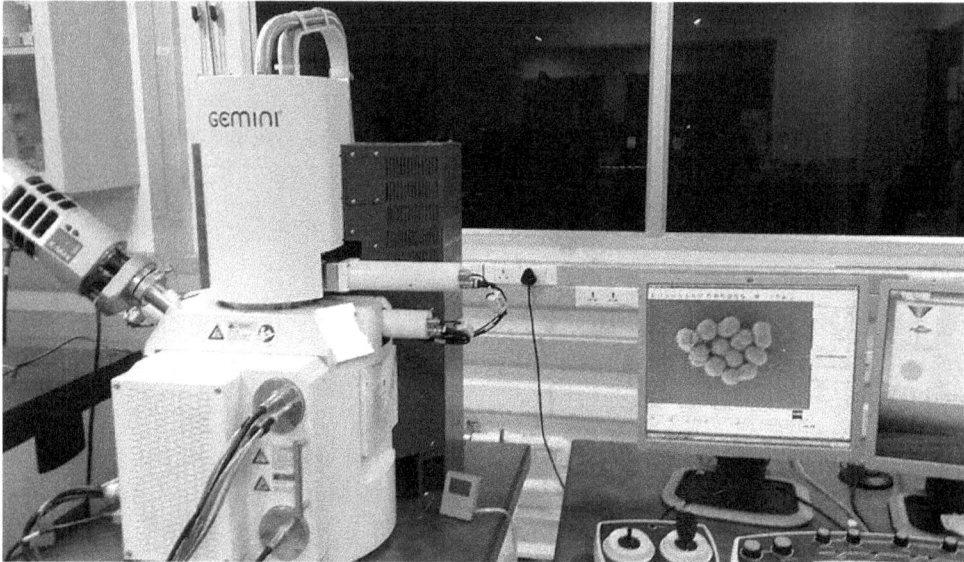

Figure 21: Carl Zeiss Ultra 55 field emission scanning electron microscope (FESEM, EDS) scanning electron microscope at the Center for Nanoscience and Engineering (CeNSE), IISc, Bangalore. Vader1941, CC BY-SA 4.0, via Wikipedia.

Using Scanning Electron Microscopy (SEM):

Scanning Electron Microscopy (SEM) is a powerful technique for imaging the surface and near-surface features of materials at high magnifications. The process involves scanning a focused electron beam across the specimen's surface, producing detailed topographical and compositional information.

Step-by-Step Procedure

1. Sample Preparation

Preparing the sample is critical for obtaining clear and accurate images. SEM samples need to be small enough to fit in the microscope's chamber, typically a few millimetres to a few centimetres in size. Non-conductive samples require a thin conductive coating (such as gold, carbon, or platinum) to prevent charging under the electron beam. The sample should also be clean and free of contamination to avoid image distortion.

2. Loading the Sample

Place the prepared sample onto a specimen holder or stub using adhesive or a clamp. Insert the holder into the SEM chamber and ensure it is securely mounted. Close the chamber door and create a vacuum, as SEM operates in a low-pressure environment to avoid electron scattering by air molecules.

3. Setting Up the Instrument

Power on the SEM and set the desired operating conditions, including:

- **Accelerating Voltage:** Adjust based on the sample material and imaging needs. Low voltages (1-5 kV) are suitable for delicate samples, while higher voltages (10-30 kV) provide better resolution and depth of field.

- **Electron Beam Current:** Optimize the beam intensity for the level of detail required.

- **Magnification and Resolution:** Select an appropriate magnification level for the features you aim to study.

4. Aligning the Electron Beam

Align the electron beam to ensure it is focused and accurately aimed at the sample. Adjust the beam spot size and working distance (distance between the sample and the objective lens) to achieve optimal imaging conditions.

5. Imaging the Sample

Begin scanning the sample surface with the electron beam. The electrons interact with the sample, generating various signals such as secondary electrons (used for topography), backscattered electrons (used for compositional contrast), and X-rays (used for elemental analysis). Use the SEM software interface to view and capture images.

6. Adjusting Image Parameters

Refine imaging parameters like focus, contrast, and brightness to enhance the quality of the image. Employ features such as tilt or rotation to view different aspects of the sample.

7. Analytical Capabilities

Many SEM instruments are equipped with additional tools for chemical and compositional analysis:

- **Energy Dispersive X-ray Spectroscopy (EDS):** Identifies elemental composition by analysing X-rays emitted during electron-sample interaction.

- **Electron Backscatter Diffraction (EBSD):** Determines crystallographic orientation.

8. Data Analysis and Storage

Save the acquired images and analytical data for further analysis. Use the SEM software to annotate, measure, or compare features on the sample.

9. Shutting Down the SEM

Once imaging is complete, safely shut down the instrument:

- Vent the chamber to return to atmospheric pressure.

- Remove the sample carefully.

- Turn off the electron beam and power down the SEM according to the manufacturer's instructions.

Scanning Electron Microscopy (SEM) finds extensive applications across various scientific and industrial fields due to its ability to provide detailed images of surfaces and compositional information. In materials science, SEM is instrumental in studying surface morphology, analysing fracture patterns, and examining grain structures. This makes it an invaluable tool for understanding material behaviour under different conditions, aiding in the development and improvement of materials for various applications.

In biology, SEM is used to image biological specimens such as cells and tissues, providing high-resolution surface details that are crucial for understanding structural and functional characteristics. Its ability to capture the intricate textures and topographies of biological materials enhances research in cellular biology, microbiology, and tissue engineering.

The electronics industry heavily relies on SEM for failure analysis and surface examination of components. It enables researchers and engineers to identify defects, analyse wear patterns, and evaluate the integrity of microelectronic devices, ensuring reliability and performance in advanced technologies.

Forensic science employs SEM to investigate trace evidence such as paint chips, fibre fragments, and tool marks. The high magnification and compositional analysis capabilities of SEM are vital for linking evidence to crime scenes or suspects, making it a powerful tool in criminal investigations. Overall, SEM's versatility and precision make it a cornerstone technique in numerous scientific and industrial domains.

Atomic Force Microscopy (AFM): AFM is a versatile technique that uses a sharp probe to scan a material's surface at the nanoscale. It provides information on surface roughness, mechanical properties, and topography without requiring conductive samples, unlike electron microscopy techniques. AFM is particularly effective for studying polymers, thin films, and biological materials. For instance, AFM can measure the nanoscale roughness of a polymer surface used in biomedical applications, ensuring its suitability for cell adhesion.

The process involves preparing the sample, calibrating the instrument, conducting the scan, and interpreting the data. Below is a guide on how to use AFM effectively:

1. Sample Preparation

AFM requires a clean and stable sample surface for accurate imaging. The sample should be:

- **Clean:** Contaminants can interfere with the probe-sample interaction. Use chemical cleaning or plasma treatment if necessary.

- **Flat:** Ensure the sample is relatively flat to avoid damaging the probe or losing resolution.

- **Mounted Securely:** Use adhesives or sample holders to ensure the sample doesn't shift during scanning.

2. Probe and Cantilever Selection

Choose the appropriate AFM probe and cantilever based on the intended application:

- **Contact Mode:** Use soft cantilevers with low spring constants for imaging surfaces directly.

- **Tapping Mode:** Select cantilevers with medium stiffness to minimize surface damage.

- **Non-Contact Mode:** Opt for stiffer cantilevers for analysing delicate or soft samples.

- **Property Analysis:** Use specialized probes for mechanical, electrical, or magnetic property measurements.

3. Instrument Setup

Set up the AFM by:

- Aligning the laser onto the cantilever's reflective surface. Adjust the position to maximize signal strength on the photodetector.

- Calibrating the instrument for sensitivity, including the cantilever's deflection and spring constant.

- Ensuring the piezoelectric scanner is operational and within its range of motion.

Figure 22: Using variable temperature atomic force microscopy within an ultrahigh vacuum chamber. Oak Ridge National Laboratory, CC BY 2.0, via Flickr.

4. Scanning Parameters

Define the scanning parameters to optimize imaging:

- **Scan Area:** Choose the size of the area to be imaged, typically in micrometres.

- **Resolution:** Set the number of pixels per line to balance resolution and scan time.

- **Scan Rate:** Adjust the rate to avoid probe damage or inaccurate imaging, with slower rates for high-resolution scans.

- **Interaction Force:** Minimize the force to prevent damage to both the probe and sample.

5. Scanning

Perform the scan:

- Lower the cantilever gradually toward the sample surface until the probe interacts with the sample.

- Begin the scan using the chosen mode (contact, tapping, or non-contact).

- Monitor the real-time feedback to ensure consistent and accurate imaging.

- Adjust scanning parameters if artifacts or noise appear in the data.

6. Data Analysis

Once the scan is complete, analyse the data:

- **Topographical Imaging:** Generate 3D surface profiles to visualize surface roughness and features.

- **Force Spectroscopy:** Use force curves to determine mechanical properties like stiffness and adhesion.

- **Property Maps:** Analyse electrical or magnetic responses for material-specific applications.

7. Maintenance and Troubleshooting

After completing the scan:

- Remove the sample and clean the stage to prevent contamination.

- Inspect the probe for wear or damage and replace it if necessary.

- Perform regular maintenance on the AFM to ensure optimal performance.

Atomic Force Microscopy (AFM) finds extensive applications across diverse fields due to its ability to provide high-resolution imaging and detailed surface property measurements. In materials science, AFM is a critical tool for investigating surface roughness, mechanical properties, and nanostructures. It enables researchers to visualize and quantify surface features at the nanoscale, contributing to advancements in coatings, composite materials, and nanotechnology.

In biology, AFM is employed to image biomolecules, cells, and tissues with unparalleled detail. Its ability to operate in liquid environments makes it particularly valuable for studying biological samples in their native states. Researchers use AFM to explore the structure and behaviour of proteins, DNA, and cellular membranes, aiding in developments in biophysics and drug delivery.

In the electronics industry, AFM is essential for analysing thin films, semiconductor surfaces, and device structures. It provides precise measurements of surface morphology and layer thickness, which are critical for optimizing the performance and reliability of electronic components. AFM also plays a significant role in failure analysis and quality control.

For polymer research, AFM is used to measure elasticity, adhesion, and surface morphology. It helps scientists understand polymer behaviour under various conditions, enabling the design of materials with tailored mechanical and surface properties for applications in packaging, biomedical devices, and advanced manufacturing. Across these fields, AFM's versatility and precision make it an indispensable tool for nanoscale analysis and innovation.

Neutron Scattering: Neutron scattering is a structural characterization technique that probes materials at the atomic and molecular levels. Unlike X-rays, neutrons interact strongly with light elements and are sensitive to magnetic properties, making this technique ideal for studying hydrogen storage materials, magnetic alloys, and soft matter systems. Neutron scattering provides unique insights into dynamic processes and structural features that are challenging to observe with other methods. For example, it is used to understand the hydrogen absorption behaviour of materials used in fuel cells.

These structural characterization techniques are foundational in materials research and development, offering insights into the arrangement and interactions of atomic and molecular structures. By applying these techniques, researchers can optimize material properties for applications in industries ranging from energy and aerospace to electronics and healthcare.

Neutron scattering is a powerful technique for investigating the structure and dynamics of materials at the atomic and molecular levels. Its application involves specific steps and considerations to ensure accurate and meaningful results.

1. Understand the Objective

Define the purpose of the neutron scattering experiment. The technique is versatile and can be used for a variety of studies:

- **Structural Analysis**: To determine atomic arrangements in crystals or amorphous materials.

- **Magnetic Properties**: To study magnetic structures and behaviours in materials.

- **Dynamic Studies**: To investigate phonons, molecular vibrations, or diffusion processes.

2. Select the Appropriate Technique

Neutron scattering encompasses various methods, each suited to specific research goals:

- **Neutron Diffraction**: Used for structural studies to determine atomic arrangements and phase compositions.

- **Small-Angle Neutron Scattering (SANS)**: Suited for studying large-scale structures like polymers, biological macromolecules, and nanomaterials.

- **Inelastic Neutron Scattering**: Focused on understanding vibrational modes, lattice dynamics, and energy transitions.

- **Reflectometry**: Used for analysing thin films and interfaces.

3. Prepare the Sample

Sample preparation is crucial for a successful experiment:

- **Material State**: The sample can be in solid, liquid, or gas form, depending on the research question.

- **Sample Size**: Ensure that the sample is appropriately sized for the neutron flux and beamline specifications.

- **Environment Control**: Some experiments require specific conditions, such as low temperatures, high pressures, or magnetic fields.

4. Access a Neutron Source

Neutron scattering experiments require access to specialized facilities with neutron sources, such as nuclear reactors or spallation sources. Examples include the Institut Laue-Langevin (ILL) in France, the Oak Ridge National Laboratory (ORNL) in the U.S., or the ISIS Neutron and Muon Source in the U.K.

5. Align the Sample and Instrument

The sample is mounted in the neutron beamline, and precise alignment is critical:

- **Beam Characteristics**: Ensure the beam energy and flux are suitable for the chosen technique.

- **Instrument Settings**: Adjust detectors and sample holders to match the experiment's requirements, such as scattering angles or wavelength resolutions.

6. Conduct the Experiment

Once alignment is complete, neutrons are directed at the sample, and scattered neutrons are detected:

- **Data Collection**: Detectors record the scattering patterns, intensities, and energy changes of neutrons.

- **Control Systems**: Modern neutron scattering instruments are equipped with automated controls and software for real-time monitoring and adjustments.

7. Analyse the Data

Data analysis is essential to extract meaningful information:

- **Scattering Patterns**: Interpret diffraction peaks, intensity distributions, or reflectivity curves.

- **Model Fitting**: Use theoretical models to match the observed data and deduce structural, magnetic, or dynamic parameters.

- **Specialized Software**: Tools like GSAS, Mantid, or SasView are often used to process and analyse neutron scattering data.

8. Validate and Report Findings

Verify results through comparisons with complementary techniques, such as X-ray diffraction or spectroscopy. Report findings with detailed descriptions of experimental conditions, data interpretation, and implications.

Neutron scattering is a complex yet invaluable technique for probing material properties at fundamental levels. Proper planning, execution, and interpretation of neutron scattering experiments enable researchers to address critical questions in materials science, physics, chemistry, and biology.

Compositional Characterization Techniques

Compositional analysis is a critical aspect of material characterization, focusing on understanding the elemental and chemical makeup of materials. This knowledge is essential for tailoring material properties, detecting impurities, and ensuring quality in various applications. Several advanced techniques are employed to achieve precise compositional insights, each suited to specific materials and contexts.

Energy Dispersive X-ray Spectroscopy (EDS or EDX): EDS is widely used to determine the elemental composition of materials and is often integrated with Scanning Electron Microscopy (SEM) or Transmission Electron Microscopy (TEM). This technique works by detecting characteristic X-rays emitted from a material when it is bombarded with high-energy electrons. EDS provides a rapid and non-destructive means of identifying and quantifying elements within a sample. For example, it is frequently employed in the semiconductor industry to detect impurities or dopants in silicon wafers, which can significantly impact the performance of electronic devices.

Steps to Use EDS:

1. Preparing the Sample The first step is ensuring the sample is properly prepared for analysis:

- The sample must have a clean and smooth surface to ensure consistent signal collection.

- For non-conductive materials, a conductive coating (e.g., gold or carbon) may be applied to prevent charging effects during SEM or TEM operation.

2. Setting Up the Instrument

- Mount the sample onto the SEM or TEM stage.

- Align the EDS detector to optimize its position relative to the sample. The detector is typically positioned at a specific angle to capture the emitted X-rays effectively.

3. Generating X-rays

- The instrument's electron beam is focused on the sample surface.

- The interaction between the high-energy electrons and the atoms in the sample excites the inner-shell electrons of the atoms.

- As these electrons return to their lower-energy states, they emit X-rays with characteristic energies unique to each element.

4. Collecting Data

- The EDS detector collects the emitted X-rays and generates a spectrum. Each peak in the spectrum corresponds to a specific element, and the peak height or area indicates the element's relative abundance.

- Adjust parameters such as beam current, acceleration voltage, and dwell time to optimize signal intensity and resolution.

5. Analysing the Spectrum

- Use the EDS software to identify the elements present in the sample. The software automatically matches the characteristic X-ray energies to known elemental fingerprints.

- Quantify the elemental composition by analysing the peak intensities.

6. Mapping and Line Scans

- Perform elemental mapping to visualize the spatial distribution of elements across the sample. This is useful for detecting inclusions, impurities, or layered structures.

- Use line scans for detailed analysis of compositional changes along a specific line in the sample.

7. Interpreting the Results

- Evaluate the spectrum and mappings to understand the sample's composition. For example, in alloy analysis, EDS can confirm the distribution of alloying elements and detect any segregation.

EDS is widely used across various fields for detailed elemental analysis and material characterization. In material science, it is employed to identify inclusions, analyse phase compositions, and detect surface contaminations, providing insights into material integrity and processing outcomes. In the semiconductor industry, EDS plays a critical role in detecting dopant distributions and impurities within silicon wafers, ensuring the reliability and performance of electronic devices. Geologists use EDS to analyse mineral compositions, helping to classify and study geological samples. In biology, the technique is applied to examine trace elements in biological specimens such as bone or tissue, offering valuable data for research in medicine and environmental science.

To achieve accurate and meaningful results with EDS, several best practices must be followed. Choosing the appropriate acceleration voltage is critical; lower voltages are preferred for surface analysis to minimize penetration depth, while higher voltages allow for deeper bulk analysis. Calibration of the detector using standard samples ensures the accuracy of elemental identification and quantification. It is also essential to understand the limitations of the technique, particularly its inability to detect elements lighter than beryllium ($Z < 4$) with standard detectors. Adhering to these practices enhances the reliability and effectiveness of EDS in diverse applications.

X-Ray Photoelectron Spectroscopy (XPS): XPS is a surface-sensitive technique that analyses the chemical composition of materials by measuring the kinetic energy of electrons ejected from the surface by X-ray photons. This method provides information about the elemental composition, chemical states, and surface bonding of materials. XPS is particularly useful for studying surface phenomena, such as the formation of oxides, corrosion layers, and coatings. For instance, it is commonly applied to analyse the surface chemistry of protective coatings in aerospace and automotive industries, ensuring that these coatings meet performance standards.

Figure 23: XPS system at Institute of Materials Research and Engineering, Singapore. Mehdi-rou, CC BY-SA 3.0, via Wikimedia Commons.

The process involves several steps, each critical to obtaining accurate and meaningful results.

1. Sample Preparation

Preparing the sample is the first and most important step. The material to be analysed should have a clean and flat surface to ensure uniform X-ray interaction. Contaminants, such as grease, dust, or adsorbed moisture, are removed to avoid interfering with the analysis. If necessary, the sample may undergo vacuum cleaning or sputtering to eliminate surface impurities.

2. Vacuum Environment

XPS operates under ultra-high vacuum (UHV) conditions to prevent interaction of emitted electrons with air molecules. This ensures a clear path for electrons from the sample to the detector and minimizes signal loss.

3. X-Ray Source Selection

The sample is irradiated with a monochromatic X-ray beam, typically generated using an aluminium (Al Kα) or magnesium (Mg Kα) X-ray source. The X-ray beam's energy excites electrons in the material, causing them to be emitted from their atomic orbitals.

4. Electron Detection

The emitted photoelectrons are collected by an electron energy analyser, which measures their kinetic energy. This energy, along with the known X-ray photon energy, allows for the calculation of the electron's binding energy using the photoelectric equation:

$$E_B = h\nu - E_K - \phi$$

where E_B is the binding energy, $h\nu$ is the photon energy, E_K is the kinetic energy, and ϕ is the spectrometer work function.

5. Data Acquisition

The spectrometer records the binding energy and intensity of the emitted electrons, producing an XPS spectrum. Peaks in the spectrum correspond to specific elements and their chemical states. The relative intensity of these peaks provides quantitative data on the elemental composition of the sample's surface.

6. Data Interpretation

The acquired spectrum is analysed to identify elements, chemical bonds, and oxidation states. High-resolution scans may be performed on specific peaks to obtain detailed information about chemical states. Peak deconvolution is used to resolve overlapping signals and refine the analysis.

7. Depth Profiling (Optional)

For layered or coated materials, depth profiling can be performed by sputtering the sample surface with an ion beam and repeatedly analysing the exposed layers. This provides information about compositional variations as a function of depth.

XPS is widely used in materials science, chemistry, and electronics. It is instrumental in analysing surface oxides, coatings, thin films, and nanostructures. In catalysis research, it provides insights into active sites and reaction mechanisms. XPS is also essential for studying chemical bonding and interactions in semiconductors, polymers, and biomaterials.

Fourier Transform Infrared Spectroscopy (FTIR): FTIR is a technique used to identify molecular bonds within a material by measuring the absorption of infrared light at specific wavelengths. Each type of bond absorbs light at characteristic frequencies, creating a unique spectral fingerprint. This method is especially effective for analysing organic materials and polymers, providing insights into their chemical structure and composition. For example, FTIR is used to study the degradation of polymers in harsh environments or to identify additives in plastic materials.

The process involves the following steps, each essential for obtaining accurate and meaningful data.

1. Sample Preparation

FTIR accommodates various sample types, including solids, liquids, and gases. The preparation method depends on the sample's physical state:

- **Solids:** Ground into a fine powder and mixed with potassium bromide (KBr) to form a pellet, or applied directly onto an Attenuated Total Reflectance (ATR) crystal.

- **Liquids:** Placed in a liquid sample cell with IR-transparent windows, such as calcium fluoride or sodium chloride.

- **Gases:** Introduced into a gas cell with a longer path length to increase interaction with the IR beam.

Proper preparation ensures uniform IR light interaction and minimizes interference from contaminants or moisture.

2. Instrument Setup

The FTIR spectrometer is calibrated to ensure accuracy. A laser (commonly a HeNe laser) is used to monitor the moving mirror's position in the interferometer, ensuring precise data acquisition. The system may require a background spectrum collection, typically taken with an empty sample chamber or a clean ATR crystal, to remove environmental and instrumental contributions.

3. IR Light Interaction

The spectrometer generates a broad spectrum of IR light, which passes through the sample. Molecules in the sample absorb specific wavelengths of IR light, causing vibrations in their chemical bonds. These vibrations are unique to the molecular structure and bond types, such as stretching, bending, or twisting.

4. Interferogram Collection

The IR light interacts with an interferometer, which splits and recombines the light to create an interferogram. This pattern represents the combined intensity of all IR wavelengths after interaction with the sample. The interferogram contains information about all molecular vibrations.

5. Fourier Transformation

A computer applies a mathematical Fourier Transform to the interferogram, converting it into an IR spectrum. This spectrum displays absorbance or transmittance as a function of wavenumber (cm^{-1}), providing a molecular fingerprint of the sample.

6. Data Interpretation

The resulting spectrum is analysed to identify functional groups and molecular structures. Key steps include:

- **Peak Assignment:** Identifying characteristic absorption peaks corresponding to specific functional groups (e.g., C-H, O-H, C=O).

- **Quantification:** Measuring peak intensities to determine concentration in quantitative analysis.

- **Comparison:** Comparing the spectrum to reference databases for material identification.

FTIR finds extensive applications across a wide range of disciplines. In materials science, it is used for characterizing polymers, composites, and coatings, providing insights into molecular composition and structural features critical for material performance. In the pharmaceutical industry, FTIR plays a crucial role in identifying active pharmaceutical ingredients (APIs) and excipients, ensuring product quality and compliance with regulatory standards. Environmental science relies on FTIR to analyse pollutants and organic compounds in water or air, facilitating the monitoring and mitigation of environmental contaminants. In biology, FTIR is used to study proteins, lipids, and nucleic acids, offering valuable information about biomolecular structures and interactions.

National Aeronautics and Space Administration
John H. Glenn Research Center at Lewis Field

Figure 24: Fourier Transform Infrared Spectroscopy. Defense Visual Information Distribution Service, Public Domain, via Picryl.

To achieve accurate and reliable results, several best practices should be followed when using FTIR. Cleanliness is essential; IR-transparent windows, sample holders, and ATR crystals must be free of contaminants to prevent spectral interference. Regular acquisition of background spectra is necessary to correct for environmental and instrumental contributions, improving the accuracy of the analysis. Routine instrument maintenance, including calibration and inspection of optical components, ensures the spectrometer operates at peak performance. By adhering to these practices, FTIR users can optimize the technique's effectiveness for diverse applications.

Raman Spectroscopy: Raman spectroscopy offers detailed information about molecular vibrations, crystal structures, and bonding in materials. This technique involves the inelastic scattering of light, which shifts in wavelength due to interactions with molecular vibrations. Raman spectroscopy is particularly effective for studying carbon-based materials like graphene, carbon nanotubes, and other nanostructures. It provides valuable data on structural integrity, defect density, and electronic properties, which are crucial for developing advanced nanomaterials and composites.

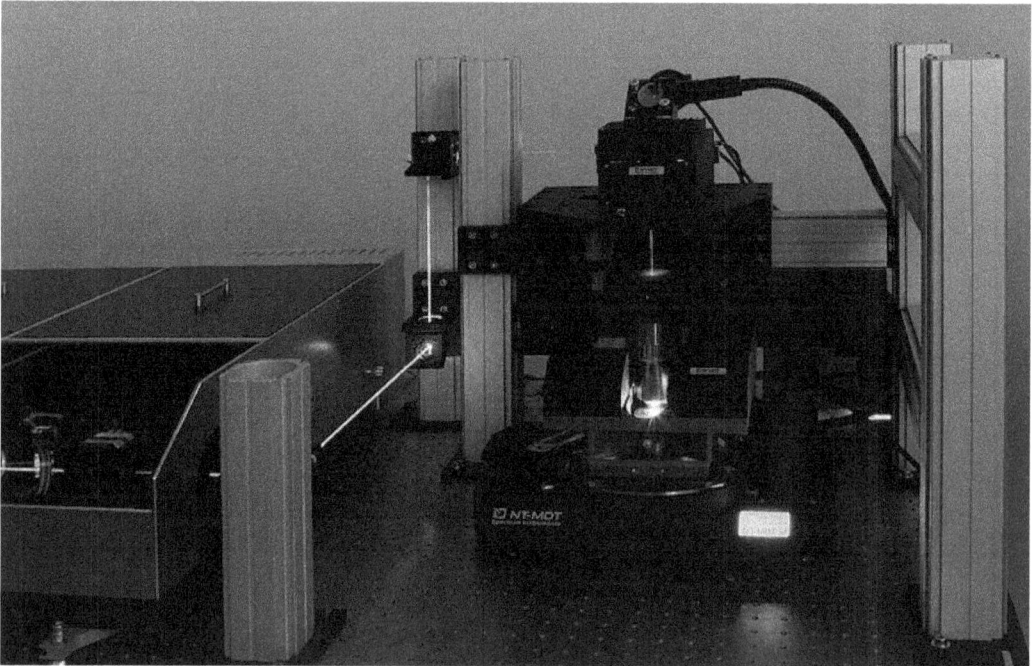

Figure 25: AFM-Raman integrates the best of AFM and Raman microscopy. AAMonitor96, CC BY-SA 3.0, via Wikimedia Commons.

Raman spectroscopy relies on the inelastic scattering of light (Raman scattering) to provide information about the material's vibrational modes.

1. Setting Up the Equipment

- **Laser Selection**: Choose a laser source with a wavelength suitable for your sample. Common options include visible (e.g., 532 nm), near-infrared (e.g., 785 nm), or ultraviolet lasers.

- **Spectrometer Alignment**: Ensure that the spectrometer is correctly aligned with the laser beam for optimal signal collection.

- **Sample Mounting**: Place the sample on the microscope stage or sample holder. The surface should be clean and flat for consistent measurements.

2. Adjusting Instrument Parameters

- **Laser Power**: Set the laser power to a level that minimizes sample damage or heating while ensuring a detectable signal.

- **Integration Time**: Adjust the integration time based on the sample's fluorescence and signal strength. Longer times increase signal-to-noise but may require stability.

- **Objective Lens**: Select an appropriate objective lens for focusing the laser on the sample, typically 10x, 50x, or 100x magnifications.

3. Acquiring Spectra

- **Focusing**: Use the microscope optics to focus the laser beam precisely on the area of interest in the sample.

- **Spectral Range**: Define the spectral range to capture relevant Raman peaks. For most materials, the range from 100 to 4000 cm^{-1} suffices.

- **Background Subtraction**: Capture a background spectrum without the sample to subtract environmental or instrument-related contributions.

4. Data Analysis

- **Peak Identification**: Analyse the Raman peaks to identify molecular vibrations, bond types, or crystal structures. Software tools often assist in this process.

- **Quantitative Analysis**: If necessary, perform quantitative analysis using calibration curves or comparison with known standards.

- **Mapping**: For spatial distribution studies, use Raman mapping to collect spectra across a defined area of the sample.

5. Interpreting Results

- Correlate the observed Raman shifts with molecular vibrations or structural information. Reference databases or theoretical models can aid interpretation.

- Compare the intensity and position of peaks to detect variations in composition, strain, or phase.

Secondary Ion Mass Spectrometry (SIMS): SIMS is a highly sensitive technique for analysing surface composition and depth profiling in thin films. It works by bombarding the material's surface with primary ions and detecting the secondary ions ejected from the surface. SIMS provides detailed elemental and isotopic information, making it indispensable for applications requiring precise depth analysis. For instance, it is used in the production of thin-film solar cells to analyse doping concentrations and ensure uniformity across layers.

Secondary Ion Mass Spectrometry involves sputtering the sample surface with a focused ion beam and analysing the ejected secondary ions.

1. Preparation of the Sample

- **Clean the Sample**: Ensure the sample surface is free of contaminants to avoid interference with the SIMS analysis.

- **Size and Compatibility**: Ensure the sample size fits the SIMS chamber and that it can withstand vacuum conditions.

- **Mounting**: Securely mount the sample on a holder to ensure stability during the analysis.

2. Instrument Setup

- **Select the Primary Ion Beam**: Choose an appropriate ion source (e.g., cesium, oxygen, or gallium) based on the material being analysed. For instance:

  o Cesium enhances the detection of negative ions.

  o Oxygen enhances the detection of positive ions.

- **Adjust Beam Energy and Current**: Set the energy and current of the ion beam to balance surface erosion rate and spatial resolution.

- **Set Sputtering Conditions**: Optimize the sputtering rate based on the analysis requirements (e.g., surface composition vs. depth profiling).

3. Acquiring Data

- **Surface Composition Analysis**:

  o Direct the ion beam onto the sample surface.

  o Collect and analyse secondary ions ejected from the top atomic layers.

  o Record the mass-to-charge ratios using a mass spectrometer.

- **Depth Profiling**:

  o Gradually sputter the sample surface layer by layer.

- Monitor secondary ion signals over time to determine changes in composition with depth.

- **Imaging**:

  - Raster the ion beam across the sample surface.

  - Generate spatially resolved maps of elemental or isotopic distributions.

4. Data Analysis

- **Mass Spectra Interpretation**:

  - Identify peaks in the mass spectrum corresponding to specific elements, isotopes, or molecular fragments.

  - Quantify relative concentrations using calibration standards or known sensitivity factors.

- **Depth Profiling**:

  - Plot the signal intensity of secondary ions as a function of sputtering time or depth to analyse concentration gradients.

- **Spatial Mapping**:

  - Use imaging data to visualize elemental or isotopic distributions across the sample.

5. Best Practices

- **Calibration**:

  - Use reference materials to calibrate the instrument for accurate quantification.

- **Minimize Artifacts**:

  - Control sputtering conditions to reduce damage or mixing of layers.

  - Ensure the vacuum system is clean to avoid contamination.

- **Selectivity**:

  - Adjust beam settings to enhance the detection of specific ions or elements of interest.

- **Avoid Over-Sputtering**:

○ Monitor the sputtering process carefully to prevent loss of critical surface information.

SIMS finds extensive applications across various fields due to its ability to provide detailed compositional and depth profiling data at high sensitivity and resolution. In materials science, it is widely used to analyse thin films, interfaces, and coatings, offering insights into their elemental composition and layer integrity. This makes it an essential tool for developing advanced materials and quality control in industrial processes.

Figure 26: Expert performing isotopic measurements on uranium particles with Secondary Ion Mass Spectrometer (SIMS). IAEA Imagebank, CC BY-SA 2.0, via Flickr.

In the semiconductor industry, SIMS is crucial for depth profiling dopants in silicon wafers, ensuring precise control over doping concentrations and distributions. This is vital for optimizing the performance and reliability of electronic devices. Geology benefits from SIMS through isotopic analysis, which helps in dating geological samples and determining their provenance, providing valuable information for studies in earth sciences and resource exploration.

In the field of biology, SIMS is used to image biomolecular distributions in tissues and cells. This capability enables researchers to map the spatial distribution of elements and isotopes within biological specimens, contributing to advancements in biomedical research and diagnostics. Environmental science also leverages SIMS for surface analysis of pollutants and contaminants. By identifying and quantifying surface-bound compounds, SIMS aids in monitoring environmental health and developing remediation strategies.

Compositional analysis techniques are vital tools in modern materials science. They provide critical information for the development, optimization, and quality control of materials across diverse industries, including electronics, aerospace, energy, and healthcare. By combining these techniques, researchers can achieve a comprehensive understanding of material composition, enabling innovations that address complex engineering challenges and improve material performance.

Property Measurement Techniques

Property measurement techniques are essential tools in materials science, enabling the detailed evaluation of a material's mechanical, thermal, electrical, magnetic, and optical properties. These methods provide critical insights for designing and optimizing materials for specific applications.

Mechanical testing focuses on assessing properties like strength, elasticity, and toughness. Techniques such as tensile testing, hardness testing, and fatigue testing are employed to quantify these attributes. Tensile testing, for instance, applies a uniaxial force to a sample to determine its response, providing valuable data on materials like metals and polymers. Hardness testing is more applicable to ceramics and metals, offering a quick and effective way to evaluate resistance to localized deformation.

Thermal analysis investigates a material's behaviour under temperature changes. Differential Scanning Calorimetry (DSC) measures heat flow during phase transitions, such as melting or glass transitions, making it a vital tool for polymer characterization. Differential Scanning Calorimetry (DSC) is a thermal analysis technique used to measure heat flow associated with phase transitions, chemical reactions, or other thermal events in a material. Following is an explanation of how to use DSC effectively.

1. Prepare the Sample

Proper sample preparation is crucial for accurate results.

- **Select the Material**: Choose a representative sample of the material to be analysed.

- **Weigh the Sample**: Use a precision balance to measure the sample mass, typically ranging from a few milligrams to about 20 mg, depending on the type of DSC and the material being tested.

- **Encapsulate the Sample**: Place the sample into a crucible or pan. For standard analyses, aluminium pans are commonly used. Seal the pan if necessary (e.g., in volatile materials or when using a sealed atmosphere).

2. Calibrate the Instrument

Calibrate the DSC instrument to ensure accurate measurements.

- **Temperature Calibration**: Use standard materials with known melting points (e.g., indium or tin) to calibrate the temperature scale.

- **Heat Flow Calibration**: Perform calibration using materials with known enthalpy changes.

3. Load the Sample and Reference

- Place the prepared sample in one of the DSC instrument's compartments.

- Place an empty, identical reference crucible in the other compartment. The reference is critical for comparing heat flow between the sample and the baseline.

4. Select the Experimental Conditions

- **Temperature Range**: Define the starting and ending temperatures based on the expected thermal events (e.g., room temperature to 500°C for polymers or higher for ceramics and metals).

- **Heating or Cooling Rate**: Set the rate of temperature change, typically between 1–20°C/min, depending on the sensitivity required.

- **Atmosphere**: Choose an inert gas (e.g., nitrogen or argon) for non-reactive environments or specific gases for reactive studies. Ensure a constant flow rate.

5. Perform the Measurement

- Start the experiment, and the DSC will record the heat flow as the sample is heated or cooled.

- The instrument will generate a thermogram, a plot of heat flow (y-axis) versus temperature or time (x-axis), showing endothermic (heat absorption) or exothermic (heat release) events.

6. Analyse the Data

- **Identify Peaks**: Peaks in the thermogram correspond to phase transitions, such as melting (endothermic) or crystallization (exothermic).

- **Measure Enthalpy Changes**: The area under the peaks represents the enthalpy change (ΔH), which can be calculated using the calibration data.

- **Determine Transition Temperatures**: Identify the onset, peak, and end temperatures of transitions like glass transition (Tg), melting (Tm), or decomposition.

7. Interpret and Report Results

- Compare the results with known reference data to identify material properties or anomalies.

- For complex materials, combine DSC with other techniques like TGA (Thermogravimetric Analysis) for comprehensive thermal analysis.

Thermogravimetric Analysis (TGA) quantifies weight changes during thermal decomposition, providing insights into material stability and composition. These techniques are indispensable for understanding the thermal properties of ceramics, polymers, and composites.

Thermogravimetric Analysis (TGA) is a thermal analysis technique used to measure changes in the mass of a material as a function of temperature or time. It is widely used to study material properties such as thermal stability, composition, and decomposition behaviour. The process involves:

1. Prepare the Sample

Proper sample preparation is essential for accurate and reproducible results.

- **Select the Material**: Choose a representative sample of the material to be analysed.

- **Weigh the Sample**: Use a precision balance to measure a small amount of the sample, typically between 2–50 mg. The sample size depends on the instrument and the material type.

- **Sample Form**: Ensure the sample is in a form suitable for analysis, such as a powder, thin film, or small solid piece. Homogeneity is key to obtaining reliable results.

2. Calibrate the Instrument

Regular calibration ensures the accuracy of TGA measurements.

- **Weight Calibration**: Use standard weights to verify the balance's accuracy.

- **Temperature Calibration**: Use standard materials with known decomposition or phase change temperatures (e.g., calcium oxalate or nickel sulphate).

3. Load the Sample

- Place the sample into a clean TGA sample pan or crucible. Materials like platinum or alumina are commonly used for crucibles due to their thermal stability.

- Carefully place the sample holder in the TGA furnace or measurement chamber to avoid spilling or uneven distribution.

4. Define Experimental Conditions

- **Temperature Range**: Choose a range suitable for the material, such as room temperature to 800°C for organic materials or up to 1500°C for ceramics and metals.

- **Heating Rate**: Select a heating rate, typically 5–20°C/min, depending on the resolution required.

- **Atmosphere**: Choose the appropriate gas environment:

    o **Inert Atmosphere**: Use nitrogen or argon to prevent oxidation or combustion.

    o **Oxidizing Atmosphere**: Use air or oxygen to study oxidation or combustion reactions.

- **Flow Rate**: Set the gas flow rate to ensure consistent thermal and atmospheric conditions, usually 20–50 mL/min.

5. Start the Experiment

- Begin the heating process, and the TGA will measure the mass of the sample continuously as the temperature increases.

- Monitor the thermogram, a plot of mass change (y-axis) versus temperature or time (x-axis), in real time.

6. Analyse the Data

- **Identify Key Points**: Examine the thermogram for events such as:

    o **Mass Loss Steps**: Indicate decomposition, volatilization, or moisture loss.

    o **Plateaus**: Represent thermal stability or the completion of a reaction.

- **Calculate Thermal Properties**: Measure the percentage of mass lost at specific temperature intervals.

- **Determine Residue**: Assess the remaining mass to analyse ash content or non-volatile residue.

7. Interpret and Report Results

- Compare the results with known standards or reference materials to identify decomposition pathways, thermal stability, or material composition.

- For complex materials, complement TGA with other techniques like Differential Scanning Calorimetry (DSC) for comprehensive thermal analysis.

Thermogravimetric Analysis (TGA) has diverse applications across multiple fields, leveraging its ability to analyse material weight changes under controlled temperature conditions. In the polymer industry, TGA is invaluable for assessing degradation temperatures, thermal stability, and the filler content in polymeric materials. These insights help in optimizing formulations and ensuring the durability of polymer products.

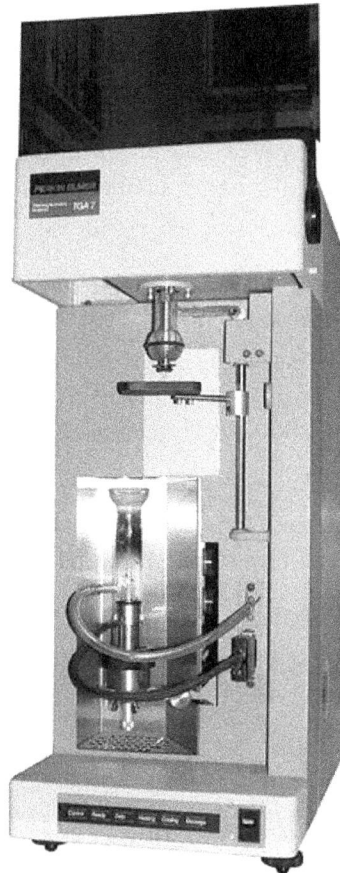

Figure 27: Thermogravimetric analyser (TGA). Luigi Chiesa, CC BY 3.0, via Wikimedia Commons.

In the pharmaceutical sector, TGA is used to analyse moisture content, thermal stability, and the purity of active ingredients and excipients. This ensures that pharmaceuticals meet stringent quality standards and remain stable under storage conditions. For metals and ceramics, TGA plays a critical role in studying oxidation behaviour and performing compositional analysis. These applications are essential in developing high-performance materials for aerospace, automotive, and industrial uses.

In the field of composites, TGA helps characterize the decomposition of individual components, providing crucial data for designing lightweight, strong, and thermally stable composite materials. Environmental science also benefits significantly from TGA, as it enables the analysis of carbon content in soils and waste materials. This information is vital for monitoring pollution levels, assessing soil health, and developing sustainable waste management practices. These applications illustrate the versatility of TGA as a tool for advancing research and quality assurance across industries.

Electrical and magnetic testing evaluates a material's conductivity and magnetic behaviour. Hall Effect measurements determine carrier concentration and mobility in semiconductors, while Vibrating Sample Magnetometry (VSM) assesses magnetic properties like saturation and coercivity in materials such as magnetic alloys. These techniques are pivotal in developing electronic and magnetic devices.

Optical characterization techniques, such as UV-Vis spectroscopy and photoluminescence (PL) spectroscopy, examine how materials interact with light. UV-Vis spectroscopy analyses light absorption and transmission in the ultraviolet and visible spectrum, while PL spectroscopy studies light emission upon excitation. These methods are especially valuable in the analysis of optoelectronic materials like quantum dots and perovskites.

Nanoindentation provides a precise way to measure hardness and elastic modulus at the nanoscale. This technique uses a sharp indenter to probe the surface of a material, making it particularly effective for studying thin films, nanocomposites, and other small-scale materials.

These property measurement techniques, when applied individually or in combination, offer a comprehensive understanding of material behaviour, enabling researchers and engineers to tailor materials for advanced technological applications across industries.

Emerging Techniques

Emerging techniques in materials characterization are revolutionizing the field by enabling unprecedented insights into the atomic and molecular structures of materials, their behaviours under various conditions, and their internal features. These advancements are crucial for fostering innovation across diverse scientific and industrial domains, including energy, healthcare, aerospace, and electronics.

One of the most significant advancements in this area is the application of Synchrotron Radiation. This technique utilizes high-energy X-rays generated in synchrotron facilities to probe the atomic and molecular structures of materials with exceptional resolution and sensitivity. For instance, synchrotron radiation has been instrumental in analysing the intricate structures of proteins and enzymes, which is vital for drug discovery [127]. Additionally, it has been employed to investigate the microstructural evolution of high-performance alloys under operational stresses, thereby informing the design of next-generation aerospace and automotive materials [128]. The ability of synchrotron radiation to provide atomic-scale resolution makes it an invaluable tool in materials science.

Figure 28: Synchrotron Radiation Source (SRS), Daresbury Laboratory. Rudi Winter, CC BY-SA 2.0, via Wikimedia Commons.

The process of utilizing synchrotron radiation for material analysis includes several key steps.

1. Understanding the Synchrotron Source Synchrotron radiation is generated when charged particles, typically electrons, are accelerated to near-light speeds and forced to change direction by magnetic fields. The resulting radiation covers a broad spectrum, including X-rays, UV, and infrared light, which can be tailored to specific experimental needs. This unique property makes synchrotron radiation ideal for probing materials at atomic and molecular levels.

2. Preparing the Sample Samples for synchrotron analysis must be carefully prepared to suit the requirements of the chosen technique. Depending on the type of analysis, the sample may need to be thin, transparent to X-rays, or mounted on specific holders. For example:

- **Crystalline materials** may require precise cutting or polishing for diffraction studies.

- **Biological samples** often need cryogenic preparation to maintain structural integrity.

- **Thin films or layered materials** may require deposition on substrates compatible with X-ray transmission.

3. Selecting the Appropriate Beamline and Technique Synchrotron facilities house multiple beamlines, each optimized for specific experimental techniques. Examples include:

- **X-ray diffraction (XRD)** for determining crystal structures.

- **X-ray absorption spectroscopy (XAS)** for probing electronic and chemical states.

- **Small-angle X-ray scattering (SAXS)** for studying nanoscale structures.

- **X-ray computed tomography (XCT)** for three-dimensional imaging of material interiors.

Researchers must consult with facility experts to select the appropriate beamline and tailor experimental conditions.

4. Conducting the Experiment The sample is placed in the beamline, where it interacts with the synchrotron radiation. Depending on the chosen technique, detectors measure various interactions:

- **Diffraction patterns** for structural analysis.

- **Absorption edges** for elemental and chemical state information.

- **Scattered radiation** for studying nanostructures and material morphology.

Data acquisition often involves automated systems for precision and efficiency. Experimental parameters, such as beam energy, angle, and exposure time, are adjusted based on the specific material and research goals.

5. Data Processing and Analysis The raw data collected during the experiment undergoes extensive processing using specialized software. This step includes:

- **Indexing and refining diffraction patterns** to solve crystal structures.

- **Analysing absorption spectra** to determine oxidation states and bonding environments.

- **Reconstructing 3D images** from tomography scans.

Researchers interpret the processed data to extract meaningful insights about the material's properties, structure, and behaviour.

Synchrotron radiation has a broad spectrum of applications across multiple disciplines, making it an indispensable tool in scientific research and technological development. In materials science, synchrotron radiation is widely employed for characterizing the structural and chemical properties of alloys, ceramics, and nanomaterials. By providing detailed insights into atomic arrangements and material phases, it supports advancements in high-performance materials and industrial applications.

In biology, synchrotron radiation plays a critical role in solving the structures of complex proteins and studying biomolecular interactions. This capability enables a deeper understanding of biological processes and facilitates the development of pharmaceuticals and therapeutic strategies. Similarly, the energy sector benefits from synchrotron radiation through its ability to investigate catalysts, batteries, and solar cell materials. These studies are crucial for designing efficient energy storage systems and renewable energy technologies.

Environmental science also leverages synchrotron radiation to analyse pollutants and mineral compositions. Its sensitivity to trace elements and compounds allows researchers to study environmental contamination and develop effective remediation strategies. Across these fields, synchrotron radiation provides unmatched precision and versatility, enabling groundbreaking discoveries and innovations.

Another emerging technique is 3D Imaging, particularly through X-ray computed tomography (XCT). XCT facilitates the non-destructive visualization of materials' internal structures in three dimensions, which is particularly beneficial for analysing porous materials, composites, and complex geometries. For example, XCT is widely used to study the pore networks in geopolymers and foams, which are critical for understanding their mechanical and thermal performance [129]. In composite materials, XCT allows researchers to examine fibre alignment, voids, and delaminations, which helps optimize manufacturing processes and ensure product reliability [130]. This non-invasive approach provides a comprehensive view of material integrity and performance.

X-Ray Computed Tomography (XCT) is a non-destructive imaging technique used to visualize the internal structures of materials in three dimensions. It relies on X-ray absorption differences within a sample to create detailed cross-sectional images, which are reconstructed into a 3D representation. This method is particularly valuable for studying materials with complex internal geometries, such as composites, porous structures, and biological specimens. The process comprises:

1. Sample Preparation

Ensure the sample is compatible with XCT in terms of size and density.

- Small and medium-sized samples (from microns to centimetres) are ideal for lab-based XCT systems.

- Dense or heavy samples may require higher-energy X-rays, typically found in synchrotron facilities.

- Samples should be free of contaminants that could affect imaging accuracy.

2. System Setup

Select the appropriate XCT system based on the desired resolution and sample characteristics.

- **Micro-CT systems** offer high-resolution imaging for small samples.

- **Industrial CT systems** are designed for larger or denser materials.
  Adjust the X-ray source and detector parameters:

- **Voltage and current**: Control the X-ray beam intensity. Higher voltages penetrate denser materials.

- **Detector resolution**: Ensure it matches the required imaging detail.

3. Scanning the Sample

Place the sample on the rotation stage.

- The stage rotates the sample incrementally while the X-ray source and detector capture projections from multiple angles.
  Set the scanning parameters:

- **Number of projections**: More projections yield higher-resolution images but increase scan time.

- **Exposure time**: Longer exposure enhances signal quality but may increase noise.
  Perform the scan and acquire raw projection data.

4. Image Reconstruction

Reconstruct the 2D projections into a 3D dataset using specialized software.

- Algorithms like filtered back-projection or iterative reconstruction convert the projections into a volumetric dataset.

- Calibrate the dataset to correct for distortions and ensure dimensional accuracy.

5. Visualization and Analysis

Analyse the reconstructed 3D data using visualization software.

- **Volume rendering**: Create a full 3D visualization of the internal structure.

- **Cross-sectional analysis**: Inspect individual layers to study specific features.

- **Quantitative analysis**: Measure parameters such as porosity, crack size, or density distribution.

In-Situ Characterization techniques represent a transformative approach by enabling real-time observation of material changes under operational conditions. These techniques simulate environments such as high temperatures, mechanical stress, or electrochemical processes, offering dynamic insights into material behaviour. For instance, in-situ studies of battery materials during charging and discharging cycles have elucidated the mechanisms of lithium-ion transport and electrode degradation, which are crucial for developing batteries with longer lifespans and improved safety profiles [131]. Furthermore, in-situ characterization techniques have been applied to understand the degradation mechanisms in perovskite solar cells, highlighting their importance in enhancing module stability [132]. The ability to track dynamic changes in structure and morphology in real time is a significant advancement in materials characterization.

These emerging techniques collectively push the boundaries of materials science, providing comprehensive data that drive advancements across various sectors. Their capabilities to offer atomic-scale resolution, real-time observation, and three-dimensional insights equip researchers and engineers with the necessary tools to develop materials that meet the challenges of modern technology and sustainability. As these techniques continue to evolve, they will undoubtedly play a critical role in shaping the future of materials science and engineering.

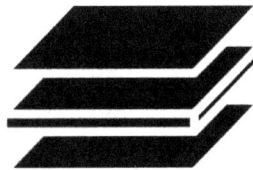

Chapter 3

Composite Materials

Definition and Types of Composites

Composites are engineered materials made by combining two or more constituent materials with different physical or chemical properties. These constituents work synergistically to create a composite with enhanced properties not found in the individual components. Typically, composites consist of a matrix material that binds and protects the reinforcement material, which provides strength and rigidity. The resulting material exhibits superior characteristics, such as high strength-to-weight ratios, improved durability, and tailored properties, making composites essential for various industrial applications.

Polymer Matrix Composites (PMCs)

Polymer matrix composites (PMCs) represent a significant advancement in material science, characterized by a polymer resin matrix reinforced with fibres or particles. This combination enhances mechanical properties such as strength and stiffness while providing a protective medium for the reinforcements, making PMCs suitable for a wide range of applications in industries where lightweight and high-performance materials are critical. The matrix is predominantly made from polymers like epoxy, polyester, or thermoplastics such as polyether ether ketone (PEEK). Epoxy resins are particularly favoured due to their excellent adhesive properties, chemical resistance, and compatibility with various reinforcements, while thermoplastics offer advantages in terms of recyclability and toughness, which are increasingly important in modern applications [133-135].

The matrix in polymer matrix composites (PMCs) serves a crucial role in bonding the fibres together and facilitating load transfer between them. This function ensures that the composite material behaves as a unified structure, allowing it to withstand applied stresses effectively. The matrix also protects the reinforcing fibres from environmental degradation, abrasion, and mechanical damage. PMCs matrices are typically classified into thermosets and thermoplastics, with thermosets being the predominant choice in advanced composite applications today.

Thermoset resins are widely used in PMCs because of their excellent mechanical properties, thermal stability, and chemical resistance. These resins require the addition of a curing agent or hardener, followed by a curing process to solidify the composite into its final form. Once cured, thermoset materials become rigid and cannot be reshaped, except for post-curing finishing processes. Common thermoset resins include epoxies, polyurethanes, phenolic resins, and bismaleimides (BMIs). Among these, epoxy resins dominate the advanced composite industry due to their versatility and superior performance characteristics.

Epoxy resins, which have been in industrial use for over 40 years, are also referred to as glycidyl compounds. These resins can be chemically modified by cross-linking with other molecules to produce a wide range of resin products, each tailored to specific performance requirements. They are available in forms ranging from low-viscosity liquids to high-molecular-weight solids, with high-viscosity liquids being the most common. This adaptability makes epoxy resins suitable for diverse applications, including aerospace, automotive, and construction.

Curing agents or hardeners are essential in thermoset systems, controlling the reaction rate and determining the final properties of the cured composite. These agents act as catalysts, facilitating the chemical cross-linking necessary for the resin to harden into a durable material. Aromatic amines, such as methylene-dianiline (MDA) and sulfonyldianiline (DDS), are among the most commonly used curing agents in advanced composite manufacturing.

Other types of curing agents include aliphatic and cycloaliphatic amines, polyaminoamides, amides, and anhydrides. The choice of curing agent depends on the specific cure characteristics and performance requirements of the composite. For example, aliphatic amines are often chosen for their lower curing temperatures, while anhydrides are preferred for high-temperature applications.

Polyurethanes represent another significant group of thermoset resins. These materials are formed by reacting a polyol component with an isocyanate compound, such as toluene diisocyanate (TDI), methylene diisocyanate (MDI), or hexamethylene diisocyanate (HDI). Polyurethanes are valued for their toughness, flexibility, and versatility in composite applications.

Phenolic and amino resins are also used in PMCs, particularly in applications requiring high thermal resistance and flame retardancy. While bismaleimides and polyamides are relatively new to the advanced composites industry, they offer promising properties, such as superior temperature resistance, and are being increasingly explored for high-performance applications.

A specialized category of thermoset-based composites involves SiC–SiC matrix systems. These high-temperature ceramic matrix composites are processed from preceramic polymers, which are used to infiltrate a fibrous preform and create a silicon carbide (SiC) matrix. SiC–SiC composites are notable for their exceptional thermal stability, making them ideal for high-temperature environments, such as aerospace and energy applications.

The matrix in PMCs is fundamental to the functionality and durability of composite materials. Thermoset matrices, especially epoxy resins, dominate the industry due to their exceptional mechanical properties and adaptability. The curing process, facilitated by various curing agents, plays a pivotal role in defining the performance characteristics of the final composite. Other thermoset systems, such as polyurethanes, phenolics, and bismaleimides, provide additional options for tailoring PMCs to meet specific application needs. Advances in resin chemistry and curing technology continue to enhance the capabilities and expand the applications of PMCs in diverse industries.

The reinforcements in PMCs typically consist of high-performance fibres, including carbon, glass, and aramid fibres. Carbon fibres are noted for their exceptional tensile strength, low density, and high stiffness, making them ideal for applications in the aerospace and automotive sectors. For instance, carbon fibre-reinforced polymers (CFRPs) are extensively used in aircraft structures, where their strength-to-weight ratio significantly contributes to fuel efficiency and structural integrity [28, 136, 137]. Glass fibres provide a more cost-effective reinforcement option, offering good strength and corrosion resistance, which is advantageous in marine and construction applications. Aramid fibres, such as Kevlar, are recognized for their impact resistance and toughness, commonly utilized in protective gear and advanced sporting equipment [138, 139].

Fibre-reinforced polymer matrix composites (PMCs) are among the most widely used advanced materials due to their excellent mechanical properties and versatility. These composites typically contain approximately 60% reinforcing fibres by volume, such as fiberglass, graphite, and aramid. Fiberglass is particularly popular because of its competitive tensile strength, relatively low stiffness, and significantly lower cost compared to other fibres. This affordability makes fiberglass one of the most commonly used fibres in industrial and commercial applications.

The mechanical properties of reinforcing fibres are highest along their lengths rather than their widths. This anisotropic property enables designers to orient fibres in various configurations to optimize the composite's physical properties for specific applications. For instance, fibres can be arranged in unidirectional layers, woven into fabrics, or randomly oriented, depending on the desired balance of strength, stiffness, and toughness. Applications of fibre-reinforced PMCs span aerospace, automotive, sports equipment, and marine industries, where their lightweight and high-strength characteristics are particularly advantageous.

Nanomaterial-reinforced PMCs represent a significant advancement in composite technology, achieving remarkable improvements in mechanical properties with minimal reinforcement loading, often less than 2% by volume. Among these materials, carbon nanotubes (CNTs) have garnered intense interest due to their exceptional tensile stiffness and strength, attributed to the strong sp^2 covalent bonds between carbon atoms. However, realizing the full potential of CNTs in composites requires effective load transfer between the nanotubes and the polymer matrix.

The dispersion and size of CNTs are critical factors in determining the final properties of the composite. Long CNTs, with their high aspect ratios, enhance tensile stiffness and strength by enabling efficient stress transfer and mitigating crack propagation. In contrast, short CNTs require interfacial adhesion to improve the composite's properties, which can be achieved through surface modifications. Functionalization techniques, both covalent and non-covalent, are employed to enhance the bond between CNTs and the polymer matrix. Covalent functionalization involves direct chemical bonding, such as oxidation or free radical reactions, while non-covalent methods rely on physical interactions like van der Waals forces or π-stacking.

The manufacturing process for CNT-reinforced PMCs depends on the polymer matrix and the desired properties. Thermoset polymers are typically processed using solution methods, where CNTs and polymers are mixed in solvents, sonicated, and cast. However, this process can damage CNTs and create structural inconsistencies. Thermoplastic polymers are processed through melt-processing, where CNTs are mixed with melted polymers, though high CNT loading increases viscosity. In-situ polymerization, suitable for polymers incompatible with solvents or heat, involves mixing CNTs with monomers, which react to form the polymer matrix, offering strong interfacial bonding.

Graphene, like carbon nanotubes, is renowned for its remarkable mechanical properties and is increasingly explored as a reinforcement in PMCs. While the mechanical enhancements achieved with graphene-reinforced PMCs are generally less than those with CNTs, graphene offers other advantages. Graphene oxide, a functionalized derivative of graphene, is easier to process due to its inherent defects, facilitating better interfacial bonding with polymer matrices. Additionally, three-dimensional graphene structures show potential for isotropic enhancement of mechanical properties, expanding their applicability.

The processing methods for graphene-reinforced PMCs mirror those used for CNT composites. Solution processing, melt-processing, and in-situ polymerization are commonly employed, with each method tailored to the specific matrix material and application requirements. The ease of functionalizing graphene oxide and its compatibility with various polymers make it an attractive alternative for developing advanced PMCs with tailored properties.

The applications of polymer matrix composites (PMCs) are extensive, particularly in sectors where performance and weight reduction are critical. In the aerospace industry, carbon fibre-reinforced polymers (CFRPs) are essential for constructing aircraft components such as fuselages and wings. Their lightweight nature significantly contributes to reduced fuel consumption and enhanced durability, making them a preferred choice in modern aircraft design. For instance, the Boeing 787 Dreamliner utilizes CFRPs extensively, with these composites constituting approximately 50% of the airframe, underscoring their pivotal role in aerospace engineering [140, 141]. The continuous advancements in materials science further enhance the capabilities of CFRPs, solidifying their position as indispensable materials in high-performance sectors [142, 143].

Similarly, the automotive industry leverages PMCs to manufacture lightweight body panels and structural components, which leads to improved fuel efficiency and safety. The integration of CFRPs in automotive applications is driven by their high strength-to-weight ratio, which allows for the design of safer and more efficient vehicles [144]. Research indicates that the use of PMCs in automotive manufacturing not only enhances performance but also contributes to sustainability efforts by reducing the overall weight of vehicles, thereby lowering emissions during operation [142, 143].

In addition to aerospace and automotive applications, the sports equipment industry benefits significantly from PMCs. Products such as tennis rackets and bicycles utilize these materials for their enhanced performance and durability, which are essential for high-level athletic performance [144]. The lightweight and strong characteristics of PMCs allow for the creation of equipment that can withstand the rigors of competitive sports while providing athletes with a performance edge.

Marine applications also extensively employ glass fibre-reinforced composites (GFRPs), particularly in constructing hulls and decks. GFRPs are favoured for their resistance to corrosion and degradation in harsh marine environments, ensuring longevity and reliability in marine structures [145]. The durability of these composites under exposure to seawater and other environmental factors makes them suitable for various marine applications, further illustrating the versatility of PMCs across different industries [143].

Metal Matrix Composites (MMCs)

Metal Matrix Composites (MMCs) represent a significant advancement in materials science, characterized by a metallic matrix—commonly aluminium, magnesium, or titanium—reinforced with various materials such as ceramic particles, whiskers, or fibres. The matrix provides essential properties such as ductility and thermal conductivity, while the reinforcement enhances stiffness, wear resistance, and overall strength, making MMCs suitable for high-performance applications.

The matrix material, typically aluminium, is favoured due to its low density, excellent strength-to-weight ratio, and corrosion resistance, which are critical in applications like aerospace and automotive industries [146, 147]. For instance, aluminium matrix composites (AMCs) are extensively utilized in automotive brake discs, where the incorporation of silicon carbide (SiC) particles significantly improves wear resistance and thermal stability [146, 147]. The reinforcement phase, often consisting of ceramic materials, plays a crucial role in enhancing the mechanical properties of the composites. Studies indicate that the inclusion of reinforcements such as SiC and alumina can lead to substantial improvements in tensile strength and hardness, thereby extending the operational capabilities of the materials under extreme conditions [148-150].

MMCs are formed by dispersing a reinforcing material within a continuous metal matrix. The reinforcement can be in the form of fibres, particles, or whiskers, and is often coated to prevent undesirable chemical reactions with the matrix material. For instance, carbon fibres are frequently utilized in aluminium matrices due to their low density and high strength. However, a significant challenge arises from the reaction between carbon and aluminium, which leads to the formation of a brittle and water-soluble compound known as Al_4C_3 at the fibre-matrix interface [151]. To mitigate this issue, carbon fibres are often coated with materials such as nickel or titanium boride, which effectively inhibit the formation of Al_4C_3 and enhance the mechanical properties of the composite [152, 153].

The matrix in MMCs serves as the continuous phase that supports the reinforcement. Typically, lighter metals like aluminium, magnesium, or titanium are used for structural applications due to their favourable strength-to-weight ratios [154, 155]. The matrix not only provides structural integrity but also influences the overall thermal and mechanical properties of the composite. In high-temperature applications, cobalt and cobalt-nickel alloy matrices are preferred due to their superior thermal stability [154]. The continuous nature of the matrix ensures that there is a path through the material, which is essential for load transfer and overall performance [156].

Reinforcement materials in MMCs can be categorized into continuous and discontinuous forms. Continuous reinforcements, such as carbon fibres or silicon carbide fibres, are aligned in a specific direction, resulting in anisotropic properties where the strength varies with direction [157]. This alignment can significantly enhance the tensile strength of the composite, as evidenced by studies showing that carbon fibre reinforced aluminium composites exhibit remarkable strength improvements when properly coated [158, 159].

On the other hand, discontinuous reinforcements, which include short fibres or particles, can be isotropic and are often easier to process using conventional metalworking techniques like extrusion or forging [155, 160]. The use of polycrystalline diamond tooling is common for machining these composites due to their enhanced hardness and wear resistance [160]. The choice of reinforcement not only affects the mechanical properties but also the thermal and wear resistance of the composite, making it crucial for specific applications [154, 161].

The processing methods for MMCs, such as powder metallurgy and stir casting, are pivotal in determining the final properties of the composites. For example, the hot pressing and extrusion techniques can enhance the interfacial bonding between the matrix and the reinforcement, thereby improving mechanical properties [162, 163]. However, it is essential to optimize processing parameters, as excessive temperatures can lead to a reduction in the strength of the matrix material, adversely affecting the composite's performance [162].

Applications of MMCs are diverse and include engine components, aerospace structures, and thermal management systems. Their ability to withstand high temperatures and mechanical stress makes them ideal for these demanding environments [164, 165]. The development of hybrid metal matrix composites (HMMCs), which incorporate multiple types of reinforcements, further expands the potential applications by combining the benefits of different materials to achieve tailored properties [160].

MMC manufacturing can be broken into three types—solid, liquid, and vapor. Semi-solid state processing is a unique approach in manufacturing metal matrix composites (MMCs) that utilizes a combination of solid and liquid phases to achieve superior material properties. One common technique is semi-solid powder processing, where a mixture of powdered metal and reinforcement material is heated to a semi-solid state. At this stage, the material exists in a partially molten condition, enabling a balance between fluidity and structural integrity. Pressure is then applied to consolidate the mixture into a dense composite material.

This method is particularly advantageous because it allows for better distribution of the reinforcement within the matrix compared to fully liquid methods. The semi-solid state minimizes segregation and settling of the reinforcement particles, resulting in a more uniform microstructure. This process is often used to produce MMCs with improved mechanical properties, such as higher strength and enhanced wear resistance, making them suitable for applications in aerospace and automotive industries.

Vapor deposition techniques involve the deposition of material from a vapor phase onto a substrate, forming a composite material. In the case of physical vapor deposition (PVD), fibres are passed through a dense cloud of vaporized metal. This vapor condenses on the fibres, coating them with a uniform metallic layer that serves as the matrix. The process can be fine-tuned to control the thickness and properties of the coating, allowing for precise engineering of the composite's final characteristics.

PVD is particularly useful for producing thin, high-performance coatings with excellent adhesion and tailored properties. This method is widely used in applications where surface properties, such as hardness, wear resistance, or corrosion resistance, are critical. For example, PVD is employed in creating lightweight, high-strength materials for aerospace components and advanced tools with enhanced durability.

In-situ fabrication techniques create MMCs directly within the material system through controlled solidification processes. One such approach involves the controlled unidirectional solidification of eutectic alloys. During solidification, the alloy separates into two distinct phases, with one phase forming lamellar or fibrous structures distributed within the matrix.

This in-situ formation of reinforcement eliminates the need for external addition of reinforcement particles or fibres, ensuring excellent bonding and uniform distribution.

This technique is particularly advantageous for producing composites with highly aligned microstructures, resulting in exceptional directional properties. For instance, the lamellar or fibrous reinforcement significantly enhances the material's strength and stiffness along the aligned direction. Applications include high-temperature structural components and specialized materials for thermal management, where tailored properties in specific orientations are essential.

Each manufacturing method, including semi-solid state processing, vapor deposition, and in-situ fabrication, offers unique benefits for producing MMCs tailored to specific applications. Semi-solid processing ensures uniform reinforcement distribution, vapor deposition enables precision coating for advanced properties, and in-situ techniques naturally integrate reinforcement during solidification. These advanced manufacturing approaches play a critical role in enabling the development of high-performance MMCs for cutting-edge technologies.

The addition of ceramic particles to metals in the formation of metal-matrix composites (MMCs) significantly impacts their mechanical properties, offering notable benefits and trade-offs. One of the most pronounced effects is the enhancement of material strength and stiffness. For instance, incorporating aluminium oxide (Al_2O_3) into aluminium (Al) matrices can elevate the yield strength of cast Al 6061 alloys from 105 MPa to 120 MPa and increase the Young's modulus from 70 GPa to 95 GPa. These improvements are attributed to the high specific stiffness of the ceramic reinforcement. However, this comes at the cost of reduced ductility, with the composite's ductility decreasing from 10% to just 2%. This trade-off reflects the balance between leveraging the stiffness of ceramics and retaining some degree of metal flexibility.

In addition to strength, MMCs exhibit significant enhancements in wear resistance and hardness. Aluminium alloys reinforced with Al_2O_3 particles demonstrate markedly improved wear resistance, making them suitable for applications in light, wear-resistant components. For example, Al-Si alloys with ceramic reinforcements show enhanced durability under frictional stress, while Al-Mg alloys reinforced with SiO_2 particles exhibit increased hardness. These properties make MMCs ideal for automotive applications, such as piston liners in lightweight aluminium engines, which often face challenges due to the softness of aluminium alloys. By incorporating ceramic particles, manufacturers can reduce reliance on heavier cast-iron liners, achieving better weight efficiency without compromising wear resistance.

Fracture toughness in MMCs is a complex property influenced by the composition of the composite and the interaction between the metal and ceramic phases. Typically, fracture toughness is dominated by the metal phase; however, in systems with significant thermal expansion mismatches, such as Cu/Al_2O_3, localized stresses can encourage crack propagation through delamination. This results in lower fracture toughness. On the other hand, systems like Al/Al$_2$O$_3$ that feature co-continuous phases exhibit improved toughness. In such systems, cracks deflect at the interfaces between ceramic and metallic phases, requiring additional energy for propagation. This deflection mechanism contributes to toughening the composite, particularly when thermal mismatches are minimized.

MMCs also strengthen materials against plastic deformation through several mechanisms. The first mechanism is direct load transfer to the stronger ceramic particles, which bear a significant portion of the applied load. Secondly, the disparity in plastic deformation between the matrix and ceramic particles leads to the pinning of dislocations at the particle-matrix interface. Dislocations, which are lower-energy pathways for plastic deformation, must bypass these obstacles by bowing around the particles, a process that requires significantly higher stress and energy. This phenomenon is analogous to precipitation hardening. Lastly, stress fields created by thermal and coherency mismatches between the ceramic and metal phases further trap dislocations, creating pileups that inhibit their movement. This collective behaviour increases the energy required for plastic deformation, effectively strengthening the composite against plastic flow.

Overall, MMCs leverage ceramic reinforcements to achieve remarkable enhancements in strength, stiffness, wear resistance, and hardness while facing challenges in ductility and fracture toughness. These effects are critical in tailoring MMCs for demanding applications such as aerospace components, automotive parts, and other high-performance environments. By understanding and controlling these mechanical property trade-offs, researchers and engineers can optimize MMC compositions for specific use cases.

Ceramic Matrix Composites (CMCs)

Ceramic Matrix Composites (CMCs) represent a significant advancement in materials science, combining the inherent properties of ceramics with the toughness provided by fibre reinforcement. The matrix in CMCs is typically composed of ceramics such as silicon carbide (SiC) or aluminium oxide (Al$_2$O$_3$), which confer exceptional thermal stability and high-temperature strength. These materials allow CMCs to withstand extreme temperatures without significant degradation, making them suitable for demanding applications in industries such as aerospace and power generation [166-168].

The reinforcement in CMCs is often achieved using ceramic fibres or whiskers, which play a crucial role in enhancing the toughness and fracture resistance of the composite. The presence of these reinforcements helps to arrest crack propagation, distribute stress more evenly, and absorb energy during deformation, thereby preventing catastrophic failure [169-171]. For instance, SiC fibres are particularly prominent in CMCs due to their high strength, thermal stability, and performance in oxidative environments, making them ideal for applications in gas turbines and aerospace components [170, 172, 173].

CMCs are particularly advantageous in high-temperature environments where conventional materials may fail. Their excellent thermal stability and resistance to oxidation make them invaluable in applications such as turbine blades and heat shields in jet engines, where they must endure extreme thermal and mechanical stresses while maintaining structural integrity [174, 175]. The lightweight nature of CMCs, combined with their ability to operate at temperatures significantly higher than traditional metal alloys, contributes to improved engine efficiency and reduced fuel consumption [176-178].

Figure 29: CVI-SiC/SiC-shaft sleeves for big pumps, diameter 100 to 300 mm, (SiC-fibre reinforced SiC material manufactured via chemical vapour infiltration). MT Aerospace AG, Augsburg, Germany, CC BY-SA 3.0, via Wikimedia Commons.

A prime example of CMCs in application is the SiC/SiC composite, which consists of a silicon carbide matrix reinforced with silicon carbide fibres. These composites are extensively utilized in the aerospace industry for components that require high-temperature resilience and lightweight properties. In power generation, SiC/SiC composites are employed in gas turbines and nuclear reactors, where they must sustain high thermal loads and resist oxidative degradation over extended periods [168, 172, 179]. The unique combination of properties offered by CMCs, including high strength, low density, and chemical inertness, positions them as critical materials for future advancements in various high-performance applications [173, 175].

The manufacturing of CMCs involves intricate processes that integrate ceramic materials with fibres or other reinforcements to achieve superior mechanical and thermal properties. The procedure generally comprises three primary steps: fibre lay-up and fixation, matrix infiltration, and final machining. These processes are tailored based on the type of matrix material and the desired properties of the final composite.

The first step involves shaping and fixing the reinforcing fibres, often referred to as rovings, into a preform that resembles the intended component shape. Techniques like filament winding, fabric lay-up, braiding, and knotting—commonly used in fibre-reinforced plastic production—are adapted for CMCs. The resulting preform forms the structural foundation for the composite, dictating its mechanical performance and load distribution.

The second and most crucial step is infiltrating the matrix material into the fibre preform. Multiple methods are employed based on the type of ceramic matrix:

1. **Deposition from a Gas Phase**: Chemical Vapor Deposition (CVD) or Chemical Vapor Infiltration (CVI) is used to deposit materials like carbon or silicon carbide into the fibre structure. For example, argon and hydrocarbons can decompose at elevated temperatures to deposit carbon, while silicon carbide can form from methyl-trichlorosilane. This process produces a porous matrix with about 10-15% porosity, as access to the inner regions of the preform diminishes with external deposition.

2. **Matrix Formation via Pyrolysis**: In this approach, a polymer precursor infiltrates the fibre preform, followed by curing and pyrolysis. Polymers like polycarbosilanes decompose to form ceramics such as silicon carbide, silicon oxycarbide, or silicon oxynitride. Repeated cycles of infiltration and pyrolysis (typically five to eight) are needed to reduce porosity and enhance matrix quality.

3. **Chemical Reaction of Elements**: Reactive methods involve chemical reactions within the preform. For instance, liquid silicon infiltration (LSI) uses molten silicon to react with porous carbon, forming silicon carbide. This technique yields a highly dense material with porosity as low as 3%.

4. **Sintering**: For oxide-based CMCs, precursors like alumina powders combined with tetra-ethyl-orthosilicate undergo sintering at relatively low temperatures (1000–1200 °C). This method is particularly suited for materials with temperature-sensitive fibres, resulting in a matrix with about 20% porosity.

5. **Electrophoresis**: In this developing technique, electrically charged ceramic particles are deposited into a preform using an electric field. While still under research, it shows potential for creating innovative CMCs with controlled properties.

After the matrix infiltration, the composite undergoes finishing steps like grinding, drilling, or lapping, often with diamond tools due to the hardness of ceramic materials. Advanced techniques such as laser, water jet, or ultrasonic machining may also be used for precise shaping. Additional treatments, including coating or impregnation, can be applied to enhance specific properties like oxidation resistance or thermal conductivity.

Carbon-Carbon Composites (C/C Composites)

Carbon-carbon composites (C/C composites) are advanced materials that consist of carbon fibres embedded within a carbon matrix, which endows them with exceptional mechanical and thermal properties. The unique all-carbon structure of these composites allows them to perform exceptionally well in high-performance applications, particularly those that require extreme heat resistance and strength. The composition and structure of C/C composites are critical to their functionality; the carbon fibres serve as the reinforcement, providing high tensile strength and stiffness, while the carbon matrix binds the fibres together and facilitates load transfer, ensuring mechanical integrity under thermal and mechanical stress [180, 181].

The matrix and reinforcement in C/C composites are specifically optimized to achieve distinct functionalities. By manipulating the fibre architecture—such as employing woven, braided, or unidirectional configurations—engineers can tailor the composite for specific applications, enhancing strength in particular directions or providing multidirectional load-bearing capabilities [180, 182]. This adaptability is crucial for applications in demanding environments, where the mechanical properties of the composite must be preserved even under severe conditions [183].

One of the most defining characteristics of C/C composites is their exceptional thermal resistance. These materials can withstand extreme temperatures, often exceeding 2000 °C in non-oxidizing environments, without melting or degrading [184]. This property is particularly advantageous in aerospace and defence applications, where materials are subjected to intense thermal and mechanical stresses. The high thermal conductivity of C/C composites also aids in the even distribution of heat, minimizing the risk of localized overheating, which can lead to material failure [181, 183].

C/C composites find extensive applications in industries that demand superior thermal and mechanical performance. In aerospace, they are critical for manufacturing components such as rocket nozzles and re-entry heat shields, where they must endure the extreme heat generated during atmospheric re-entry or rocket propulsion [185, 186]. In aviation, these composites are favoured for aircraft brake systems, providing reliable performance under the extreme friction and heat encountered during landing [180, 186]. Furthermore, in motorsports and advanced automotive engineering, C/C composites are utilized in high-performance braking systems due to their lightweight nature and excellent thermal resistance [184, 185].

A prominent example of the application of C/C composites is in space exploration. These materials are integral to the thermal protection systems (TPS) of spacecraft, which protect vehicles and their occupants from the extreme temperatures experienced during re-entry into Earth's atmosphere. Their ability to withstand and dissipate heat efficiently ensures that spacecraft components remain intact and operational under harsh conditions [183, 185, 186]. The ongoing advancements in technology are likely to expand the versatility and performance of C/C composites, paving the way for new engineering innovations and applications in high-tech industries [184, 185].

The production of carbon-carbon (C/C) composites is a complex and meticulous process designed to create materials with exceptional mechanical strength, thermal stability, and resistance to extreme temperatures. The manufacturing involves several stages, including the selection and preparation of carbon fibres, impregnation with a carbonaceous matrix, and repeated heat treatments to optimize the material's properties.

1. Fiber Selection and Preform Preparation

The process begins with the selection of high-quality carbon fibres, which serve as the reinforcement in the composite. These fibres are typically arranged into a desired architecture, such as unidirectional, woven, or braided forms, depending on the intended application. The fibre architecture significantly influences the composite's mechanical properties, including tensile strength, stiffness, and thermal conductivity.

The arranged fibres are shaped into a preform that closely resembles the final geometry of the component. Techniques such as filament winding, braiding, or manual lay-up are used to create this preform, ensuring the fibres are uniformly distributed to maximize strength and load-bearing capabilities.

2. Impregnation with Carbonaceous Matrix

The next step involves impregnating the fibre preform with a carbonaceous matrix precursor. Common precursors include resins (such as phenolic resin) or pitch, which serve as sources of carbon. The preform is soaked in the liquid precursor under vacuum or pressure to ensure thorough penetration and minimize voids within the structure. This stage is critical as it forms the basis for creating a strong bond between the fibres and the matrix.

3. Carbonization

Once impregnated, the preform undergoes carbonization, where it is heated to high temperatures (typically 900–1200 °C) in an inert atmosphere, such as argon or nitrogen. During this process, the organic components of the precursor decompose, leaving behind a carbon matrix. This step also removes volatile impurities, creating a dense carbon-carbon composite. However, this initial carbon matrix is still porous and lacks the desired mechanical and thermal properties.

4. Densification

To improve the density and strength of the composite, densification is performed. This involves repeating the impregnation and carbonization steps multiple times, typically five to ten cycles, depending on the application. Each cycle fills the remaining pores with additional carbon, gradually increasing the composite's density and reducing porosity. This iterative process ensures the final material achieves the desired mechanical strength and thermal resistance.

5. Graphitization

In some applications, the carbon matrix is further treated through a graphitization process, where the material is heated to extremely high temperatures (above 2000 °C). This step enhances the crystalline structure of the carbon, improving its thermal conductivity and oxidation resistance. Graphitization also reduces internal stresses, increasing the composite's durability in high-temperature environments.

6. Final Machining and Coating

The densified composite is then machined into its final shape using diamond-tipped tools, as the material's hardness can make conventional machining difficult. If the application requires protection against oxidation, the component is often coated with a layer of silicon carbide (SiC) or other protective materials to prevent degradation in oxygen-rich environments.

Figure 30: Carbon/carbon fibre panels, panel A112 #25 post test isometric view of back side. Defense Visual Information Distribution Service, Public Domain, via Picryl.

Chemical Vapor Infiltration (CVI), Liquid Pitch Infiltration (LPI), and Resin Transfer Molding (RTM) are particularly notable for their effectiveness in forming a dense carbon matrix and enhancing composite properties.

Chemical Vapor Infiltration (CVI): Chemical Vapor Infiltration is a gas-phase process designed to deposit carbon into the pores of the fibre preform. In this technique, the fibre preform is placed in a reaction chamber, and hydrocarbon gases, such as methane or propane, are introduced under controlled pressure and temperature conditions. The hydrocarbons decompose upon contact with the heated preform, depositing carbon on and between the fibres.

CVI is advantageous because it allows precise control over the carbon deposition process, resulting in uniform infiltration and minimal residual porosity. Additionally, the process can be customized to optimize the properties of the carbon matrix, such as its density and crystallinity. However, CVI is time-consuming and requires specialized equipment, making it suitable for applications demanding high-performance materials, such as aerospace components and rocket nozzles.

Liquid Pitch Infiltration (LPI): Liquid Pitch Infiltration is a liquid-phase process that uses pitch—a viscous, carbon-rich material—as the carbon precursor. In this method, the fiber preform is submerged in molten pitch, ensuring that the pitch infiltrates the pores and fills the voids within the structure. After infiltration, the preform undergoes a pyrolysis process, where it is heated to high temperatures in an inert atmosphere. This process decomposes the pitch, leaving behind a carbon matrix.

LPI is particularly effective for producing composites with a high degree of densification, as repeated cycles of infiltration and pyrolysis can significantly reduce porosity. This technique is also more cost-effective than CVI, making it suitable for applications that do not require the extreme precision of gas-phase processes. However, the challenge of achieving complete infiltration in highly complex geometries and the need for multiple cycles are notable limitations.

Resin Transfer Molding (RTM): Resin Transfer Molding is a technique that involves injecting liquid resin into a fibre preform under pressure. The resin serves as the precursor for the carbon matrix. Once the resin is fully infused into the preform, the material is cured to solidify the resin and provide initial structural integrity. Subsequently, the cured composite undergoes pyrolysis to convert the resin into carbon, forming the matrix.

RTM is highly advantageous for its ability to produce near-net-shape components with minimal post-processing. The use of molds allows precise control over the geometry and dimensions of the final component, reducing waste and machining requirements. RTM is also more adaptable for large-scale production compared to CVI and LPI. However, like LPI, RTM often requires multiple cycles of infiltration and pyrolysis to achieve the desired density and mechanical properties.

Hybrid Composites

Hybrid composites are advanced materials engineered by combining two or more types of reinforcements or matrices to achieve an optimal mix of mechanical, thermal, or chemical properties. This engineering approach allows for leveraging the strengths of different components, enabling hybrid composites to meet specific performance requirements that single-component composites cannot achieve. For instance, the combination of carbon fibres, known for their high stiffness and tensile strength, with glass fibres, which offer cost-effectiveness and flexibility, exemplifies how hybrid composites can optimize performance characteristics [187, 188].

The unique structural design of hybrid composites is a critical feature that distinguishes them from traditional composites. These materials can incorporate various types of fibres, such as carbon, glass, or aramid (e.g., Kevlar), within a single polymer matrix. The arrangement of these reinforcements can vary; they may be layered separately, woven into hybrid fabrics, or mixed within the matrix, depending on the desired properties [189, 190]. For example, a hybrid composite that combines carbon and glass fibres can achieve a balance of high stiffness from the carbon fibres while benefiting from the flexibility and lower cost of glass fibres [191]. The matrix material, typically a thermoset resin like epoxy or a thermoplastic such as polyether ether ketone (PEEK), plays a crucial role in binding the reinforcements, ensuring load transfer, and protecting the fibres from environmental damage [192, 193].

Hybrid composites find extensive applications across various industries where tailored performance is essential. In the sporting goods sector, they are utilized in products like tennis rackets, bicycles, and skis, where a combination of lightweight properties with durability and impact resistance is critical [194, 195]. The automotive industry employs hybrid composites in body panels to reduce vehicle weight without compromising safety or structural integrity, thus enhancing fuel efficiency and overall performance [196]. In aerospace, these materials are favoured for their strength, toughness, and impact resistance, making them suitable for components such as aircraft panels and rotor blades [197, 198]. Additionally, hybrid composites are increasingly used in energy storage devices, including battery casings, due to their ability to withstand high mechanical and thermal loads [199].

Examples of hybrid composites illustrate their versatility and effectiveness. One notable example is the combination of carbon and Kevlar fibres in a polymer matrix, which provides high stiffness and tensile strength alongside excellent impact resistance and toughness. This specific hybrid is widely used in protective equipment, such as helmets and body armour, as well as in aerospace components requiring lightweight yet durable materials [188, 197]. Another example is the use of glass and basalt fibres in wind turbine blades, where glass fibres contribute cost-effectiveness and good tensile strength, while basalt fibres enhance thermal resistance and fatigue performance, ensuring durability in harsh environmental conditions [194, 195].

The primary advantage of hybrid composites lies in their ability to optimize material properties. By carefully selecting and combining reinforcements, manufacturers can tailor the composite to meet specific mechanical, thermal, or environmental challenges. This flexibility also allows for cost-performance trade-offs, making hybrid composites accessible for a broader range of applications [190, 196]. Furthermore, hybridization can effectively address the deficiencies of individual materials, such as the brittleness of certain fibres, by combining them with more ductile options [189, 200].

Natural Fibre Composites (NFCs)

Natural Fiber Composites (NFCs) are increasingly recognized for their potential in sustainable materials engineering, primarily due to their composition of natural fibres such as jute, hemp, and sisal embedded within a polymer matrix. This combination not only enhances mechanical properties but also promotes environmental sustainability, making NFCs a viable alternative to traditional synthetic composites.

The matrix materials used in NFCs are often biodegradable or thermoplastics, with polypropylene being a common choice. Polypropylene, due to its favourable mechanical properties and processability, serves as an effective matrix that can be combined with various natural fibres to create composites with enhanced performance characteristics. For instance, studies have shown that the orientation and stacking sequence of fibres, such as jute and sisal, significantly affect the mechanical behaviour of the resulting composites, leading to improved interlaminar shear strength and overall structural integrity [201, 202]. The mechanical properties of these composites can be further optimized through the careful selection of fibre types and their respective orientations within the matrix [203, 204].

Natural fibres provide several advantages, including sustainability and cost-effectiveness. They are renewable resources that contribute to reducing the carbon footprint associated with composite production. Research indicates that natural fibres like hemp not only offer good mechanical properties but also enhance the biodegradability of the composites [205, 206]. For example, hemp fibre composites have been successfully utilized in automotive applications, such as car door panels, due to their lightweight nature and sustainable manufacturing processes [206, 207]. This application exemplifies the growing trend of using NFCs in various industries, including automotive interiors, construction materials, and consumer products, where environmental benefits are increasingly prioritized [208].

In terms of specific applications, hemp fibre composites have demonstrated notable performance in automotive interiors, where weight reduction and sustainability are critical factors. The incorporation of hemp fibres into polymer matrices has been shown to improve mechanical properties such as tensile strength and impact resistance, making them suitable for structural applications [206, 207]. Furthermore, the development of hybrid composites that combine natural fibres with synthetic fibres has been explored to enhance specific properties, thereby broadening the application scope of NFCs [209].

Matrix and Reinforcement Components

Composite materials are engineered by combining two or more distinct components to create a material with superior properties compared to its individual constituents. The two primary components of a composite material are the matrix and the reinforcement. Each plays a critical role in determining the composite's overall performance, structure, and application suitability.

The matrix is the continuous phase in a composite material that binds the reinforcement together. It acts as the glue that holds the reinforcing components in place, ensuring load transfer between them and protecting them from environmental damage, such as moisture, corrosion, or UV radiation. The matrix also provides shape and structural integrity to the composite.

Matrix materials can be classified into three main types:

1. **Polymeric Matrices**: These include thermosetting resins like epoxy, polyester, and vinyl ester, and thermoplastics like polypropylene (PP) and polyether ether ketone (PEEK). Polymeric matrices are lightweight, corrosion-resistant, and versatile, making them suitable for applications ranging from aerospace components to sporting goods.

2. **Metallic Matrices**: Metals such as aluminium, titanium, and magnesium are used as matrices in metal matrix composites (MMCs). These matrices provide high strength, thermal conductivity, and durability, often used in automotive and aerospace industries where strength-to-weight ratio is crucial.

3. **Ceramic Matrices**: Materials like silicon carbide (SiC) and aluminium oxide (Al_2O_3) serve as matrices in ceramic matrix composites (CMCs). These matrices are ideal for high-temperature environments and applications requiring excellent thermal stability, such as jet engines and turbine components.

The choice of matrix material depends on the desired properties of the composite, including thermal resistance, chemical resistance, flexibility, and cost.

Different matrix materials are selected based on the desired properties they impart to composite materials, ensuring they meet specific application requirements effectively.

For thermal resistance, epoxy resin is a commonly used matrix in aerospace and automotive applications due to its high thermal stability and ability to endure extreme temperatures without degradation. Silicon carbide (SiC), a ceramic matrix, is ideal for components like turbine blades and heat shields, offering exceptional resistance to high temperatures and oxidation. Polyimide, a thermoset polymer, is often used in electronics and aerospace for its ability to perform in environments exceeding 300°C.

When chemical resistance is critical, polyvinyl ester (PVE) is employed in corrosion-resistant coatings and marine applications because of its excellent resistance to acids and alkalis. Teflon, also known as polytetrafluoroethylene (PTFE), is used in chemical processing equipment and non-stick coatings due to its inertness and resistance to harsh chemicals. Aluminium oxide (Al_2O_3), a ceramic matrix, is chosen for use in chemical reactors and other environments exposed to aggressive chemicals.

Flexibility is another key consideration. Polypropylene (PP), a thermoplastic matrix, is widely used in consumer goods and automotive interiors for its flexibility and impact resistance. Thermoplastic polyurethane (TPU) is found in sports equipment and medical devices due to its flexibility, durability, and abrasion resistance. Elastomeric matrices, such as silicone rubber, are utilized in flexible electronics and soft robotics for their high elasticity.

Cost efficiency is often prioritized in composite design. Polyester resin is a cost-effective choice for boat hulls and construction materials due to its adequate mechanical properties. Glass fibre-reinforced plastic (GFRP) is used in wind turbine blades and water tanks as a low-cost composite matrix. Magnesium alloys, as metal matrices, strike a balance between cost and lightweight properties, making them suitable for automotive applications.

For applications requiring a high strength-to-weight ratio, carbon fibre-reinforced polymer (CFRP) is extensively used in aerospace and high-performance automotive parts due to its remarkable strength and low weight. Aluminium matrix composites are employed in lightweight components such as bicycle frames and electronic heat sinks, benefiting from their strength and low density. Polyether ether ketone (PEEK), a high-performance thermoplastic, is used in medical implants and aerospace components for its superior mechanical properties.

Durability and wear resistance are essential for certain applications. Phenolic resins are used in brake pads and clutch plates due to their ability to withstand friction and wear. Zirconia matrix composites are utilized in cutting tools and dental implants for their hardness and wear resistance. Polycarbonate (PC), a tough thermoplastic, is employed in bulletproof windows and safety helmets.

In applications requiring electrical and thermal conductivity, copper matrix composites are used in heat exchangers and electrical connectors for their excellent conductivity. Graphite matrix materials are applied in high-temperature furnaces and electrodes, offering thermal shock resistance and conductivity. Conductive polymers, such as PEDOT, are employed in flexible electronic devices and sensors for their conductivity and flexibility.

Finally, for applications needing high toughness, Kevlar-reinforced epoxy composites provide excellent impact resistance and are used in ballistic armour and helmets. Aramid fibres combined with polyurethane matrices are used in flexible protective gear and industrial belts. Boron fibre in metallic matrices is chosen for aerospace applications requiring high fracture toughness and stiffness.

By aligning the matrix material with the specific requirements of the application, engineers can design composites that achieve an optimal balance of performance, cost, and environmental suitability.

Reinforcements play a crucial role in enhancing the mechanical properties of composite materials, contributing significantly to their strength, stiffness, and overall performance. These reinforcements, which can take various forms such as fibres, particles, whiskers, and nanomaterials, are embedded in a matrix to provide structural support and resist various forces, including tensile, compressive, and shear stresses.

Fibrous Reinforcements: Fibers are among the most commonly used reinforcements due to their exceptional strength-to-weight ratios and directional strength characteristics. Carbon fibres, for instance, are renowned for their high tensile strength and stiffness, making them ideal for applications in aerospace, automotive, and sporting goods industries [210, 211]. Glass fibres, on the other hand, offer a cost-effective solution with substantial strength, widely utilized in construction materials and consumer products [211]. Aramid fibres, such as Kevlar, are noted for their high toughness and impact resistance, which makes them suitable for protective gear like body armour and helmets [210]. The mechanical performance of carbon fibre reinforced composites (CFRPs) is particularly notable, as they exhibit a high specific strength and modulus, which is essential for lightweight structural applications [212, 213].

Particulate Reinforcements: Particulate reinforcements, such as silicon carbide and aluminium oxide, are dispersed within the matrix to enhance properties like wear resistance, thermal stability, and hardness. These materials are particularly effective in applications requiring high durability, such as cutting tools and thermal coatings [214]. The incorporation of nano-scale particulates has been shown to significantly improve the compressive strength of titanium composites, indicating the effectiveness of particulate reinforcements in enhancing mechanical properties [214].

Whiskers: Whiskers, which are short, single-crystal fibres, provide exceptional strength and stiffness. Silicon carbide whiskers, for example, are utilized in high-performance ceramics, contributing to improved mechanical properties and thermal stability [215]. Their unique structure allows for effective load transfer within the composite, enhancing overall performance.

Nanomaterials: The advent of nanomaterials, such as carbon nanotubes and graphene, has revolutionized the field of composite materials. These materials can significantly enhance properties at lower volume fractions compared to traditional reinforcements. For instance, carbon nanotubes have been shown to improve tensile strength and electrical conductivity in nanocomposites, making them highly desirable for advanced applications [216]. The integration of nanomaterials into composites not only improves mechanical properties but also opens avenues for multifunctional applications, such as self-healing composites [217].

The interaction between the matrix and reinforcement in composite materials is crucial for determining their overall performance. The matrix serves as a medium that transfers external loads to the reinforcements, which are responsible for bearing the majority of the stress. A strong interfacial bond between the matrix and the reinforcement is essential for effective load transfer, and this bond can be enhanced through various surface treatments or the use of chemical coupling agents [142, 218]. For instance, the mechanical properties of composites can be significantly improved by optimizing the bonding at the interface, which directly influences the stress distribution within the material [219].

The alignment and orientation of reinforcements within the matrix also play a pivotal role in defining the mechanical properties of composites. Unidirectional fibres are known to provide maximum strength in a single direction, making them ideal for applications where directional strength is critical. Conversely, woven or randomly oriented fibres can impart isotropic properties, which are beneficial in applications requiring uniform strength in multiple directions [220, 221]. The choice of reinforcement orientation is thus a key factor in tailoring the mechanical performance of composite materials to meet specific application requirements [222, 223].

Applications of matrix and reinforcement combinations are diverse and tailored to meet specific performance criteria. Carbon fibre-reinforced polymers (CFRPs) are widely used in the aerospace industry due to their lightweight and high-strength characteristics, which are essential for improving fuel efficiency and performance [138, 212]. Similarly, aluminium matrix composites are employed in automotive applications, such as brake rotors, where improved wear resistance is critical [224, 225]. Ceramic matrix composites are particularly advantageous in high-temperature environments, such as jet engines, due to their ability to withstand extreme thermal conditions without significant degradation [226]. The selection of appropriate matrix and reinforcement combinations is thus vital for optimizing the performance of composite materials in various engineering applications [227].

The matrix and reinforcement components in composite materials work in tandem to create advanced materials with tailored properties. By selecting the appropriate combination of matrix and reinforcement, engineers can design composites that meet specific performance requirements across industries such as aerospace, automotive, construction, and healthcare. This ability to customize makes composite materials an indispensable part of modern engineering and innovation.

Processing Techniques for Composites

Processing techniques for composites are pivotal in defining the structural integrity, performance, and quality of the final product. These methods are tailored to align with the properties of the matrix and reinforcement materials, the desired application, and economic considerations. The techniques can be broadly categorized into methods for thermoset and thermoplastic composites, as well as specialized approaches for advanced materials like metal matrix composites (MMCs) and ceramic matrix composites (CMCs).

Hand Lay-Up and Spray-Up Techniques

Hand lay-up and spray-up techniques are widely recognized methods in the composite manufacturing industry, particularly for producing large components with cost efficiency.

Hand lay-up is characterized by the manual placement of reinforcement fibres, typically woven mats, into a mold, followed by the application of resin through brushing or spraying. This method is particularly advantageous for low-volume production, as it requires minimal tooling and equipment investment, making it a cost-effective choice for manufacturers [142, 228]. However, one of the significant drawbacks of the hand lay-up technique is the tendency to produce voids within the laminate, which can compromise the mechanical properties of the final product [229, 230].

The spray-up technique, a variant of hand lay-up, involves spraying a mixture of resin and chopped fibres onto the mold surface. This method allows for a more uniform distribution of materials and can be particularly beneficial in applications where complex shapes are required [231]. Both techniques are commonly employed in industries such as marine, automotive, and construction, where large composite structures like boat hulls, automotive panels, and storage tanks are produced [232, 233]. The choice between hand lay-up and spray-up often depends on the specific requirements of the application, including the desired mechanical properties, production volume, and cost considerations.

Research has shown that the mechanical properties of composites produced via hand lay-up can be significantly influenced by factors such as the type of reinforcement used, the resin system, and the processing conditions [234, 235]. For instance, studies indicate that while hand lay-up can yield composites with good fatigue properties, the presence of voids often leads to reduced stiffness and overall structural integrity [236]. In contrast, spray-up can enhance the impact resistance of composites, making it a suitable choice for applications requiring high durability [231].

The Lay-Up process is a molding technique used to manufacture composite materials, where the final product is formed by layering multiple sheets of fibre reinforcements impregnated with a matrix material. These layers are meticulously arranged to achieve desired structural properties. The process can be categorized into Dry Lay-Up, commonly used in the aerospace industry for its ability to create complex shapes with excellent mechanical properties, and Wet Lay-Up, which is more suited to applications with less demanding performance requirements.

Cutting: The cutting phase is the initial step where fibre fabrics are shaped according to the specifications of the composite part. Despite the high tensile strength of fibres, their relatively low shear strength makes them easy to cut. Cutting can be manual, semi-automatic, or fully automated. Manual cutting involves tools like scissors, cutters, or knives, whereas automated methods include die-cutting, laser cutting, and water jet cutting. Automated processes allow for higher precision, reduced material wastage, and increased production rates. However, advanced techniques like laser and water jet cutting generate localized high temperatures, potentially altering the material properties near the cut edges. To optimize material usage, the nesting layout, which digitally arranges shapes to minimize waste, is employed.

Lamination: Lamination involves stacking the fibre layers in the correct sequence and orientation to form the composite's structural framework. For Wet Lay-Up, this stage includes preparing and applying the resin, as the fabrics are not pre-impregnated. To maintain the integrity of the composite, the process is often carried out in cleanrooms to prevent contamination. A mold is used to shape the laminate, and its material is chosen based on factors such as thermal expansion, stiffness, and compatibility with the composite material.

The lamination process starts with applying a release agent to the mold to prevent adhesion between the resin and the mold. Additional layers, such as peel-ply for surface protection and breather layers for even vacuum distribution, are added during this stage. A vacuum bag is used to remove trapped air and excess resin, improving the laminate's mechanical properties. Depending on the application, lamination may be performed manually, semi-automatically, or entirely automatically using machines like tape-laying systems for precise placement and orientation of fibres.

Polymerization: Polymerization, the final stage, solidifies the composite structure and determines its mechanical properties. This step can be performed at room temperature, in industrial ovens, or in autoclaves. Autoclave polymerization combines temperature, vacuum, and hydrostatic pressure to enhance bonding between layers and expel any trapped air or volatile compounds, producing composites with superior mechanical performance. However, this method is expensive and generally suited for high-performance applications. Industrial ovens provide temperature and vacuum control but lack pressure application, making them suitable for components requiring moderate strength and stiffness. For simpler geometries, matched-die molding uses male and female molds to compress the laminate into the desired shape. While this method offers excellent dimensional control and surface finish, it can result in fibre misalignment and is less versatile than autoclave processing.

The Lay-Up process is highly versatile, allowing for customization of composite properties by varying fibre orientation, layer sequence, and resin type. This flexibility makes it ideal for small production runs and prototypes, particularly in aerospace, automotive, and marine industries. However, the process is labour-intensive, resulting in higher costs and lower production rates compared to automated composite manufacturing techniques. Polymerization methods like autoclave and industrial ovens add further costs but enhance the quality and mechanical properties of the final product.

Compression Molding

Compression molding is a widely adopted technique for the high-volume production of thermoset composites, particularly due to its ability to produce components with complex geometries and high mechanical strength. This method involves placing pre-impregnated reinforcement materials, known as prepregs, or bulk molding compounds into a heated mold. The application of heat and pressure facilitates the flow and curing of the resin, resulting in a rigid structure that is essential for various applications, including automotive parts, appliance housings, and electrical components.

The advantages of compression molding are particularly evident in the production of thermoset composites. For instance, it allows for the use of longer fibres, which significantly enhance the mechanical properties of the final product compared to other methods like injection molding [237]. Compression molding is noted for its cost-effectiveness and efficiency, making it a preferred choice for manufacturing fibre-reinforced composites, especially in the automotive industry where lightweight and high-strength components are critical [238]. The process also minimizes internal stresses and reduces the likelihood of defects such as porosity, which can compromise the integrity of the final product [239].

Moreover, the versatility of compression molding enables the production of complex shapes and structures that are essential in modern engineering applications. Research indicates that this method is particularly effective for creating components with intricate designs, as it allows for uniform distribution of materials under pressure, thus enhancing the mechanical stability of the composites [240]. The ability to tailor the mechanical properties through the selection of reinforcement materials and the molding process itself further underscores the significance of compression molding in composite manufacturing [241].

In addition to its mechanical advantages, compression molding is also associated with environmental benefits. The process can be optimized to reduce waste and energy consumption, aligning with the growing demand for sustainable manufacturing practices [238, 242]. The use of eco-friendly materials in compression molding, such as natural fibres and recycled thermosets, is gaining traction, particularly in automotive applications where reducing the environmental footprint is a priority [238, 243].

Figure 31: Compression Molding Machine. Konglhask, CC BY-SA 4.0, via Wikimedia Commons.

Compression molding is a manufacturing process that uses heat and pressure to shape materials into desired forms. It is a versatile method suitable for high-strength and complex products, often involving thermoset resins in their partially cured state. The process begins by placing the preheated material into an open, heated mold cavity. The mold is then closed, applying pressure that forces the material to fill the mold, and heat is maintained until the material cures. For rubber materials, this process is referred to as vulcanization. Compression molding is commonly used with materials like fiberglass reinforcements and advanced composite thermoplastics.

Compression molding is highly advantageous for producing large, intricate parts with minimal material wastage, making it cost-effective when working with expensive compounds. It is particularly suited for creating flat or moderately curved parts, as well as ultra-large basic shapes that exceed the capacity of extrusion techniques. The method is widely used in automotive industries to manufacture components such as hoods, fenders, and spoilers. It is also employed for producing high-performance parts for aerospace, sports equipment, and even consumer goods like water bottles and golf ball cores.

The materials used in compression molding include bulk molding compound (BMC), sheet molding compound (SMC), and advanced thermoplastic materials like Poly(p-phenylene sulfide) (PPS) and Polyether ether ketone (PEEK). SMC materials are often precut to match the surface area of the mold, ensuring uniform coverage and reducing flow orientation during compression. Compression molding can also create sandwich structures by incorporating core materials such as honeycomb or polymer foam.

The compression molding process involves multiple stages, starting with the placement of the molding material into the mold cavity. The material, often in the form of pellets or sheets, may be preheated to facilitate flow and molding. Once placed, the mold is closed using a hydraulic press that applies pressure, ensuring the material conforms to the mold's shape. The mold is maintained at high temperatures to cure the material, and then cooled before the final product is ejected.

Key process parameters include molding time, temperature, and pressure, with typical clamp pressures ranging from 300 to 400 tons. The molds, which resemble clam shells, are designed with cavities that shape the material. Excess material or flash is removed after molding to ensure a clean finish.

There are three primary mold types used in compression molding: flash plunger-type, straight plunger-type, and landed plunger-type molds. The flash plunger-type mold requires precise material measurements and produces a horizontal flash, while the straight plunger-type mold allows for slight inaccuracies in material charge, producing vertical flash. The landed plunger-type mold requires an accurate charge and produces no flash.

In addition to standard molds, compression molding machines often feature preforms made by extruders or roller dies, which are cut to length before being placed in the mold. This ensures uniform material distribution and reduces defects. The process may include hydraulic or manual assistance for loading and unloading molds.

Compression molding is widely used for products requiring specific geometries or performance characteristics. For example, preforms for water bottles are die-cut sheets that are placed around a core and molded under pressure. Similarly, golf ball centers are produced using preforms that are loaded into molds and cured under heat and pressure. In both cases, the process ensures precise shaping and high-quality finishes.

The method is also used to manufacture thermoplastic composites for automotive applications, including long fiber-reinforced thermoplastics (LFT) and glass fiber mat-reinforced thermoplastics (GMT). These materials offer high strength-to-weight ratios, making them ideal for lightweight yet durable components.

Despite its advantages, compression molding has some limitations. It may result in poor product consistency and difficulties in controlling flashing. Additionally, while it is suitable for high-strength applications, it is less effective for parts requiring extreme precision or highly complex geometries. Engineers must carefully consider factors such as material amount, heating techniques, and mold design to optimize the process and ensure quality.

Compression molding remains a cornerstone of modern manufacturing, balancing cost efficiency, material utilization, and product performance across a wide range of industries.

Resin Transfer Molding (RTM)

Resin Transfer Molding (RTM) is a prominent manufacturing technique for producing fibre-reinforced composites, characterized by the injection of liquid resin into a closed mold containing a fibre preform. This process allows for excellent control over fibre volume fractions and results in high-quality surface finishes on both sides of the composite component. The RTM process is particularly advantageous in industries such as aerospace, automotive, and wind energy, where precision and performance are critical [244-246].

The RTM process operates by injecting low-viscosity thermosetting resin into a mold cavity that has been pre-filled with dry reinforcement fabric. The resin permeates the fibres, and the composite cures under controlled pressure and temperature conditions, leading to enhanced mechanical properties and structural integrity [246]. This method not only allows for the creation of complex geometries but also facilitates the integration of metallic inserts within the composite, thereby enhancing functionality without compromising structural performance [247]. The ability to achieve uniform resin distribution and optimal fibre wetting is a significant advantage of RTM, as it minimizes defects such as voids and dry spots, which are common in other composite manufacturing techniques [248].

Moreover, RTM is often compared to other liquid composite molding processes, such as Vacuum Assisted Resin Transfer Molding (VARTM) and Compression Resin Transfer Molding (CRTM). While VARTM utilizes a flexible vacuum bag to assist in resin infusion, RTM employs rigid tooling, which can lead to differences in cost and production efficiency [249, 250]. The RTM process has been recognized for its cost-effectiveness and high production rates, making it a preferred choice for large-scale applications in various sectors, including automotive and aerospace [251, 252].

Recent advancements in RTM technology have focused on improving the flow characteristics and curing kinetics of the resin, which are critical for optimizing the manufacturing process and ensuring the quality of the final product [253, 254]. Innovations such as high-pressure resin transfer molding (HP-RTM) and thermoplastic resin transfer molding (T-RTM) have emerged, combining the benefits of traditional RTM with enhanced material properties and recyclability [255]. These developments underscore the ongoing evolution of RTM as a versatile and efficient method for producing high-performance composite materials.

Pultrusion

Pultrusion is a highly efficient continuous manufacturing process that produces long, constant cross-section profiles, such as rods, tubes, and beams. This technique involves the pulling of reinforcement fibres through a resin bath, followed by their passage into a heated die where the composite material cures. The pultrusion process is characterized by its ability to create profiles with consistent dimensions and properties, making it an attractive option for various applications in construction, infrastructure, and electrical sectors.

The fundamental mechanics of pultrusion involve the impregnation of continuous fibres with a resin, which is then shaped and cured in a heated die. This process allows for the production of composite materials with a high strength-to-weight ratio, excellent corrosion resistance, and good mechanical properties [256, 257]. The efficiency of pultrusion is further enhanced by its minimal labour requirements and the automation potential it offers, which leads to reduced production costs and increased output [258, 259]. Moreover, the process is adaptable to various types of fibres and resins, including thermosetting and thermoplastic composites, which broadens its applicability across different industries [260, 261].

Pultruded composites are particularly valued for their lightweight nature and resistance to environmental degradation, making them suitable for demanding applications. For instance, glass fibre-reinforced polymer (GFRP) composites produced via pultrusion are increasingly used in construction due to their mechanical performance and durability [262, 263]. These materials are non-conductive and non-magnetic, which adds to their utility in electrical applications [257]. Furthermore, the ability to achieve high fibre volume fractions during the pultrusion process enhances the mechanical properties of the resulting composites, making them ideal for structural applications [264].

The pultrusion process incorporates several critical components. A continuous roll of reinforced fibres or woven mats feeds into the system, guided by tension rollers to ensure uniformity. The fibres are then passed through a resin impregnator, which saturates them with resin. The resin-impregnated fibres are subsequently pulled through a heated die, which shapes and hardens the composite. After exiting the die, the finished fibre-reinforced polymer is continuously pulled forward using a pull mechanism, creating long, consistent profiles.

Figure 32: Pultrusion system. JSortimo, CC BY-SA 3.0, via Wikimedia Commons.

Resins used in pultrusion include thermosetting polymers like polyester, polyurethane, vinyl ester, and epoxy, as well as thermoplastics such as polybutylene terephthalate (PBT) and polyethylene terephthalate (PET). The resin provides resistance to environmental factors like corrosion, UV radiation, and impact, while the glass or fibre reinforcement imparts strength and durability. Surface veils can be added for additional protection against erosion and fibre bloom.

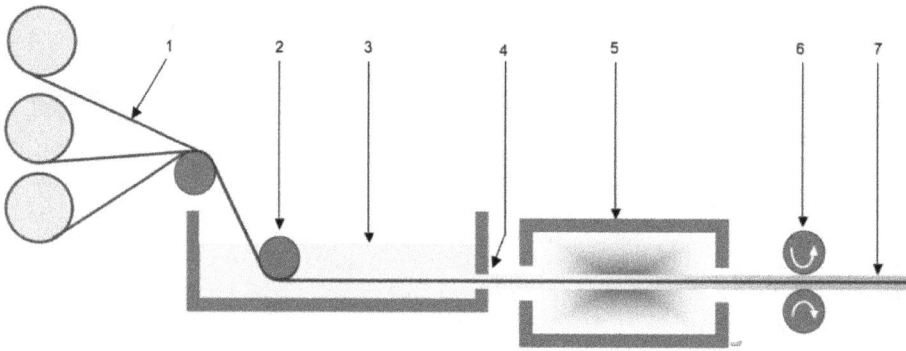

1. Continuous roll of reinforced fibres/woven fibre mat
2. Tension roller
3. Resin impregnator
4. Resin-soaked fibre
5. Die and heat source
6. Pull mechanism
7. Finished hardened fibre reinforced polymer

Figure 33: Visualization of the pultrusion process. LaurensvanLieshout assumed (based on copyright claims), Public domain, via Wikimedia Commons.

Pultrusion is highly energy-efficient and resource-saving, making it an attractive option for various industries. It produces profiles with enhanced chemical, thermal, and mechanical properties. For instance, pultruded composites are used in agriculture for chemically resistant slatted floors, in construction for reinforcement bars and window stiffeners, and in aerospace for lightweight structural components. Other applications include sports equipment like skis and golf flagsticks, electrical engineering components such as fiberglass rods, and automotive structural parts.

Pultrusion is particularly suited to the production of straight profiles, which makes it ideal for beams, rods, and similar structures. Recent advancements, however, have expanded its capabilities to include curved profiles, broadening its application range.

The standard pultrusion process is limited to straight profiles due to the fixed nature of the die. A notable modification, known as "Radius-Pultrusion," was developed to overcome this limitation. In this process, the die moves back and forth along the profile, enabling the manufacture of two- and three-dimensional curved shapes. This innovation is particularly valuable for applications requiring complex geometries or low-distortion textile reinforcements.

Radius-pultrusion relies on specialized equipment with movable stages. Linear movement is sufficient for large radii, while circular movements are required for smaller radii. These advancements have expanded the versatility of pultrusion, allowing it to meet the demands of modern industries.

Pultrusion machines are designed in different configurations to suit specific manufacturing needs. The two main types are reciprocating machines, which operate in a hand-over-hand motion, and continuous machines, also known as cat-track systems. For radius pultrusion, the equipment includes two moving stages: one for pulling and the other for holding the mold. The choice of equipment depends on the complexity of the profiles being manufactured and the required production efficiency.

Pultrusion technology aligns with ecological and economic priorities. Products made using thermoplastic matrices are recyclable, making them more environmentally friendly compared to thermosetting composites. Additionally, the high production efficiency and material consistency of pultrusion contribute to cost savings and sustainable manufacturing practices.

Filament Winding

Filament winding is a sophisticated manufacturing technique primarily utilized for producing hollow, cylindrical structures such as pipes, pressure vessels, and rocket casings. This process involves the impregnation of continuous fibres with resin, which are then wound around a rotating mandrel in specific patterns. Upon curing, the mandrel is removed, resulting in a robust and lightweight composite structure. The filament winding technique is particularly advantageous due to its ability to provide excellent control over fibre orientation, which is crucial for optimizing the mechanical properties of the final product. This method is widely adopted in aerospace and chemical storage applications due to its efficiency and effectiveness in producing high-performance composite materials.

The filament winding process is recognized for its high forming efficiency and quality, especially in the fabrication of rotary composite components. Huang [265] emphasizes that this technique enhances the performance of composite materials and contributes to reducing production costs, making it a valuable method in various industrial applications. Furthermore, Srebrenkoska et al. [266] highlight the effectiveness of filament winding in producing composite structures specifically designed for gas storage and transportation, underscoring its relevance in industries that require durable and lightweight materials. The ability to tailor fibre orientation during the winding process allows for the optimization of mechanical properties, which is essential in applications such as pressure vessels and aerospace components [267].

In addition to its mechanical advantages, filament winding is also noted for its automation capabilities, which enhance production efficiency. Jois discusses the challenges associated with high-speed and high-throughput requirements in filament winding, particularly in the context of manufacturing composite pressure vessels [268]. This shift towards more efficient manufacturing processes is critical as industries increasingly demand faster production times without compromising quality. Moreover, the integration of advanced materials, such as carbon fibre-reinforced composites, further enhances the structural integrity and performance of the final products, as evidenced by the work of Kadri, who explores the optimization of pressure vessels for hydrogen storage [269].

The versatility of filament winding extends to various applications, including defence, space exploration, and chemical storage. Demirci and Sahin [267] note that this technique is extensively utilized in the production of composite pipes for various sectors, emphasizing its broad applicability and importance in modern engineering. Additionally, the dynamic modelling of the winding system, as discussed by Hou et al. [270], reveals the complexities involved in maintaining balance and precision during the winding process, which is crucial for ensuring the quality of the final composite structure.

Filament winding is a specialized fabrication technique primarily used for creating hollow or cylindrical structures such as pressure vessels and tanks. The process involves winding continuous fibres under controlled tension around a rotating mandrel to achieve specific patterns and properties. This method is widely employed for its ability to produce high-strength, lightweight components with an optimal fibre-to-resin ratio.

The process begins with the preparation and impregnation of fibres, which are typically glass, carbon, or aramid. These fibres are impregnated with resin to enhance bonding and performance characteristics. Impregnation can be achieved through a resin bath (wet winding) or by using pre-impregnated fibres (dry winding). The choice of resin system is critical, as it determines the composite's strength, corrosion resistance, and durability. Common resins include epoxy, polyurethane, polyester, and vinyl ester.

Once impregnated, the fibres are wound around a mandrel, which acts as the mold core. The mandrel rotates around a central axis, while the delivery system moves horizontally to lay fibres in patterns such as circumferential, helical, or polar. The winding angle and tension significantly influence the mechanical properties of the final product. High-angle winding provides circumferential strength, while low-angle winding enhances axial tensile strength.

After achieving the desired thickness, the resin is cured using heat from an autoclave, oven, or radiant heaters. Once cured, the mandrel is removed, leaving the hollow composite structure. In some cases, such as gas storage tanks, the mandrel may remain as a functional liner.

There are variations in the filament winding process to suit different applications. Continuous winding is used for low-pressure pipes and simple components, employing a hoop pattern to enhance structural properties. Discontinuous winding is applied to high-pressure components such as pressure vessels and complex shapes, allowing for customized fibre orientation. Advanced robotic and CNC filament winding systems offer greater precision and flexibility, enabling the creation of complex geometries like curved profiles.

Filament winding utilizes various materials to achieve specific properties. Glass, carbon, and aramid fibres are commonly used for their mechanical and thermal characteristics. Resins such as epoxy and polyurethane are preferred for high-strength aerospace components, while polyester and vinyl ester are cost-effective options for less demanding applications. The mandrel material may vary depending on whether it will be removed or integrated into the final product.

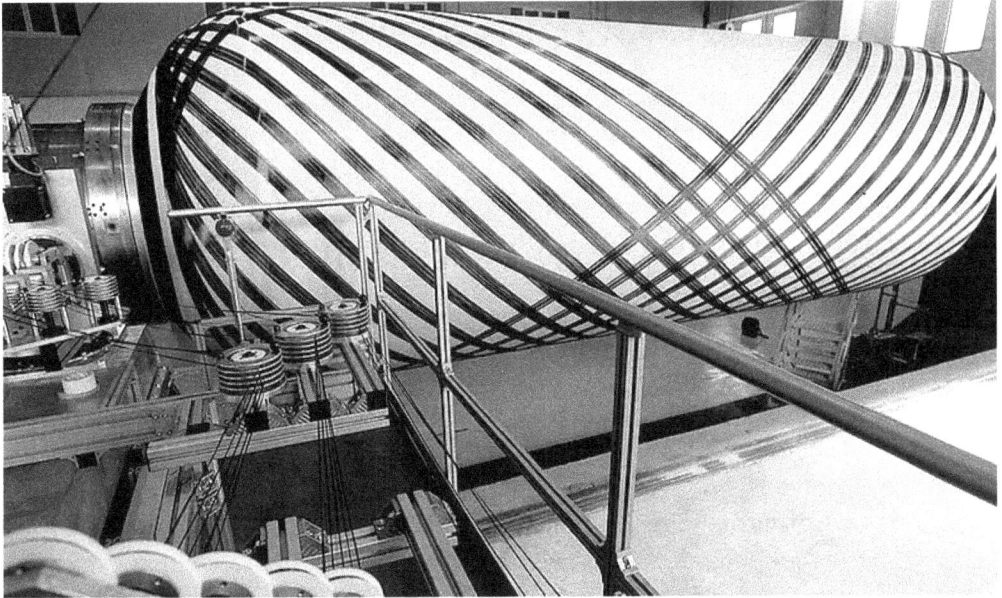

Figure 34: Filament winding of the P80 solid rocket motor. Gdipasquale1, CC0, via Wikimedia Commons.

This technique is highly versatile and finds applications across numerous industries. In aerospace, it is used for rocket casings, fuselage components, and heat-resistant structures. The automotive sector employs filament winding for lightweight drive shafts and structural components. Industrial applications include pipes for water, sewage, and chemical transportation, while sports and leisure industries use the technique for golf club shafts, oars, and bicycle components. In the energy sector, it is critical for manufacturing pressure vessels for compressed natural gas (CNG) and hydrogen storage.

While filament winding offers significant advantages, such as superior mechanical properties, flexibility in fibre orientation, and cost efficiency for large-scale production, it also poses challenges. These include the need for specialized machinery and expertise, difficulty in achieving smooth outer surfaces without secondary operations, and potential health risks associated with resin systems. The release of volatile organic compounds (VOCs) like styrene and concerns over bisphenol A (BPA) in epoxy resins have prompted manufacturers to adopt safer, low-emission resins.

Despite these challenges, filament winding remains a cornerstone of advanced composite manufacturing, enabling the production of high-performance components for a wide array of critical applications.

Automated Fibre Placement (AFP) and Tape Laying

Automated Fibre Placement (AFP) and Automated Tape Laying (ATL) are advanced manufacturing techniques that leverage robotic systems to achieve high precision in the placement of fibre tows or tapes onto molds. These methods are particularly advantageous for the production of large and complex components, such as aircraft fuselages and wind turbine blades, where the need for consistency and repeatability is paramount. The automation inherent in AFP and ATL not only enhances the accuracy of the manufacturing process but also significantly reduces labour costs and material waste, thereby improving overall manufacturing efficiency.

The AFP and ATL processes utilize advanced robotics to lay down composite materials with high precision. According to Liu et al. [271], these automated techniques provide clear advantages over traditional hand lay-up methods, including higher efficiency and improved laying accuracy, which are critical for producing high-quality composite structures. Yassin and Hojjati [272] further emphasize that these methods are capable of in situ consolidation of thermoplastic composites, which allows for the production of advanced composite laminates at high throughput. The ability to automate the placement of materials leads to a reduction in human error and variability, which is crucial for maintaining the integrity of complex geometries [273].

Moreover, the economic benefits of AFP and ATL are substantial. Bahar and Sinapius [274] highlight that these technologies not only enhance the quality of the laminate but also contribute to reduced production times and costs, making them increasingly attractive for the aerospace industry. Rakhshbahar and Sinapius [275] also note that AFP and ATL are well-suited for manufacturing large, geometrically complex components, which is essential in industries where performance and weight savings are critical. The efficiency of these processes is further supported by the findings of Hirsch [276], who discusses the integration of ATL and AFP in various industries beyond aerospace, indicating their versatility and effectiveness in reducing production costs and waste.

In addition to economic advantages, the technological advancements in AFP and ATL have led to improved material utilization rates. Miao [277] points out that AFP allows for independent control of multiple prepreg fibres, which is particularly beneficial for complex surface placements. This capability not only enhances the design freedom for engineers but also ensures that the mechanical properties of the final product are optimized [278]. Furthermore, the integration of in situ monitoring technologies, as discussed by Oromiehie et al. [279], enhances the accuracy and quality of the manufacturing process by merging multiple stages such as cutting, lay-up, and curing into a single automated system.

Fiber placement is an advanced automated composites manufacturing process designed to enhance precision, efficiency, and versatility in the production of composite materials. This method involves heating and compacting pre-impregnated synthetic resin fibres onto complex tooling mandrels. The fibres, referred to as "tows," are bundles of carbon fibres impregnated with epoxy resin, typically measuring approximately 0.500 inches (12.7 mm) wide and 0.005 inches (0.13 mm) thick. These tows are supplied on spools and fed into specialized fibre placement machines (FPM) capable of handling between 12 and 32 tows simultaneously. Depending on the number of tows placed per course, the resulting width can range from 1.5 inches to 4 inches.

In the fibre placement process, the tows are directed through a heating and compaction roller situated on the machine's head. Using robotic or automated machine movements, the tows are laid in courses across a mandrel or tooling surface. These courses are typically arranged in orientations of 0°, +45°, -45°, and 90°, enabling the creation of composite plies with enhanced properties in all directions. This layering strategy ensures that the final product possesses superior mechanical strength, stiffness, and durability.

Fiber placement machines are often rated by their deposition rate, measured in weight per unit of time, such as pounds per hour (lb/h) or pounds per minute (lb/min). This rating reflects the efficiency and productivity of the machine, particularly in high-volume manufacturing environments.

Automated Fibre Placement (AFP) machines represent a significant advancement in composite manufacturing technology. These machines are specifically designed to automate the placement of fibre reinforcements on molds or mandrels. They utilize multiple narrow-width tows, typically 8 millimetres (0.31 inches) or less, made from thermoset or thermoplastic pre-impregnated materials. This allows for the precise and efficient formation of composite layups.

Figure 35: Automated Fibre Placement Robot. Alidarejeh, CC BY-SA 4.0, via Wikimedia Commons.

AFP technology offers several advantages over traditional composite manufacturing methods. It provides higher precision and increased deposition rates compared to manual laminators, ensuring consistent quality and reducing material waste. While AFP allows for the production of more complex layup geometries than Automated Tape Laying (ATL), its deposition rates may not match those of ATL. Nevertheless, AFP is uniquely suited for manufacturing intricate structures that would be impossible to create using other techniques.

Fibre placement is particularly valuable in industries requiring high-performance composite materials, such as aerospace, automotive, and defence. It enables the production of lightweight yet durable components with tailored properties to meet specific structural and mechanical demands. The automated nature of the process ensures high repeatability, reduced labour costs, and minimized errors, making it an essential technology for advanced manufacturing.

This technique has opened new possibilities for designing and fabricating complex composite parts, pushing the boundaries of what can be achieved with traditional manufacturing methods. By combining high precision, flexibility, and efficiency, fibre placement is shaping the future of composite material production.

Injection Molding

Injection molding is a widely adopted manufacturing process primarily utilized for thermoplastic composites, which are materials that combine thermoplastic resins with reinforcement fibres. This method involves mixing reinforcement fibres with thermoplastic pellets, which are then melted and injected into a mold. Upon cooling, the composite solidifies into the desired shape, making it an efficient technique for producing intricate parts in large volumes. This process is particularly advantageous for applications in various industries, including automotive, electronics, and consumer goods.

The injection molding process is characterized by its ability to create complex geometries with high precision and excellent surface finishes. For instance, Radzi et al. [280] emphasize that plastics injection molding is one of the most prevalent cyclic processes for manufacturing thermoplastic materials, particularly for automotive components where dimensional accuracy is critical. Furthermore, the use of injection molding allows for the production of lightweight and strong parts, as noted by Jiang et al. [281], who highlight the high lightweight potential of hybrid thermoplastic composites produced through injection overmolding techniques. This capability is essential in the automotive industry, where reducing weight can lead to improved fuel efficiency.

Moreover, the integration of reinforcement fibres into thermoplastic matrices enhances the mechanical properties of the final products. For example, Azenha et al. [282] discuss the direct impregnation of reinforcement textiles using thermoplastic injection molding, which can yield high-strength composite parts. This is corroborated by Xu et al. [283], who note that fibre-reinforced thermoplastics (FRTPs) offer superior toughness and recyclability compared to traditional fibre-reinforced plastics (FRPs). The ability to achieve high fibre volume fractions in these composites further contributes to their mechanical performance, making them suitable for demanding applications such as electronic housings and automotive components [284].

The versatility of injection molding extends to its application in producing components with varying geometries and sizes. For instance, the process is not only limited to large automotive parts but also extends to smaller, intricate components used in electronics, as highlighted by the work of Chen et al. [285], which discusses the production of automotive trim components using fibre-reinforced polypropylene composites. Additionally, the process parameters, such as mold temperature and injection speed, significantly influence the adhesion strength and overall quality of the molded parts, as discussed by Bex et al. [286].

Advanced Processing for MMCs and CMCs

Metal matrix composites (MMCs) and ceramic matrix composites (CMCs) are advanced materials that necessitate specialized fabrication techniques due to their unique material properties. For MMCs, methods such as stir casting, powder metallurgy, and pressure infiltration are commonly employed. Stir casting is a widely utilized technique that allows for the uniform distribution of reinforcements within the metallic matrix, enhancing the mechanical properties of the composite [287]. Powder metallurgy is another effective method that facilitates the production of MMCs with tailored microstructures and properties, particularly when reinforced with materials like silicon nitride [288]. Pressure infiltration, which involves injecting molten metal into a porous preform, is crucial for achieving a homogeneous distribution of reinforcements, thereby improving the overall performance of the composite [156, 289].

In the context of CMCs, specialized techniques such as chemical vapor infiltration (CVI), polymer infiltration and pyrolysis (PIP), and reaction bonding are employed to enhance properties like high-temperature stability and oxidation resistance. CVI is particularly effective for fabricating CMCs, as it allows for the deposition of a ceramic phase within a porous preform, resulting in improved mechanical properties and thermal stability [290]. PIP involves the infiltration of a polymer into a ceramic preform, followed by pyrolysis to convert the polymer into a ceramic matrix, which is beneficial for achieving a dense and strong composite [290]. Reaction bonding, on the other hand, utilizes chemical reactions to form a ceramic matrix around the reinforcements, which is advantageous for producing composites with superior thermal and mechanical properties [290].

Polymer Infiltration and Pyrolysis (PIP) is a sophisticated fabrication process utilized to create high-performance ceramic matrix composites (CMCs). This technique involves infiltrating pre-ceramic polymers into a fibre-reinforced preform, followed by a pyrolysis step where the polymer is transformed into a ceramic matrix through high-temperature heat treatment. PIP is especially valuable for producing composites with intricate geometries and tailored properties, making it ideal for applications in aerospace, automotive, and various industrial sectors.

The process begins with the preparation of a fibre-reinforced preform, which serves as the structural framework of the composite. These preforms are typically composed of fibres such as silicon carbide (SiC), carbon, or alumina, arranged in configurations like woven fabrics, braided forms, or filament-wound structures. The preform provides the composite with its foundational mechanical properties and shape.

In the polymer infiltration stage, the preform is saturated with a liquid polymer precursor, such as polycarbosilane or polysiloxane, which will later convert into a ceramic matrix. This infiltration is typically achieved using vacuum or pressure-assisted methods to ensure the polymer thoroughly penetrates the fibre structure, filling voids and ensuring uniform distribution. This step is critical for achieving a homogeneous matrix in the final composite.

Once infiltrated, the polymer precursor undergoes curing to solidify and bond with the fibres. This involves heating the infiltrated preform at moderate temperatures to induce cross-linking reactions in the polymer. Curing ensures that the polymer maintains its structure and establishes strong bonds with the fibres, which is essential before the pyrolysis stage.

During pyrolysis, the cured preform is subjected to high temperatures in an inert or controlled atmosphere, such as argon or nitrogen. This process decomposes the polymer, releasing volatile byproducts and forming a ceramic phase. Depending on the polymer precursor, the resulting matrix typically comprises amorphous silicon carbide, silicon nitride, or carbon. However, pyrolysis leaves some residual porosity in the matrix, which can compromise the composite's properties. To address this, multiple cycles of polymer infiltration and pyrolysis are performed until the desired matrix density and characteristics are achieved, with five to eight cycles being standard for reducing porosity to acceptable levels.

PIP offers several advantages. It provides exceptional design flexibility, enabling the creation of complex shapes and geometries. The process allows for tailoring the matrix composition by selecting specific polymer precursors, enabling customization of thermal, mechanical, and chemical properties. Furthermore, PIP operates at relatively lower processing temperatures than sintering or chemical vapor infiltration (CVI), reducing thermal stresses on the fibres.

Despite its benefits, PIP has some limitations. The process often leaves residual porosity in the matrix, requiring additional cycles to minimize. The production times are lengthy due to the repeated cycles of infiltration and pyrolysis, which increases costs. Moreover, the volume reduction during pyrolysis can cause shrinkage and potential cracking, necessitating meticulous process control.

PIP is widely employed in industries requiring materials with exceptional thermal and mechanical properties. In aerospace, it is used for components like turbine blades, nozzles, and heat shields operating in extreme temperatures. The automotive industry uses PIP for producing brake discs and other wear-resistant parts, while the energy sector relies on it for high-temperature seals and gas turbine components. Additionally, PIP finds applications in creating corrosion-resistant parts for chemical processing equipment.

Polymer Infiltration and Pyrolysis remains a versatile and effective technique for manufacturing high-performance ceramic matrix composites. By combining fibre reinforcement with a ceramic matrix, PIP enables the production of materials with unparalleled strength, thermal stability, and resistance to wear and corrosion. Continuous advancements in PIP technology aim to enhance its efficiency and address its challenges, ensuring its ongoing relevance in advanced material manufacturing.

Chemical Vapor Infiltration (CVI) is an advanced ceramic engineering technique employed to fabricate fibre-reinforced composites by infiltrating a fibrous preform with a ceramic matrix using reactive gases at elevated temperatures. This method has been pivotal in producing materials like carbon-carbon composites and ceramic-matrix composites (CMCs). Unlike Chemical Vapor Deposition (CVD), which deposits material on bulk surfaces, CVI focuses on porous substrates, allowing the matrix to infiltrate and bond with the fibres.

CVI involves several critical steps. Initially, the fibrous preform, made from yarns, woven fabrics, or three-dimensional shapes like filament-wound or braided structures, is supported on a porous metallic plate. A mixture of reactive gases carrying the matrix material is introduced into a reactor, where the preform undergoes chemical reactions at high temperatures. This results in the matrix material infiltrating the fibre crevices and bonding to the fibre surfaces.

During the process, unreacted gases and residual matrix material exit the reactor and are treated in an effluent plant. Induction heating is commonly used to maintain isothermal and isobaric conditions in conventional CVI. As the reaction progresses, a coating of the matrix material forms on the fibre surface, reducing the fibre diameter while increasing the composite's mechanical strength.

The traditional "hot wall" technique in CVI, though effective, is slow and time-intensive. To address this, modified approaches such as Forced-flow CVI (FCVI) with thermal gradients have been developed. In this method, a pressurized flow of gases and matrix material is forced through the preform, achieving a denser and less porous final material. This technique operates at a temperature gradient, typically ranging from 1050 °C in cooler zones to 1200 °C in furnace zones, significantly reducing processing time.

CVI is widely used in fabricating high-performance components across various industries. Examples include:

1. **Carbon/Carbon (C/C) Composites**: PAN-based carbon felts serve as preforms infiltrated with kerosene at 1050 °C under atmospheric pressure. These composites are integral to aerospace applications due to their exceptional thermal stability.

2. **Silicon Carbide/Silicon Carbide (SiC/SiC) Composites**: SiC fibres are used as preforms, and matrix infiltration occurs through reactions involving methyltrichlorosilane (CH_3SiCl_3) at approximately 1000 °C. This composite offers high-temperature resistance, making it suitable for turbine blades and heat shields.

CVI offers numerous benefits, including enhanced mechanical properties, corrosion resistance, and thermal shock resistance. The process operates at relatively low infiltration temperatures, minimizing residual stresses and preserving the geometry of the fibres. CVI also allows for the production of large, complex shapes and offers flexibility in choosing matrix and fibre combinations, such as SiC, C, BN, and ZrC. By carefully controlling gas purity, manufacturers can achieve a highly pure and uniform matrix.

Despite its advantages, CVI has notable drawbacks. The process results in residual porosity of about 10-15%, which can affect material properties. Additionally, CVI has a slow production rate and involves high capital investment and processing costs, limiting its scalability.

The unique characteristics of both MMCs and CMCs arise from their composite nature, where the interaction between the matrix and the reinforcements plays a critical role in determining the final properties of the material. For instance, the incorporation of ceramic particles or fibres into a metal matrix can significantly enhance wear resistance and mechanical strength, making these composites suitable for a wide range of applications in aerospace, automotive, and structural engineering [291, 292]. Similarly, CMCs exhibit exceptional thermal stability and oxidation resistance, making them ideal for high-temperature applications such as turbine engines and heat shields [290].

Additive Manufacturing

Additive manufacturing (AM), commonly referred to as 3D printing, has emerged as a transformative technology in the production of composites, particularly due to its ability to facilitate continuous fibre reinforcement and tailored material deposition. This capability enables the creation of highly customized components characterized by complex geometries, which are increasingly sought after in various high-performance sectors, including aerospace, medical, and sports applications. The flexibility in design and material efficiency offered by AM makes it a compelling choice for these industries, where performance and precision are paramount.

The aerospace sector, in particular, has seen significant advancements through the adoption of AM technologies. Research indicates that AM allows for the production of critical parts with enhanced geometrical freedom and reduced material waste, which is crucial for meeting the stringent requirements of aerospace applications [293, 294]. For instance, continuous carbon fibre reinforced polymer composites (CFRPCs) are gaining traction in aerospace due to their superior mechanical properties, such as high strength-to-weight ratios and corrosion resistance [295]. Furthermore, the integration of advanced materials, such as nickel-based superalloys, into AM processes has been explored to enhance component performance under extreme conditions, which is vital for gas turbine engines [296].

In the medical field, AM's ability to produce patient-specific implants and prosthetics is revolutionizing treatment approaches. The technology allows for the customization of devices to fit individual anatomical requirements, thereby improving patient outcomes [297, 298]. Similarly, in high-performance sports, AM is being utilized to create lightweight and durable equipment tailored to the specific needs of athletes, showcasing the versatility of this manufacturing technique [299].

Moreover, the efficiency of AM extends beyond just material usage; it significantly reduces lead times and costs associated with traditional manufacturing methods. The layer-by-layer approach of AM not only minimizes waste but also enables rapid prototyping and direct component development, which are critical in fast-paced industries [300, 301]. This efficiency is further enhanced by the ability to make rapid design changes without the need for extensive retooling, thereby streamlining the production process [302].

The term "additive manufacturing" (AM) gained traction in the 2000s, primarily due to its association with processes that involve the addition of material to create objects, contrasting with "subtractive manufacturing," which refers to traditional machining techniques that remove material. This distinction became increasingly relevant as industries began to adopt AM technologies for various applications, particularly in metalworking and end-use part production contexts, while "3D printing" was often limited to polymer technologies in popular discourse [303, 304]. The evolution of these terms reflects a broader shift in manufacturing paradigms, where AM encompasses a range of techniques, including but not limited to 3D printing, which is often perceived as a more accessible, consumer-oriented technology [305].

By the early 2010s, the terms "3D printing" and "additive manufacturing" began to be used interchangeably, albeit in different contexts. "3D printing" became a term favoured by consumer-maker communities and media, while "additive manufacturing" was more commonly used in formal industrial settings, including by machine manufacturers and technical standards organizations [304, 305]. This duality highlights the growing recognition of AM technologies across various sectors, with experts noting that while the terms are synonymous in casual usage, there is an emerging distinction where AM is seen as a broader category that includes additional technologies beyond traditional 3D printing methods [304, 305].

The terms "rapid manufacturing" and "on-demand manufacturing" emerged in the 2000s, reflecting a shift towards more agile production methods that contrast sharply with the lengthy lead times associated with traditional manufacturing processes [303, 306]. This shift is indicative of a broader trend towards flexibility and responsiveness in manufacturing, enabled by advancements in AM technologies. Agile tooling, which utilizes AM methods for the rapid production of tooling and fixtures, exemplifies this trend, allowing manufacturers to quickly adapt to changing market demands [303, 306]. The continued relevance of subtractive manufacturing terminology, despite the rise of AM, underscores the complementary nature of these processes in modern manufacturing environments [307].

3D printing begins with modelling, a critical step in the additive manufacturing process. The creation of 3D models can involve computer-aided design (CAD) software, 3D scanning, or photogrammetry, which converts 2D images into 3D structures. CAD-designed models are preferred for their precision and lower error rates compared to models generated by scanning or photographic methods. Errors such as holes, noise shells, and manifold issues can occur during modelling, particularly with scanned models, but these can often be addressed through "repair" processes in CAD software.

The data for 3D printing is typically saved in STL (stereolithography) format, a standard file type that represents the surface geometry of an object using triangular facets. However, due to its inefficiency in handling complex, high-resolution designs, the Additive Manufacturing File Format (AMF) has been introduced. AMF improves on STL by employing curved triangulations, reducing file size and preserving design accuracy.

Before printing, an STL or AMF file is processed by a slicer software that converts the 3D model into thin layers and generates G-code instructions for the printer. The resolution of the printer, including layer thickness and X-Y resolution, determines the detail and surface finish of the printed object. Modern printers can achieve layer thicknesses as fine as 16 micrometres, with X-Y resolutions comparable to laser printers.

Printing times vary widely based on the size and complexity of the object. Additive manufacturing significantly reduces the time required to produce intricate designs compared to traditional methods, with most prints completed within hours.

While 3D printers deliver high precision, post-processing can enhance the dimensional accuracy, surface smoothness, and functionality of printed parts. Methods such as sanding, bead blasting, and chemical vapor smoothing are used to refine surfaces. For instance, polymers like ABS can be smoothed chemically with acetone, resulting in a polished finish. Post-print annealing further improves internal bonding, enhancing properties like fracture toughness and heat resistance.

Additive-Subtractive Hybrid Manufacturing (ASHM) combines 3D printing with machining to achieve superior surface finishes and accuracy. This hybrid approach allows for precise modifications either during or after the printing process.

Challenges like the stair-stepping effect on curved or tilted surfaces can be minimized by employing adaptive layer height techniques. Painting and material-specific coatings also offer aesthetic and functional enhancements.

Multi-material 3D printing overcomes the limitations of single-material systems, enabling the creation of objects with heterogeneous properties. This technology is particularly promising in industries like healthcare, where it allows for customized pills and implants. Despite its potential, multi-material printing faces challenges in material compatibility and software development.

4D printing expands on 3D printing by incorporating time as a dimension. This advanced technique uses smart materials capable of responding to external stimuli like heat or moisture to change shape or functionality. Applications include self-assembling structures and adaptive components in aerospace, medical, and commercial industries. However, challenges such as material development, precision, and scalability remain to be addressed.

Initially, 3D printing was dominated by polymers due to their ease of use. Today, advancements have expanded its scope to include metals and ceramics, enhancing its versatility. The earliest patented 3D printing technologies utilized UV-cured resins, thermoplastics, and inkjet-based materials. Over the decades, innovations in material science and software have transformed 3D printing into a cornerstone of modern manufacturing.

From aerospace and automotive to healthcare and consumer goods, 3D printing continues to revolutionize industries by offering unparalleled design flexibility and efficiency. The integration of multi-material and 4D printing technologies promises further innovation, paving the way for adaptive, high-performance materials and components. While technical challenges remain, ongoing research and development are rapidly transforming 3D printing into a universally indispensable manufacturing technology.

3D printing involves a diverse range of processes and printer types, each suited to specific applications and materials. The ISO/ASTM52900-15 standard categorizes additive manufacturing into seven primary processes: vat photopolymerization, material jetting, binder jetting, powder bed fusion, material extrusion, directed energy deposition, and sheet lamination. These processes differ in how they layer materials to build parts and the materials used, impacting factors like cost, speed, precision, and mechanical properties.

Vat photopolymerization uses a liquid resin cured by light to form solid layers. This method is precise and ideal for detailed parts, often employing stereolithography (SLA) or digital light processing (DLP) techniques.

Material jetting, an evolution of inkjet technology, deposits droplets of photopolymers or wax in layers. Each layer is cured using ultraviolet light, enabling the creation of highly detailed and colourful parts. This process is often used for prototypes or parts requiring complex geometries.

Powder bed fusion (PBF) processes, including selective laser sintering (SLS) and selective laser melting (SLM), use a laser or electron beam to fuse powdered material into solid layers. SLS works with polymers, while SLM and electron beam melting (EBM) focus on metals, creating fully dense parts with mechanical properties similar to conventionally manufactured components. These methods allow for intricate designs and are widely used in aerospace, automotive, and medical applications.

Material extrusion, commonly referred to as fused deposition modelling (FDM) or fused filament fabrication (FFF), uses thermoplastic filaments melted and extruded through a nozzle to build layers. This process is cost-effective and suitable for a wide range of applications, from prototyping to functional parts. Recent innovations, like fused particle fabrication (FPF), allow the use of pellets instead of filaments, enabling the recycling of materials and reducing costs.

In binder jetting, a liquid adhesive binds layers of powdered material, forming a "green" part that is later cured or sintered. This method can handle materials such as ceramics, metals, and polymers. Its ability to produce full-colour prototypes and complex shapes makes it suitable for diverse industries, from art to industrial manufacturing.

Directed energy deposition uses a focused energy source, such as a laser or electron beam, to melt material—either powder or wire—as it is deposited. This method is highly versatile, allowing for the repair of existing parts and the creation of new, intricate geometries. DED is commonly used for metallic materials in aerospace and defence.

Sheet lamination bonds layers of material, often paper, metal, or polymer sheets, using adhesive or ultrasonic welding. Laminated object manufacturing (LOM) is a subset of this process that cuts cross-sections of adhesive-coated material and layers them to form parts. These techniques are cost-effective for large, simple components but less precise than other methods.

Computed axial lithography creates parts using light projections into a photopolymer resin, building entire layers at once. It is faster than conventional methods and can embed objects within the prints. Liquid additive manufacturing (LAM) uses liquid materials like silicone rubber, vulcanizing them into solid parts.

Cryogenic 3D printing solidifies liquid layers by freezing, enabling unique applications in biomedicine and other fields requiring low temperatures.

Modern 3D printing extends beyond traditional polymers to include metals, ceramics, and composites. Processes like powder bed fusion enable precise metal printing for critical applications, while binder jetting and liquid metal deposition have expanded the range of printable materials.

The choice of a 3D printing process depends on factors such as cost, material properties, and the complexity of the design. While vat photopolymerization excels in detail, processes like PBF and DED are suited for robust, industrial-grade components. Post-processing, such as annealing or surface finishing, is often required to enhance the mechanical properties and aesthetics of printed parts.

Despite advancements, challenges like limited material options, production speed, and high initial costs for industrial-grade printers persist. However, ongoing innovations in multi-material printing, precision, and material science are paving the way for broader adoption across industries, from healthcare and aerospace to consumer goods and beyond.

Hybrid and Combined Techniques

In some cases, manufacturers combine multiple processing techniques to optimize composite properties. For example, hybrid processes may use a combination of RTM and AFP to produce components with high strength, precision, and complex geometries. These combined methods are particularly valuable in advanced industries like aerospace and defence, where the demand for superior performance is paramount.

By selecting the appropriate processing technique, manufacturers can tailor the composite material's properties to meet specific application requirements, balancing cost, performance, and production efficiency. These techniques are continually evolving, driven by advancements in materials science and engineering innovation.

Applications of Composites in Aerospace, Automotive, and Construction

Composite materials have significantly transformed the aerospace, automotive, and construction industries due to their lightweight nature, high strength, and resistance to environmental factors. Each sector has adopted composites to enhance performance, durability, and efficiency, leading to innovative applications.

The aerospace industry has been at the forefront of composite adoption, particularly in structural components. Carbon fibre-reinforced polymers (CFRP) are extensively utilized in aircraft fuselages, wings, and tail sections. These materials provide an exceptional strength-to-weight ratio, which is crucial for improving fuel efficiency and reducing operational costs [308]. Furthermore, ceramic matrix composites (CMCs) are employed in jet engines, specifically in turbine blades and exhaust systems, where they can withstand extreme temperatures, thereby minimizing the need for heavy cooling systems and enhancing engine efficiency [309, 310].

In spacecraft, composites are vital for enduring harsh conditions such as extreme temperatures and radiation exposure. Carbon-carbon composites are commonly used in heat shields for re-entry vehicles, while composite honeycomb structures are integral for lightweight reinforcement in satellites [311]. Additionally, helicopter rotor blades often utilize glass or carbon fibre composites, which provide the necessary flexibility and strength to withstand aerodynamic forces, thus improving performance and safety [311, 312].

In the automotive sector, the integration of composites is essential for achieving lightweight designs that enhance fuel efficiency and overall performance. Carbon fibre composites are increasingly used in vehicle chassis and body panels, particularly in high-performance vehicles, where they reduce weight without compromising safety [313, 314]. Moreover, interior components such as dashboards and door panels often incorporate glass fibre-reinforced plastics (GFRP), which offer durability and aesthetic flexibility while contributing to weight savings [313].

The rise of electric vehicles (EVs) has further accelerated the use of composites in the automotive industry. Composites are employed in battery enclosures, motor housings, and structural reinforcements, maximizing the energy-to-weight ratio and extending driving range [314]. Additionally, ceramic composites are utilized in braking systems for their heat resistance and durability, which enhances braking performance and reduces wear [313, 315].

In construction, composites are valued for their durability and resistance to environmental degradation. Fibre-reinforced polymers (FRPs) are increasingly used to strengthen concrete structures, enhancing tensile strength and extending the lifespan of bridges, buildings, and tunnels [316]. Composite panels made from fiberglass or aluminium are commonly used for exterior cladding, providing lightweight and weather-resistant solutions with diverse design options [316].

Composites are also utilized in roofing materials and insulation panels due to their thermal resistance and lightweight properties, making them ideal for energy-efficient construction [316]. Furthermore, GFRP is employed in manufacturing pipes and tanks for water and chemical transport, offering corrosion resistance and durability [316]. In renewable energy, composites such as epoxy and carbon fibre are critical in wind turbine blades, where a high strength-to-weight ratio is necessary to withstand wind forces and maximize energy generation [317].

The advantages of composite materials span across various sectors. Their lightweight nature significantly enhances efficiency in both aerospace and automotive applications, leading to improved fuel economy and performance [313, 318]. In construction, composites' corrosion resistance reduces maintenance costs and extends service life, while their design flexibility allows for complex shapes tailored to specific requirements [316]. Overall, the enhanced performance characteristics of composites, including improved strength, durability, and thermal stability, make them indispensable in critical applications across these industries [313, 317, 318].

As material technologies continue to evolve, the applications of composites are expected to expand further, fostering sustainability and enhanced performance across aerospace, automotive, and construction sectors.

Chapter 4

Nanomaterials

Introduction to Nanotechnology and Nanomaterials

Nanotechnology

Nanotechnology is defined as the manipulation and engineering of matter at the nanoscale, specifically within the range of 1 to 100 nanometres (nm). At this scale, materials exhibit unique physical and chemical properties due to quantum mechanical effects and an increased surface area-to-volume ratio, which significantly alters their behaviour and interactions compared to their bulk counterparts [319, 320]. The term encompasses various disciplines, including surface science, molecular biology, organic chemistry, and semiconductor physics, reflecting its multidisciplinary nature [320, 321]. The initial vision for nanotechnology involved the precise manipulation of atoms and molecules to create macroscale products, a concept now referred to as molecular nanotechnology [322].

The applications of nanotechnology are vast and transformative, impacting numerous fields such as medicine, electronics, and energy. In nanomedicine, for instance, nanotechnology is being utilized for targeted drug delivery systems, enhancing the efficacy of treatments while minimizing side effects [323, 324]. Nanoelectronics aims to develop smaller, faster, and more efficient electronic devices by leveraging the unique properties of nanomaterials [322]. Additionally, nanotechnology plays a critical role in energy production, contributing to advancements in solar panels, batteries, and fuel cells [325]. However, the rapid advancement of nanotechnology has raised concerns regarding the toxicity and environmental impact of nanomaterials, prompting discussions about the need for regulatory frameworks to ensure safety and sustainability [326, 327].

Historically, the concept of nanotechnology can be traced back to Richard Feynman's 1959 lecture, "There's Plenty of Room at the Bottom," where he envisioned the potential for atomic-level manipulation of materials [322]. The term "nanotechnology" was first introduced by Norio

Taniguchi in 1974, but it gained significant traction following K. Eric Drexler's 1986 book, "Engines of Creation," which outlined the possibilities of molecular assemblers [322]. The 1980s marked significant advancements in the field, particularly with the invention of the scanning tunnelling microscope, which enabled the visualization and manipulation of individual atoms [322, 328]. The discovery of fullerenes and carbon nanotubes further propelled the field, establishing a foundation for the development of nanoscale materials [322].

The early 2000s saw a surge in interest and investment in nanotechnology, leading to its integration into commercial products, such as antibacterial silver nanoparticles and nanoparticle-based sunscreens [328]. However, this period also sparked debates regarding the feasibility and implications of molecular nanotechnology, with discussions between proponents and sceptics highlighting the theoretical and practical challenges that remain [322, 329]. Governments worldwide recognized the potential of nanotechnology, leading to initiatives like the U.S. National Nanotechnology Initiative, which aimed to formalize definitions and provide funding for research and development [325, 329].

Currently, nanotechnology is characterized by a focus on atomically precise manufacturing and advanced nanomaterials, with ongoing research aimed at integrating these technologies into renewable energy, healthcare, and electronics [325]. Despite facing challenges related to scalability, cost, and environmental safety, the field continues to evolve, promising innovative solutions to pressing global issues while also raising ethical and regulatory considerations [326, 330].

Nanomaterials

Nanomaterials are defined as materials with structural components that have at least one dimension in the nanometre range, typically between 1 to 100 nanometres (nm). At this nanoscale, materials exhibit unique properties that differ significantly from their bulk counterparts, primarily due to quantum mechanical effects and a high surface area-to-volume ratio. These distinctive properties make nanomaterials valuable for a wide range of scientific, industrial, and commercial applications [331, 332].

One of the key features of nanomaterials is their high surface area, which enhances chemical reactivity, catalytic properties, and interaction with other substances. This increased surface area relative to volume allows for greater interaction with surrounding environments, making nanomaterials particularly effective in applications such as catalysis and drug delivery [331]. Additionally, quantum effects become significant at the nanoscale, influencing electronic, optical, and magnetic behaviours. For instance, the electronic properties of nanoparticles can vary dramatically with size, affecting their conductivity and optical characteristics [331, 333]. Furthermore, many nanomaterials, such as carbon nanotubes, exhibit exceptional mechanical strength and flexibility compared to traditional materials, which opens avenues for their use in lightweight and durable applications [332, 334].

Nanomaterials can be categorized into various types, including nanoparticles, nanotubes, nanowires, nanofilms, and nanocomposites. Nanoparticles can be metallic (e.g., gold and silver) or non-metallic (e.g., silica and quantum dots) and are widely used in applications ranging from electronics to medicine [335]. Nanotubes and nanowires, such as carbon nanotubes, are cylindrical nanostructures that play a crucial role in electronics and materials science due to their unique electrical and mechanical properties [333, 334]. Nanofilms and nanocoatings are thin layers that modify surface properties for applications in optics and corrosion resistance, while nanocomposites combine nanoscale fillers with matrix materials to enhance mechanical, thermal, or electrical properties [332, 336].

The applications of nanomaterials are vast and varied, spanning fields such as medicine, energy, electronics, environmental science, materials science, and consumer products. In medicine, nanomaterials are utilized for drug delivery systems, diagnostic tools, and antimicrobial coatings [337]. In energy, they contribute to the development of improved batteries, fuel cells, and solar panels [336, 338]. The electronics industry benefits from nanomaterials through the creation of high-performance transistors, displays, and data storage devices [331, 338]. Environmental applications include water purification, air filtration, and pollutant degradation, demonstrating the versatility of nanomaterials in addressing global challenges [336, 338].

Despite their potential, the use of nanomaterials also raises challenges and concerns, particularly regarding toxicity, production costs, and regulatory standards. The potential health and environmental risks associated with exposure to nanomaterials necessitate thorough safety assessments and the establishment of guidelines to ensure safe development and usage [339, 340]. Furthermore, the high production costs of certain nanomaterials can limit their large-scale applications, highlighting the need for cost-effective synthesis methods [341].

Nanomaterials can be classified in multiple ways, including by origin, structure, or chemical composition. One fundamental way to classify nanomaterials is based on their origin.

Natural Nanomaterials occur in nature without human intervention. Examples include volcanic ash particles, fine smoke particles, and biological molecules like haemoglobin. The iridescent colours of a peacock's feathers, for instance, result from nanometre-scale structures on their surface that manipulate light. These naturally occurring nanomaterials highlight how nanoscale phenomena have been integral to the natural world long before human discovery.

Artificial Nanomaterials are produced by human activity, either unintentionally or intentionally. Unintentional artificial nanomaterials include particles emitted from fossil fuel combustion, such as vehicle exhaust, which often contains nanoscale particulates. On the other hand, intentionally produced nanomaterials are those specifically designed and engineered for advanced applications. Scientists and engineers create these materials for use in industries like electronics, medicine, and environmental science.

Nanomaterials can also be classified into two major categories: fullerenes and nanoparticles. These classifications encompass both naturally occurring and artificially created materials, each with unique properties and applications.

Fullerenes are a distinctive type of carbon-based nanomaterial and represent a specific allotrope of carbon. Allotropes are different molecular forms of the same element. While diamond and graphite are familiar examples of carbon allotropes, fullerenes introduce a nanoscale dimension that expands the possibilities of carbon's applications. Spherical fullerenes, commonly referred to as buckyballs, consist of carbon atoms arranged in a soccer-ball-like structure made of tightly bonded hexagons and pentagons. Known as buckminsterfullerenes, see Figure 36, these structures are exceptionally stable and capable of withstanding extreme pressures and temperatures. Their durability enables them to exist in harsh environments, including outer space, where they were first discovered around planetary nebulae. Buckyballs have promising potential in chemical tracing applications, where they can encapsulate elements like helium and track the movement of pollutants through ecosystems.

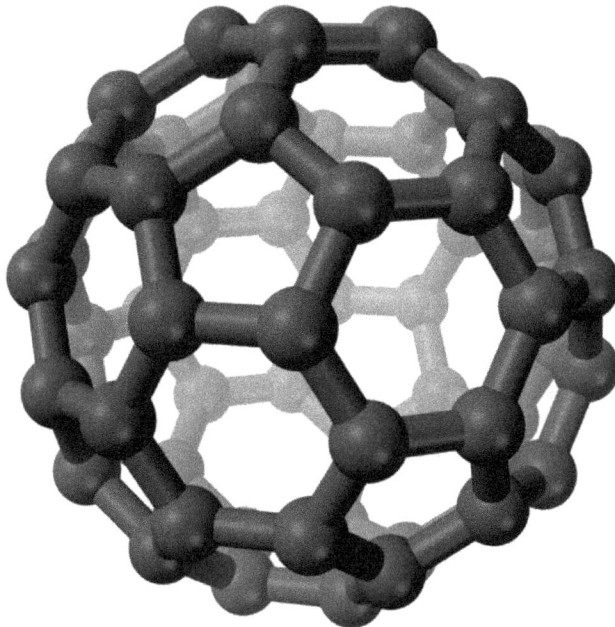

Figure 36: Ball-and-stick model of the buckminsterfullerene molecule, also known as buckyball and by its formula C60. It is notable as the most common fullerene, and for its resemblance to a football. Jynto, CC0, via Wikimedia Commons.

Another type of fullerene is the tubular variant, known as carbon nanotubes. These are cylindrical graphene structures known for their extraordinary strength and flexibility. Carbon nanotubes surpass diamond in hardness and outmatch rubber in flexibility, making them indispensable for applications requiring superior mechanical properties. NASA has been exploring their use in developing ultra-black coatings for satellites, which reduce light reflection and enhance data accuracy.

Nanoparticles, another major class of nanomaterials, represent nanometre-scale versions of various elements, including carbon, gold, silicon, and titanium. These materials are integral to numerous technological advancements. Among nanoparticles, quantum dots stand out for their unique fluorescent properties. Composed of elements such as cadmium and sulphur, quantum dots have been utilized in solar cells, fabric dyes, and cutting-edge display technologies.

Nanoparticles also play a transformative role in nanomedicine. For instance, gold nanoparticles have been designed to combat lymphoma, a form of cancer that targets cholesterol cells. These gold-core nanoparticles mimic cholesterol, attaching to lymphoma cells and preventing them from feeding on actual cholesterol. This innovative approach effectively starves the cancer cells, offering a promising avenue for targeted cancer therapies.

Intentionally produced nanomaterials are categorized into four primary types: carbon-based, metal-based, dendrimers, and nanocomposites. Each category possesses unique properties and applications that are critical in various fields, including medicine, electronics, and materials science.

Carbon-based nanomaterials, such as carbon nanotubes and fullerenes, are synthesized through methods like chemical vapor deposition. This process involves a substrate, typically silicon, where carbon nanotubes grow facilitated by a catalyst, often iron, and a heated gas containing carbon. The result is a highly structured material with exceptional mechanical and electrical properties, making it suitable for applications in nanotechnology and aerospace, such as advanced coatings for satellites [342]. The unique properties of carbon nanotubes, including their high tensile strength and electrical conductivity, have led to extensive research into their use in various applications, from drug delivery systems to advanced composite materials [342].

Metal-based nanomaterials, including gold nanoparticles and quantum dots, are synthesized using various methods that allow for precise control over their size and properties. Quantum dots, for instance, are created by forming small crystals of two different elements under high temperatures, where the size of these nanocrystals dictates their fluorescent properties [343]. This tunability is crucial for applications in medical imaging and photonics. Gold nanoparticles, in particular, have garnered attention for their photothermal properties, which can be enhanced through encapsulation in dendrimers, thus improving their efficacy in targeted drug delivery systems [344].

Dendrimers are highly branched, synthetic macromolecules characterized by a core, inner shell, and outer shell, with functional end groups that can be tailored for specific applications. They can be synthesized using either a divergent or convergent method, allowing for precise control over their architecture [345]. Dendrimers have shown promise in drug delivery due to their ability to encapsulate drugs and facilitate targeted delivery to specific cells, minimizing side effects [346]. Research indicates that dendrimers can enhance the solubility and stability of drugs, making them effective carriers in therapeutic applications [346]. Furthermore, studies have demonstrated that the surface properties of dendrimers can be modified to improve biocompatibility and reduce toxicity, which is essential for their use in biomedical applications [347].

Nanocomposites are materials that integrate nanomaterials with bulk materials to enhance their properties. They can be classified into three main types: nanoceramic matrix composites (NCMCs), metal matrix composites (MMCs), and polymer matrix composites (PMCs) [342]. NCMCs, often referred to as nanoclays, are used to improve the thermal and mechanical properties of packaging materials, while MMCs are utilized in aerospace and automotive industries for their lightweight and strength characteristics [342]. PMCs, which incorporate polymers with nanomaterials, are particularly promising in the field of tissue engineering, where they can serve as scaffolds for cell growth and tissue regeneration, potentially revolutionizing treatments for injuries and organ loss [342].

Nanomanufacturing

Nanomanufacturing refers to the creation of materials, devices, or systems at the nanometre scale, which ranges from 1 to 100 nanometres. At this scale, traditional manufacturing techniques need significant adaptation or replacement to manipulate matter with such precision. Two primary approaches define nanomanufacturing processes: top-down and bottom-up methods, each with unique principles and applications.

In top-down nanomanufacturing, bulk materials are carved, etched, or shaped to achieve nanoscale features. This approach typically involves processes such as photolithography and etching, which have been foundational in producing microchips for decades. The continuous development of this method has allowed manufacturers to create ever-smaller and more efficient transistors, enabling exponential increases in computational power as described by Moore's Law.

For example, in the semiconductor industry, top-down techniques have driven the evolution of microchips by shrinking transistors to nanometre-scale dimensions, thereby increasing chip density and performance. However, as silicon-based microchips approach their physical limits, alternatives like graphene-based microchips are gaining attention. Graphene, a one-atom-thick sheet of carbon atoms, offers superior electrical conductivity, mechanical strength, and thermal properties. The introduction of graphene into top-down manufacturing processes could revolutionize the industry by enabling faster, smaller, and more energy-efficient chips.

Bottom-up nanomanufacturing works in the opposite direction, assembling materials or devices atom-by-atom or molecule-by-molecule. This approach mimics processes observed in nature, such as molecular self-assembly. It relies on fundamental interactions at the atomic and molecular level, including chemical bonds and quantum mechanical forces, to create structures with precise nanoscale properties.

Tech companies experimenting with bottom-up approaches are pushing the boundaries of electronics by developing transistors and other components from nanomaterials like quantum dots, nanotubes, and single molecules. Quantum dots, for instance, are semiconductor particles that exhibit unique optical and electronic properties due to their nanoscale dimensions. These materials can be used to create atom-thick transistors, which promise unparalleled speed, efficiency, and miniaturization in electronic devices.

The bottom-up approach offers a pathway to overcoming the physical limitations of traditional materials and methods. By leveraging the properties of individual molecules, bottom-up nanomanufacturing holds the potential to transform industries ranging from electronics to medicine.

The integration of top-down and bottom-up approaches could define the future of nanomanufacturing. While top-down methods excel at creating complex structures with established scalability, bottom-up techniques offer the precision and versatility needed for next-generation materials and devices. Together, these methods could enable breakthroughs in computing, energy storage, drug delivery, and more.

Nanomanufacturing remains a rapidly evolving field, with advancements continuously reshaping the possibilities for innovation at the nanoscale. By combining traditional manufacturing expertise with cutting-edge nanoscale engineering, industries can harness the extraordinary potential of nanotechnology to address challenges and unlock new opportunities.

Synthesis Methods for Nanomaterials

To produce nanomaterials, researchers utilize a variety of synthesis methods broadly categorized into "top-down" and "bottom-up" approaches. These methods are tailored to create nanomaterials with specific sizes, shapes, and properties, optimizing them for various applications.

Top-Down Synthesis

Top-down synthesis is a crucial approach in nanotechnology, involving the creation of nanoscale structures by breaking down larger bulk materials through mechanical, chemical, or physical methods. This technique is particularly advantageous for applications in electronics and microfabrication, where precise structuring at the nanoscale is essential. However, challenges such as achieving uniformity, precision, and control over the nanoscale features persist in these methods [348, 349].

Mechanical Milling: Mechanical milling is a prominent top-down synthesis method that utilizes high-energy ball mills to grind bulk materials into fine powders with nanoscale grains. The process involves repeated impacts and friction between the milling balls and the material, leading to particle size reduction. This method is effective for producing nanocrystalline metals, alloys, and ceramics, such as aluminium and titanium, which exhibit enhanced mechanical and thermal properties when refined to nanoscale dimensions [350, 351]. Mechanical milling is favoured for its simplicity, cost-effectiveness, and scalability for industrial production. However, it can introduce defects or impurities into the final material, often due to contamination from the milling equipment. Furthermore, achieving precise control over particle shape and size distribution remains a significant challenge in this method [349, 352].

Lithography: Lithography is another critical technique in top-down synthesis, employed to create nanoscale patterns on substrates. Techniques such as photolithography, electron beam lithography, and nanoimprint lithography utilize different mechanisms to define nanoscale features. Photolithography, for instance, uses light to transfer patterns from a mask onto a photosensitive material, while electron beam lithography employs focused electron beams for higher resolution [353, 354]. Lithography is extensively used in fabricating electronic circuits and sensors, making it indispensable in semiconductor industries due to its high precision and reproducibility. However, the high costs associated with specialized equipment and materials, along with limitations to primarily two-dimensional structures, pose challenges for broader applications [352, 354].

Etching: Etching, a subtractive process, employs chemical or plasma-based methods to remove material and create nanoscale features. This process can be isotropic, removing material uniformly, or anisotropic, targeting specific directions for material removal to achieve well-defined shapes. Etching is vital in microelectronics for creating intricate patterns on silicon wafers and developing nanostructured surfaces for applications like photonic crystals [353, 354]. The primary advantage of etching lies in its ability to produce highly defined shapes with nanoscale precision. Nevertheless, stringent control of process conditions is essential to prevent over-etching, which can compromise the dimensions and functionality of the final structures [352, 354].

While top-down synthesis methods are invaluable for creating nanoscale features across various materials and applications, they face limitations such as inconsistent particle size, shape, and purity, along with significant material wastage. Ongoing research is focused on refining these techniques, integrating hybrid approaches, and developing cost-effective and environmentally friendly solutions [352]. As advancements continue, top-down synthesis is poised to remain a cornerstone of nanotechnology and microfabrication, addressing current challenges and expanding its applications in emerging fields [352, 354].

Bottom-Up Synthesis

Bottom-up synthesis is a crucial approach in nanomaterials fabrication, characterized by the assembly of materials atom by atom or molecule by molecule. This technique allows for exceptional control over the structure, composition, and properties of the resulting nanomaterials, making it particularly suitable for high-performance applications across various fields, including electronics, medicine, and catalysis. The versatility of bottom-up synthesis methods, such as Chemical Vapor Deposition (CVD), sol-gel processing, hydrothermal methods, electrochemical deposition, and molecular self-assembly, underscores their significance in advancing nanotechnology.

Chemical Vapor Deposition (CVD) is a prominent bottom-up synthesis method where gaseous precursors react at elevated temperatures on a substrate, leading to the formation of thin layers of nanomaterials. CVD is widely utilized for producing high-quality nanostructures, including carbon nanotubes and graphene, which are essential for next-generation electronic devices due to their outstanding electrical conductivity and mechanical properties [355, 356]. The advantages of CVD include its ability to yield uniform and high-purity nanomaterials, which is critical for applications in electronics and coatings [355, 356]. However, the process is energy-intensive and requires sophisticated equipment, which can pose challenges for large-scale production [355, 356].

The sol-gel process is another versatile bottom-up synthesis technique that transforms liquid precursors into a colloidal solution (sol) and subsequently into a gel. This method allows for precise control over the chemical composition and structural properties of nanomaterials, enabling the production of nanoparticles, thin films, and porous materials [357]. The sol-gel process is particularly valuable in applications such as catalysts and sensors, with nanostructured silica and titania being used in anti-reflective coatings and high-performance catalysts [357]. Despite its flexibility, the sol-gel process demands meticulous control of reaction conditions to ensure uniformity and reproducibility [357].

Hydrothermal and solvothermal methods involve chemical reactions conducted in sealed vessels under high pressure and temperature, making them effective for synthesizing nanomaterials with controlled shapes and sizes, such as semiconductor nanoparticles and metal oxides [358]. These methods are environmentally friendly and suitable for a broad range of materials, enhancing their appeal for sustainable nanotechnology applications [358]. For instance, zinc oxide and titanium dioxide nanoparticles produced through hydrothermal methods are utilized in photocatalysis and UV-blocking agents [358]. However, these methods can be time-consuming and require specialized equipment, such as autoclaves, to maintain the necessary high-pressure environment [358].

Electrochemical deposition is a technique that allows for the controlled deposition of nanomaterials onto conductive substrates via electrochemical reactions. This method is particularly advantageous for creating nanostructured films and coatings with tailored properties, making it suitable for applications in batteries, fuel cells, and corrosion-resistant coatings [358]. The cost-effectiveness and scalability of electrochemical deposition make it an attractive option, although its applicability is limited to conductive substrates [358].

Molecular self-assembly is a process that relies on the spontaneous organization of molecules or nanoparticles into well-defined structures through intermolecular forces. This method is capable of producing intricate and highly ordered nanostructures, which have significant applications in drug delivery, nanostructured membranes, and photonic crystals [357]. For example, self-assembled liposomes are employed to enhance targeted drug delivery while minimizing side effects [357]. The ability of molecular self-assembly to create complex structures with atomic precision is a notable advantage; however, it is sensitive to environmental factors, necessitating careful control to achieve consistent results [357].

Hybrid and Emerging Techniques

Innovative nanomaterial synthesis methods have emerged as a crucial area of research, particularly through the integration of top-down and bottom-up approaches. These hybrid strategies enhance the control over material properties, improve environmental compatibility, and broaden the applications of nanotechnology. Among these methods, Atomic Layer Deposition (ALD), laser ablation, and biological synthesis stand out for their unique capabilities and applications.

Atomic Layer Deposition (ALD) is a precise bottom-up synthesis technique that allows for the deposition of thin films one atomic layer at a time. This process involves alternating chemical reactions where a substrate is exposed to specific precursors, resulting in the formation of a single atomic layer. ALD is particularly valued in industries requiring high precision and uniformity, such as microelectronics and catalysis, where it is used to deposit dielectric layers in semiconductors and create protective coatings on advanced materials [359]. The atomic-scale precision of ALD ensures consistent layer thickness even on complex geometries, making it indispensable in advanced manufacturing. However, the slow deposition rate poses challenges for scalability in large-scale production [360].

Laser ablation represents a powerful top-down approach for synthesizing nanoparticles by focusing a high-energy laser beam onto a bulk material, causing it to vaporize and subsequently condense into nanoparticles. This technique is particularly effective for producing high-purity metal, ceramic, and composite nanoparticles without the need for chemical solvents, thus ensuring minimal contamination [361]. The ability to control particle size through laser parameters is a significant advantage, making laser ablation suitable for applications in electronics and catalysis where purity is paramount [362]. Despite its advantages, the high energy requirements and specialized equipment necessary for laser ablation can increase operational costs and limit its broader adoption [363].

Biological synthesis of nanomaterials utilizes natural processes involving microorganisms, enzymes, or plant extracts to produce nanoparticles. This eco-friendly method avoids toxic chemicals and minimizes environmental impact, making it an attractive alternative for sustainable manufacturing [364]. For instance, gold and silver nanoparticles synthesized using plant extracts have shown promise in medical applications, including cancer therapies and diagnostic tools [365]. However, challenges such as slower reaction rates and limited control over particle size and uniformity hinder the scalability of biological synthesis for industrial applications [365].

Considerations for Nanomaterial Synthesis

To ensure their effectiveness and functionality, several critical factors must be considered during the synthesis stage, ranging from precise control over physical characteristics to environmental sustainability.

One of the most significant challenges in nanomaterial synthesis is achieving precise control over the size and shape of the particles. Many applications, particularly in electronics, catalysis, and medicine, require nanoparticles with specific dimensions and morphologies to optimize their performance. For instance, in drug delivery, the size of nanoparticles influences their ability to penetrate cellular membranes, while the shape can affect circulation time and cellular uptake. Achieving uniformity in size and shape also enhances material properties like reactivity and mechanical strength. Advanced synthesis methods such as chemical vapor deposition (CVD), hydrothermal techniques, and molecular self-assembly are often employed to meet these stringent requirements.

The purity of nanomaterials is a crucial consideration, as impurities can dramatically alter their properties and compromise their intended functionality. In fields such as nanoelectronics and biomedical applications, even trace amounts of contaminants can disrupt electrical conductivity or biocompatibility. Achieving high purity often requires precise control over raw materials, reaction conditions, and post-synthesis processing. Techniques such as laser ablation and biological synthesis are particularly valued for their ability to produce highly pure materials. Post-synthesis purification steps, including centrifugation and filtration, are also commonly employed to eliminate residual byproducts.

While many synthesis methods excel in laboratory settings, their scalability is a critical factor for industrial applications. The ability to produce nanomaterials in large quantities without compromising quality is essential for their adoption in commercial markets. Methods like sol-gel processes and hydrothermal synthesis are relatively scalable, but other advanced techniques, such as atomic layer deposition (ALD), may face challenges in meeting industrial-scale demands due to slow production rates. Engineers and researchers are continuously working to refine these methods, developing hybrid techniques and automation systems to bridge the gap between small-scale experimentation and large-scale production.

As nanotechnology becomes more pervasive, its environmental implications are under increasing scrutiny. The use of toxic chemicals, high energy consumption, and waste generation in traditional synthesis methods can pose significant environmental risks. Sustainable nanomaterial production focuses on eco-friendly processes that minimize these impacts. Biological synthesis, for instance, leverages natural agents like plant extracts and enzymes, reducing the need for harmful chemicals. Similarly, advancements in green chemistry aim to develop processes that use renewable resources, reduce waste, and operate under milder conditions. Addressing the environmental footprint of nanomaterial synthesis is not only a matter of regulatory compliance but also a step toward aligning the industry with global sustainability goals.

Current Utilisation of Nanomaterials

Nanomaterials are widely used across various industries, ranging from healthcare to energy, due to their unique properties. Carbon-based nanomaterials are among the most significant, with applications such as carbon nanotubes (CNTs), which are used in lightweight composites for aerospace, automotive components, and sports equipment. They are also utilized in conductive films for touch screens and batteries, with products offered by companies like Arkema and Nanocyl. Graphene is another critical carbon-based material, commonly found in flexible electronics, high-performance batteries, and advanced coatings. Companies such as Graphenea and Haydale offer graphene-based inks. Additionally, fullerenes, including buckyballs, are incorporated into solar cells, lubricants, and drug delivery systems, with some being used in antioxidant skincare creams.

Metal-based nanomaterials also play a crucial role in modern applications. Gold nanoparticles are employed in medical diagnostics, drug delivery systems, and chemical catalysts, with providers like NanoComposix leading the market. Silver nanoparticles are widely used for their antimicrobial properties in coatings for medical devices, clothing, and food packaging, with products like Polygiene's antimicrobial fabrics. Titanium dioxide nanoparticles are a common component in sunscreens, self-cleaning glass, and paints, with companies such as Evonik and

PPG Industries marketing NanoTiO$_2$. Similarly, zinc oxide nanoparticles are used in sunscreens, cosmetics, and rubber production, with BASF supplying UV-blocking ZnO nanoparticles.

Ceramic nanomaterials, such as silicon dioxide (SiO$_2$) nanoparticles, find applications in anti-caking agents, scratch-resistant coatings, and toothpaste, with products like Cab-O-Sil® by Cabot Corporation. Aluminium oxide (Al$_2$O$_3$) nanoparticles are utilized in polishing, thermal insulation, and wear-resistant coatings, with Nanodur by Evonik Industries as a prominent example.

Polymer-based nanomaterials include dendrimers, which are used in drug delivery, imaging, and water treatment, as seen in Starpharma's products. Polymeric nanoparticles are applied in controlled drug release systems and biodegradable implants, with companies like AdvaNano offering these materials. Composite nanomaterials such as nanoclay composites reinforce plastics in packaging, automotive parts, and construction materials, with BYK's Cloisite® nanoclay additives leading the market. Metal-organic frameworks (MOFs) are another category, used in gas storage, separation technologies, and catalysis, with companies like MOF Technologies producing these materials.

Semiconductor nanomaterials, such as quantum dots, are integral to high-resolution displays, solar cells, and bioimaging. For instance, Samsung integrates quantum dot technology into its QLED TVs. Biological and green nanomaterials also have significant applications. Cellulose nanocrystals (CNCs) are used to reinforce bioplastics, create lightweight materials, and improve food packaging, with products offered by Celluforce and American Process Inc. Chitosan nanoparticles are found in wound dressings, antimicrobial coatings, and drug delivery systems, with NovaMatrix providing these materials.

Magnetic nanomaterials, such as iron oxide nanoparticles, are utilized in magnetic resonance imaging (MRI), drug delivery, and data storage. Companies like Ocean Nanotech offer SPIONs (Superparamagnetic Iron Oxide Nanoparticles) for these purposes.

These nanomaterials continue to drive innovation, addressing critical challenges and enabling advancements in consumer goods, healthcare, and environmental sustainability. Their versatile properties and applications make them indispensable in shaping future technologies.

Properties and Behaviour at the Nanoscale

Nanomaterials exhibit unique properties and behaviours that significantly differ from their bulk counterparts, primarily due to their nanoscale dimensions. This leads to the dominance of quantum mechanical effects and an increased surface area-to-volume ratio, resulting in remarkable physical, chemical, and mechanical characteristics.

At the nanoscale, the surface area-to-volume ratio of materials increases dramatically, which means a larger proportion of atoms are located on the surface. This phenomenon significantly influences the material's reactivity and interactions with the environment. Enhanced reactivity is a notable characteristic of nanomaterials, as the abundance of surface atoms facilitates more efficient chemical reactions compared to bulk materials. For instance, nanoparticles used as catalysts can accelerate reactions more effectively than their bulk counterparts due to their high surface area [366, 367]. Additionally, the increased surface area enhances properties like energy absorption, making nanomaterials particularly suitable for applications in batteries and supercapacitors [368].

Quantum mechanical phenomena play a crucial role in the behaviour of materials at the nanoscale. Quantum confinement occurs when the dimensions of a material are reduced below a critical size, leading to discrete energy levels and size-dependent optical properties. This effect is prominently observed in quantum dots, which can emit different colours based on their size [369]. Furthermore, tunnelling effects at the nanoscale allow electrons to "tunnel" through barriers, impacting electrical conductivity and enabling advancements in technologies such as quantum computing [366].

Nanomaterials exhibit unique optical behaviours due to their interaction with light at the nanoscale. One significant phenomenon is Localized Surface Plasmon Resonance (LSPR), which occurs in metal nanoparticles like gold and silver. This effect, where conduction electrons resonate with light waves, is exploited in various sensing technologies and imaging applications [370]. Moreover, nanostructures can absorb and scatter light more efficiently, enhancing their application in solar cells and display technologies [371].

The mechanical properties of nanomaterials often surpass those of their bulk counterparts. For example, materials such as carbon nanotubes and nanocrystalline metals exhibit exceptional tensile strength and hardness due to a reduced presence of structural defects [372]. Additionally, nanomaterials like graphene are noted for their remarkable flexibility and toughness, making them ideal for lightweight and durable applications [373].

Thermal behaviour also changes at the nanoscale. Nanomaterials such as carbon nanotubes and graphene demonstrate enhanced thermal conductivity, which is beneficial for heat dissipation applications [374]. Conversely, nanoparticles often exhibit lower melting points than bulk materials due to the increased mobility of surface atoms [375].

Magnetism at the nanoscale is influenced by size and quantum effects. Nanoparticles of ferromagnetic materials can exhibit superparamagnetism, where magnetization can be easily reversed, making them suitable for applications in magnetic resonance imaging (MRI) and data storage [371, 376]. Additionally, magnetic nanoparticles can display enhanced magnetic moments due to surface spin effects, further broadening their application potential [370].

Nanomaterials demonstrate unique chemical properties resulting from their surface structure and reactivity. Their high surface energy allows nanoparticles to act as efficient catalysts in chemical reactions [375]. The ability to tailor surface functionalization enables precise chemical interactions, which is crucial in applications such as drug delivery and sensing [371].

The interaction of nanomaterials with biological systems is markedly different from that of bulk materials. Nanoparticles can penetrate cellular membranes more easily due to their size, making them suitable for targeted drug delivery applications [377]. Additionally, functionalized nanomaterials can detect biomolecules with high specificity, advancing diagnostics and therapeutic strategies [373].

Nanomaterials are highly sensitive to environmental changes, such as temperature, pressure, and chemical surroundings. This sensitivity is harnessed in sensors designed to detect minute changes in environmental conditions, enhancing their utility in monitoring applications [378].

Applications in Electronics, Medicine, and Energy Storage

Nanomaterials have emerged as pivotal components in various fields, particularly in electronics, medicine, and energy storage. Their unique properties at the nanoscale have facilitated advancements that were previously unattainable with conventional materials.

In the realm of electronics, nanomaterials have significantly enhanced device performance through miniaturization and efficiency improvements. For instance, semiconductors and transistors have benefited from materials such as graphene and carbon nanotubes, which enable the creation of transistors with higher speeds and lower energy consumption, thus paving the way for next-generation nanoelectronics [379, 380]. Quantum dots, another class of nanomaterials, are utilized in high-resolution displays and photodetectors, providing vibrant colours and energy efficiency in devices like televisions and monitors [381, 382]. Furthermore, the large surface area and quantum effects of nanomaterials allow for the development of highly sensitive sensors capable of detecting gases, chemicals, and biomolecules [383].

Memory devices have also seen advancements with the incorporation of magnetic nanoparticles in spintronics, which enhances data storage density and retrieval speeds [384]. The flexibility of electronics has been improved through the use of graphene and conductive polymers, leading to the development of flexible screens and wearable technology [385]. Additionally, thermal management in electronic devices has been enhanced by nanomaterials like carbon nanotubes and graphene, which improve heat dissipation [386].

Nanomaterials are transforming the medical field by improving diagnostics, therapeutics, and regenerative medicine. In drug delivery, nanoparticles such as liposomes and polymeric nanoparticles are employed to target specific tissues, enhancing drug efficacy while minimizing side effects [387, 388]. Imaging techniques have also benefited from nanomaterials; for example, gold and iron oxide nanoparticles are utilized in MRI and CT scans to provide better contrast and precision [389, 390].

In cancer therapy, nanomaterials like gold nanoshells and quantum dots facilitate targeted treatments through photothermal and photodynamic therapies, which selectively destroy cancer cells [391, 392]. Regenerative medicine has seen the use of nanostructured scaffolds that support cell growth and tissue repair [393]. Moreover, functionalized nanoparticles serve as biosensors for the early detection and monitoring of diseases [394]. In the realm of vaccines, nanoparticles act as adjuvants, enhancing immune responses and improving vaccine delivery [395].

The role of nanomaterials in energy storage technologies is critical, as they enhance capacity, efficiency, and lifespan. In lithium-ion batteries, nanostructured electrodes, such as silicon nanowires and graphene, significantly improve energy density and charging speed [396, 397]. Solid-state batteries benefit from nanomaterials that create safer and more efficient solid electrolytes [398]. Supercapacitors leverage nanomaterials like graphene and carbon nanotubes to achieve high energy density and rapid charge-discharge cycles [399].

Nanocatalysts, particularly platinum nanoparticles, enhance the efficiency and reduce the costs of hydrogen fuel cells [400]. In solar energy applications, quantum dots and perovskite nanomaterials improve the efficiency of photovoltaic cells by optimizing light absorption and conversion [401, 402]. Furthermore, nanostructured materials such as metal-organic frameworks (MOFs) and carbon nanotubes facilitate high-capacity hydrogen storage, which is essential for fuel applications.

Chapter 5

Bioplastics and Sustainable Materials

Overview of Bioplastics and Their Types

Bioplastics, defined as materials derived from renewable biomass sources, present a sustainable alternative to conventional fossil-fuel-based plastics. Historically, bioplastics have roots in early materials like shellac and cellulose, which were among the first plastics utilized before the petroleum industry emerged in the late 19th century, leading to the dominance of fossil-fuel-based plastics due to their cost-effectiveness and versatility [403, 404]. In recent years, there has been a resurgence of interest in bioplastics, driven by the frameworks of the bioeconomy and circular economy, which aim to reduce reliance on non-renewable resources and address pressing environmental concerns [404, 405].

The production of bioplastics involves various methods, including direct processing from natural biopolymers such as polysaccharides (e.g., corn starch and cellulose) and proteins (e.g., soy protein and gelatin) [406, 407]. Additionally, chemical synthesis from sugar derivatives and lipids, as well as fermentation processes utilizing microorganisms, have emerged as significant pathways for bioplastic production, notably for advanced bioplastics like polyhydroxyalkanoates (PHA) [408, 409]. These diverse production methods underscore the adaptability of bioplastics to meet various industrial needs, highlighting their potential in a range of applications from packaging to agricultural uses [410].

One of the primary advantages of bioplastics is their independence from fossil fuels, which are finite and unevenly distributed globally, contributing significantly to environmental degradation [403, 405]. Many bioplastics are produced from agricultural and industrial waste materials, such as straw and food waste, promoting waste valorization and sustainability [407, 410]. Life cycle analyses indicate that certain bioplastics can have a lower carbon footprint compared to their fossil-based counterparts, particularly when biomass serves as both the raw material and energy source [405, 411]. However, it is crucial to note that not all bioplastic production processes are efficient; some may even result in a higher carbon footprint than traditional plastics, emphasizing the need for optimized production methods [405, 412].

The degradability of bioplastics is influenced more by their molecular structure than by their biomass origin. Durable bioplastics, such as bio-PET and bio-polyethylene, are designed to mimic fossil-based polymers and can be recycled through existing systems [411, 413]. Conversely, biodegradable options like polylactic acid (PLA) and polybutylene succinate (PBS) offer unique disposal pathways but may disrupt recycling systems due to their differing properties [413, 414]. While biodegradability is advantageous for specific applications, such as agricultural mulch, it necessitates careful consideration of environmental conditions, as not all bioplastics decompose readily in natural settings [413, 414].

Despite their growing significance, bioplastics still represent a small fraction of global plastic production, accounting for about 2% of the over 380 million tons produced in 2018 [404, 405]. The most commercially significant bioplastics include PLA and starch-based products, which have seen increasing demand [404, 405]. However, challenges such as cost, scalability, and performance continue to hinder broader adoption [405, 412]. As research progresses and investments in bioplastic technologies increase, their market presence is anticipated to expand, although fossil-based plastics continue to dominate production [404, 405].

The International Union of Pure and Applied Chemistry (IUPAC) defines biobased polymers as those derived from biomass, emphasizing the distinction between bioplastics and fossil-derived polymers [404]. However, the term "bioplastic" can be misleading, as it does not inherently guarantee environmental friendliness. Comprehensive life cycle analyses are essential to assess the environmental benefits of bioplastics compared to their fossil-based counterparts [405, 411]. Applications for bioplastics are currently limited, primarily to disposable items like packaging and straws, but they also serve as coatings for paper, replacing petrochemical alternatives [404, 405]. "Drop-in" bioplastics, which are chemically identical to fossil-based plastics but derived from renewable resources, offer a seamless integration into existing infrastructure, facilitating a transition to more sustainable alternatives [404, 405].

Bioplastics are classified into several types based on their sources and chemical composition, each with unique features and applications.

Starch-based bioplastics are the most widely used, derived from natural starches such as corn, potato, or rice. These materials often incorporate thermoplastic starch blended with biodegradable polyesters to enhance mechanical properties and flexibility. Commonly used in packaging materials, compostable bags, and food containers, starch-based bioplastics are a staple in the bioplastics market.

Cellulose-based bioplastics, made from cellulose—a primary component of plant cell walls—include cellulose esters like cellulose acetate. These bioplastics exhibit improved mechanical properties, gas permeability, and water resistance compared to starch-based options. Their applications include packaging, films, and coatings, making them valuable in various industries.

Protein-based bioplastics are derived from proteins such as wheat gluten, casein, or soy protein. These materials are biodegradable and renewable, though they are often blended with other biodegradable polymers to improve water resistance and mechanical strength. Protein-based bioplastics find applications in agricultural films, packaging, and edible films.

Polyhydroxyalkanoates (PHAs) are linear polyesters naturally produced through bacterial fermentation of sugar or lipids. These bioplastics are biodegradable, versatile, and highly customizable. PHAs are widely used in medical sutures, drug delivery systems, and packaging due to their adaptability.

Polylactic acid (PLA) is a transparent plastic made from lactic acid derived from fermented plant-based sugars like corn or sugarcane. It is biodegradable under industrial composting conditions and is commonly used for bottles, food packaging, and 3D printing filaments. While PLA has limitations in impact strength and thermal robustness, it remains a popular choice for various applications.

Figure 37: Four polylactic acid cups. Cmglee, CC BY-SA 4.0, via Wikimedia Commons.

Bio-polyethylene (Bio-PE) is chemically identical to conventional polyethylene but is produced from renewable sources like sugarcane. Although non-biodegradable, it is recyclable and has a lower carbon footprint than its fossil-based counterpart. Bio-PE is utilized in packaging, bottles, and consumer goods.

Bio-polyamides (Bio-PA), derived from renewable resources such as castor oil, include options like Polyamide 11 and Polyamide 410. These bioplastics exhibit high thermal resistance and mechanical strength, making them suitable for automotive components, electrical insulation, and sports equipment.

Drop-in bioplastics are chemically identical to their fossil-based equivalents but are derived from renewable resources. Examples include Bio-PET, Bio-PP, and Bio-PE. These materials can be seamlessly integrated into existing production infrastructure and are used in bottles, textiles, and packaging.

Lipid-derived polymers are synthesized from plant and animal fats and oils. These bioplastics are comparable to crude oil-based materials and offer high customizability. Applications include coatings, adhesives, and flexible films, showcasing their versatility in industrial use.

Genetically engineered bioplastics are created using genetically modified microorganisms or plants to improve synthesis efficiency. These bioplastics can optimize production processes and generate novel polymers. They are used in advanced medical materials, packaging, and textiles, addressing complex industrial needs.

Some specific types of bioplastics include:

Polysaccharide-Based Bioplastics: Starch-Based Plastics: Thermoplastic starch represents one of the most widely used bioplastics, accounting for approximately 50% of the bioplastics market. It can be produced by gelatinizing starch and solution casting, making it simple enough to be made at home. In industrial applications, starch-based bioplastics are used for drug capsules due to their ability to absorb humidity. However, pure starch-based plastics are brittle, and plasticizers such as glycerol, glycol, and sorbitol are added to enhance processability and flexibility. These additives allow starch to be processed using conventional polymer techniques like extrusion, injection molding, and compression molding. The properties of starch-based bioplastics are influenced by the amylose-to-amylopectin ratio, with higher amylose content providing superior mechanical properties but posing challenges in processing due to higher gelatinization temperatures and melt viscosity.

To overcome limitations, starch-based bioplastics are often blended with biodegradable polyesters, such as polylactic acid (PLA), polycaprolactone, or Ecoflex. These blends enhance mechanical and composting properties, making them suitable for industrial applications. Starch/polyester films are commonly used in packaging, including bakery bags, magazine wrappings, and compostable waste bags. Recent advances in starch-based nanocomposites have improved mechanical properties, thermal stability, and resistance to moisture and gas permeability, broadening their utility.

Cellulose-Based Plastics: Cellulose bioplastics, such as cellulose esters (e.g., cellulose acetate and nitrocellulose), are derived from cellulose and its derivatives. While cellulose acetate exhibits thermoplastic properties, its high cost limits its use in packaging. However, incorporating cellulose fibres into starch-based bioplastics enhances mechanical strength and water resistance, expanding their range of applications.

Protein-Based Plastics: Proteins like wheat gluten and casein offer potential as raw materials for bioplastics. Soy protein has been used historically in plastic production, including in automotive body panels. However, soy protein-based plastics face challenges related to water sensitivity and cost. Blending soy protein with biodegradable polyesters improves water resistance and reduces costs, paving the way for broader adoption.

Aliphatic Polyesters: Polyhydroxyalkanoates (PHAs): PHAs, such as poly-3-hydroxybutyrate (PHB), are linear polyesters naturally produced by bacterial fermentation of sugar or lipids. PHAs are biodegradable, ductile, and suitable for medical applications. PHB, in particular, has characteristics similar to polypropylene, with the ability to form transparent, heat-resistant films. Industrial-scale production of PHB is expanding, particularly in regions like South America.

Polylactic Acid (PLA): PLA is a transparent plastic derived from renewable sources such as maize and dextrose. Although biodegradable under industrial composting conditions, PLA has lower impact strength and thermal robustness compared to conventional plastics. PLA is widely used for films, fibres, and 3D printing filaments, making it a versatile bioplastic for various industries.

Polyamide 11 and Polyamide 410: Polyamide 11 (PA 11), derived from natural oils, offers high thermal resistance and reduced greenhouse gas emissions compared to its fossil-based counterparts. It is used in automotive fuel lines, electrical sheathing, and flexible pipes. Similarly, Polyamide 410 (PA 410), derived 70% from castor oil, combines high melting points and chemical resistance, making it suitable for demanding applications.

Bio-Derived Polyethylene: Bio-derived polyethylene is chemically identical to traditional polyethylene but is produced from renewable sources like sugarcane ethanol. It is non-biodegradable but recyclable, with a lower carbon footprint compared to fossil-based polyethylene. Companies like Braskem produce this material for applications ranging from packaging to consumer goods.

Genetically Modified Feedstocks and Lipid-Derived Polymers: Advancements in genetically modified crops and bacteria enable efficient bioplastic production. Lipid-derived polymers, synthesized from plant and animal fats, have properties comparable to petroleum-based plastics. Recent developments, such as oleogels based on vegetable oils, have introduced novel bioplastics like OleoPlast, which are recyclable, biodegradable, and adaptable to various processing techniques.

These diverse types of bioplastics highlight their growing relevance in industries ranging from packaging and agriculture to automotive and healthcare. With continued innovation and investment, bioplastics have the potential to address environmental challenges while meeting industrial demands.

Polymer Chemistry and Biodegradability

Polymer chemistry is a vital field that encompasses the study of polymers, which are large molecules formed by the repetition of smaller units known as monomers. This discipline focuses on the synthesis of these macromolecules as well as their properties and applications across various industries. Polymers can be categorized into natural and synthetic types, with natural polymers including cellulose and proteins, while synthetic polymers encompass materials like polyethylene and nylon [415]. The classification of polymers extends to their structural characteristics, which include thermoplastics, thermosetting polymers, and elastomers. Each type exhibits distinct properties that are influenced by their chemical structure, molecular weight, and the nature of the bonding within the polymer chains [416, 417].

The properties of polymers are crucial for their applications, and they can be tailored through various polymerization techniques, such as addition and condensation polymerization. These methods allow for the manipulation of polymer characteristics to meet specific requirements in fields ranging from materials science to biomedical applications [418, 419]. For instance, the development of biodegradable elastomers has gained attention due to their potential in drug delivery and tissue engineering, where the degradation rate can be controlled to match the biological environment [420]. The incorporation of specific functional groups, such as amide bonds, can enhance the stability and biodegradability of these materials, making them suitable for various applications [416, 417].

Biodegradability is a significant aspect of polymer chemistry, particularly in the context of environmental sustainability. The increasing awareness of plastic pollution has spurred research into biodegradable polymers that can decompose naturally without harming the ecosystem. Studies have shown that synthetic biodegradable elastomers can be designed to degrade through mechanisms such as hydrolysis and enzymatic activity, which are critical for their application in medical devices and environmental solutions [416, 420]. Moreover, the exploration of bio-based copolyesters has demonstrated promising results in achieving both elasticity and biodegradability, indicating a shift towards more sustainable polymer materials [421].

Biodegradability of polymers is a critical area of research due to the increasing environmental concerns associated with synthetic polymers and plastic waste. Biodegradability refers to the ability of materials to decompose into simpler, non-toxic components through the action of microorganisms, which is essential for reducing pollution and promoting sustainable materials. The biodegradability of polymers is influenced by their chemical structure, physical properties, and environmental conditions.

The chemical structure of polymers significantly impacts their biodegradability. Polymers that contain hydrolytically unstable bonds, such as ester, amide, or glycosidic bonds, are more susceptible to microbial degradation. For instance, natural biopolymers like starch and cellulose, as well as synthetic biodegradable polymers such as polylactic acid (PLA) and polycaprolactone (PCL), possess these types of bonds, facilitating their breakdown by microbial enzymes [422-424]. Conversely, polymers with stable C–C bonds, such as polyethylene and polypropylene, resist biodegradation and persist in the environment for extended periods [424]. This highlights the importance of selecting appropriate chemical structures in the design of biodegradable materials.

Several factors influence the biodegradability of polymers, including polymer structure, crystallinity, molecular weight, and the presence of functional groups.

1. Polymer Structure: The arrangement of monomer units affects how easily microorganisms can access and degrade the polymer. Linear and less cross-linked polymers, such as aliphatic polyesters, are generally more biodegradable than highly cross-linked or aromatic polymers, which present a more rigid structure that hinders enzymatic attack [425, 426].

2. Crystallinity: The degree of crystallinity in polymers plays a crucial role in biodegradation. Amorphous regions, which are less densely packed, are more accessible to water and enzymes, allowing for easier microbial degradation. In contrast, highly crystalline polymers are more resistant to biodegradation due to their dense structure [427]. For example, studies have shown that the biodegradation rate of poly(lactic acid) can be significantly influenced by its crystallinity [427].

3. Molecular Weight: Lower molecular weight polymers tend to be more biodegradable because their shorter chain lengths facilitate microbial breakdown. High molecular weight polymers require initial cleavage into smaller fragments before they can be metabolized by microorganisms, which can slow down the biodegradation process [422, 423, 428].

4. Functional Groups: The presence of polar functional groups, such as hydroxyl (-OH) and carboxyl (-COOH) groups, enhances the biodegradability of polymers by increasing their hydrophilicity. This property allows for better water absorption and swelling, which facilitates enzymatic and microbial attack [422-424]. For instance, the polar nature of starch and cellulose makes them more susceptible to hydrolytic degradation compared to nonpolar synthetic polymers [422, 423].

The environment in which polymers are placed also significantly affects their biodegradation rates. Composting conditions, characterized by high moisture and microbial activity, have been shown to enhance the degradation of biodegradable polymers like polybutylene succinate (PBS) and its copolymers [429]. In contrast, polymers placed in anaerobic or less active environments may degrade at much slower rates, underscoring the importance of environmental factors in the biodegradation process [429].

Biodegradable Polymers

Biodegradable polymers represent a significant advancement in materials science, primarily due to their ability to decompose into natural substances such as carbon dioxide, water, and biomass under specific environmental conditions. These polymers can be categorized into natural and synthetic types, each with unique properties and applications.

Natural biodegradable polymers are derived from renewable biological sources and are inherently compatible with natural decomposition processes. Cellulose, a major component of plant cell walls, is one of the most abundant natural polymers. Its biodegradation is facilitated by cellulolytic microorganisms that produce enzymes like cellulase, breaking down cellulose into glucose [430]. This property makes cellulose widely utilized in products such as paper and biodegradable films, reinforcing its role in sustainable industries [423].

Proteins such as collagen and gelatin also exemplify natural biodegradable polymers. These materials degrade into amino acids through enzymatic activity, making them particularly valuable in medical applications, including tissue engineering and drug delivery systems, due to their biocompatibility and biodegradability in biological environments [423, 431]. Furthermore, polysaccharides like starch and chitosan are notable examples of biodegradable polymers. Starch, sourced from plants such as corn and potatoes, is frequently employed in biodegradable plastics, while chitosan, derived from chitin in crustacean exoskeletons, exhibits antimicrobial properties and finds applications in wound healing and water purification [430, 431].

Synthetic biodegradable polymers are engineered to mimic the degradability of natural polymers while offering enhanced mechanical properties and tunable degradation rates. Polylactic acid (PLA) is a prominent example, synthesized from lactic acid derived from renewable resources like corn starch or sugarcane. PLA is widely used in packaging and disposable cutlery due to its biodegradability under industrial composting conditions, although its degradation requires specific conditions such as elevated temperatures and controlled humidity [432, 433].

Polyhydroxyalkanoates (PHA) are another class of synthetic biodegradable polymers produced through microbial fermentation of renewable resources. PHAs are fully biodegradable and exhibit a diverse range of mechanical properties, making them suitable for applications in packaging, agriculture, and medicine [434, 435]. Their ability to tailor properties by varying monomer composition enhances their versatility, although high production costs remain a challenge [430, 431].

Polycaprolactone (PCL), a synthetic aliphatic polyester, is characterized by a relatively slow degradation rate, making it ideal for long-term biomedical applications such as drug delivery devices and tissue scaffolds. The biodegradation of PCL primarily occurs through hydrolysis of its ester bonds, followed by microbial assimilation, and it is often blended with other polymers to enhance its properties for various applications [423, 436].

Chemical Structure and Biodegradability

The chemical structure of a polymer plays a pivotal role in determining its susceptibility to biodegradation. This process primarily depends on the presence of specific chemical bonds and functional groups that can be hydrolysed or enzymatically cleaved, enabling the breakdown of polymer chains into smaller, more manageable fragments. Several key bond types and structural features influence the biodegradability of polymers.

Ester Bonds are among the most common features in biodegradable polymers. Found in polyesters like polylactic acid (PLA) and polycaprolactone (PCL), ester bonds are particularly susceptible to hydrolysis and enzymatic action. During hydrolysis, water molecules break these bonds, fragmenting the polymer into smaller oligomers or monomers. Enzymes such as esterases further catalyse this process, facilitating a faster and more efficient breakdown. For example, PLA, widely used in packaging and medical applications, degrades into lactic acid under industrial composting conditions, where moisture and enzymes are abundant.

Ester bonds are a type of chemical bond formed between a carboxylic acid group ($-COOH$) and an alcohol group ($-OH$) through a condensation reaction, which results in the release of a water molecule (H_2O). This bond is characterized by the functional group $R-COOR'$, where R and R' are organic groups. Ester bonds are a key structural component in many biological and synthetic molecules, including lipids, polyesters, and other polymers.

The process of forming an ester bond is known as esterification. It typically occurs through the following reaction:

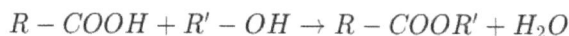

$$R - COOH + R' - OH \rightarrow R - COOR' + H_2O$$

This reaction requires acidic or basic conditions as a catalyst to speed up the process. The reverse reaction, where an ester is broken down into its original alcohol and acid components, is called hydrolysis.

Ester bonds exhibit distinct properties that make them essential in both natural and synthetic contexts. These bonds are polar due to the presence of electronegative oxygen atoms, which contribute to their moderate solubility in water, particularly in smaller ester molecules. Ester bonds are also characterized by their susceptibility to hydrolysis, a process that can occur chemically under acidic or basic conditions or enzymatically through the action of specific enzymes like esterases. While ester bonds are stable under neutral conditions, their sensitivity to pH changes makes them prone to cleavage in acidic or basic environments.

In nature, ester bonds play a pivotal role in biological systems. Triglycerides, the primary storage form of fat in animals, consist of three fatty acids esterified to a glycerol backbone, forming a critical energy reserve. Similarly, the structure of DNA and RNA relies on phosphodiester bonds, a specialized form of ester linkage, to connect nucleotides and maintain the integrity of genetic material. These biological examples highlight the significance of ester bonds in maintaining life processes.

In industrial applications, ester bonds are integral to the production of synthetic polymers such as polyesters. Polyethylene terephthalate (PET), a polymer composed of repeating ester units, is extensively used in textiles and plastic packaging. Biodegradable polyesters like polylactic acid (PLA) and polycaprolactone (PCL) also incorporate ester bonds, offering environmentally friendly alternatives for applications in compostable packaging and biomedical devices. Furthermore, esters are prominent in the fragrance and flavour industries, where their volatility and pleasant aromas make them valuable for perfumes and food flavourings, such as ethyl acetate and isoamyl acetate.

The biodegradability of ester bonds further underscores their environmental importance. These bonds can be hydrolysed by water or enzymatic activity, making ester-containing polymers ideal for sustainable applications. Polymers like PLA and PCL degrade efficiently under appropriate conditions, contributing to reduced environmental impact in products designed for composting or medical use.

Amide Bonds are another significant feature influencing biodegradability. These bonds are present in natural polymers like proteins and synthetic polyamides. Proteolytic enzymes, such as proteases, specifically target amide bonds, breaking them into peptides or individual amino acids. This enzymatic degradation is crucial in the natural recycling of biological materials and is also harnessed in biomedical applications involving protein-based biodegradable materials.

Amide bonds are a type of covalent bond formed between a carboxylic acid group *(–COOH)* and an amine group (*–NH$_2$*) through a condensation reaction, resulting in the release of a water molecule (*H$_2$O*). These bonds are characterized by the functional group *R–CONH–R'*, where *R* and *R'* represent organic groups. Amide bonds are fundamental in both biological and synthetic molecules, particularly in proteins and synthetic polyamides.

Amide bonds are created through a process called amidation, typically occurring as follows:

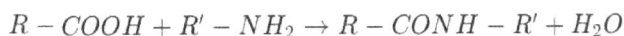

$$R - COOH + R' - NH_2 \rightarrow R - CONH - R' + H_2O$$

This reaction requires activation, as the direct reaction between a carboxylic acid and an amine is often slow. In biological systems, enzymes like aminoacyl-tRNA synthetases catalyse amide bond formation, while in industrial or laboratory settings, coupling agents (e.g., carbodiimides) are used to accelerate the reaction.

Amide bonds exhibit distinct properties that make them fundamental to both natural and synthetic systems. These bonds are inherently polar due to the electronegative oxygen and nitrogen atoms, which create a partial double-bond character. This feature imparts rigidity and planarity to the bond, contributing to the structural stability of the molecules they form. Despite their polarity, amide bonds are highly resistant to hydrolysis under normal physiological conditions. Their cleavage typically requires enzymatic action by proteases or exposure to harsh chemical environments, such as strong acids or bases. The partial double-bond character further enhances their stability, making them robust in various biological and synthetic contexts.

In nature, amide bonds are indispensable. They are the backbone of proteins and peptides, often referred to as peptide bonds in biochemical contexts. These bonds link amino acids, forming the primary structure of proteins, which are essential for the structure and function of living organisms. Additionally, amide-like linkages are present in certain components of DNA and RNA, playing a role in the complex architecture of nucleic acids.

In industrial applications, amide bonds have significant utility. Synthetic polyamides like nylon and Kevlar, formed through the polymerization of diamines and dicarboxylic acids, are well-known for their strength, durability, and wide-ranging applications, from textiles to bulletproof materials. In medicinal chemistry, amide bonds contribute to the stability and specificity of many pharmaceuticals, enabling precise interactions within biological systems.

The biodegradability of amide bonds depends on their context. In natural systems, proteolytic enzymes like proteases hydrolyse amide bonds, breaking proteins into smaller peptides or individual amino acids. This process is critical for nutrient recycling and cellular metabolism. However, in synthetic polyamides like nylon, the high stability of amide bonds makes them less prone to biodegradation. While certain microorganisms or extreme environmental conditions can facilitate their breakdown, the development of biodegradable polyamides represents an important area of research for creating environmentally sustainable materials.

Glycosidic Bonds are integral to the structure of polysaccharides like cellulose and starch. These bonds connect sugar units in the polymer chain and are readily broken down by specific enzymes such as amylases (for starch) and cellulases (for cellulose). The enzymatic cleavage of glycosidic bonds releases glucose or smaller sugar molecules, which microorganisms can easily assimilate and metabolize. This characteristic makes polysaccharide-based polymers highly biodegradable and suitable for applications like compostable packaging and agricultural films.

A glycosidic bond is a covalent bond that links a carbohydrate (sugar molecule) to another molecule, which may be another carbohydrate, a protein, a lipid, or a non-carbohydrate moiety. These bonds are fundamental in forming complex carbohydrates and various biomolecules. They are established through a dehydration reaction, where a hydroxyl group (–OH) from one sugar reacts with a hydrogen atom from another molecule, resulting in the removal of a water molecule.

The structure and formation of glycosidic bonds typically involve the hydroxyl group of the anomeric carbon (carbon-1 in glucose, for instance) of one sugar molecule reacting with a hydroxyl group or another functional group from a second molecule. The orientation of the bond determines its classification. Alpha (α) glycosidic bonds form when the hydroxyl group on the anomeric carbon is below the plane of the sugar ring, as seen in maltose, which features an α(1→4) glycosidic bond between two glucose molecules. In contrast, beta (β) glycosidic bonds occur when the hydroxyl group on the anomeric carbon is above the plane of the sugar ring, exemplified by lactose, which is formed by a β(1→4) glycosidic bond between galactose and glucose.

Glycosidic bonds can also be categorized by their chemical composition. O-glycosidic bonds involve the oxygen atom of the hydroxyl group on the anomeric carbon linking to another hydroxyl group. N-glycosidic bonds, found in nucleotides, connect the anomeric carbon of a sugar to a nitrogen atom in a base, as seen in DNA and RNA. C-glycosidic bonds, although less common, form directly between the carbon atom of the sugar and another carbon atom. S-glycosidic bonds are rare and involve sulphur replacing oxygen in the linkage.

Glycosidic bonds play essential roles in biological systems and synthetic applications. In polysaccharides like starch, glycogen, and cellulose, glycosidic bonds serve as the primary linkages. Starch and glycogen contain α(1→4) and α(1→6) glycosidic bonds, while cellulose has β(1→4) bonds, making it resistant to enzymatic degradation by most organisms. In nucleotides and nucleic acids, N-glycosidic bonds link sugar molecules to nitrogenous bases, forming the backbone of DNA and RNA. Glycoproteins and glycolipids rely on glycosidic bonds to attach sugars to proteins or lipids, which are critical for cell signalling, immunity, and structural integrity. Disaccharides like sucrose, lactose, and maltose also feature glycosidic bonds between two monosaccharides.

The hydrolysis of glycosidic bonds is a vital process for breaking down carbohydrates into simpler forms. This reaction involves the addition of water to cleave the bond and is facilitated by specific enzymes. Amylase catalyses the breakdown of α-glycosidic bonds in starch and glycogen, lactase hydrolyses the β-glycosidic bond in lactose, and cellulase cleaves β-glycosidic bonds in cellulose, although this enzyme is not produced by humans.

The Carbon Backbone vs. Heteroatoms distinction is critical in determining a polymer's degradability. Polymers with purely carbon backbones, such as polyethylene and polypropylene, are highly resistant to biodegradation because the absence of reactive sites makes them inert to hydrolytic or enzymatic attack. In contrast, polymers incorporating heteroatoms like oxygen or nitrogen in their backbone—such as polyesters and polyamides—are more readily degradable. These heteroatoms introduce polar functional groups, increasing the polymer's affinity for water and facilitating enzymatic interactions that accelerate the breakdown process.

The distinction between carbon backbones and heteroatoms in polymer chemistry is crucial for understanding the structural and functional diversity of polymers, particularly their chemical properties and biodegradability. A carbon backbone refers to a polymer chain primarily composed of carbon-carbon (C-C) bonds. These backbones are characteristic of non-polar, hydrophobic polymers such as polyethylene, polypropylene, and polystyrene. The simplicity and stability of carbon-carbon bonds make these polymers resistant to chemical and biological degradation. Their lack of reactive sites and non-polar nature reduces their interactions with water or enzymes, contributing to their durability. However, this also leads to environmental persistence, posing significant challenges for biodegradability.

In contrast, heteroatoms are atoms other than carbon that are incorporated into the polymer backbone or side chains, such as oxygen, nitrogen, sulphur, or phosphorus. Polymers containing heteroatoms, such as polyesters, polyamides, and polyurethanes, often feature polar groups or bonds like ester, amide, or ether groups. These polar bonds introduce chemical reactivity and susceptibility to enzymatic attack, significantly influencing the polymer's physical and chemical properties.

The differences between these types of polymers have several key implications. In terms of chemical reactivity, carbon backbone polymers are chemically inert due to their stable C-C bonds, making them less reactive to acids, bases, and enzymes. On the other hand, polymers with heteroatoms often contain functional groups that enable chemical reactions such as hydrolysis or oxidation. For example, ester bonds in polyesters can undergo hydrolysis, while amide bonds in polyamides can be cleaved by proteolytic enzymes.

Polarity and solubility also differ significantly. Carbon backbone polymers are generally non-polar and hydrophobic, which contributes to their insolubility in water. In contrast, heteroatom-containing polymers are more polar, making them compatible with water and other polar solvents. This property is particularly advantageous in applications like biodegradable plastics and hydrogels.

Mechanical properties also vary based on polymer structure. Carbon backbone polymers, such as polyethylene, are typically flexible and exhibit high tensile strength, making them suitable for applications like packaging and structural materials. Polymers with heteroatoms may have varied mechanical properties depending on the specific heteroatoms and functional groups present. For instance, polyamides like nylon are strong and durable, while polyesters like polylactic acid tend to be more brittle.

Biodegradability is another critical distinction. Polymers with purely carbon backbones are generally non-biodegradable because they lack bonds that can be readily cleaved by biological processes. Heteroatom-containing polymers, however, are more likely to be biodegradable due to the presence of polar groups, such as ester or amide bonds, that are susceptible to enzymatic or chemical breakdown. This characteristic makes these polymers particularly important in environmentally friendly materials and biomedical applications.

The applications of these polymers reflect their structural differences. Carbon backbone polymers are used in applications requiring durability and chemical resistance, such as plastic bags, containers, and insulation materials. Heteroatom-containing polymers, by contrast, are preferred in applications demanding biodegradability or specific chemical functionalities, including medical implants, biodegradable packaging, and high-performance engineering materials.

Mechanisms of Polymer Biodegradation

The biodegradation of polymers is a complex, sequential process that transforms synthetic or natural polymers into environmentally benign substances like carbon dioxide, water, methane, and biomass. This degradation involves a combination of chemical, enzymatic, and biological mechanisms that work together to break down polymer structures.

Hydrolysis is often the initial step in polymer biodegradation. During this stage, water molecules interact with the polymer chains, breaking them into smaller fragments. This process can occur chemically or be facilitated by biological catalysts such as enzymes. Hydrolysis specifically targets chemical bonds within the polymer, such as ester, amide, or glycosidic bonds, which are particularly susceptible to cleavage in the presence of water. Chemical hydrolysis often takes place under acidic or basic conditions, where protons or hydroxyl ions catalyse the bond cleavage, accelerating degradation. Enzymatic hydrolysis involves enzymes like esterases, proteases, and glycosidases, which specifically target and cleave particular bonds. For instance, esterases break ester bonds in polyesters, while proteases act on amide bonds in protein-based polymers. This step reduces the molecular weight of the polymer, increasing its surface area and making it more accessible for further degradation.

The next stage, enzymatic degradation, is a biologically driven process where enzymes secreted by microorganisms attack polymer chains and break them into smaller fragments such as oligomers, dimers, or monomers. This stage is particularly crucial for polymers that do not degrade efficiently through chemical hydrolysis alone. Specific enzymes like lipases, cellulases, and amylases play vital roles in this process. For example, lipases degrade aliphatic polyesters like polyhydroxyalkanoates (PHAs), cellulases break glycosidic bonds in cellulose-based polymers, and amylases hydrolyse starch into maltose and glucose. The specificity of these enzymes ensures that degradation occurs only at targeted functional groups or regions within the polymer, making the process efficient while preserving other materials in mixed environments. Enzymatic degradation produces fragments small enough for microorganisms to assimilate, setting the stage for the next phase.

Microbial assimilation completes the biodegradation process. Once the polymer chains are fragmented into smaller molecules, microorganisms such as bacteria and fungi absorb these degradation products as sources of carbon and energy. These smaller molecules are transported across microbial cell membranes and metabolized through pathways like glycolysis, the tricarboxylic acid cycle (TCA cycle), or the β-oxidation pathway. These metabolic processes convert the breakdown products into carbon dioxide, methane (under anaerobic conditions), water, and microbial biomass. The byproducts of microbial assimilation are non-toxic and naturally occurring, making this stage environmentally friendly. In composting environments, the primary byproducts are carbon dioxide and water vapor, while anaerobic conditions produce methane.

The stages of polymer biodegradation—hydrolysis, enzymatic degradation, and microbial assimilation—are highly interdependent. Microbial assimilation relies on the efficiency of enzymatic degradation to produce bioavailable molecules. Similarly, enzymatic degradation depends on hydrolysis to expose susceptible bonds and functional groups in the polymer matrix. The efficiency and speed of these mechanisms are influenced by several factors, including the chemical structure, molecular weight, crystallinity, and functional groups of the polymer. Environmental conditions like temperature, pH, moisture, and microbial diversity also play crucial roles. Polymers with larger surface areas degrade faster due to increased exposure to water and enzymes. Additives in polymers can enhance or inhibit degradation depending on their composition and interactions with the polymer.

Production Methods for Bioplastics

The production of bioplastics involves a range of processes, depending on the type of bioplastic and its intended application. Bioplastics can be derived from natural sources such as plants, microorganisms, or animal products, and their production typically involves biological, chemical, or industrial synthesis.

Biological Fermentation

Biological fermentation is a pivotal process in the production of bioplastics, particularly through the conversion of renewable feedstocks into biopolymers such as Polyhydroxyalkanoates (PHAs) and Polylactic Acid (PLA). This process utilizes microorganisms to transform substrates like sugarcane, corn, and vegetable oils into valuable bioplastics, thereby promoting sustainability and reducing reliance on fossil fuels.

PHAs are a family of biodegradable polyesters synthesized by bacteria as storage compounds during fermentation. Various studies have demonstrated that bacteria such as Alcaligenes sp. can effectively utilize sugar industry waste to produce PHAs, with nutrient concentration playing a crucial role in maximizing yield [434]. Additionally, the metabolic engineering of microorganisms has been explored to enhance PHA production from alternative substrates, including waste frying oils, which can serve as a sustainable feedstock [437]. This highlights the versatility and potential of microbial fermentation in producing environmentally friendly materials.

On the other hand, PLA is produced through the fermentation of plant-based sugars into lactic acid, which is then polymerized to form PLA. The predominant method for lactic acid production is microbial fermentation, accounting for nearly 90% of the total production [438]. Research has shown that various strains of Lactobacillus can be immobilized to improve lactic acid yield from renewable feedstocks, such as agricultural waste [439]. The fermentation process can be optimized by controlling parameters like temperature and pH, which are critical for maximizing lactic acid production and subsequent PLA synthesis [440].

The environmental benefits of biological fermentation are significant, as it utilizes renewable resources and generates fewer toxic byproducts compared to traditional petrochemical processes. However, achieving optimal fermentation conditions requires precise control over various factors, including nutrient supply and microbial strain selection [441]. For instance, the production of butyric acid from renewable feedstocks has been shown to support rural economies while providing a sustainable alternative to petroleum-derived products [442]. This underscores the importance of fermentation technology in advancing bioplastic production and contributing to a circular economy.

Starch Processing

Starch processing plays a crucial role in the production of starch-based bioplastics, which are derived from natural sources such as corn, potatoes, and rice. These bioplastics are increasingly favoured due to their biodegradability and sustainability compared to traditional petroleum-based plastics. The initial step in starch processing involves gelatinization, where starch is mixed with plasticizers like glycerol or sorbitol and subjected to heat. This process disrupts the crystalline structure of starch, transforming it into a thermoplastic material suitable for various applications [443]. The gelatinization process is critical as it enhances the starch's ability to be molded and shaped, making it more versatile for manufacturing [443].

Following gelatinization, blending is often employed to enhance the mechanical properties and flexibility of thermoplastic starch (TPS). Blending TPS with other biodegradable polymers, such as polylactic acid (PLA) or polycaprolactone (PCL), can significantly improve its performance. For instance, studies have shown that the mechanical properties of starch-based blends are influenced by the dispersion of starch within the polymer matrix, which affects the microstructure and interfacial interactions of the materials [444, 445]. The incorporation of starch into polymers like low-density polyethylene (LDPE) creates a polar phase within a nonpolar matrix, which can lead to improved properties if strong interfacial interactions are established [444]. However, without proper blending, starch-based bioplastics can exhibit brittleness, limiting their practical applications [446].

The simplicity and cost-effectiveness of starch processing are significant advantages, making it an attractive option for producing biodegradable materials. However, the inherent brittleness of unblended starch-based bioplastics poses challenges. Research indicates that the mechanical performance of these materials can be enhanced through the careful selection of blending ratios and the use of additives [445]. For example, the ratio of starch to other polymers can influence both the mechanical properties and the biodegradability of the final product, with higher starch content often leading to increased biodegradability due to enhanced microbial accessibility [447].

Chemical Synthesis

Chemical synthesis is a critical process in the conversion of monomers derived from biological sources into polymers, particularly in the production of bioplastics. This synthesis often involves complex industrial chemical reactions that can be finely controlled to yield high-quality materials with tailored properties. Two notable examples of bioplastics produced through chemical synthesis are polylactic acid (PLA) and bio-based polyethylene (Bio-PE).

PLA is synthesized from lactic acid, which is typically produced through the fermentation of carbohydrates. Following fermentation, lactic acid undergoes polymerization through either condensation or ring-opening methods to form PLA. This process allows for the creation of biodegradable plastics that can serve as alternatives to conventional petroleum-based plastics. The polymerization process is crucial as it determines the molecular weight and properties of the resulting PLA, impacting its application in various fields such as packaging and biomedical devices [448].

On the other hand, bio-based polyethylene (Bio-PE) is produced from ethanol derived from renewable sources such as sugarcane or corn. The production process involves the dehydration of ethanol to form ethylene, which is then polymerized to create polyethylene. Notably, Bio-PE is chemically identical to its fossil-fuel-based counterpart, making it an attractive option for reducing the carbon footprint associated with plastic production. The conversion of ethanol to ethylene can be achieved through various catalytic processes, with zeolite catalysts like H-ZSM-5 being particularly effective in enhancing the yield and efficiency of this transformation [449-451].

The precision in controlling the polymerization process is a significant advantage of chemical synthesis. This control enables the production of bioplastics with specific mechanical and thermal properties, which can be tailored for particular applications. For instance, the choice of catalyst and reaction conditions can influence the molecular weight distribution and crystallinity of the polymers, thereby affecting their biodegradability and performance [452, 453]. Furthermore, advancements in metabolic engineering and synthetic biology are expanding the potential for producing a wider variety of bioplastics from renewable resources, enhancing the sustainability of plastic production [453].

Extrusion and Injection Molding

The shaping and processing of bioplastics into finished products commonly utilize industrial techniques such as extrusion and injection molding. These methods are integral to the production of various bioplastic applications, including packaging materials and consumer goods.

Extrusion is a prevalent technique where bioplastic pellets are melted and formed into continuous shapes such as films, sheets, or fibres. This process is particularly advantageous for producing packaging materials due to its efficiency and ability to create uniform products. For instance, Coltelli et al. [454] highlight that extrusion is widely accepted in the plastics industry and is essential for the thermo-mechanical shaping of bioplastics, including those derived from proteins and other renewable resources. Additionally, the review by Jiménez-Rosado et al. [455]confirms that extrusion is a conventional method used for transforming bioplastics like soy protein into various forms, emphasizing its role in the mass production of bioplastic products. Moreover, the versatility of extrusion allows for the incorporation of additives that enhance the properties of the final product, making it suitable for a range of applications [456].

Injection molding is another critical technique employed in the bioplastics industry, where molten bioplastics are injected into molds to create complex three-dimensional objects. This method is particularly effective for mass production, allowing for the creation of intricate shapes such as containers and automotive parts. Aversa et al. [457] discuss the use of injection molding for producing bioplastic bottles, demonstrating its application in creating functional and sustainable packaging solutions. Furthermore, the study by Santana [407] emphasizes that injection molding can be adapted for various bioplastics, including those derived from seaweed, showcasing its flexibility in processing different materials. The efficiency of injection molding is underscored by its lower energy demands compared to other methods, making it a preferred choice for industrial applications [458].

Both extrusion and injection molding require optimized processing conditions to ensure the quality of bioplastic products. Factors such as temperature, pressure, and material composition must be carefully controlled to achieve the desired mechanical properties and performance characteristics. For example, the research by Capezza [459] indicates that the processing techniques for bioplastics must align with those used for traditional plastics to ensure compatibility and performance. Additionally, the work by Gurram et al. [458] highlights that the choice of processing method can significantly affect the final properties of bioplastic films, further emphasizing the need for optimization in processing conditions.

Biocatalysis

Biocatalysis has emerged as a pivotal technique in the synthesis and modification of bioplastics, leveraging the unique properties of enzymes to catalyse polymerization reactions and modify existing polymers. Enzymes such as lipases play a crucial role in the polymerization of monomers into biodegradable polyesters, including polycaprolactone (PCL). For instance, lipase-catalysed reactions have been extensively studied for their ability to facilitate the synthesis of polyesters through ring-opening polymerization and condensation reactions, showcasing the potential of enzymatic processes in producing environmentally friendly materials [460, 461]. The enzymatic approach not only enhances the efficiency of polymer synthesis but also contributes to the development of bioplastics that are biodegradable and derived from renewable resources [462].

Biocatalysis is instrumental in the modification of existing polymers to improve their properties for specific applications. Enzymes can alter the functional groups of natural polymers such as cellulose and starch, enhancing their functionality and applicability in various fields. For example, enzymatic modifications have been successfully employed to improve the dyeability of polyester fabrics, demonstrating the ability of enzymes to introduce new functional groups without compromising the integrity of the base polymer [463]. Additionally, the enzymatic modification of starch has been highlighted as a method to enhance its properties for bioplastic applications, allowing for tailored functionalities that meet specific industrial requirements [464].

Despite the advantages of biocatalysis, including high specificity and eco-friendliness, challenges remain regarding the cost and stability of enzymes. The economic feasibility of enzyme-catalysed processes is often hindered by the high costs associated with enzyme production and the need for stable enzyme formulations that can withstand industrial conditions [465]. Nevertheless, ongoing research is focused on optimizing enzyme use, such as through immobilization techniques that can reduce costs and enhance enzyme stability during polymerization and modification processes [466].

Genetically Engineered Organisms

Advances in genetic engineering have significantly impacted the production of bioplastics through the modification of microorganisms and plants. Microbial production of bioplastics, particularly polyhydroxyalkanoates (PHAs) and polylactic acid (PLA), has been enhanced by genetically modifying bacteria and yeast. These engineered microorganisms can exhibit improved yield and reduced production costs, making them viable alternatives to traditional petroleum-based plastics. For instance, Al-Khairy et al. [467] highlight that microbial fermentation products, such as PHAs and PLA, are among the most popular bioplastics produced through genetic engineering, demonstrating the potential for sustainable production methods. Additionally, Dey et al. [468] emphasize the sustainability benefits of bioplastics derived from renewable resources, including those produced by genetically modified microorganisms.

In the realm of plant-based bioplastics, genetic engineering has enabled crops to be modified to produce biopolymers such as starch and cellulose in greater quantities and with improved quality. Yali [469] discusses the implications of genetically modified organisms (GMOs) in agriculture, noting that these modifications can lead to enhanced traits that support biopolymer production. Furthermore, Liu et al. [470] point out that genetically modified plants can be engineered to confer increased resistance to pests and diseases, which can indirectly enhance the production of bioplastics by ensuring higher crop yields and better quality raw materials. The use of genetically modified crops for bioplastic production is further supported by Haile et al. [471], who discuss the various advantages of GM crops, including improved nutritional value and enhanced agronomic traits.

While the potential of genetically engineered systems for bioplastic production is substantial, these advancements come with significant investment requirements and ethical and regulatory challenges. Turnbull et al. [472] provide an overview of the regulatory landscape surrounding genetically modified crops, highlighting the complexities involved in assessing the risks and benefits of such technologies. Additionally, Wadood et al. [473] discuss the economic benefits of genetic engineering in agriculture, while also acknowledging the cultural and ethical considerations that accompany the adoption of GMOs. These factors underscore the need for careful consideration of the implications of genetic engineering in the context of bioplastic production.

Thermochemical Conversion

Thermochemical conversion is a critical process in the production of bioplastics from biomass, utilizing methods such as pyrolysis and gasification to break down organic materials into valuable monomers and intermediates. Pyrolysis involves the thermal decomposition of biomass in the absence of oxygen, resulting in the formation of bio-oils that can be further processed into monomers like lactic acid or ethylene. Research indicates that the yield and properties of bio-oil can vary significantly based on the type of biomass and the pyrolysis conditions employed. For instance, studies have shown that fast pyrolysis can yield bio-oil from various feedstocks, such as rice husks and sugarcane bagasse, with bio-oil yields reaching approximately 70% of the biomass weight [474, 475].

The composition of bio-oil is influenced by several factors, including the type of biomass, moisture content, and pyrolysis temperature. For example, the bio-oil derived from different biomass sources exhibits varying acidity levels, which can affect its suitability for further chemical processing [474, 476]. Additionally, the use of catalysts in pyrolysis has been shown to enhance both the yield and quality of bio-oil, making catalytic pyrolysis an attractive option for improving the efficiency of biomass conversion [477, 478].

Gasification, another thermochemical conversion method, transforms biomass into synthesis gas (syngas), a mixture of carbon monoxide and hydrogen. This syngas can then be utilized to produce various chemicals and fuels, including those necessary for bioplastic synthesis. The gasification process allows for the conversion of non-food biomass and waste materials into valuable products, thereby contributing to sustainable production practices [479, 480]. The flexibility of gasification in handling different biomass types and its ability to produce high-energy syngas make it a promising technology for the future of renewable energy and bioplastics [479].

Testing Procedures

Testing procedures for bioplastics are essential for certifying their environmental claims, such as compostability, biodegradability, and biobased content. These tests ensure that bioplastics meet specific regulatory standards and provide credible information to manufacturers, consumers, and environmental agencies.

Industrial Compostability - EN 13432 and ASTM D6400: In the European Union, the EN 13432 standard governs the compostability of plastic products. To claim industrial compostability, plastics must meet several stringent criteria, including the disintegration of the material within 12 weeks and the biodegradation of polymeric ingredients into carbon dioxide within 180 days. Additional tests assess plant toxicity and the presence of heavy metals to ensure safety and environmental compliance. The United States has a similar standard, ASTM D6400, which outlines comparable requirements. These certifications are commonly achieved by starch-based plastics, PLA-based plastics, and certain aliphatic-aromatic co-polyesters like succinates and adipates. However, additive-based bioplastics marketed as photodegradable or oxo-degradable do not currently comply with these standards.

Compostability - ASTM D6002: ASTM D6002 was a method for determining the compostability of plastics, defining compostable materials as those that biologically decompose into carbon dioxide, water, inorganic compounds, and biomass at a rate consistent with known compostable materials. This definition faced criticism for focusing on the disappearance of material rather than the production of humus or compost, which is traditionally associated with composting. Due to these concerns, ASTM D6002 was withdrawn in 2011 and has not been replaced, leaving a gap in compostability standards for bioplastics in some contexts.

Biobased Content - ASTM D6866: The ASTM D6866 method certifies the biologically derived content of bioplastics by measuring the percentage of renewable carbon present in a material. This is achieved by analyzing the levels of carbon-14, a radioactive isotope that decays over geological timeframes. Biomass-based products contain carbon-14, whereas fossil fuel-derived materials do not. This method helps distinguish biobased plastics from petrochemical products. Importantly, biobased content is not synonymous with biodegradability. For instance, high-density polyethylene (HDPE) can be entirely biobased but is not biodegradable. Despite this, biobased non-biodegradable plastics play a crucial role in reducing greenhouse gas emissions, especially when used in energy recovery processes.

Anaerobic Biodegradability - ASTM D5511 and ASTM D5526: The ASTM D5511-12 and ASTM D5526-12 methods evaluate the biodegradability of plastics in anaerobic environments, such as landfills. These tests comply with international standards like ISO DIS 15985, providing a framework to assess how bioplastics degrade in oxygen-free conditions. Anaerobic biodegradation is essential for understanding the environmental impact of bioplastics in waste management systems where aerobic composting is not feasible.

These standardized testing methods are critical for ensuring the credibility of environmental claims made by bioplastics manufacturers. They provide benchmarks for comparing different materials, help guide product design, and inform consumers about the environmental performance of bioplastics. Additionally, these certifications foster trust and support the adoption of bioplastics in various industries, including packaging, agriculture, and construction.

Industrial Applications and Environmental Impact

Bioplastics, made from renewable materials such as starch, cellulose, wood, and sugar, offer a promising alternative to fossil-fuel-derived plastics. Their production is more sustainable, addressing global concerns about plastic pollution and environmental degradation. The environmental impact of bioplastics, however, is complex and multifaceted, encompassing both benefits and challenges.

One of the key advantages of bioplastics is their ability to significantly reduce greenhouse gas emissions and non-renewable energy use. Replacing fossil-based plastics with bioplastics in manufacturing and packaging can contribute to a smaller carbon footprint. Companies worldwide are adopting bioplastics to enhance the sustainability of their products. However, the environmental benefits of bioplastics are nuanced. While they generally perform better in terms of carbon emissions and energy use, they can contribute to eutrophication and acidification. The farming practices required to produce biomass, including the use of chemical fertilizers, lead to nitrate and phosphate runoff into water bodies. This runoff causes harmful algal blooms, creating oxygen-depleted dead zones in aquatic ecosystems and disrupting marine life.

Bioplastics also increase acidification due to emissions from fertilizer application and related agricultural activities. The production of bioplastics is linked to other environmental concerns, such as the use of pesticides, water consumption, soil erosion, and loss of biodiversity. Additionally, land use changes for biomass cultivation can lead to significant carbon emissions, reducing the environmental advantages of bioplastics. Despite these drawbacks, bioplastics exert lower human and terrestrial toxicity and carcinogenic potentials than conventional plastics. However, they may increase aquatic toxicity and stratospheric ozone depletion due to emissions of nitrous oxide during fertilizer use.

Another concern is the competition between bioplastics and food production. First-generation feedstock bioplastics rely on edible crops like corn, which could otherwise be used for food. Second-generation bioplastics address this issue by using non-food crops or waste materials, while third-generation bioplastics utilize algae, providing a more sustainable solution. The biodegradability of bioplastics is another significant aspect of their environmental impact. Bioplastics can decompose in various environments, including soil, compost, and aquatic systems, making them more acceptable than non-biodegradable plastics. Composting bioplastics is particularly effective, as it not only facilitates biodegradation but also reduces greenhouse gas emissions. However, biodegradation in water bodies can harm aquatic ecosystems, posing additional challenges.

In the construction industry, bioplastics have gained attention for their potential as eco-friendly materials. Recent advancements have improved their durability and performance, making them viable for applications like insulation, flooring, panels, and structural components. For instance, polylactic acid (PLA) and polyhydroxyalkanoates (PHA) are used in insulation materials due to their thermal properties and biodegradability. Bioplastic composites offer sustainable alternatives for flooring, wall cladding, and even formwork for concrete casting.

The benefits of bioplastics in construction are notable. They reduce carbon emissions, minimize waste, and promote energy efficiency. As the bioplastics market grows, it presents economic opportunities for manufacturers and suppliers. However, challenges such as higher production costs and limited performance in certain applications must be addressed. Ongoing research aims to enhance the properties of bioplastics, making them competitive with traditional materials.

The future of bioplastics appears promising as industries increasingly prioritize sustainability. Despite current limitations, advancements in bioplastic technology are expected to expand their applications and reduce costs, fostering their adoption across sectors. As the construction industry embraces greener practices, bioplastics are set to play a pivotal role in developing sustainable materials, contributing to a more environmentally conscious future.

Chapter 6

Material Development, Testing and Characterization

Developing a New Advanced Material

Creating a new advanced material involves a multidisciplinary approach that combines scientific research, engineering principles, and practical applications. This process typically unfolds in several stages, from conceptualization and experimentation to testing and implementation. The following outlines an approach to develop an advanced material.

1. Identifying the Need and Setting Objectives

The first step in developing an advanced material is to define its purpose. This involves identifying a specific problem or unmet need that the material will address. Common drivers for new materials include improving performance, reducing costs, enhancing sustainability, or enabling entirely new technologies.

For example:

- Developing lightweight yet strong materials for aerospace applications.

- Creating materials with enhanced thermal conductivity for electronics.

- Designing materials with biocompatibility for medical implants.

In the development of advanced materials, the initial step involves identifying specific needs and setting clear objectives that the material aims to address. This process is critical as it lays the foundation for subsequent research and development efforts. Common drivers for the creation of new materials include the need to improve performance, reduce costs, enhance sustainability, and enable innovative technologies. For instance, the aerospace industry often seeks lightweight yet strong materials to enhance fuel efficiency and performance [481], while the electronics sector requires materials with improved thermal conductivity to manage heat dissipation effectively [482]. Additionally, the medical field demands biocompatible materials for implants to ensure patient safety and promote healing [483, 484].

The identification of unmet needs can be guided by sustainability considerations, as highlighted by Danso [485], who emphasizes the importance of environmental, social, and economic indicators in the selection of construction materials. This approach is echoed in the work of Gunansyah [486], who discusses the broader implications of sustainable development in material selection, suggesting that economic growth must be balanced with environmental stewardship. Such frameworks are essential in guiding the development of materials that not only meet performance criteria but also contribute positively to sustainability goals.

Moreover, specific applications illustrate the diverse objectives that advanced materials can fulfill. For example, the development of biodegradable magnesium alloys for orthopaedic implants aims to enhance biocompatibility and promote bone healing while minimizing long-term foreign body reactions [487]. Similarly, the exploration of polymeric materials, such as polymethyl methacrylate (PMMA), demonstrates the need for coatings that maintain the functionality of medical devices while ensuring biocompatibility [488]. These examples underscore the necessity of aligning material properties with the intended application, which is a critical aspect of setting objectives in material development.

2. Researching Existing Materials and Technologies

Understanding the limitations of existing materials and exploring related technologies is crucial. This stage includes a review of scientific literature, patents, and current market offerings to ensure the new material fills a unique niche.

For instance:

- Studying carbon fibre composites before attempting to enhance their properties with nanomaterials.

- Reviewing polymers used in flexible electronics to identify shortcomings like low durability.

Researching existing materials and technologies is a critical step in the development of innovative materials, particularly in fields such as composites and flexible electronics. This process involves a thorough review of scientific literature, patents, and current market offerings to identify gaps that new materials can fill. For instance, when considering enhancements to carbon fibre composites with nanomaterials, it is essential to understand the existing properties and limitations of carbon fibre itself. Studies have shown that the mechanical properties of carbon fibre composites can be significantly influenced by factors such as fibre orientation and the ratio of different fibre types used in hybrid composites. For example, Bakar et al. [489] demonstrated that varying the carbon fibre ratio affects the impact properties of hybrid kenaf/carbon fibre reinforced epoxy composites, highlighting the importance of fibre characteristics in composite performance. Furthermore, Swolfs et al. [490] developed a global load-sharing model that predicts a hybrid effect in terms of failure strain for carbon/glass hybrids, indicating that understanding the interactions between different fibre types is crucial for optimizing composite materials.

In the realm of flexible electronics, reviewing the limitations of polymers is equally important. Hu et al. [491] noted that hydrogel-based flexible electronics currently face challenges related to their functionality, often being limited to singular applications. This limitation underscores the necessity for further research to enhance the versatility and durability of these materials. Additionally, the work of Wang et al. [492] emphasizes the need for innovative design concepts in flexible electronics, particularly in developing stretchable conductors and semiconductors that do not compromise electrical performance. The integration of conductive polymers into flexible devices has been a focal point, as these materials offer tunable electrical conductivity and flexibility, which are essential for wearable technologies [493]. However, the durability of these polymers remains a concern, as highlighted by Cheng et al. [494], who explored the integration of conductive fillers into hydrogels to improve their mechanical properties and conductivity.

The exploration of new materials must also consider the recyclability and thermal stability of substrates used in flexible electronics. Chen et al. [495] discussed the importance of selecting polymer substrates that can withstand various fabrication processes while maintaining their structural integrity. This aspect is crucial as the demand for sustainable materials in electronics continues to grow. The development of self-healing materials, as investigated by Wang et al. [496], also presents a promising avenue for enhancing the longevity and functionality of flexible electronics.

3. Conceptual Design

Based on the identified objectives and research, scientists and engineers propose theoretical models for the new material. This includes defining its composition, structure, and desired properties. Computational tools, such as materials simulation software, are often used to predict how different combinations of components will behave.

In the conceptual design phase of new materials, scientists and engineers develop theoretical models that define the composition, structure, and desired properties of the materials. This process is crucial as it lays the groundwork for subsequent experimental validation and application. The integration of computational tools, particularly materials simulation software, plays a pivotal role in predicting the behaviour of various material combinations under different conditions.

Computational materials discovery has become a cornerstone of modern materials science, enabling researchers to explore a vast array of compounds and their properties. Asta [497] highlights that high-throughput computations and first-principles techniques facilitate the creation of materials databases, allowing designers to screen for materials with optimal properties for specific applications. This computational approach not only accelerates the discovery process but also enhances the understanding of how different compositions and structures influence material performance.

The application of machine learning techniques in materials science has gained traction, particularly in analysing microstructural features and their correlation with material properties. For instance, Exl et al. [498] discuss the use of machine learning to predict stress hot-spots in polycrystalline metals by correlating microstructural features with mechanical behaviour. This data-driven approach complements traditional computational methods and provides deeper insights into the material design process.

Theoretical modelling is essential for understanding the mechanical properties of composite materials. Durbaca et al. [499] emphasize the importance of defining theoretical models for composite plates, which involves evaluating their structural elements and physical-mechanical characteristics. Such models are vital for optimizing the design and application of composite materials in various industrial contexts. Additionally, El-Azab notes that the development of rigorous models for materials structure and behaviour has been integral to advancing materials science, particularly in conjunction with experimental methods [500].

Furthermore, numerical simulations, including finite element analysis (FEA), are widely employed to predict the mechanical behaviour of composite materials. For instance, Tao et al. [501]propose a three-dimensional micromechanical model to calculate the mechanical properties of cemented particulate composites, showcasing the utility of theoretical methods in practical applications. The integration of numerical simulations with theoretical models allows for more accurate predictions of material performance, which is crucial for effective material design.

4. Synthesis and Fabrication

Once the conceptual design is finalized, the material is synthesized. This may involve:

- **Chemical Methods**: Creating new molecules or compounds through chemical reactions (e.g., polymerization, sol-gel processes).

- **Physical Methods**: Using physical deposition or mechanical processing techniques like melt processing, extrusion, or 3D printing.

- **Hybrid Methods**: Combining multiple synthesis techniques to create complex structures, such as using nanomaterials to reinforce traditional composites.

During this stage, material samples are fabricated for evaluation. Techniques like additive manufacturing or molecular self-assembly may be employed for novel designs.

The synthesis and fabrication of materials is a critical phase in material science, involving various methodologies that can be categorized into chemical, physical, and hybrid methods. Each approach has unique advantages and applications, which are essential for the development of novel materials.

Chemical Methods: Chemical synthesis techniques are fundamental in creating new molecules or compounds through reactions such as polymerization and sol-gel processes. For instance, the wet chemical method is particularly favoured for its ability to produce high-purity nanocrystalline powders with homogeneous molecular mixing at low processing temperatures, which is crucial for applications requiring specific material properties [502]. Additionally, the soft chemical synthesis methods, such as the water-assisted solid-state reaction (WASSR), allow for the synthesis of nanosized materials with minimal mechanical load, enhancing the efficiency and quality of the produced materials [503]. The versatility of chemical methods is further exemplified by their application in synthesizing nanoparticles, where biological methods have emerged due to their cost-effectiveness and environmental friendliness [504].

Physical Methods: Physical synthesis techniques, including melt processing, extrusion, and 3D printing, are pivotal for material fabrication. The advent of additive manufacturing has revolutionized the production of complex geometries and structures, allowing for the precise control of material properties [505]. For example, carbon nanostructures can be produced using various 3D printing technologies, which facilitate the creation of high-quality nanoproducts [505]. Furthermore, mechanical processing techniques such as chemical vapor deposition (CVD) are widely used for synthesizing materials like graphene, which is essential for numerous applications due to its outstanding electrical and mechanical properties [506].

Hybrid Methods: Hybrid synthesis methods combine multiple techniques to enhance material properties and functionalities. For instance, the integration of nanomaterials into traditional composites can significantly improve their mechanical strength and thermal stability [507]. The controlled synthesis of organic-inorganic hybrid materials has gained traction due to their applicability in fields such as biosensing and advanced materials [508]. Moreover, the use of machine learning to predict optimal synthesis conditions and material properties exemplifies the innovative approaches being adopted in material synthesis [509].

The synthesis and fabrication of materials involve a diverse array of methods, each contributing uniquely to the development of advanced materials. Chemical methods provide precision and control over material properties, physical methods enable the production of complex structures, and hybrid methods leverage the strengths of multiple techniques to create superior materials.

5. Characterization

The material is analysed to determine if it meets the desired properties. Various characterization techniques are employed to assess its physical, chemical, mechanical, thermal, and electrical properties. Common tools include:

- Scanning Electron Microscopy (SEM) and Transmission Electron Microscopy (TEM) for structural analysis.

- X-Ray Diffraction (XRD) for crystallographic studies.

- Dynamic Mechanical Analysis (DMA) for mechanical properties.

- Thermogravimetric Analysis (TGA) for thermal stability.

Characterization of materials is a critical aspect of materials science, as it allows researchers to determine whether a material meets the desired properties through various analytical techniques. The characterization process typically involves assessing physical, chemical, mechanical, thermal, and electrical properties using a range of sophisticated tools.

One of the primary techniques employed in structural analysis is Scanning Electron Microscopy (SEM) and Transmission Electron Microscopy (TEM). SEM provides high-resolution images of the surface morphology of materials, while TEM is invaluable for examining the internal structure at atomic resolution. For instance, Cha et al. [510] highlight the significance of TEM in investigating delicate materials, emphasizing its role in providing detailed imaging and analytical capabilities essential for materials characterization. Furthermore, advancements in high-resolution scanning/transmission electron microscopy (HR-S/TEM) have enabled researchers to visualize atomic arrangements and determine the position of individual atoms with remarkable precision, as noted by Leer et al. [511]. This capability is crucial for understanding the microstructural characteristics of materials, as demonstrated by Wang et al. [512], who utilized TEM to study the morphology of graphene and other two-dimensional materials at atomic scales.

In addition to structural analysis, X-Ray Diffraction (XRD) is a fundamental technique used for crystallographic studies. XRD allows researchers to determine the crystalline structure of materials by analysing the diffraction patterns produced when X-rays interact with the material's atomic lattice. This technique is essential for identifying phase compositions and understanding the crystallographic properties of materials, as noted by Fultz and Howe, who discuss the importance of XRD in the context of materials science [513].

Dynamic Mechanical Analysis (DMA) is another critical technique used to assess the mechanical properties of materials. DMA measures the material's response to oscillatory stress, providing insights into its viscoelastic behaviour. This technique is particularly useful for polymers and composites, where understanding the mechanical response under varying temperatures and frequencies is vital for applications in engineering and materials design, as discussed by Velez et al. [514].

Thermogravimetric Analysis (TGA) is employed to evaluate the thermal stability of materials by measuring weight changes as a function of temperature. This technique is crucial for understanding the thermal degradation and stability of materials under different conditions. For example, Anber et al. [515] discuss the use of in situ TEM heating experiments to assess the thermal stability of high entropy alloys, illustrating the importance of combining TGA with advanced microscopy techniques for comprehensive material characterization.

6. Testing and Optimization

After initial characterization, the material undergoes application-specific testing to evaluate its performance under real-world conditions. For example:

- Testing a new alloy under extreme temperatures for aerospace applications.

- Measuring the wear resistance of a polymer in automotive parts.

Based on test results, the material may require further optimization. This iterative process involves tweaking the synthesis methods or composition to fine-tune its properties.

In the field of materials science, the iterative process of testing and optimization is crucial for ensuring that materials meet specific performance criteria for various applications. After initial characterization, materials undergo rigorous testing to evaluate their performance under real-world conditions. For instance, aerospace applications often require testing of new alloys under extreme temperatures to assess their thermal stability and mechanical integrity. Sorgente and Tricarico [516] highlight that the characterization of superplastic aluminium alloys involves iterative optimization techniques that minimize discrepancies between experimental results and numerical simulations, demonstrating the importance of precise material parameter identification in high-stakes applications such as aerospace. Similarly, Dai and Ni [517] emphasize that reliable testing methods are essential for accurately assessing material properties, particularly in extrusion processes where flow stress and friction factors are critical for performance.

In automotive applications, the wear resistance of polymers is a significant concern, as these materials are subjected to harsh conditions that can lead to degradation. The iterative nature of testing allows for the refinement of material compositions to enhance their durability. For example, Iyer et al. [518] discuss a data-centric Bayesian optimization framework that integrates experimental data to guide the design of materials with improved properties, which is particularly relevant for automotive components where wear resistance is paramount. Additionally, the work of Schmaltz and Willner [519] illustrates the use of finite element model updating strategies to identify optimal material behaviours, which can be applied to enhance the wear resistance of polymers through iterative testing and optimization

The optimization process itself often involves tweaking synthesis methods or compositions based on test results. This iterative refinement is essential for achieving desired material properties. For instance, the research by Cao [520] introduces a sequential density-based approach for layout optimization of multi-material systems, which can be adapted to refine material properties through iterative adjustments. Furthermore, the development of advanced optimization algorithms, such as those discussed by Zhang et al. [521], demonstrates the potential for improving material performance through adaptive iterative learning control methods. These methodologies not only enhance the material properties but also streamline the testing process, making it more efficient and effective.

7. Scaling Up Production

Once the material meets the desired specifications, the next step is to scale up its production. This involves designing manufacturing processes that are cost-effective, scalable, and environmentally sustainable. Key considerations include:

- Availability of raw materials.

- Energy requirements.

- Waste management and environmental impact.

Scaling up production is a critical phase in the manufacturing process that requires careful consideration of various factors to ensure efficiency, cost-effectiveness, and environmental sustainability. The transition from small-scale to large-scale production involves designing manufacturing processes that not only meet market demands but also adhere to sustainable practices. Key considerations in this process include the availability of raw materials, energy requirements, and waste management.

The availability of raw materials is a fundamental aspect of scaling up production. Sustainable manufacturing practices emphasize the importance of sourcing materials that are not only cost-effective but also environmentally friendly. For instance, Alayón et al. [522] highlight that the adoption of sustainable production principles is essential for manufacturers to align their operations with environmental goals, which includes the use of renewable resources and eco-friendly materials. Furthermore, the role of government regulations in promoting sustainable sourcing is significant, as stricter policies can drive manufacturers to adopt more sustainable practices [523].

Energy requirements are another critical consideration during the scale-up process. Manufacturers must evaluate their energy consumption and seek ways to optimize it to reduce costs and environmental impact. Research indicates that integrating energy efficiency into manufacturing processes can lead to significant improvements in sustainability performance [524]. For example, the implementation of energy-efficient technologies not only lowers operational costs but also minimizes the carbon footprint of manufacturing activities [525]. Additionally, the use of renewable energy sources is increasingly recognized as a viable strategy for enhancing sustainability in manufacturing [526].

Waste management is integral to scaling up production sustainably. Effective waste management strategies can significantly reduce the environmental impact of manufacturing processes. The literature suggests that sustainable manufacturing practices, such as recycling and waste minimization, are essential for reducing the ecological footprint of production [527]. Moreover, a comprehensive approach to waste management involves not only the reduction of waste at the source but also the development of systems for reusing materials and recycling by-products [528]. This holistic view aligns with the principles of life cycle assessment (LCA), which evaluates the environmental impacts of a product throughout its life cycle, from raw material acquisition to end-of-life disposal [528].

8. Integration and Application

The material is then integrated into prototypes or systems to assess its functionality in practical applications. Collaborating with industry partners is often essential at this stage, as they can provide insights into implementation challenges and market needs.

The integration of materials into prototypes or systems is a critical phase in the development process, allowing for the assessment of functionality in practical applications. This stage often necessitates collaboration with industry partners, who provide valuable insights into implementation challenges and market needs. Such partnerships can enhance the innovation process by leveraging external knowledge and resources, thereby facilitating more effective integration of new technologies into existing systems.

Collaboration between universities and industry is particularly significant in this context. Gretsch et al. [529] highlight that partnerships with both industrial and academic entities can support the success of front-end innovation, especially when the degree of innovativeness is high. This assertion is reinforced by Moellers et al. [530], who discuss the advantages of internal collaborations within multi-business firms, emphasizing that these collaborations can enhance knowledge flows and address implementation challenges more effectively than external open innovation practices. The importance of understanding the dynamics of these collaborations is further supported by the findings of Hydle et al. [531], who illustrate that the relationships between entrepreneurial firms and incumbents are complex and require relational capabilities for successful digital transformation.

Prototyping serves as a vital tool in this integration process, allowing designers to explore and refine concepts before full-scale implementation. Deininger et al. [532] note that novice designers often face challenges in utilizing prototypes effectively, which can hinder the identification of design flaws. This is echoed by Camburn et al. [533], who provide a comprehensive review of design prototyping methods, emphasizing the importance of iterative prototyping in addressing design objectives and refining concepts. The role of prototypes as a means of communication and proof of concept is also highlighted by Coutts et al. [534], who assert that prototypes embody key functionalities that can be tested and validated.

The iterative nature of prototyping is crucial for fostering collaboration among multiple stakeholders. Mahtani et al. [535] emphasize that involving various stakeholders throughout the design process enhances collaborative prototyping and co-creation, ultimately leading to more effective design outcomes. This collaborative approach is further supported by the work of Suteja and Hadiyat [536], who discuss the optimization of rapid prototyping methods to improve material removal rates and dimensional accuracy, underscoring the technical challenges that can arise during the prototyping phase.

9. Safety and Regulatory Compliance

Before commercialization, the material must undergo rigorous safety testing to ensure it complies with industry standards and regulatory requirements. This is particularly critical for materials used in medical, aerospace, or consumer products.

Before commercialization, materials intended for use in sensitive applications such as medical devices, aerospace components, or consumer products must undergo rigorous safety testing to ensure compliance with industry standards and regulatory requirements. This process is critical to mitigate risks associated with material failure or adverse reactions in end-users, particularly in medical applications where biocompatibility is paramount.

The ISO 10993 standards play a significant role in ensuring the biocompatibility of medical devices. These standards outline the necessary tests and evaluations to confirm that materials used in medical devices do not elicit harmful biological responses when in contact with the body. The increasing use of medical devices necessitates stringent adherence to these standards to prevent adverse health effects, thus ensuring patient safety and regulatory compliance [537]. Furthermore, the mechanical properties of materials, such as Nitinol used in stents and surgical implants, must be consistent and reliable, as variations can lead to device failure. Manufacturers are therefore required to implement strict quality control measures throughout the production process to meet these critical safety standards [538].

In the aerospace sector, standardization of testing methods is equally important. The development of shape memory alloys (SMAs) for aerospace applications necessitates a consensus on testing protocols to certify these materials for high-stress environments. This involves collaboration among various stakeholders, including engineers and regulatory bodies, to establish agreed-upon properties and testing methods that ensure the materials can withstand operational demands [539]. The aerospace industry is particularly sensitive to material failure, making compliance with rigorous testing standards essential for safety and performance.

Moreover, the commercialization of innovative materials, such as those derived from biofabrication processes, presents unique challenges in meeting established safety and regulatory frameworks. The integration of new technologies often requires the adaptation of existing standards or the development of new ones to address the specific characteristics and risks associated with these materials [540]. As the biofabrication industry evolves, it is crucial to ensure that these materials undergo thorough evaluation to align with safety regulations and industry standards.

10. Commercialization

Once all testing and compliance are complete, the material is ready for market launch. This involves developing marketing strategies, identifying potential customers, and establishing supply chains. Continuous feedback from end-users may lead to further refinements.

The commercialization of new materials and technologies is a multifaceted process that encompasses various strategic elements, including market launch, marketing strategies, customer identification, and supply chain establishment. Once testing and compliance are completed, the transition to market readiness involves a systematic approach to ensure successful commercialization.

The development of effective marketing strategies is crucial for the successful launch of new products. Firms can adopt different commercialization strategies, such as product-based or licensing-based approaches, depending on their market context and technological capabilities. Morricone et al. [541] highlight that these strategies can vary significantly across and within industries, indicating the need for tailored approaches to maximize market impact. Additionally, understanding customer experience is vital, as it directly influences marketing effectiveness. Stein and Ramaseshan [542] emphasize the importance of identifying customer experience touchpoints to enhance the overall customer journey, which can lead to improved marketing outcomes.

Identifying potential customers is another critical aspect of commercialization. Lilja et al. [543] propose that commercial experiences should be defined as memorable events that customers are willing to pay for, suggesting that understanding customer perceptions can guide firms in targeting their marketing efforts effectively. Furthermore, continuous feedback from end-users is essential for refining products and services post-launch. This iterative process allows companies to adapt their offerings based on real-world usage and customer satisfaction, which is supported by the findings of Huang and Xu [544], who discuss the importance of evaluating both internal technological development and external market conditions to assess commercialization potential.

Establishing robust supply chains is equally important in the commercialization process. The integration of digital technologies into business models can facilitate this aspect, as noted by Llanes [545], who discusses how e-commerce can enhance the operational capabilities of businesses, particularly in agribusiness contexts. This digital transformation is increasingly relevant in today's market landscape, where efficiency and responsiveness are critical for maintaining competitive advantage.

Further, the role of intellectual property (IP) in commercialization strategies cannot be overlooked. Pererva [546] emphasizes the need for well-defined strategies that address market, technical, and legal challenges associated with IP commercialization, which can significantly affect the efficiency of innovative activities. This strategic alignment is essential for ensuring that technological innovations can be effectively translated into marketable products.

Key Tools and Technologies in Advanced Material Development

The development of advanced materials involves leveraging cutting-edge tools and technologies to design, model, and produce materials with superior properties tailored to specific applications. These tools enable researchers and engineers to explore the fundamental behaviours of materials, predict outcomes, and create innovative solutions with unprecedented precision and efficiency. Below is a detailed explanation of some of the most influential tools and technologies in this field.

High-Performance Computing (HPC): High-performance computing has revolutionized the way materials are studied and developed. By harnessing the power of supercomputers, researchers can model the behaviour of materials at the atomic or molecular level, which provides valuable insights into their properties and potential applications. Computational techniques such as density functional theory (DFT), molecular dynamics simulations, and finite element analysis are widely used to explore material behaviours under different conditions.

HPC allows for the virtual testing of materials before physical prototypes are created, significantly reducing the time and cost associated with experimentation. For example, the simulation of stress-strain responses in composites or the electronic properties of semiconductors can be achieved with high accuracy. This predictive capability accelerates the discovery of materials for industries such as aerospace, energy, and electronics.

Artificial Intelligence (AI): Artificial intelligence is becoming an indispensable tool in material development due to its ability to analyse vast datasets and identify patterns that might be difficult for humans to discern. Machine learning algorithms can predict material properties, optimal compositions, and processing conditions based on historical data and theoretical models.

AI-driven tools enable the design of materials with specific attributes, such as strength, flexibility, or thermal resistance, by identifying the most promising combinations of elements and compounds. Additionally, AI can optimize manufacturing processes by predicting potential defects and recommending adjustments in real time. This capability not only speeds up material discovery but also ensures higher quality and efficiency in production.

Nanotechnology: Nanotechnology is a cornerstone of advanced material development, focusing on engineering materials at the nanoscale to achieve unique properties and functionalities. By manipulating structures at the atomic or molecular level, nanotechnology enables the creation of materials with enhanced strength, conductivity, thermal resistance, or optical properties.

Applications of nanotechnology in material science are vast. For example, nanocomposites with carbon nanotubes or graphene exhibit remarkable mechanical strength and electrical conductivity, making them suitable for aerospace and electronic applications. Similarly, nanoparticles are used to improve the performance of coatings, catalysts, and medical devices. The precision offered by nanotechnology allows for the creation of materials with tailored properties that were previously unattainable.

Additive Manufacturing: Additive manufacturing, commonly known as 3D printing, is a transformative technology in material development, enabling the rapid prototyping and production of complex structures. Unlike traditional subtractive manufacturing methods, additive manufacturing builds objects layer by layer, allowing for intricate designs that are often impossible to achieve using conventional techniques.

This technology is particularly valuable for testing new materials and creating customized solutions for industries such as healthcare, automotive, and aerospace. For instance, metal additive manufacturing is used to produce lightweight yet strong components for aircraft, while bioprinting is advancing the field of tissue engineering. Additive manufacturing also supports the efficient use of materials, reducing waste and lowering production costs.

Integration of Tools and Technologies: The true potential of advanced material development lies in the integration of these tools and technologies. For instance, HPC can be used to model material behaviours, while AI analyses the resulting data to identify the best material formulations. Nanotechnology can then be employed to engineer the material at the desired scale, and additive manufacturing can bring the design to life with precision.

This interdisciplinary approach not only streamlines the material development process but also opens up new possibilities for innovation. As these tools and technologies continue to evolve, they will play a pivotal role in addressing global challenges, from sustainable energy solutions to advanced healthcare applications, driving the future of material science and engineering.

Developing a new advanced material is a complex but rewarding process that requires collaboration across multiple disciplines. By combining theoretical understanding, innovative techniques, and rigorous testing, scientists and engineers can create materials that push the boundaries of technology and address critical challenges in industries ranging from healthcare to energy and aerospace.

Mechanical Testing: Tensile, Compression, and Fatigue

Mechanical testing is a critical aspect of evaluating advanced materials, providing insights into their behaviour under various forces and conditions. Among the most common and essential tests are tensile testing, compression testing, and fatigue testing. These tests enable engineers and researchers to understand material strength, deformation characteristics, and durability, ensuring the material meets the demands of its intended applications.

Tensile Testing

Tensile testing measures a material's ability to resist deformation and failure under a uniaxial tensile load. In this test, a specimen is clamped at both ends and pulled apart at a controlled rate until it fractures. Key parameters measured include ultimate tensile strength (UTS), yield strength, elongation at break, and Young's modulus.

For advanced materials, tensile testing provides crucial data on their elasticity, ductility, and toughness. For instance, in composites or nanostructured materials, tensile tests reveal how the inclusion of reinforcements or nanoscale structures enhances strength and stiffness. High-performance materials like carbon fibre-reinforced polymers (CFRPs) exhibit exceptional tensile strength-to-weight ratios, making them ideal for aerospace and automotive applications. By analysing stress-strain curves obtained from tensile testing, engineers can also predict how materials will perform under real-world conditions, such as load-bearing or high-stress environments.

Figure 38: Tensile testing on a coir composite. Kerina yin, CC0, via Wikimedia Commons.

Ultimate tensile strength (UTS), often referred to simply as tensile strength, is a critical mechanical property that defines the maximum stress a material can endure while being stretched or pulled before it ultimately fails or breaks. This property is particularly significant for understanding the behaviour of materials under tensile loads and plays an essential role in material selection and design.

The UTS is determined by performing a tensile test, where a material specimen is subjected to a uniaxial pulling force until fracture occurs. During the test, the stress (force per unit area) and strain (deformation relative to the original length) are measured, and the results are plotted on a stress-strain curve. The highest point on this curve represents the ultimate tensile strength, measured in units of stress such as pascals (Pa) in the SI system or pounds per square inch (psi) in the U.S. customary system. For compression rather than tension, the equivalent parameter is called compressive strength.

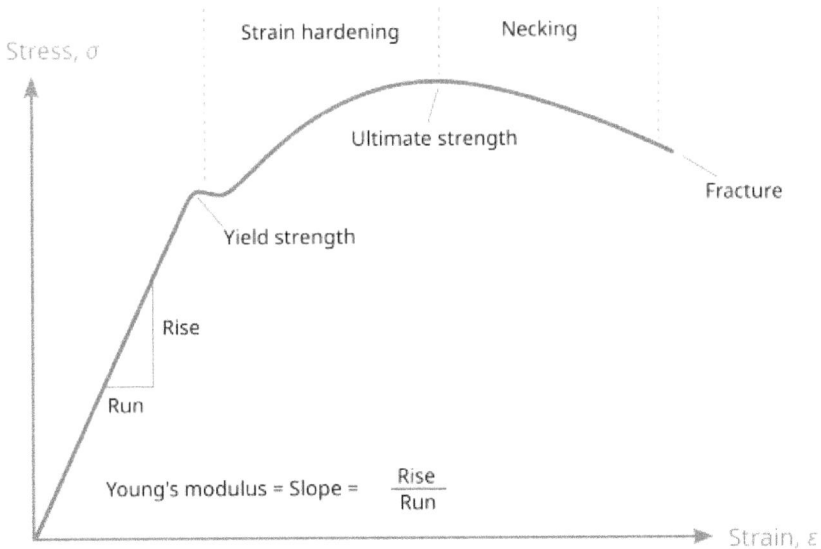

Figure 39: Typical stress vs. strain diagram for a ductile material (e.g. steel). Nicoguaro, CC BY 4.0, via Wikimedia Commons.

Unlike the yield strength, which marks the transition from elastic to plastic deformation, the UTS occurs after significant plastic deformation in ductile materials. In brittle materials, the UTS closely coincides with the yield point, as they typically fracture with minimal plastic deformation.

UTS is an intensive property, meaning it does not depend on the size of the test specimen. However, it can be influenced by factors such as the material's microstructure, surface defects, and the testing environment, including temperature. Materials with sharp fractures and no significant plastic deformation are classified as brittle, while those capable of undergoing plastic deformation and necking before failure are classified as ductile.

The tensile strength is critical for applications involving brittle materials, as these materials lack a distinct yield point. For ductile materials, yield strength is more relevant for design purposes because it ensures that the material remains in the elastic deformation range under working loads.

In ductile materials like aluminium and structural steel, the stress-strain relationship initially follows a linear path, where deformation is entirely elastic and reversible. The elastic region continues until the material reaches its yield point, beyond which it undergoes plastic deformation. At this stage, permanent changes in shape occur even after the load is removed.

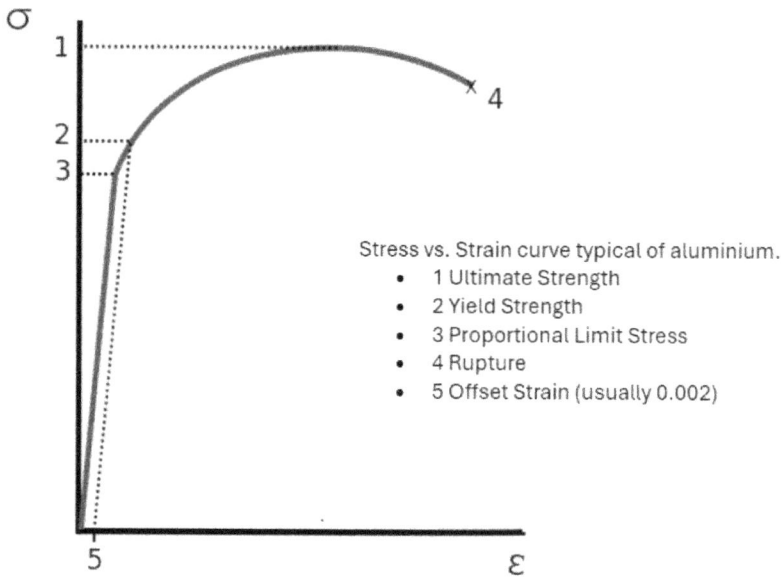

Figure 40: Stress vs. Strain curve typical of aluminium. Toiyabe, CC-BY-SA-3.0, via Wikimedia Commons.

As strain increases beyond the yield point, the material enters a strain-hardening region, where it resists further deformation due to internal structural changes. Eventually, necking begins, characterized by a localized reduction in the specimen's cross-sectional area. This phenomenon causes the stress to appear to decrease when calculated using the original cross-sectional area, leading to a peak in the engineering stress-strain curve. This peak corresponds to the ultimate tensile strength.

In practical applications, UTS is not typically used as a design parameter for ductile materials, as design standards prioritize the yield strength to prevent unacceptable plastic deformation. However, UTS is valuable for quality control and material identification.

For brittle materials, UTS is a fundamental parameter for design and analysis, as these materials do not exhibit a clear yield point. It is particularly crucial in applications where sudden failure under tensile loads must be avoided. Examples include ceramics, glass, and some composites.

In quality control, UTS serves as an indicator of material consistency and performance. It is also employed in identifying unknown materials, providing a rough estimation of their mechanical properties based on standardized testing procedures.

The yield point in materials science marks the transition between elastic and plastic behaviour in a material. It is a critical concept in understanding how materials respond to stress and deformation. When a material is stressed below its yield point, it deforms elastically, meaning it will return to its original shape once the load is removed. However, once the stress exceeds the yield point, the material enters the plastic region, where some of the deformation becomes permanent and irreversible. This transition is crucial for assessing material performance and designing mechanical components.

The yield strength or yield stress corresponds to the stress level at which the material begins to deform plastically. This property is vital in engineering and design, as it represents the maximum load a material can endure without undergoing permanent deformation. For ductile materials like aluminium and steel, the yield point is not always sharply defined, as these materials exhibit a gradual onset of plastic deformation. In such cases, an offset yield point, commonly set at 0.2% plastic strain, is used. This offset yield point provides a practical measure of a material's yield strength, especially for engineering applications.

For brittle materials, the yield strength is often indistinguishable from the ultimate tensile strength since these materials fail without significant plastic deformation. In contrast, for ductile materials, the yield strength and ultimate tensile strength are distinct values. The ratio of these two parameters plays an essential role in applications like pipeline design, influencing material selection and performance evaluation.

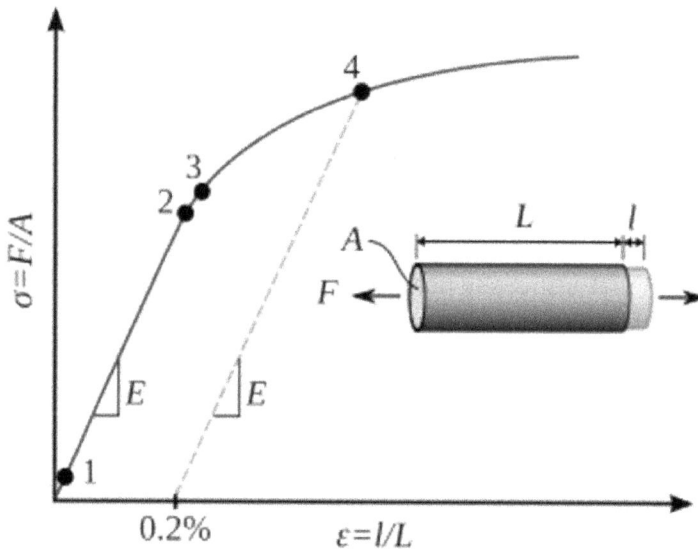

Figure 41: Illustration of offset yield point. Sigmund, Public domain, via Wikimedia Commons.

Yield behaviour can also be described in terms of three-dimensional principal stresses using yield criteria and yield surfaces. Different yield criteria are developed for various materials, such as metals, polymers, or composites, to account for their unique stress-strain behaviours.

Testing for yield strength involves controlled tensile testing, where a specimen is pulled until deformation or fracture occurs. The stress-strain relationship is recorded, and the yield point is identified on the curve. In some materials, particularly mild steel, both upper and lower yield points may be observed. The upper yield point represents the initial stress at which yielding begins, followed by a drop to the lower yield point, which is often used for conservative engineering calculations.

Yield strength testing can also be approximated using hardness tests, where indentation hardness correlates roughly with tensile strength. However, for precise applications, direct tensile testing remains the preferred method. Advanced methods like indentation plastometry enable stress-strain analysis from indentation tests, offering new possibilities for evaluating material properties with minimal sample preparation.

In structural engineering, yielding has practical implications. Yielded structures exhibit reduced stiffness, increased deflection, and decreased buckling strength. Permanent deformation after yielding may introduce residual stresses, which can affect the material's subsequent performance. Nonetheless, strain hardening in many engineering metals increases the yield stress after unloading, enhancing the material's resistance to future deformation.

In conclusion, the yield point and yield strength are foundational concepts in materials science and engineering, influencing material selection, component design, and performance analysis. Their accurate determination through testing and modelling ensures the reliability and safety of structures and mechanical systems.

Crystalline materials can be engineered to increase their yield strength through various mechanisms that modify their internal structure. Yield strength, which measures the stress required to deform a material plastically, can be enhanced by introducing defects or altering the material's microstructure. These modifications typically require a higher applied stress to move dislocations—linear defects in the crystal lattice—thereby increasing the material's yield strength. Unlike some properties that depend solely on a material's composition, yield strength is highly sensitive to the methods of material processing.

Work hardening, or strain hardening, occurs when a material is deformed plastically, increasing the density of dislocations within its crystal lattice. As these dislocations multiply and interact, they form a network that resists further movement. The result is an increase in the material's yield strength. Dislocations may become entangled, further impeding their motion and requiring greater stress for additional deformation. This mechanism is particularly effective in metals and is governed by the relationship:

$$\Delta\sigma_y = Gb\sqrt{\rho}$$

Here, $\Delta\sigma_y$ represents the increase in yield stress, G is the shear elastic modulus, b is the magnitude of the Burgers vector (which quantifies dislocation motion), and ρ is the dislocation density.

Solid solution strengthening involves alloying a material with impurity atoms that occupy positions in the crystal lattice. These impurity atoms relieve localized strains caused by dislocations, creating obstacles to their movement. The impurity concentration and the strain induced in the lattice by these atoms are critical factors in this strengthening process. The effect is described by:

$$\Delta\tau = Gb\sqrt{C_s}\epsilon^{\frac{3}{2}}$$

In this equation, $\Delta\tau$ represents the change in shear stress (related to yield stress), C_s is the concentration of solute atoms, and ϵ represents the strain induced in the lattice.

Particle or precipitate strengthening involves introducing a secondary phase within the material. These particles act as barriers to dislocation motion. When a dislocation encounters a particle, it can either shear the particle or bypass it through a process known as bowing or Orowan looping. The effectiveness of this mechanism depends on the size, spacing, and interaction between the particles and the dislocations. The stress required for shearing and bowing is described by:

$$\Delta\tau = \frac{r_{\text{particle}}}{l_{\text{interparticle}}}\gamma_{\text{particle-matrix}}$$

$$\Delta\tau = \frac{Gb}{l_{\text{interparticle}} - 2r_{\text{particle}}}$$

Here, $r_{particle}$ is the particle radius, $l_{interparticle}$ is the distance between particles, and $\gamma_{particle}$- is the interfacial surface tension.

Grain boundary strengthening, also known as Hall-Petch strengthening, relies on the presence of grain boundaries, which act as barriers to dislocation motion. As the grain size decreases, the number of grain boundaries increases, enhancing the material's resistance to plastic deformation. Dislocations accumulate at grain boundaries, creating repulsive forces that hinder further movement. The relationship governing this mechanism is:

$$\sigma_y = \sigma_0 + kd^{-1/2}$$

Here, σ_y is the yield stress, σ_0 is the stress required to move dislocations, k is a material-specific constant, and d is the grain size.

Practical Example: Strengthening Aluminium Alloys for Aerospace Applications

Aluminium alloys are widely used in aerospace applications due to their lightweight and good mechanical properties. However, to meet the stringent requirements of yield strength and durability, the material must be engineered using various strengthening mechanisms. Let's explore how the described formulas apply in practice.

Work Hardening

During the processing of aluminium alloys, such as through rolling or forging, the material undergoes plastic deformation. This increases the dislocation density, enhancing its yield strength.

Suppose the initial dislocation density (ρ_0) is 10^{10} m^{-2}, and after work hardening, it increases to 10^{14} m^{-2}. The Burgers vector (b) for aluminium is approximately 2.86×10^{-10} m, and the shear modulus (G) is 26 GPa.

Using the formula:

$$\Delta \sigma_y = Gb\sqrt{\rho}$$

Calculate the increase in yield stress:

$$\Delta \sigma_y = 26 \times 10^9 \times 2.86 \times 10^{-10} \times \sqrt{10^{14}}$$

$$\Delta \sigma_y = 26 \times 10^9 \times 2.86 \times 10^{-10} \times 10^7$$

$$\Delta \sigma_y = 7.44 \times 10^7 \, \text{Pa} = 74.4 \, \text{MPa}$$

This increase in yield strength makes the alloy suitable for load-bearing components in aircraft.

Solid Solution Strengthening

Aluminium is alloyed with magnesium to improve its strength through solid solution strengthening. The addition of magnesium atoms introduces lattice strain, impeding dislocation motion.

If the magnesium concentration (C_s) is 3%, and the strain induced (ϵ) is 0.1, we use the formula:

$$\Delta \tau = Gb\sqrt{C_s}\epsilon^{3/2}$$

Assuming G=26 GPa and b=2.86×10^{-10} m:

$$\Delta \tau = 26 \times 10^9 \times 2.86 \times 10^{-10} \times \sqrt{0.03} \times (0.1)^{3/2}$$

$$\Delta \tau = 26 \times 10^9 \times 2.86 \times 10^{-10} \times 0.173 \times 0.0316$$

$$\Delta \tau \approx 3.94 \, \text{MPa}$$

This added strength enhances the alloy's ability to withstand operational stresses.

Particle/Precipitate Strengthening

Aluminium alloys can be strengthened by precipitates, such as Al$_3$Mg, that block dislocation motion. The particle radius ($r_{particle}$) is 50 nm, and the interparticle distance ($l_{interparticle}$) is 200 nm. Assume G=26 GPa and b=2.86×10^{-10} m.

Using the bowing formula:

$$\Delta\tau = \frac{Gb}{l_{\text{interparticle}} - 2r_{\text{particle}}}$$

$$\Delta\tau = \frac{26 \times 10^9 \times 2.86 \times 10^{-10}}{200 \times 10^{-9} - 2 \times 50 \times 10^{-9}}$$

$$\Delta\tau = \frac{26 \times 10^9 \times 2.86 \times 10^{-10}}{100 \times 10^{-9}}$$

$$\Delta\tau = 7.44\,\text{MPa}$$

The additional yield strength from precipitates makes the alloy suitable for high-stress environments, such as aircraft wings.

Grain Boundary Strengthening

Grain boundary strengthening is achieved by refining the grain size of aluminium alloys. If the initial grain size (*d*) is 10 µm and reduced to 5 µm, the Hall-Petch relation applies:

$$\sigma_y = \sigma_0 + kd^{-1/2}$$

Assume σ_0=100 MPa and $k = 0.5\,\text{MPa} \cdot \mu\text{m}^{1/2}$:

$$\sigma_y = 100 + 0.5 \times 5^{-1/2}$$

$$\sigma_y = 100 + 0.5 \times 0.447$$

$$\sigma_y = 100 + 0.224 = 100.224\,\text{MPa}$$

This refinement significantly enhances the alloy's strength, making it more resistant to plastic deformation.

By combining these mechanisms—work hardening, solid solution strengthening, particle strengthening, and grain boundary refinement—the yield strength of aluminium alloys can be precisely engineered to meet the rigorous demands of aerospace applications. Each formula provides a quantitative basis for optimizing the material's properties during processing and application.

Elongation at break is a key mechanical property of materials, particularly in materials science and engineering. It measures the extent to which a material can deform plastically before it ultimately fractures. This property provides insight into a material's ductility and flexibility, indicating how much strain a material can endure before failure.

Elongation at break is defined as the percentage increase in a material's original length when subjected to a tensile force until it breaks. It is typically expressed as a percentage of the original length. This property is measured during a tensile test, where a specimen is stretched under a controlled force until it fractures. The elongation is calculated using the formula:

$$\text{Elongation at Break } (\%) = \left(\frac{\Delta L}{L_0} \right) \times 100$$

Where:

- ΔL is the change in length at the point of fracture.

- L_0 is the original gauge length of the specimen.

Elongation at break is a direct indicator of a material's ductility. Ductile materials, such as metals like aluminium and copper, exhibit high elongation values, deforming significantly before breaking. In contrast, brittle materials, such as ceramics and glass, show low elongation values, fracturing with minimal deformation.

For structural applications, high elongation at break is desirable in materials that need to absorb energy or withstand deformation without immediate failure. For example, components subjected to dynamic loads or impacts benefit from materials with high elongation at break to prevent catastrophic failure.

In manufacturing, elongation at break is used to evaluate the quality and consistency of materials. Deviations from expected values can indicate issues in material processing or impurities in the composition.

Ductile metals like copper and aluminium demonstrate elongation at break values ranging from 20% to 50%. This characteristic allows them to undergo significant plastic deformation before ultimately fracturing, making them ideal for applications requiring flexibility and resilience. On the other hand, high-strength steels, which are known for their increased hardness and reduced ductility, typically exhibit lower elongation values in the range of 5% to 15%.

Flexible polymers, such as polyethylene, stand out with elongation at break values that often exceed 100%, highlighting their capacity for exceptional plastic deformation. This makes them suitable for applications like packaging and films. In contrast, brittle polymers like polystyrene show much lower elongation values, typically between 1% and 2%, reflecting their limited ability to stretch before breaking.

Ceramics and glass, known for their inherent brittleness, exhibit very low elongation at break values, often less than 1%. These materials are prone to fracture with minimal deformation, limiting their use in applications requiring significant flexibility.

Fibre-reinforced composites present variable elongation at break values, largely influenced by the composition of the matrix and the reinforcement material. While the polymer matrix may exhibit higher elongation, the embedded fibres contribute strength but generally reduce the overall ductility of the composite. This balance between flexibility and strength makes composites versatile for a range of structural and high-performance applications.

Materials with a homogeneous and ductile microstructure generally exhibit higher elongation at break. The composition of the material plays a critical role in determining its ability to undergo plastic deformation before failure, as ductile materials tend to distribute stress more evenly, allowing for greater elongation.

Temperature is another significant factor influencing elongation at break. Elevated temperatures typically enhance a material's ductility by reducing its resistance to plastic deformation. At higher temperatures, atomic mobility increases, facilitating the movement of dislocations and enabling the material to stretch further before breaking.

Strain rate, or the speed at which a material is deformed, also impacts elongation at break. Materials subjected to higher strain rates often exhibit lower elongation values. Under rapid loading, materials behave more brittly, as there is insufficient time for dislocations to move and accommodate the applied stress, leading to premature failure.

Defects and impurities within the material's structure, such as voids or inclusions, act as stress concentrators. These imperfections can significantly reduce elongation at break by providing localized points of weakness, which initiate cracks and lead to earlier fracture.

Processing techniques play a crucial role in modifying a material's properties, including elongation at break. Methods such as annealing can improve ductility by relieving residual stresses and refining the microstructure. A well-refined microstructure allows for more uniform stress distribution and enhances the material's ability to elongate before breaking.

Young's modulus, often denoted as E, is a fundamental material property in the field of mechanics and materials science. It quantifies a material's stiffness, or its ability to resist deformation under tensile or compressive stress. Specifically, Young's modulus measures the ratio of stress (force per unit area) to strain (relative deformation) within the elastic (non-permanent) deformation range of a material. This property is pivotal in characterizing the mechanical behaviour of materials under load.

Mathematically, Young's modulus is expressed as:

$$E = \frac{\sigma}{\epsilon}$$

Where:

- E = Young's modulus (measured in units of pressure, such as pascals or megapascals in SI units)

- σ = Stress (force per unit area, typically measured in pascals)

- ϵ = Strain (dimensionless, defined as the ratio of the change in length to the original length)

Young's modulus applies to the elastic region of a material's stress-strain curve, where the material returns to its original shape upon unloading. In this range, the relationship between stress and strain is linear, as described by Hooke's Law:

$$\sigma = E \cdot \epsilon$$

This linear relationship holds until the material reaches its yield point, beyond which permanent deformation occurs.

Young's modulus is typically expressed in units of pascals (Pa), with common values ranging in the gigapascal (GPa) range for most structural materials. For instance, metals like steel and aluminium have high Young's modulus values, reflecting their resistance to elastic deformation, whereas polymers and elastomers exhibit much lower values, indicating higher flexibility. Materials such as ceramics and glass, although possessing high Young's modulus values, are brittle and fail at relatively low strains.

When materials undergo elastic deformation, they revert to their original shape upon removal of the applied stress. This behaviour is represented by the linear region of the stress-strain curve, governed by Hooke's law, where stress is directly proportional to strain. The slope of this linear portion is Young's modulus. The higher the modulus, the greater the stress required to produce a given strain. An idealized perfectly rigid material would exhibit an infinite Young's modulus, while fluids, which deform without significant force, would have a modulus of zero.

Young's modulus is distinct from other mechanical properties such as strength, which measures the maximum stress a material can withstand before failure, and hardness, which quantifies a material's surface resistance to deformation. It is also independent of geometric stiffness, which relates to the shape and structural design of a material, such as the enhanced bending stiffness in I-beams compared to simple rods of the same material.

In practical applications, Young's modulus is critical for calculating deformations in materials subjected to tensile or compressive loads. For example, in structural engineering, it helps predict how beams or columns will deflect under load. It is particularly applicable in cases of uniaxial stress, where force is applied in one direction, and is often used alongside other elastic constants such as the shear modulus, bulk modulus, and Poisson's ratio to describe the complete elastic behaviour of isotropic materials.

The value of Young's modulus varies significantly among materials, reflecting their inherent stiffness and elastic properties. Metals such as steel and aluminium exhibit high Young's modulus values, which indicate a strong resistance to elastic deformation under applied stress. These properties make them ideal for applications requiring strength and rigidity. In contrast, polymers and elastomers possess much lower Young's modulus values, allowing for greater flexibility and deformation under stress, which is advantageous in applications requiring resilience or elasticity. Ceramics and glass, while characterized by high Young's modulus values indicating stiffness, are inherently brittle and have very low strain limits, making them prone to fracture under tension.

Young's modulus plays a crucial role in engineering and design, as it helps predict how materials respond to mechanical forces. In structural design, it is used to determine the deflection of beams, columns, and bridges under applied loads, ensuring stability and safety. For mechanical components, it guides the design of machine parts to ensure they can withstand expected stresses without deforming excessively. Furthermore, Young's modulus is a key criterion in material selection for applications that require specific stiffness or flexibility, allowing engineers to choose materials best suited to the demands of a project.

Several factors influence the value of Young's modulus for a given material. Temperature is a significant factor; higher temperatures generally reduce Young's modulus, making materials more ductile and less stiff. The microstructure of a material, including its grain size, impurities, and phase composition, can also alter its stiffness. Additionally, for anisotropic materials such as composites or wood, the orientation of stress application affects Young's modulus, as stiffness can vary depending on the direction of the load.

Young's modulus is fundamental in material science, serving as a critical link between theoretical understanding and practical application. By quantifying a material's stiffness, it enables precise analysis of how structures and components behave under applied forces, ensuring that engineering systems are safe, efficient, and reliable.

Compression Testing

Compression testing determines a material's behaviour when subjected to compressive forces. Unlike tensile testing, where the material is stretched, compression testing involves squeezing the material to measure its ability to withstand shortening or crushing forces. Parameters such as compressive strength, modulus of elasticity, and deformation are assessed.

Advanced materials, such as ceramics, foams, and composites, are often tested under compression to evaluate their suitability for structural applications. For example, ceramics used in armour systems or high-temperature components must demonstrate excellent compressive strength to endure extreme forces without failure. Similarly, polymer foams used in packaging or insulation require high compressive resilience to protect against impacts. Compression testing also highlights the material's behaviour under buckling or densification, providing insights into its energy absorption capabilities.

Compressive strength is a critical property in mechanics that quantifies a material or structure's ability to withstand forces tending to reduce its size, known as compressive loads. This property contrasts with tensile strength, which measures a material's resistance to forces causing elongation or tension. Both compressive and tensile strengths, along with shear strength, are analysed independently to understand a material's behaviour under different loading conditions.

When subjected to compressive stress, materials may respond in one of two ways: they may fracture at their compressive strength limit, or they may deform irreversibly without immediate failure. For materials that undergo irreversible deformation, the acceptable limit of deformation often defines the compressive load capacity. This makes compressive strength a vital parameter in designing structures, where failure under compression must be avoided.

Compressive strength is typically determined using a universal testing machine. The test involves subjecting a specimen, often cylindrical, to a steadily increasing uniaxial compressive load until failure occurs. The machine records the stress and corresponding strain during the test, generating a stress-strain curve. The compressive strength corresponds to the maximum stress the material can withstand, represented by the peak of the curve. In this process, the specimen undergoes axial shortening and lateral expansion as the load increases, providing insights into both elastic and plastic deformation behaviours.

In the elastic region of the stress-strain curve, the material follows Hooke's law, where stress is proportional to strain ($\sigma = E\varepsilon$), with E being the Young's modulus for compression. Within this region, the material deforms elastically, returning to its original shape and size once the stress is removed. Beyond the elastic limit, the material transitions into plastic deformation, where permanent changes in shape occur, even after the load is released. This transition is marked by the yield point, a key indicator of the material's behaviour under compressive stress.

It is essential to distinguish between engineering stress and true stress in the analysis of compressive strength. Engineering stress is calculated using the original cross-sectional area of the specimen ($\sigma_e = F/A_0$), while true stress accounts for the changing area of the specimen during deformation. Similarly, engineering strain ($\varepsilon_e = (l - l_0)/l_0$) is based on initial and current lengths, whereas true strain ($\epsilon' = \ln(l/l_0)$) provides a more accurate measure of large deformations, particularly in ductile materials.

The compressive strength is defined by the specific point on the engineering stress-strain curve where the material reaches its maximum load capacity without failure. This point is characterized by the load (F*) and the corresponding strain (ε_e^*) just before the material undergoes crushing or fracturing. Understanding compressive strength and its relationship with strain is fundamental in designing materials and structures that can reliably endure compressive forces without catastrophic failure.

When a material is subjected to uniaxial compressive loads, its dimensions change, causing deviations between the calculated engineering stress and true stress. Engineering stress is computed using the original cross-sectional area of the specimen, while true stress accounts for the actual, continually changing cross-sectional area during deformation. This distinction becomes crucial in applications where precise stress-strain relationships are required, such as in material research or advanced engineering applications.

Under compressive loading, a material's length decreases, and its cross-sectional area increases. This lateral expansion leads to differences between the true stress ($\sigma'=F/A$, where A is the current area) and engineering stress ($\sigma_e=F/A_0$, where A_0 is the original area). For many routine tests, such as quality control in concrete production, measuring engineering stress at failure suffices. However, true stress calculations are critical for understanding a material's properties during plastic deformation.

In compressive tests, the geometry of the specimen and the friction between the specimen and the testing machine significantly impact stress distribution. Friction at the contact interfaces can restrict lateral expansion at the specimen's ends, resulting in a barrelling effect. This bulging at the centre leads to non-uniform stress distribution, with higher stresses concentrated at the core and lower stresses near the edges. Consequently, friction can artificially increase the apparent compressive strength, complicating the interpretation of results.

In an ideal test with no friction at the contact points, the lateral expansion of the specimen occurs uniformly along its length, maintaining consistent stress and strain. In such cases, the relationship between true stress and engineering stress can be described as:

$$\sigma' = \sigma_e(1 + \epsilon_e)$$

Here, ϵ_e is the engineering strain, and the true stress is generally lower than the engineering stress due to compressive strain being negative. High-bulk modulus materials like metals maintain near-constant volume during compression, allowing the true stress to be more accurately represented by this equation under frictionless conditions.

In real-world scenarios, friction at the specimen ends creates constraints that distort stress distribution and influence test outcomes. This friction restricts lateral expansion, intensifying the barrelling effect. To mitigate this, lubricants such as MoS_2 or PTFE sheets can be applied to the interfaces, reducing friction and promoting more uniform stress distribution. Additionally, spherical or self-aligning fixtures can minimize uneven loading.

If friction cannot be entirely eliminated, correction formulas or geometric extrapolation methods are employed to adjust for its effects. For example, the true compressive stress in a barrelled specimen can be calculated using a correction factor *C*, which incorporates parameters like bulge radius, specimen dimensions, and applied load:

$$\sigma' = C\sigma_a$$

where $\sigma_a = \dfrac{4F}{\pi d_2^2}$ is the apparent stress, and CCC is a coefficient accounting for geometric distortions caused by friction.

In cases where high aspect ratios lead to buckling before achieving true compressive strength, geometric extrapolation offers an alternative. Tests with specimens of varying aspect ratios are conducted, and the results are analysed to estimate the true compressive strength. By plotting compressive stress values against aspect ratios and extrapolating to the theoretical condition where aspect ratio approaches zero, the true stress value can be determined.

Understanding the deviation between true and engineering stress is essential for accurately assessing material behaviour under compressive loads. Friction and geometric effects must be carefully managed or corrected to obtain reliable data. These considerations are critical in designing materials and structures subjected to compressive forces, ensuring their performance and safety under real-world conditions.

Different materials exhibit varying relationships between their compressive and tensile strengths, dictated by their atomic structure, bonding, and mechanical behaviour. Concrete and ceramics, for instance, display much higher compressive strengths compared to tensile strengths. This disparity arises because these materials excel at resisting forces that compress their atomic structure but fracture easily when subjected to tensile forces that pull their atomic bonds apart. On the other hand, composite materials, such as glass fibre epoxy matrix composites, tend to have higher tensile strengths than compressive strengths. The fibre reinforcement provides excellent resistance to stretching forces, but the matrix material's compressive behaviour often limits overall performance. Metals, which are inherently ductile, show relatively balanced compressive and tensile strengths. However, under compressive loads, metals fail differently — through mechanisms like buckling, crumbling, or shear — as opposed to the necking or defect-driven fractures seen under tensile loads.

The way materials fail under compression depends on their geometry, ductility, and constraints during loading. For slender specimens with a high length-to-radius ratio, buckling often governs failure. This mode occurs when the material's structure bends under compressive forces rather than undergoing direct axial compression. Ductile materials, when loaded axially in compression, typically yield, displaying the barrelling effect, where the material bulges at the centre due to lateral expansion. Brittle materials exhibit distinct failure modes based on external constraints. Without confining pressure, brittle materials like ceramics or concrete often fail by axial splitting, where cracks develop parallel to the load direction. Under moderate confining pressure, these materials are more likely to experience shear fractures. High confining pressure, however, can induce ductile-like failure, even in brittle materials, as the constraints suppress crack propagation.

Axial splitting is a common failure mechanism in brittle materials subjected to compressive forces without lateral constraint. As the material compresses, elastic energy builds up and seeks release. Cracks form parallel to the applied load, relieving the tensile strain created in directions perpendicular to the compressive force. This phenomenon is influenced by the material's Poisson's ratio, which defines the relationship between axial compression and lateral expansion. Once axial splitting begins, the material divides into micro-columns, which may fail further due to frictional forces and stress redistribution. Variations in material properties, such as Young's modulus, can exacerbate this behaviour, leading to stress shielding and further crack propagation.

Microcracking is a significant contributor to compressive failure in brittle and quasi-brittle materials. These small cracks originate at stress concentrators like pores, inclusions, or weak interfaces. Under compressive loads, local tensile stresses near these anomalies promote crack formation and growth. Microcracks typically nucleate perpendicular to pre-existing cracks or flaws, growing into secondary cracks. These secondary cracks can extend far beyond the length of the original flaw, particularly in uniaxial compression. As microcracking progresses throughout the material, it can lead to the formation of shear bands, which are localized regions of intense deformation. These shear bands signify the onset of material instability and are often precursors to catastrophic failure under compressive loads.

In construction, compressive strength is a vital property of concrete, influencing design and quality control. The compressive strength of concrete is typically measured using standardized tests, such as ASTM C39, where cylindrical or cubic specimens are subjected to axial compressive loads until failure. The strength is expressed as the characteristic compressive strength (e.g., f_{ck}) of a 150 mm cube after 28 days of curing. Interim tests, such as those performed after 7 days, help predict the final strength and identify potential issues early.

Design values for compressive strength incorporate a safety factor, ensuring that structural elements remain within acceptable performance limits. Ultra-high performance concrete (UHPC), with compressive strengths exceeding 150 MPa, represents an advanced class of materials used in specialized applications like high-rise buildings and bridges. Testing for compressive strength often involves careful control of loading rates, specimen preparation, and environmental conditions to ensure reliable and reproducible results.

Understanding compressive strength is crucial for designing structures subjected to heavy loads, such as bridges, foundations, and dams. In addition to standard compression tests, other mechanical properties like modulus of elasticity, tensile strength, and fracture toughness are often measured to provide a comprehensive understanding of material behaviour. Standards such as ASTM C469 (modulus of elasticity) and ASTM C1609 (flexural strength) complement compressive strength tests, enabling engineers to design more efficient and resilient structures. For advanced materials like UHPC, additional tests may include microstructural analysis and fatigue testing to explore long-term performance under varying loads and environmental conditions.

Deformation is the change in shape, size, or volume of a material or structure due to an applied force or load. It can result from external stresses such as tension, compression, shear, bending, or torsion. Deformation is fundamental to understanding material behaviour under various mechanical conditions and is vital in designing structures and components to ensure safety and functionality.

Deformation is classified into two main types based on the material's response to the applied force. Elastic deformation involves temporary and reversible changes in shape or size. When the force is removed, the material returns to its original form. On the other hand, plastic deformation results in permanent and irreversible changes in shape or size, occurring when the stress exceeds the material's yield strength.

Elastic deformation occurs within the elastic limit of a material, governed by Hooke's Law, which states that stress is proportional to strain in this region. The extent of elastic deformation is influenced by the material's modulus of elasticity or stiffness. For instance, stretching a rubber band or bending a spring within its elastic limit are examples of elastic deformation. Elastic deformation is temporary, proportional to stress, and typically occurs at lower stress levels.

Plastic deformation takes place when the stress applied surpasses the yield strength of the material, causing permanent changes in shape. In this phase, the stress-strain relationship becomes non-linear, and the material does not revert to its original configuration once the force is removed. This type of deformation is common in ductile materials like metals. Plastic deformation is characterized by its permanence, non-linear stress-strain relationship, and occurrence beyond the yield point.

Various stresses can lead to deformation. Tensile stress stretches a material, causing elongation. Compressive stress compresses a material, making it shorter or more compact. Shear stress leads to layers of material sliding relative to each other. Torsional stress twists the material along its axis, while bending stress causes one side of a material to experience tension and the opposite side compression.

Several factors influence deformation. Material properties such as the elastic modulus, yield strength, and ductility play a significant role. A higher modulus of elasticity corresponds to greater stiffness and less deformation under a given load. The yield strength determines the transition point from elastic to plastic deformation, while ductile materials can sustain more plastic deformation before failure. The characteristics of the applied load, including its magnitude, direction, and duration, also affect deformation. Higher temperatures typically lower a material's resistance to deformation by reducing its yield strength and elastic modulus. Additionally, rapid loading can make materials appear stronger and less deformable due to strain rate sensitivity. Finally, the geometry of the material, such as its shape and size, influences deformation. For example, thin beams are more prone to bending than thicker ones under the same load.

Deformation is quantified using strain, which measures the relative change in size or shape of a material. Strain is a dimensionless quantity expressed as:

$$\varepsilon = \frac{\Delta L}{L_0}$$

where:

- ε is the strain,
- ΔL is the change in length,
- $L0$ is the original length.

Deformation can also be represented on a stress-strain curve, where:

- The linear region represents elastic deformation.
- The non-linear region represents plastic deformation.

Fatigue Testing

Fatigue testing assesses a material's durability under cyclic loading, mimicking conditions where materials experience repeated stress over time. This test is essential for evaluating the long-term performance and safety of materials in dynamic or fluctuating environments, such as those found in aircraft, bridges, or biomedical implants.

In fatigue testing, a material is subjected to cyclic loading at varying stress levels, often below its ultimate tensile strength. The test continues until the material fails, and results are used to create an S-N curve (stress versus the number of cycles). Advanced materials like superalloys, composites, and nanostructured metals are rigorously tested for fatigue to ensure they can withstand millions of cycles without failure.

Fatigue testing is particularly critical for advanced materials exposed to extreme environments. For instance, turbine blades in jet engines must endure high-cycle fatigue due to constant stress from thermal and mechanical loads. By identifying the material's fatigue limit, engineers can design components that operate safely within their endurance limits, reducing the risk of catastrophic failure.

Integration and Relevance

The integration of tensile, compression, and fatigue testing provides a comprehensive understanding of an advanced material's mechanical performance. Tensile testing highlights a material's strength and ductility, compression testing evaluates its load-bearing and energy absorption capabilities, and fatigue testing ensures long-term reliability under cyclic loading. Together, these tests form the foundation for material selection and design in applications requiring high performance, safety, and durability.

As advanced materials continue to evolve, these mechanical testing methods are augmented with computational modelling and real-time monitoring technologies, allowing for more precise and efficient evaluations. This synergy between traditional testing and modern technologies ensures that advanced materials can meet the increasing demands of cutting-edge industries, from aerospace and automotive to energy and biomedical engineering.

Thermal Analysis: DSC, TGA, and DMA

Thermal analysis encompasses techniques used to study the properties of materials as they change with temperature. Key methods include Differential Scanning Calorimetry (DSC), Thermogravimetric Analysis (TGA), and Dynamic Mechanical Analysis (DMA), each with distinct functionalities for material characterization.

Differential Scanning Calorimetry (DSC)

Differential Scanning Calorimetry (DSC) is a pivotal thermoanalytical technique that quantifies the difference in heat flow between a sample and a reference as a function of temperature. This method is essential for investigating phase transitions and thermal reactions, providing critical insights into processes such as melting, crystallization, glass transitions, and curing. In a typical DSC experiment, both the sample and reference are maintained at nearly identical temperatures, allowing for precise tracking of heat flow associated with thermal events. This capability enables scientists to identify endothermic and exothermic transitions, which are crucial for understanding material behaviour under varying thermal conditions [547, 548].

The operational principles of DSC involve subjecting the sample and reference to a controlled temperature program, typically involving linear heating. The reference material, which is often a metal like indium or tin, is selected for its stable and well-defined heat capacity across the temperature range of interest, ensuring accurate comparisons [547, 548]. The evolution of DSC since its inception by Watson and O'Neill in 1962 has led to significant advancements, including the introduction of adiabatic differential scanning calorimeters by Privalov and

Monaselidze in 1964, which have broadened the applications of this technique in fields such as biochemistry [548, 549].

DSC instruments can be categorized into two primary types: Heat-flux DSC and Power Differential DSC. Heat-flux DSC measures the heat flux difference between the sample and reference, utilizing a setup where both are placed in crucibles on a sample holder with integrated temperature sensors. This configuration allows for the calculation of heat flow by integrating the temperature difference between the crucibles [547, 548]. In contrast, Power Differential DSC employs separate thermally insulated furnaces for the sample and reference, directly recording the electrical power needed to maintain equal temperatures, thus providing a more straightforward measurement of thermal energy changes [547, 548].

Recent innovations in DSC technology, such as Fast-scan DSC (FSC) and Temperature Modulated DSC (TMDSC), have further enhanced the capabilities of this technique. FSC utilizes micromachined sensors to achieve ultrahigh scanning rates, facilitating the analysis of rapid phase transitions and thermophysical properties of thermally labile compounds [548, 549]. TMDSC, on the other hand, combines linear heating with superimposed sinusoidal temperature variations, allowing for the differentiation of overlapping thermal events. This technique is particularly valuable for studying complex materials, as it enables the separation of specific heat capacity changes associated with glass transitions from time-dependent phenomena like curing or dehydration [550, 551].

The detection of phase transitions using Differential Scanning Calorimetry (DSC) is based on the principle that when a sample undergoes a physical transformation, such as melting or crystallization, it requires either more or less heat flow to maintain the same temperature as the reference. This difference in heat flow allows DSC to detect and quantify thermal events. For example, during the endothermic process of melting, a sample absorbs heat, necessitating additional heat flow to maintain temperature parity with the reference. Conversely, exothermic processes, like crystallization, release heat, reducing the need for additional heat flow. By observing these differences, DSC provides precise measurements of heat absorbed or released during phase transitions, offering valuable insights into material properties such as glass transition temperatures and crystallization behaviours. This makes it an essential tool in quality control and polymer studies.

Differential thermal analysis (DTA) is a related technique, differing primarily in how it measures thermal changes. While DSC focuses on maintaining equal temperatures between the sample and reference and measures the heat flow required, DTA involves maintaining equal heat flow to the sample and reference and measuring the temperature difference that results. Both techniques yield similar information about phase transitions and thermal properties, but DSC's measurement of heat flow is particularly suited for quantifying the enthalpy of transitions. The results from DSC experiments are typically presented as curves of heat flux versus temperature or time. These curves reveal the enthalpy of transitions by integrating the peaks corresponding to specific thermal events, with the enthalpy calculated using a calorimetric constant specific to the instrument.

DSC has diverse applications in characterizing thermal properties of materials. It is widely used to determine fusion and crystallization events, glass transition temperatures, and oxidation behaviours. For amorphous solids, glass transitions appear as a step in the baseline of the DSC signal, representing changes in heat capacity without formal phase changes. Crystallization events, which are exothermic, result in peaks in the DSC curve, while melting processes are endothermic and also produce distinct peaks. These transitions are critical for creating phase diagrams and studying the thermal behaviour of various chemical systems. Additionally, DSC can provide thermodynamic insights into proteins, revealing information about their stability, structural changes, and interactions with ligands or lipids. This data is particularly valuable in understanding mutations, drug-ligand interactions, and protein denaturation processes.

Experimental considerations play a significant role in obtaining reliable DSC data. The sample condition, choice of crucibles, and scan rates all influence the quality and accuracy of the results. For instance, using a properly sealed crucible ensures sample stability and prevents contamination, but care must be taken to avoid pressure-induced artifacts. Finer powders may enhance signal strength due to increased contact surface, and the sample mass must be carefully chosen to ensure adequate thermal response without compromising resolution. Scan rates can affect the sharpness and visibility of peaks; faster rates amplify peak height but reduce temperature resolution, potentially overlapping closely occurring transitions.

DSC also allows for environmental control using purge gases like nitrogen, argon, or oxygen, depending on the temperature range and desired test conditions. For high-temperature applications, inert gases like argon minimize heat loss, while oxygen is ideal for oxidative studies. These factors, combined with the simplicity and reliability of DSC instrumentation, make it an indispensable tool for both routine quality control and advanced research. Applications extend to polymers, where DSC helps analyse thermal transitions like glass transition, crystallization, and melting. It also aids in identifying polymer degradation and detecting impurities or plasticizers. For liquid crystals, DSC captures subtle energy changes during phase transitions, enabling a detailed understanding of mesomorphic states.

The technique is equally valuable in studying oxidative stability and safety. Oxidative induction time (OIT) and oxidative-onset temperature (OOT) tests provide insights into the resistance of materials to oxidation, helping define optimal storage conditions. Additionally, DSC can serve as an initial safety screening tool to evaluate the thermal stability of substances, guiding decisions about safe processing temperatures. In pharmaceuticals, DSC supports the development and characterization of drug compounds, particularly in ensuring stability in amorphous forms or refining processing conditions to prevent crystallization. It can also determine compound purity through freezing-point depression, offering a reliable method for identifying the presence of impurities.

Thermogravimetric Analysis (TGA)

Thermogravimetric Analysis (TGA) is a powerful analytical technique that measures the weight changes in a material as a function of temperature or time. This method provides valuable insights into the thermal stability, composition, and decomposition behavior of a material. By monitoring the weight loss or gain of a sample under controlled conditions, TGA enables the characterization of processes such as moisture evaporation, thermal decomposition, and oxidation. The data obtained from TGA can be used to understand the behavior of materials under various thermal and environmental conditions, making it a versatile tool for research and quality control.

The working principle of TGA involves heating a sample in a controlled environment, which can be an inert, oxidizing, or reducing atmosphere, depending on the nature of the analysis. The atmosphere is crucial in determining the type of reactions or processes that the material undergoes during heating. For instance, an inert atmosphere like nitrogen prevents oxidation and focuses on thermal decomposition, while an oxidizing atmosphere like oxygen or air allows for the study of combustion or oxidative stability. As the sample is heated, the weight change is continuously recorded, revealing key thermal events. Weight loss typically indicates processes such as the evaporation of moisture, the loss of volatile components, or the decomposition of organic matter, while weight gain can indicate reactions like oxidation.

TGA has a wide range of applications. One of its primary uses is in determining decomposition temperatures, which help identify the thermal stability limits of a material. This is critical for materials that will be exposed to high temperatures in industrial processes or end-use applications. TGA is also invaluable for measuring the moisture and volatile content of materials, such as water, solvents, or plasticizers. This information is essential for applications like polymer processing, where moisture content can affect product quality. Additionally, TGA is used for composition analysis to differentiate between organic and inorganic components. This is particularly useful in materials research, where understanding the proportion of each component is necessary for optimizing performance. In oxidative stability studies, TGA can assess how materials react under oxidative conditions, providing data that is crucial for evaluating the longevity and safety of materials.

Derivative Thermogravimetry (DTG) is an extension of TGA that enhances its analytical capabilities. DTG calculates the rate of weight change as a function of temperature or time, which is particularly useful for identifying overlapping thermal events. For example, in a material with multiple decomposition steps, DTG can distinguish between these steps by analyzing the peaks corresponding to the rate of weight loss. This derivative approach provides clearer and more detailed information about complex thermal processes, making it easier to interpret the data.

The versatility and precision of TGA make it an indispensable tool in many fields, including materials science, pharmaceuticals, polymers, and environmental studies. Its ability to provide detailed information about thermal stability, composition, and reactivity under controlled conditions ensures its continued relevance in both research and industrial applications. Whether it is determining the thermal limits of a polymer, assessing the moisture content of a pharmaceutical compound, or analysing the oxidative stability of a material, TGA offers a reliable and effective method for understanding material behaviour under thermal stress.

Dynamic Mechanical Analysis (DMA)

Dynamic Mechanical Analysis (DMA) is a sophisticated technique used to evaluate the mechanical properties of materials under varying conditions, such as temperature, time, frequency, or applied stress. This method is essential for understanding the viscoelastic behaviour of materials, which exhibit characteristics of both elastic solids and viscous liquids. By probing the interplay between these properties, DMA provides valuable insights into how materials respond to dynamic forces, making it a key tool in materials research and development.

The working principle of DMA involves applying a sinusoidal stress or strain to a material and measuring its corresponding response. The relationship between the applied force and the material's deformation is analysed to determine properties such as modulus, damping, and phase transitions. This sinusoidal approach allows the separation of elastic and viscous responses, which are central to viscoelastic materials. Elastic behaviour corresponds to the material's ability to store energy and recover its shape, while viscous behaviour reflects energy dissipation as heat. By systematically varying parameters like temperature or frequency, DMA can map out the material's performance across a range of conditions.

Key properties measured in DMA include the storage modulus (E'), loss modulus (E''), and tan delta (δ). The storage modulus represents the elastic component of the material, indicating its ability to store energy during deformation. It reflects the stiffness of the material and provides insights into its structural integrity. The loss modulus, on the other hand, represents the viscous component, capturing the energy dissipated as heat during deformation. This parameter is critical for understanding the damping behaviour of the material. Tan delta, the ratio of loss modulus to storage modulus, is a measure of the material's damping efficiency. It indicates how well the material can dissipate energy, and peaks in the tan delta curve often correspond to critical transitions, such as the glass transition temperature (Tg).

One of the primary applications of DMA is the determination of the glass transition temperature, a critical property for polymers and other viscoelastic materials. The Tg is identified as a peak in the tan delta curve and marks the temperature at which the material transitions from a rigid, glassy state to a soft, rubbery state. This transition has significant implications for material performance, particularly in applications involving variable temperatures. DMA is also used to study time-dependent behaviours such as creep and stress relaxation, which are essential for understanding long-term material performance under load. Frequency-dependent measurements further reveal how materials respond to dynamic forces, providing insights into their behaviour in real-world applications where oscillatory forces are present.

Another key application of DMA is the determination of crosslink density in elastomers and thermosetting polymers. Crosslink density is a measure of the interconnectedness of polymer chains, which directly affects the mechanical properties of the material. DMA provides a precise method for quantifying this parameter, helping optimize materials for specific applications. Additionally, the technique is valuable for studying the effect of additives, fillers, and processing conditions on material properties.

DMA's versatility and precision make it indispensable for material scientists and engineers. By providing a detailed understanding of viscoelastic behaviour, it helps in the design and optimization of materials for a wide range of applications, from automotive components and medical devices to adhesives and coatings. Its ability to measure mechanical properties across varying conditions ensures that materials can be tailored to meet specific performance requirements, enhancing their functionality and reliability in diverse environments.

Integration in Material Research

The integration of Differential Scanning Calorimetry (DSC), Thermogravimetric Analysis (TGA), and Dynamic Mechanical Analysis (DMA) in material research offers a comprehensive framework for characterizing materials by elucidating their thermal and mechanical behaviours. Each technique contributes unique insights, and their combined application is particularly effective in revealing the complexities of material properties and performance.

DSC is pivotal in identifying phase transitions, such as melting, crystallization, and glass transition, which are critical for understanding material behaviour under varying thermal conditions. For instance, the glass transition temperature (Tg) is crucial for polymers, as it indicates the temperature range where materials transition from a rigid to a flexible state, impacting their application in environments with fluctuating temperatures [552]. This is particularly relevant in the design of materials that require precise thermal control, as evidenced by studies that utilize DSC to evaluate thermal transitions in polymeric materials [552, 553].

TGA complements DSC by focusing on the thermal stability of materials through the measurement of weight changes as a function of temperature or time. This technique is essential for assessing decomposition, moisture content, and volatile components, thereby providing insights into the durability and performance of materials under thermal stress. For example, TGA has been effectively used to differentiate between organic and inorganic components in composite materials, which aids in optimizing formulations for enhanced thermal stability [554]. The ability of TGA to reveal compositional changes is invaluable in material development, particularly in applications where thermal stability is paramount [554].

DMA evaluates the mechanical performance of materials by analysing their response to dynamic forces across various temperatures and frequencies. This technique is instrumental in understanding viscoelastic behaviour, which includes energy storage and dissipation—key factors that influence material performance in practical applications. DMA has been employed to study phenomena such as creep and stress relaxation in polymers, allowing researchers to fine-tune mechanical properties to meet specific requirements [554, 555]. The insights gained from DMA are critical for industries where mechanical performance under dynamic loading conditions is essential, such as in aerospace and automotive applications [555].

The synergistic application of DSC, TGA, and DMA provides a holistic view of material properties. For instance, in polymers, DSC can identify thermal transitions, TGA can assess thermal stability and composition, and DMA can measure mechanical properties like stiffness and damping. This integration allows researchers to correlate thermal events with mechanical behaviour, leading to a deeper understanding of how materials respond to real-world conditions [553, 555]. The combined use of these techniques is widely applied across various materials, including polymers, composites, metals, and ceramics, ensuring high-quality control and facilitating failure analysis in critical industries [554, 555].

Relationship Between Mechanical Testing and Thermal Analysis

The relationship between mechanical testing and thermal analysis is rooted in their complementary roles in comprehensively understanding a material's performance. Mechanical testing, which includes tensile, compression, and fatigue tests, evaluates how materials respond to applied forces, measuring properties such as strength, deformation, and durability. In contrast, thermal analysis techniques such as Differential Scanning Calorimetry (DSC), Thermogravimetric Analysis (TGA), and Dynamic Mechanical Analysis (DMA) focus on how materials respond to temperature changes, thermal decomposition, and viscoelastic behaviour. By combining these approaches, researchers gain a holistic understanding of a material's behaviour under both mechanical and thermal conditions, which are often interdependent in real-world applications.

One key area where mechanical and thermal properties intersect is in the thermal effects on mechanical behaviour. Materials frequently exhibit changes in stiffness, tensile strength, and fatigue resistance when exposed to varying temperatures. DSC and DMA provide critical data on thermal transitions, such as the glass transition temperature (Tg), melting points, and crystallization events. These transitions are pivotal in understanding how a material's mechanical properties evolve with temperature. For instance, polymers often lose stiffness and tensile strength above their Tg, which can limit their effectiveness in load-bearing applications.

Thermal decomposition also directly impacts a material's structural integrity and mechanical performance. TGA is essential for evaluating a material's thermal stability by measuring weight loss during decomposition. This analysis identifies the temperature limits within which a material can maintain its mechanical properties. For example, a composite material exposed to high temperatures might experience weight loss due to the breakdown of organic components, leading to a significant reduction in tensile or compressive strength. Such information is crucial for determining the operational limits of materials in high-temperature environments.

Viscoelastic behaviour and fatigue are additional areas where thermal and mechanical testing overlap. DMA measures viscoelastic properties such as storage modulus, loss modulus, and damping behaviour (tan delta), providing insights into how materials perform under dynamic loading conditions. Fatigue testing evaluates a material's ability to endure cyclic stresses over time, which often involves hysteresis and energy dissipation. DMA helps quantify these effects by analysing damping properties, enabling predictions of a material's resistance to fatigue under varying thermal conditions.

The effects of additives and modifications further illustrate the synergy between mechanical and thermal analyses. Techniques like DSC and TGA can characterize how additives, fillers, or reinforcements influence a material's thermal behaviour, such as enhancing thermal stability or altering thermal transitions. These changes often correlate with improved mechanical properties, such as increased tensile strength or better fatigue resistance. For example, fibre-reinforced polymers might demonstrate both superior thermal stability and enhanced mechanical performance, which can be precisely quantified using these complementary testing methods.

Material selection for applications requiring combined mechanical and thermal performance highlights the importance of integrating these methods. In industries like aerospace, automotive, and electronics, materials must withstand both mechanical stress and thermal challenges. Mechanical testing ensures that materials possess adequate strength, ductility, and fatigue resistance, while thermal analysis confirms that these properties remain stable within the operational temperature range. For instance, composite materials in aerospace applications must endure both mechanical loads and thermal cycling, and the combined use of tensile testing and TGA ensures their reliability under these conditions.

Practical applications further illustrate this relationship. In aerospace, composite materials must maintain load-bearing capabilities and thermal stability, with mechanical testing and thermal analysis working together to ensure performance. In the automotive sector, tires are subjected to cyclic loading and temperature variations, requiring DMA to evaluate viscoelastic behaviour and fatigue testing to confirm durability. In electronics, circuit board materials must withstand thermal expansion and cycling while retaining mechanical integrity, which is assessed through a combination of TGA, DSC, and tensile or compression testing.

The relationship between mechanical testing and thermal analysis is one of interdependence. Understanding how materials respond to both mechanical forces and thermal changes enables engineers and scientists to design materials that meet structural and environmental demands. By integrating these methodologies, it becomes possible to predict and optimize material performance under diverse and often extreme conditions, ensuring reliability and functionality in various applications.

Chemical and Spectroscopic Characterization: XRD, SEM, and TEM

Chemical and spectroscopic characterization techniques such as X-Ray Diffraction (XRD), Scanning Electron Microscopy (SEM), and Transmission Electron Microscopy (TEM) focus primarily on analysing the structural, chemical, and morphological aspects of materials. In contrast, mechanical testing (tensile, compression, and fatigue) and thermal analysis (DSC, TGA, and DMA) evaluate a material's performance under mechanical loads and thermal conditions, respectively. These approaches serve different purposes and operate on distinct principles, complementing one another to provide a comprehensive understanding of materials.

Chemical and spectroscopic characterization techniques investigate the material's composition, structure, and micro/nanoscale features. XRD identifies crystalline phases, lattice structures, and defects, offering insights into how atomic arrangements influence material properties. SEM provides high-resolution surface imaging and elemental composition using electron interactions with the material. TEM extends this capability to the atomic level, providing detailed structural and chemical information about materials at the nanoscale.

Mechanical testing, on the other hand, evaluates how materials respond to external forces, measuring properties like strength, elasticity, ductility, and durability. These tests assess the practical load-bearing capacity and deformation behaviour of materials. Thermal analysis complements mechanical testing by examining how materials respond to temperature changes, identifying critical thermal transitions, decomposition behaviour, and viscoelastic properties. Together, these techniques address material performance under real-world operating conditions.

Chemical and spectroscopic techniques provide qualitative and quantitative information about material structure and composition. XRD reveals crystallographic data, such as phase identification and lattice parameters, which influence mechanical and thermal properties. SEM provides topographical and compositional data, while TEM offers atomic-scale imaging and chemical mapping, which can explain property variations due to defects, grain boundaries, or interfaces.

In contrast, mechanical and thermal analyses provide macroscopic property data. Mechanical testing yields stress-strain relationships, fatigue life, and failure characteristics, directly related to material application. Thermal analysis provides data on heat capacity, thermal transitions, and decomposition, crucial for materials exposed to thermal cycling or elevated temperatures.

Chemical and spectroscopic techniques often focus on the micro to atomic scale, examining structures down to individual atoms or small material regions. For instance, TEM can resolve atomic lattice arrangements, and SEM can highlight surface features at high magnifications.

Mechanical and thermal analyses typically operate on a macroscopic scale, studying bulk material properties. Mechanical testing evaluates whole samples under controlled loads, and thermal analysis assesses thermal behaviour over a sample's entirety. These methods focus on how the entire material performs as a unit, rather than its microscopic details.

Chemical and spectroscopic characterization is crucial for understanding the foundational structure and composition of materials, which often dictate their mechanical and thermal behaviour. For example, XRD can identify a material's crystalline phase, which may correlate with its tensile strength or thermal stability. SEM and TEM can pinpoint defects or inclusions that cause early mechanical failure or thermal instability.

Mechanical testing and thermal analysis focus on application-driven performance metrics. Industries like aerospace, automotive, and electronics rely on mechanical and thermal tests to ensure materials meet practical requirements, such as withstanding loads, resisting thermal degradation, or maintaining dimensional stability during temperature changes.

While these techniques differ, they are highly complementary. Structural and compositional insights from XRD, SEM, and TEM can explain phenomena observed in mechanical and thermal tests. For instance, an XRD analysis revealing a phase transformation at high temperatures can help interpret changes in modulus measured by DMA. Similarly, SEM imaging of fatigue cracks can elucidate failure mechanisms identified during fatigue testing. TEM can reveal nanoscale features like dislocations or precipitates that influence tensile strength or thermal stability.

Chemical and spectroscopic characterization techniques focus on understanding a material's atomic and microstructural features, providing foundational knowledge of its makeup and morphology. In contrast, mechanical testing and thermal analysis assess how materials perform under external forces and thermal conditions, offering practical insights into their usability and reliability. Together, these techniques offer a multi-scale, multi-disciplinary approach to material characterization, connecting fundamental structure with performance to meet diverse research and industrial needs.

Non-Destructive Testing Techniques

Non-Destructive Testing (NDT) techniques are essential for evaluating the properties, integrity, and performance of advanced materials without causing damage or altering their structure. These methods are widely used in industries such as aerospace, automotive, energy, and construction to ensure material reliability, identify defects, and assess performance under operational conditions. For advanced materials like composites, ceramics, and high-performance alloys, NDT methods play a crucial role due to their unique properties and specialized applications.

Key Non-Destructive Testing Techniques

Ultrasonic Testing (UT): This technique employs high-frequency sound waves to identify internal flaws or measure material properties. A transducer emits ultrasonic waves into the material, and the reflected waves are analysed to detect inconsistencies. For instance, carbon fibre-reinforced composites, which have complex internal structures, can be effectively inspected for delaminations and voids using UT. Variants such as phased-array ultrasonic testing (PAUT) enhance imaging capabilities, allowing for detailed flaw characterization [556].

Ultrasonic waves, typically in the range of 0.1–15 MHz and occasionally up to 50 MHz, are introduced into the material via a transducer, and their behaviour as they propagate through the material provides critical information about its integrity. A common application of UT is ultrasonic thickness measurement, often used to monitor corrosion or erosion in pipelines. Ultrasonic testing is extensively employed across industries for inspecting welds and ensuring their structural integrity.

This technique is versatile and effective for testing metals and alloys like steel and aluminium, as well as non-metallic materials like concrete, wood, and composites, although resolution diminishes with less homogeneous materials. Industries such as aerospace, automotive, manufacturing, and construction rely on UT for its ability to provide fast, accurate, and reliable results. For example, aerospace manufacturers use UT to inspect critical components like turbine blades and fuselage panels, while automotive sectors employ it to evaluate welds and structural parts.

The origins of ultrasonic testing date back to the 1930s, with significant advancements following in the 1940s. Dr. Floyd Firestone of the University of Michigan was granted the first patent for a practical ultrasonic testing device in 1942. His invention introduced the concept of using high-frequency vibrations to detect inhomogeneities such as cracks or voids within solid materials, even when these flaws were entirely internal. This innovation laid the foundation for modern ultrasonic testing.

Further advancements came in the 1960s, particularly with the development of piezoelectric transducers. These devices convert electrical energy into ultrasonic waves, which propagate through the material being tested. Reflected waves from discontinuities within the material are then converted back into electrical signals and analysed. This basic principle remains central to UT today, with modern enhancements such as phased-array systems and electromagnetic acoustic transducers (EMATs) improving resolution and efficiency.

The fundamental process of ultrasonic testing involves an ultrasound transducer connected to a diagnostic machine. The transducer emits ultrasonic waves into the material, and its response is monitored using either reflection or attenuation methods. In the reflection, or pulse-echo mode, the same transducer generates and receives the ultrasonic waves. Reflections occur at material interfaces, such as the back wall or internal flaws, and the diagnostic machine interprets these reflections as signals. The signal amplitude represents the intensity of the reflection, while its position corresponds to the location of the interface or flaw.

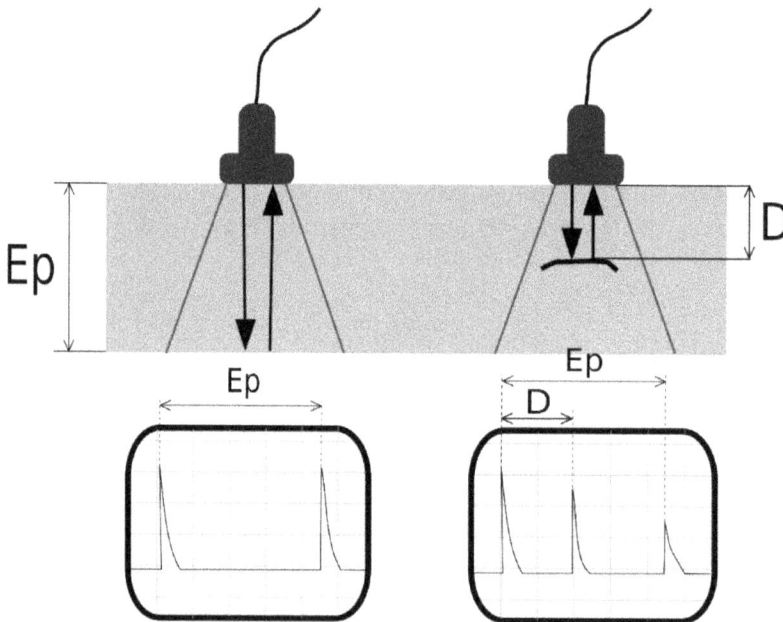

Figure 42: Ultrasonic Inspection principle of a material. Romary, CC BY-SA 3.0, via Wikimedia Commons.

In attenuation, or through-transmission mode, a transmitter sends ultrasonic waves through one side of the material, while a receiver on the opposite side detects the transmitted waves. Imperfections or variations in the material reduce the intensity of the transmitted waves, signalling the presence of flaws or changes in the material's properties. For both methods, the use of a couplant, such as oil or water, enhances the process by minimizing energy loss at the interface between the transducer and the material.

Ultrasonic testing is capable of much more than flaw detection. For instance, it can be used to measure grain size in metals. This non-destructive method evaluates ultrasonic velocities, attenuation, and backscatter features to estimate grain size. Theoretical models for scattering attenuation, developed by researchers like Stanke and Kino, provide the basis for correlating ultrasonic data with grain size. For example, in pure niobium, grain size can be linearly correlated with attenuation due to grain boundary scattering. This allows for non-destructive determination of material properties using relatively simple instruments.

Another advanced application is the characterization of early-stage fatigue or creep damage in materials. Nonlinear ultrasonic methods leverage the distortion of intensive ultrasonic waves when they encounter microstructural damage. This distortion is quantified using the acoustic nonlinearity parameter (β), calculated from the harmonic components of the ultrasonic signal. By identifying these subtle changes, UT provides critical insights into material health and potential failure mechanisms.

UT is highly sensitive and versatile, capable of detecting flaws deep within materials while requiring access to only one side of the object. It is equally effective for evaluating material properties, thickness, and internal structures. Modern advancements, such as phased-array instruments, have enhanced its precision, enabling detailed imaging and defect characterization. The non-destructive nature of UT makes it particularly valuable for inspecting expensive or critical components, such as those used in aerospace or nuclear applications, where material integrity is paramount.

Ultrasonic testing remains one of the most reliable and widely used NDT techniques for advanced materials and industrial applications. Its ability to detect internal flaws, measure material properties, and assess early damage stages, combined with ongoing technological improvements, ensures its continued relevance and utility in ensuring material safety and performance.

Radiographic Testing (RT): RT utilizes X-rays or gamma rays to produce images of a material's internal structure. This method is particularly adept at identifying voids, cracks, and inclusions in dense materials like metal-matrix composites and ceramics. Advanced digital radiography and computed tomography (CT) provide three-dimensional imaging, facilitating detailed visualization of internal features without damaging the material [557].

Radiographic Testing (RT) is a widely used non-destructive testing (NDT) method that employs high-energy electromagnetic radiation, such as x-rays or gamma rays, to examine the internal structure of materials and components. This technique is instrumental in identifying internal flaws, defects, or inconsistencies within a material, ensuring quality and reliability in manufacturing processes. Radiographic Testing is extensively applied across industries such as aerospace, automotive, construction, and energy to inspect critical components and ensure structural integrity.

The principle of radiographic testing involves placing the test object between a radiation source and a detection medium, such as film or an electronic detector. As the radiation penetrates the object, variations in material density and thickness cause differential attenuation through scattering and absorption processes. These differences in radiation absorption are captured as an image, either on film or digitally, revealing internal features of the object. The resulting image highlights areas of higher or lower density, enabling the detection of flaws such as cracks, voids, inclusions, or porosity. Modern imaging technologies have significantly improved this process, offering a range of methods including Film Radiography, Real-Time Radiography (RTR), Computed Tomography (CT), Digital Radiography (DR), and Computed Radiography (CR).

The sources of radiation used in industrial radiography are x-rays and gamma rays, both of which possess short wavelengths and high energy levels. X-rays are typically generated by specialized equipment, while gamma rays are produced from radioactive isotopes such as Cobalt-60 or Iridium-192. The choice between x-rays and gamma rays depends on the application, with gamma rays often used for thicker or denser materials due to their higher penetration power. Because of the inherent hazards associated with radiation, stringent safety measures are mandatory during radiographic testing, including adherence to local rules and regulations, the use of shielding, and strict access controls to protect personnel and the environment.

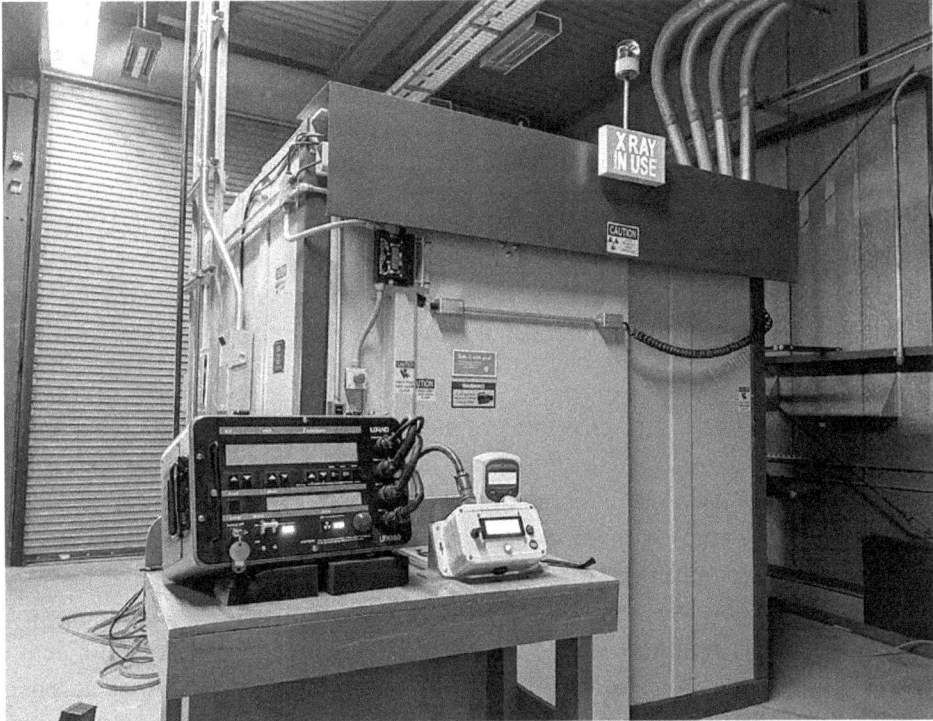

Figure 43: Xray vault used in radiography. Dandersound, CC BY-SA 4.0, via Wikimedia Commons.

Computed Tomography (CT) is an advanced radiographic technique that provides both cross-sectional and 3D volumetric images of the test object. Unlike traditional 2D radiography, CT eliminates the superimposition of features, offering unparalleled clarity and detail of the internal structure. CT works by capturing multiple x-ray images from different angles around the object, which are then reconstructed into a 3D model using advanced algorithms. This capability allows for detailed analysis of complex geometries, internal defects, and material distributions, making it particularly valuable for industries requiring high precision, such as aerospace, medical device manufacturing, and advanced material research.

The versatility of radiographic testing, particularly with advanced techniques like CT, allows for comprehensive evaluation of a wide range of materials, including metals, composites, and ceramics. This method can detect internal defects that may not be visible with other NDT methods, making it indispensable for critical applications where reliability and safety are paramount. Whether inspecting welds, castings, or high-performance components, radiographic testing provides a thorough understanding of internal features, ensuring that materials meet the required specifications and standards.

Eddy Current Testing (ECT): ECT is an electromagnetic technique primarily used to detect surface and subsurface defects in conductive materials. By inducing eddy currents through an alternating current in a coil, this method can identify flaws and variations in material properties. ECT is especially useful for inspecting thin metallic layers and coatings, making it relevant for advanced alloys in aerospace and automotive applications [558].

Eddy-current testing (ECT) is a non-destructive testing (NDT) method that employs electromagnetic induction to detect and characterize surface and sub-surface flaws in conductive materials. This technique is widely used across various industries, including aerospace, petrochemical, and manufacturing, to ensure the integrity of components without causing damage.

The foundation of ECT lies in the principles of electromagnetism. In 1831, English scientist Michael Faraday discovered electromagnetic induction, observing that a time-varying magnetic field passing through a conductor induces an electric current within it. This phenomenon, known as eddy currents, was further explored by French physicist Léon Foucault in 1855, who is credited with their discovery. In 1879, English-born scientist David Edward Hughes demonstrated how the properties of a coil change when placed in contact with metals of different conductivity and permeability, applying this knowledge to metallurgical sorting tests. The practical application of ECT as an NDT technique for industrial purposes was significantly advanced during World War II in Germany. Professor Friedrich Förster, working at the Kaiser-Wilhelm Institute, adapted eddy current technology for industrial use, developing instruments for measuring conductivity and sorting mixed ferrous components. In 1948, Förster founded a company, now known as the Foerster Group, where he made substantial progress in developing and marketing practical ECT instruments.

ECT operates on the principle of electromagnetic induction. A coil of conductive wire, known as the probe, is energized with an alternating current, producing an alternating magnetic field around it. When this probe is brought near a conductive material, the alternating magnetic field induces circulating currents, termed eddy currents, within the material. Variations in the material's electrical conductivity, magnetic permeability, or the presence of defects alter the flow of these eddy currents. These changes affect the impedance of the probe coil, which can be measured to detect and characterize flaws. The depth of penetration of eddy currents is influenced by factors such as the frequency of the alternating current and the material properties, a concept known as skin depth.

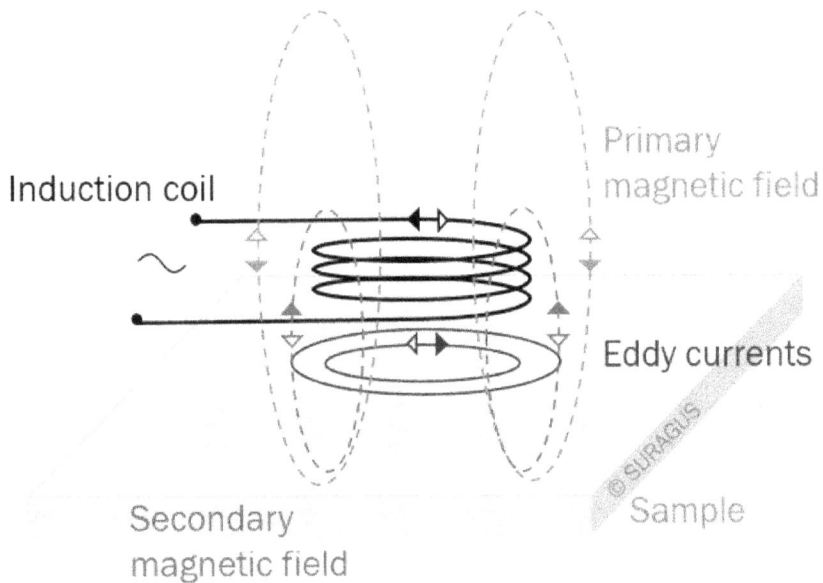

Figure 44: Visualization of Eddy Current Induction by Induction Coil. Stefan Trache, CC BY-SA 4.0, via Wikimedia Commons.

ECT is versatile, with two primary applications being surface inspection and tubing inspection. In surface inspection, ECT is extensively used in the aerospace industry to detect tight cracks and other surface defects in both ferromagnetic and non-ferromagnetic materials. For tubing inspection, conventional ECT is applied to non-ferromagnetic tubing, such as steam generator tubes in nuclear plants and heat exchanger tubes in power and petrochemical industries, to detect pits, wall loss, and corrosion. Variations of ECT, like full saturation eddy current testing, are employed for partially magnetic materials, utilizing a magnetic field to suppress permeability variations and enhance defect detection. Additionally, ECT is useful for measuring electrical conductivity, coating thickness, and assessing material properties such as hardness and grain size.

The global NDT equipment market has seen significant growth, with the magnetic and electromagnetic NDT equipment segment, which includes ECT, estimated at $220 million in 2012. This market was projected to grow at a compound annual growth rate of 7.5%, reaching approximately $315 million by 2016. Advancements in ECT technology, such as the development of eddy current array (ECA) techniques, have enhanced inspection capabilities, allowing for faster and more accurate detection of defects. The ongoing research and development in this field continue to expand the applications and effectiveness of ECT in various industries.

Eddy-current testing is a crucial non-destructive evaluation method rooted in the principles of electromagnetism. Its development over the years has led to a reliable and efficient technique for detecting and characterizing flaws in conductive materials, playing a vital role in quality assurance and safety across multiple industries.

Eddy current testing (ECT) has evolved significantly to address limitations associated with traditional methods. Among the innovative techniques developed, the Eddy Current Array (ECA) and Lorentz Force Eddy Current Testing (LET) stand out for their enhanced capabilities in detecting flaws in conductive materials.

Eddy Current Array (ECA) technology builds upon the fundamental principles of conventional ECT, utilizing an array of coils arranged in a specific topology. This configuration allows for a tailored sensitivity profile that is optimized for detecting various types of defects. The multiplexing of coils minimizes mutual inductance, enabling faster inspections and broader coverage compared to traditional methods. ECA systems are noted for their reduced operator dependence, as they yield more consistent results than manual raster scans, thus enhancing the reliability of inspections [559]. Furthermore, the simplicity of scan patterns facilitates easier data analysis and improves defect positioning and sizing through encoded data [560]. The flexibility of array probes also allows for inspections in hard-to-reach areas, making ECA a powerful tool in non-destructive testing (NDT) applications.

ECA technology is particularly beneficial in industries where time efficiency and accuracy are critical, such as in the inspection of carbon steel welds. The advantages of ECA, including faster inspections and improved detection capabilities, underscore its growing adoption in various sectors [559].

In contrast to traditional ECT, which is limited by the skin effect and typically only detects surface-level defects, Lorentz Force Eddy Current Testing (LET) offers a novel approach for identifying deeper flaws within conductive materials. LET operates by generating eddy currents through the relative motion between a conductor and a permanent magnet, which allows for the detection of defects that are not accessible through conventional methods [561, 562]. This technique modifies the traditional ECT approach by changing both the induction method of eddy currents and the detection of their perturbations. The Lorentz force generated when a magnet passes over a defect provides a clear signal that can be measured, thus indicating the presence of flaws [561, 562].

The principle of LET is grounded in electromagnetic induction and the Lorentz force law. According to this law, a charged particle moving through a magnetic field experiences a force proportional to its velocity, the magnetic field's strength, and the particle's charge. In LET, when a conductive material moves relative to a magnetic field, or vice versa, eddy currents are induced within the material. These circulating electric currents flow in closed loops and generate their own magnetic fields. When these eddy currents interact with an externally applied magnetic field, a Lorentz force is produced, acting on the conductive material. If the material is free of defects, the Lorentz force is uniform. However, the presence of defects such as cracks, voids, or inclusions disrupts the eddy current distribution, leading to localized changes in the Lorentz force. These disturbances are detected and analysed to determine the size, position, and nature of the defects.

LET comprises several key components. A strong permanent magnet or an electromagnet generates the external magnetic field required to induce eddy currents. The technique is applicable exclusively to conductive materials, as eddy currents cannot form in non-conductive substances. Relative motion between the magnet and the material is essential to generate the time-varying eddy currents, which is typically achieved through linear or rotational movement. Highly sensitive force sensors are used to measure the Lorentz force and detect minute variations caused by defects.

The working process of LET begins with positioning a magnet near the surface of the conductive material to be inspected. Relative motion is introduced, inducing eddy currents within the material. As these currents interact with the magnetic field, a Lorentz force is generated. For defect-free materials, this force exhibits a uniform profile. However, when the eddy currents encounter defects such as cracks or voids, the current distribution is disrupted, leading to localized changes in the Lorentz force. These anomalies are recorded and analysed, enabling the precise detection and characterization of defects.

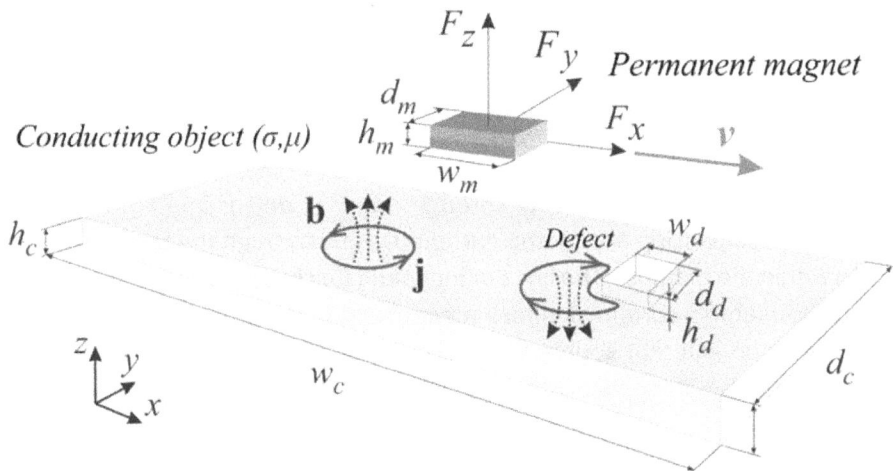

Figure 45: Lorentz force eddy current testing. Research Training Group Lorentz Force Velocimetry and Lorentz Force Eddy Current Testing, CC BY-SA 3.0, via Wikimedia Commons.

Research has demonstrated that LET can effectively detect defects at greater depths than traditional ECT, making it a valuable technique in applications where deep-seated flaws may compromise structural integrity. Studies have shown that the detection capabilities of LET can surpass those of classical ECT, particularly in scenarios involving moving conductors [563, 564]. The ability to measure the Lorentz force allows for real-time monitoring of defects, enhancing the overall efficacy of NDT practices [565].

Despite its advantages, LET has some limitations. It is restricted to conductive materials, as non-conductive materials cannot support eddy currents. Accurate defect detection requires precise calibration and detailed knowledge of the material's properties, such as conductivity and permeability. The technique's reliance on relative motion makes it sensitive to environmental factors like vibrations or uneven motion, which can introduce noise in the measurements. Additionally, the advanced sensors and force measurement systems used in LET can be costly compared to traditional eddy current testing setups.

Magnetic Particle Testing (MPT): This method is applicable to ferromagnetic materials. By applying a magnetic field, defects disrupt the field, allowing fine magnetic particles to accumulate at defect sites, which become visible under appropriate lighting. While limited to ferromagnetic materials, MPT is highly effective for detecting surface and near-surface flaws in advanced magnetic alloys [566].

Magnetic Particle Inspection (MPI) is a widely used nondestructive testing (NDT) process designed to detect surface and shallow subsurface discontinuities in ferromagnetic materials, such as iron, nickel, cobalt, and their alloys. By applying a magnetic field to the part under examination, MPI identifies imperfections that disrupt the material's magnetic flux. This disruption, referred to as flux leakage, occurs because air and non-magnetic defects cannot support as much magnetic field density as ferromagnetic materials. The resulting leakage attracts ferrous particles, which form visible indications of defects, allowing evaluators to determine their nature, location, and severity.

The process of magnetization can occur through either direct or indirect methods. Direct magnetization involves passing an electric current directly through the test object, creating a magnetic field within the material. The lines of magnetic flux in this scenario are perpendicular to the flow of the electric current. Indirect magnetization, in contrast, involves applying a magnetic field from an external source without passing current through the object. In both methods, the induced magnetic field interacts with potential discontinuities, creating areas of flux leakage.

MPI relies on different types of electrical currents to generate magnetic fields. Alternating current (AC) is commonly used for detecting surface discontinuities due to its tendency to concentrate at the material's surface, a phenomenon known as the skin effect. However, AC has limited penetration depth, making it unsuitable for detecting deeper flaws. Direct current (DC), including full-wave DC (FWDC) and half-wave DC (HWDC), is used for subsurface defect detection. FWDC can penetrate deeper into the material, but its effectiveness decreases for large cross-sectional parts. HWDC, on the other hand, offers improved particle mobility due to its pulsating nature, making it advantageous for detecting both surface and subsurface defects. An AC electromagnet is often preferred for surface defect detection, while DC is better suited for identifying subsurface flaws.

MPI equipment varies based on application requirements. A common device used in mass production is the wet horizontal MPI machine, which includes a head and tail stock for positioning the part and a coil to alter the orientation of the magnetic field. Handheld magnetic yokes are versatile tools for field inspections, especially for welds and remote locations. These yokes induce a magnetic field between two poles, but their inspection range is limited to the area between the poles. For large-scale inspections, the yoke must be rotated 90 degrees to detect horizontal and vertical discontinuities, a process that can be time-intensive.

After completing the inspection, the magnetized part must be demagnetized to remove any residual magnetism. Demagnetization requires an equal or greater magnetic field than the one used for magnetization, which is then gradually reduced to zero. This process can be performed using various methods, including pull-through AC demagnetization, AC decaying demagnetization, and reversing full-wave DC demagnetization. Pull-through AC demagnetization involves passing the part through a high magnetic field that diminishes as the part moves away from the coil. AC decaying demagnetization gradually reduces the AC current over time, leaving the magnetic domains randomized. Reversing full-wave DC demagnetization alternates the current direction while reducing its intensity, effectively neutralizing residual magnetism. Each method is selected based on the geometry and material properties of the part.

The particles used in MPI are typically composed of iron oxide and are applied either in a dry or wet suspension. Wet system particles, ranging from less than 0.5 to 10 micrometres in size, are carried in water or oil-based solutions and often fluoresce under ultraviolet (UV) light for enhanced visibility. Dry particles, ranging from 5 to 170 micrometres, are used in white-light environments and are applied using air powder applicators. Aerosol-applied particles are another option, combining the convenience of premixed particles in spray cans for small-scale or field applications.

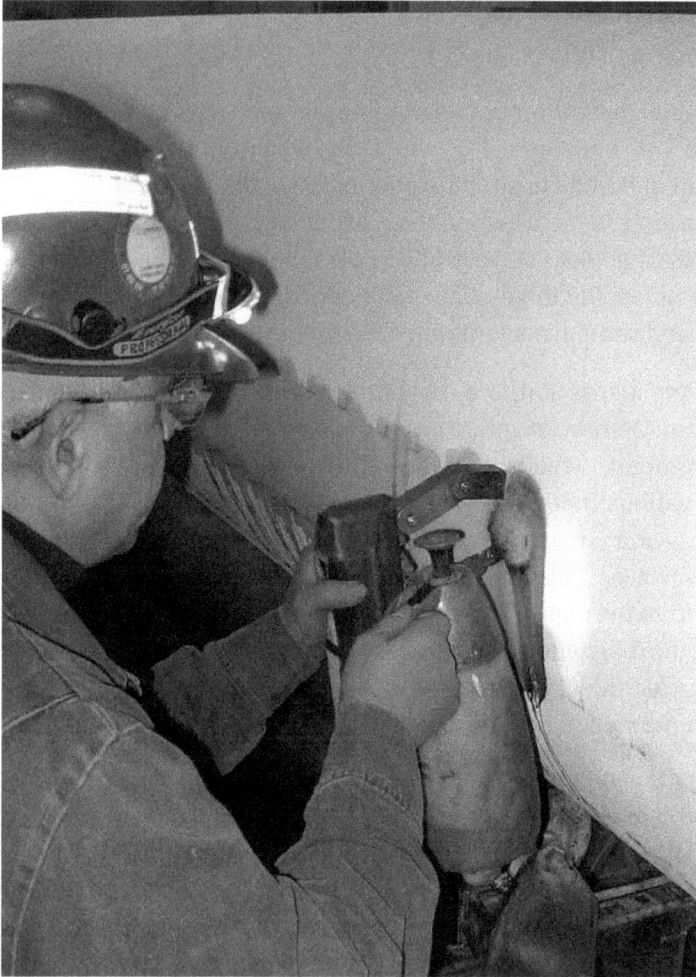

Figure 46: A technician performs magnetic-particle inspection (MPI) on a pipeline to check for stress corrosion cracking. Davidmack at English Wikipedia, CC0, via Wikimedia Commons.

Inspection involves several steps to ensure accuracy. The workpiece is first cleaned to remove contaminants, followed by calculations to determine the necessary magnetizing current. A magnetic pulse is applied while ferrous particles are sprayed onto the surface. Operators use UV light to detect indications of defects, which appear within specific angles relative to the magnetic field. Depending on the geometry of the part, the orientation of the magnetic field may need to be adjusted to inspect areas that cannot be effectively examined in a single pass. If defects are found, the workpiece is either rejected or accepted based on predefined criteria, and the part is demagnetized before use.

Magnetic Particle Inspection is a versatile and effective NDT technique for ensuring the integrity of ferromagnetic materials. Its ability to detect surface and shallow subsurface defects makes it invaluable in industries such as aerospace, automotive, and construction. By utilizing a combination of electrical currents, specialized equipment, and inspection techniques, MPI provides a reliable method for identifying defects that could compromise material performance and safety.

Infrared Thermography (IRT): IRT detects thermal anomalies on a material's surface using infrared cameras. When subjected to heat or stress, defects disrupt heat flow, creating detectable thermal patterns. This technique is particularly beneficial for composites and layered materials, where traditional methods may struggle due to complex geometries [567].

Infrared thermography (IRT), also known as thermal video or thermal imaging, is a technique that uses a thermal camera to capture and create images based on the infrared radiation emitted by objects. This process is grounded in the principles of infrared imaging science, utilizing the fact that all objects with a temperature above absolute zero emit infrared radiation according to the black body radiation law. Thermographic cameras detect radiation in the long-infrared range of the electromagnetic spectrum, typically between 9,000 and 14,000 nanometres (9–14 μm). The images produced, called thermograms, visually represent variations in temperature. Since warmer objects emit more radiation, they stand out clearly against cooler backgrounds when viewed through a thermal camera. This capability allows thermography to be effective in various applications, from military surveillance to wildlife observation, both during the day and at night.

Thermography is particularly valuable in medical diagnostics and other biological studies. It can monitor physiological changes in humans and warm-blooded animals, such as during allergy testing or veterinary assessments. However, its use in some contexts, such as breast screening, has been controversial. The U.S. Food and Drug Administration (FDA) has cautioned against relying solely on thermography for cancer detection, emphasizing that it should not replace mammography. During health crises, such as the 2009 swine flu pandemic, thermography was employed at airports and by government authorities to detect elevated body temperatures as a potential sign of infection, demonstrating its utility in public health surveillance.

Thermal imaging has a rich history and has seen a dramatic rise in commercial and industrial applications over the past half-century. Firefighters use it to see through smoke, locate individuals, and identify the source of a fire. Maintenance technicians apply thermography to detect overheating in electrical systems or machinery, signalling potential failures. In building construction, it is used to identify thermal signatures that indicate heat leaks, enabling improvements in insulation and energy efficiency. These practical uses highlight the importance of thermography in safety, maintenance, and energy conservation.

Modern thermographic cameras resemble camcorders in appearance and functionality. They use advanced focal plane arrays (FPAs) to detect longer wavelengths of infrared radiation, such as mid- and long-wavelength infrared. Common sensor materials include indium antimonide (InSb), indium gallium arsenide (InGaAs), mercury cadmium telluride (HgCdTe), and quantum well infrared photodetectors (QWIP). More recent advancements involve uncooled microbolometer sensors, which are cost-effective but offer lower resolution compared to cooled detectors. While some older or more sensitive sensors require cryogenic cooling, such as through miniature Stirling cycle refrigerators or liquid nitrogen, modern microbolometers eliminate this need, reducing complexity and cost. Despite these advances, thermal cameras remain significantly more expensive than visible-spectrum counterparts, and high-end models are often export-restricted due to their military applications.

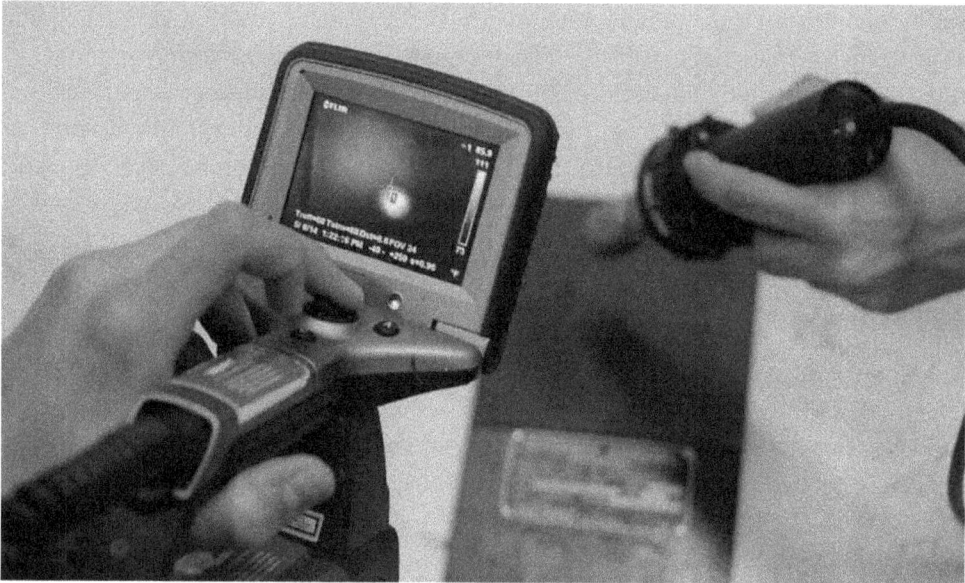

Figure 47: Airman Kawika Cadman, 509th Maintenance Squadron non-destructive inspection technician, uses infrared technology to find discrepancies in aircraft parts. Defense Visual Information Distribution Service, Public Domain, via National Archives and Defense Visual Information Distribution Service.

Thermal images, or thermograms, are visual representations of the total infrared energy emitted, transmitted, and reflected by an object. The complexity of this measurement arises from the fact that infrared radiation comes from multiple sources, including the object itself, transmitted radiation from external sources, and radiation reflected off the object's surface. A thermal imaging camera uses algorithms to process these inputs and reconstruct a temperature image. However, this reconstruction is an approximation since external factors, such as ambient radiation and object emissivity, influence the measurement. Emissivity, a material property indicating the ability to emit thermal radiation, varies by material and wavelength. For instance, metals often have low emissivity, while non-metals like asphalt or oxides exhibit higher emissivity.

Accurate temperature measurement using thermography requires proper emissivity settings in the camera. For precise readings, thermographers may apply high-emissivity materials, such as black insulation tape, to the object's surface, or they may rely on emissivity tables for approximate values. In situations where emissivity adjustments are not feasible due to accessibility or safety concerns, the accuracy of thermographic measurements may be affected. Other factors, such as absorption by the surrounding medium and reflections from nearby objects, also complicate temperature estimation, requiring careful calibration and interpretation.

Thermal imaging cameras typically produce monochrome images, as their sensors do not differentiate between wavelengths of infrared radiation. However, these images are often displayed in pseudo-colour, where variations in colour are used to represent temperature differences. Bright colours such as white or yellow indicate warmer areas, while cooler regions appear as shades of blue or black. This technique, known as density slicing, enhances the human eye's ability to interpret temperature variations and is particularly useful in applications where fine temperature differences need to be identified. A colour scale accompanying the image relates specific colours to temperature values, ensuring clarity in temperature measurement and analysis.

Thermographic cameras, also known as infrared cameras or thermal imagers, are devices designed to create images using infrared (IR) radiation, functioning similarly to visible light cameras but operating within a different spectral range. Unlike visible light cameras that capture wavelengths between 400 and 700 nanometres, thermographic cameras detect infrared wavelengths from approximately 1,000 nanometres (1 micrometre) to 14,000 nanometres (14 micrometres). This capability allows them to form thermal images, or thermograms, based on the heat emitted by objects. The technology leverages the fact that all objects with a temperature above absolute zero emit thermal infrared energy, enabling cameras to passively visualize objects regardless of ambient light conditions. Thermal cameras are particularly sensitive to temperatures warmer than -50°C (-58°F), making them valuable for a wide range of applications, from industrial diagnostics to medical imaging.

Thermal cameras translate infrared radiation into a visible image by measuring energy in the far-infrared wavelength and converting it into a visual display. This process makes temperature variations within a scene readily observable, as warmer objects emit more infrared radiation and appear distinct from cooler surroundings. Specifications for thermal cameras include parameters like pixel resolution, frame rate, noise-equivalent temperature difference (NETD), spectral band, and dynamic range. Despite their effectiveness, thermal cameras generally have lower resolutions than optical cameras, with common resolutions around 160x120 or 320x240 pixels, although high-end models can achieve resolutions up to 1280x1024 pixels. These devices are significantly more expensive than visible-spectrum cameras, though recent advancements have led to affordable smartphone-compatible models.

Thermographic cameras are categorized into two main types based on their detectors: cooled and uncooled infrared detectors. Cooled infrared detectors are housed within vacuum-sealed enclosures and require cryogenic cooling to function effectively. This cooling prevents the sensors from being overwhelmed by their own thermal radiation. Typical operating temperatures for these detectors range from 4 Kelvin (-269°C) to just below room temperature, with most modern systems functioning between 60 Kelvin (-213°C) and 100 Kelvin (-173°C). While cooled detectors offer superior image quality and sensitivity, especially for objects at or below room temperature, they are costly to produce and maintain. Cooling systems such as Stirling cryocoolers or Joule–Thomson expansion systems are commonly used but add to the complexity and energy demands of these devices.

Materials used in cooled infrared detectors include narrow-gap semiconductors like indium antimonide (InSb), mercury cadmium telluride (MCT), and lead selenide. These materials are highly sensitive to specific infrared wavelengths, making them ideal for high-performance imaging. Superconducting bolometers represent another cooled detector technology, capable of detecting individual photons. However, their use is largely confined to scientific research due to their extreme sensitivity and the need for precise environmental shielding.

Uncooled infrared detectors operate at or near ambient temperature and do not require cryogenic cooling, making them smaller, less expensive, and easier to use. These detectors rely on materials that change resistance, voltage, or current when exposed to infrared radiation, with the resulting changes measured electronically. Although uncooled detectors are less sensitive and offer lower image quality compared to their cooled counterparts, they are sufficient for many practical applications. Common technologies for uncooled detectors include microbolometers, which consist of temperature-sensitive materials like vanadium oxide or amorphous silicon. These materials form pixels that detect minute temperature differences, allowing the camera to construct a thermal image. Advances in uncooled focal plane arrays (UFPA) continue to improve their sensitivity and resolution, making them increasingly competitive in the thermal imaging market.

Thermal imaging cameras also find specialized applications through technologies like charge-coupled device (CCD) and complementary metal-oxide-semiconductor (CMOS) sensors. While these sensors are primarily designed for visible light, they can detect near-infrared radiation under specific conditions, allowing them to perform basic thermal imaging for high-temperature objects. Similarly, infrared films can capture black-body radiation within a limited temperature range but are less versatile than modern thermographic systems.

The comparison between thermal imaging and night-vision devices highlights a key distinction: starlight-type night-vision systems amplify ambient light, whereas thermal cameras detect heat emitted by objects. Some infrared cameras marketed as night vision devices can detect near-infrared radiation but are not typically used for thermography, as they rely on active illumination rather than passive heat detection.

Passive and active thermography are two approaches used in infrared thermal imaging, each serving distinct purposes based on the thermal characteristics of the target object and its environment. Both rely on the fundamental principle that all objects above absolute zero (0 K) emit infrared radiation, making thermal imaging an effective tool for detecting and analyzing temperature variations. The choice between passive and active thermography depends on the specific application and the temperature dynamics of the object or environment being inspected.

Passive thermography involves capturing infrared radiation emitted naturally by objects, without the need for an external energy source. In this approach, the features of interest are already at a higher or lower temperature compared to their surroundings, creating a natural thermal contrast. This method is widely used in scenarios where objects inherently generate heat or have temperature differences due to environmental or physiological factors. Applications of passive thermography include surveillance, where people or animals are monitored based on their body heat against cooler backgrounds, and medical diagnosis, particularly in thermology, where temperature variations on the skin can indicate underlying health conditions. Since passive thermography depends solely on naturally occurring thermal emissions, it is a non-invasive and convenient method for observing thermal anomalies in real-time.

Active thermography, on the other hand, involves the use of an external energy source to create a thermal contrast between the feature of interest and its background. This method is necessary when the object of interest is in thermal equilibrium with its surroundings, making it difficult to distinguish based on natural thermal emissions alone. By applying an external stimulus, such as a heat source, the thermal response of the object can be measured, revealing hidden features or defects. Active thermography is particularly useful in industrial applications, such as non-destructive testing, where flaws or subsurface defects in materials can be identified by analysing the thermal response to external heating or cooling. Additionally, the active approach is valuable in scientific research, as it enables enhanced imaging resolution by leveraging the non-linearities of black-body radiation. This capability allows active thermography to surpass the diffraction limit of imaging systems, making it suitable for advanced applications like super-resolution microscopy.

The effectiveness of both passive and active thermography depends on the quality of the thermal imaging equipment, typically focal plane array (FPA) infrared cameras. These cameras operate in the mid-wave infrared (MWIR, 3 to 5 µm) and long-wave infrared (LWIR, 7 to 14 µm) bands, which correspond to high-transmittance windows in the infrared spectrum. MWIR and

LWIR cameras are capable of detecting subtle thermal variations, enabling precise measurements and detailed imaging. In passive thermography, the camera captures the natural thermal emissions of the target, while in active thermography, it records the thermal changes induced by the external energy source.

Thermography offers numerous advantages, making it a valuable tool in various fields, including industrial maintenance, medical diagnostics, and safety monitoring. One of its primary benefits is the ability to provide a visual representation of temperature variations over a large area, enabling easy comparison of thermal conditions. By capturing real-time images, thermography can track moving targets and detect temperature anomalies, such as overheating components, before they fail. This capability is especially useful in preventive maintenance, allowing operators to identify potential issues early and avoid costly downtime. Furthermore, thermography is highly versatile, as it can be employed in environments that are inaccessible or hazardous for other methods. It is a non-destructive testing technique, making it suitable for inspecting delicate or critical systems without causing damage. Thermography is effective in identifying defects in various materials, such as metal or plastic parts, and can even detect objects in low-light or dark areas. In the medical field, thermography has specific applications, particularly in physiotherapy, where it can monitor temperature variations related to inflammation or circulation.

Despite its numerous advantages, thermography has some limitations and disadvantages that must be considered. High-quality thermographic cameras are often expensive, with costs typically exceeding $3,000. These prices are attributed to the advanced technology required for larger pixel arrays, such as state-of-the-art 1280x1024 resolution sensors. While less expensive models are available, their lower resolution, ranging from 40x40 to 160x120 pixels, results in reduced image quality, making it more challenging to distinguish closely positioned targets within the same field of view. Additionally, the refresh rate of thermographic cameras varies significantly between models. Some cameras may only refresh at 5–15 Hz, while high-performance models can achieve refresh rates of 180 Hz or more in non-full-window modes. This variation impacts the ability to capture fast-moving or dynamic targets effectively.

Another limitation of thermography is the dependency on proper calibration and environmental factors. Many thermographic cameras do not provide raw irradiance data, which is crucial for constructing accurate thermal images. Without proper calibration for emissivity, distance, ambient temperature, and relative humidity, the resulting images may be inaccurate. Moreover, interpreting thermal images can be challenging, particularly for objects with erratic temperature distributions. While active thermography can mitigate some of these challenges by artificially creating thermal contrast, interpretation difficulties remain a concern in certain scenarios.

Thermographic measurements are inherently influenced by the emissivity and reflectivity of the target surface. Radiant heat energy levels captured by the camera can be affected by factors such as sunlight reflection or the surface's emissivity properties, leading to potential errors. Most thermographic cameras offer a temperature measurement accuracy of ±2% or worse, which is generally less precise than contact-based methods. Additionally, thermography is limited to detecting surface temperatures, making it unsuitable for measuring internal temperatures without specific techniques or assumptions.

Acoustic Emission Testing (AET): AET monitors stress waves released from materials under load, generated by events such as crack propagation or fibre breakage. Sensors capture these signals, enabling real-time monitoring of structural health, especially in composite materials used in aerospace and wind turbine applications [568].

Acoustic emission (AE) refers to the release of transient elastic waves within a material, triggered by irreversible structural changes such as crack formation, plastic deformation, or the accumulation of stress. These emissions occur when the accumulated elastic energy in a material is suddenly released, generating stress waves that propagate through the solid. The phenomenon is commonly observed during mechanical loading, temperature gradients, or material aging, making it an essential tool for understanding material behaviour under stress. By detecting these emissions, AE provides valuable insights into the internal state of materials and structures, especially in cases where cracks, dislocations, or other imperfections develop.

The mechanism behind AE involves the generation of elastic pulses during events like microcrack nucleation or the movement of dislocations in the material. For instance, in metals with a body-centred cubic (bcc) lattice, dislocations may accumulate and break through a boundary under mechanical stress, emitting a primary elastic pulse. This phenomenon enables the early detection of micro-deformation processes, dislocation nucleation, and crack formation. AE acts like a warning system, where each crack "screams" as it grows. By analysing the acoustic signals, engineers can determine the critical size of a crack and monitor its growth rate. Once a crack reaches a critical size, catastrophic failure occurs, often at speeds close to half the speed of sound in the material. Using advanced AE equipment, it is possible to measure parameters like the intensity of events and the total number of emissions, which help estimate crack length and predict imminent failure.

AE waves are integral to several applications, particularly in structural health monitoring (SHM), quality control, and process evaluation. In SHM, AE is used to detect, locate, and characterize damage in critical structures such as bridges, pressure vessels, and pipelines. By monitoring these acoustic signals, engineers can identify stress points or areas prone to failure and schedule maintenance before catastrophic damage occurs. In materials research, AE helps evaluate mechanical performance and characterize failure mechanisms, providing insights into phenomena like crack growth, delamination, and friction. Recent advancements allow AE to characterize the nature of the source mechanism, offering detailed information on whether a defect warrants repair.

The elastic waves generated during AE events span a wide frequency range, typically from under 1 kHz to up to 100 MHz, although most energy is concentrated between 1 kHz and 1 MHz.

Signal processing techniques, such as Fast Fourier Transform (FFT), are used to analyse the frequency domain of these waves, offering detailed insights into the source characteristics. Employing statistical methods, such as Poisson stream analysis, enhances the reliability and depth of AE diagnostics, providing a robust framework for assessing structural integrity.

AE plays a critical role in non-destructive testing (NDT) by monitoring emissions caused by material failure or stress rather than externally generated waves. Unlike conventional ultrasonic testing, AE detects active processes like crack formation during mechanical loading. For example, in welding, AE can monitor crack formation during the process rather than inspecting the weld after completion. Transducers mounted on active structures, such as aircraft components or pressurized pipelines, can detect cracks as they form and pinpoint their location by analysing the time it takes for the waves to reach multiple sensors. Long-term AE monitoring is particularly valuable for high-pressure vessels and pipelines, where early detection of cracks ensures safety and prevents catastrophic failures. Standards developed by organizations like ASME and ISO guide the use of AE in such applications.

Beyond traditional structural applications, AE has proven useful in unconventional fields. It has been employed to estimate corrosion in reinforced concrete structures and diagnose historical artifacts like the Tsar Bell in Moscow. The method is actively used in the nuclear, aerospace, railway, and rocket industries for monitoring and diagnostics of critical systems. In energy storage, AE is being explored for monitoring the health of lithium-ion batteries. Piezoelectric sensors detect acoustic signals associated with mechano-electrochemical events, such as electrode grinding, phase transitions, or gas evolution, providing real-time insights into battery performance and safety.

AE also has applications in process monitoring, where it detects anomalies in industrial systems. For example, it has been successfully applied to monitor fluidized beds and determine endpoints in batch granulation processes. This versatility underscores AE's value in both research and industry, making it a critical tool for ensuring the safety and efficiency of a wide range of systems and materials.

Laser-Based Techniques: Techniques like Laser Ultrasonics and Laser Shearography provide high precision and sensitivity for advanced materials. Laser Ultrasonics generates and detects ultrasonic waves using laser pulses, while Laser Shearography detects surface deformations caused by defects. These methods are particularly valuable for thin, lightweight composite structures in aerospace [569].

Laser ultrasonics is a sophisticated non-contact non-destructive testing (NDT) method that employs laser technology to generate and detect ultrasonic waves within materials. This technique is particularly advantageous for inspecting materials that present challenges for traditional contact-based ultrasonic testing, such as those with rough surfaces, elevated temperatures, or complex geometries. The process begins with a pulsed laser that delivers a high-energy pulse to the material's surface, leading to rapid thermal expansion or ablation, which in turn generates ultrasonic waves that propagate through the material. Subsequently, a continuous-wave laser or an optical interferometer detects these waves as they return to the surface after interacting with internal features like defects or structural inhomogeneities [570, 571].

The advantages of laser ultrasonics are manifold. It operates without direct contact, allowing for high spatial resolution and making it suitable for inspecting fragile or moving materials. This method has found extensive applications in industries such as aerospace and manufacturing, where it is utilized for detecting defects, measuring thickness, and monitoring material properties [572, 573]. Specifically, laser ultrasonics is effective in identifying cracks, voids, and delaminations in both composite and metallic materials, as well as measuring the thickness of coatings and thin films [571].

In contrast, laser shearography is another advanced optical NDT method that focuses on measuring strain distribution to detect surface and subsurface defects. This technique utilizes the principle of speckle interferometry, where coherent laser light reflects off a material's surface, creating a speckle pattern. When the material is subjected to external stress—whether thermal, mechanical, or vibrational—the resulting deformation alters the speckle pattern. An optical sensor captures these changes, producing a shearogram that highlights strain anomalies indicative of defects such as delaminations or cracks [574, 575]. Laser shearography is non-contact, rapid, and capable of inspecting large areas, making it particularly useful in sectors like aerospace and automotive for assessing composite materials and bonded structures [575, 576].

While both laser ultrasonics and laser shearography utilize laser technology for material inspection, they serve different purposes and can complement each other in complex inspections. Laser ultrasonics excels in providing depth information and detecting internal flaws, while laser shearography is adept at analysing surface strain to identify near-surface defects. Together, these techniques enhance the reliability and scope of NDT in critical applications, particularly in industries where material integrity is paramount [572, 573, 577].

Visual Testing (VT): Although basic, Visual Testing remains a crucial initial step in NDT. Utilizing tools like borescopes or high-resolution cameras, inspectors can identify surface defects or anomalies. Enhanced VT techniques, such as automated visual inspection systems with AI algorithms, improve accuracy and efficiency in inspecting advanced materials [578].

NDT is indispensable for advanced materials due to its non-invasive nature, which is critical for expensive materials often fabricated into complex geometries. These techniques ensure structural integrity, optimize material performance, and mitigate the risk of in-service failures. Furthermore, modern NDT methods are increasingly automated, providing rapid and reliable results suitable for high-throughput industrial applications [579].

The unique characteristics of advanced materials, such as anisotropy and heterogeneous structures, can complicate the effectiveness of traditional NDT methods. For example, in fibre-reinforced composites, the differing properties of fibres and matrix materials can complicate test result interpretation. Similarly, the high stiffness and brittleness of ceramics may limit the applicability of certain NDT techniques. To address these challenges, advanced methods like CT imaging, phased-array UT, and laser-based techniques have been developed, offering enhanced sensitivity and precision [580].

NDT for advanced materials is vital in sectors where safety and reliability are critical. In aerospace, NDT ensures the integrity of composite airframes and turbine blades. In automotive applications, it is used to inspect lightweight materials for structural components. The energy sector employs NDT for monitoring wind turbine blades and advanced coatings in nuclear reactors. Additionally, medical device manufacturing relies on NDT to ensure the quality of biocompatible materials and implants [581].

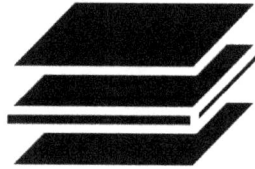

Chapter 7

Sustainability and Environmental Considerations

Lifecycle Analysis of Advanced Materials

Lifecycle Analysis (LCA), also known as lifecycle assessment, is a systematic methodology used to evaluate the environmental, economic, and social impacts of a material or product throughout its entire lifecycle. For advanced materials, such as composites, nanomaterials, and high-performance alloys, LCA provides crucial insights into their sustainability, resource efficiency, and overall impact from production to disposal or recycling. The lifecycle of advanced materials encompasses multiple stages, including raw material extraction, manufacturing, usage, and end-of-life management.

Lifecycle Assessment (LCA) has emerged as a critical tool in evaluating the environmental impacts of advanced materials across various industries, including aerospace, automotive, electronics, and renewable energy. In the aerospace sector, LCA is instrumental in assessing the trade-offs associated with carbon fibre composites. While these materials have a high production impact, they significantly contribute to fuel savings in aircraft, thus presenting a complex sustainability challenge that LCA can help navigate [582, 583]. The ability to quantify these trade-offs allows manufacturers to make informed decisions that balance performance with environmental responsibility.

In the renewable energy sector, LCA plays a pivotal role in evaluating the sustainability of materials used in technologies such as wind turbine blades and solar panels. By analysing the entire lifecycle of these materials, LCA ensures that the environmental benefits of clean energy technologies are not undermined by the impacts associated with material extraction, processing, and disposal [584, 585]. This holistic approach is essential for promoting truly sustainable energy solutions.

The electronics industry also benefits from LCA, particularly in identifying sustainable alternatives to rare materials used in high-performance devices. As the demand for electronics continues to rise, the pressure on rare material resources increases, making it crucial to explore and evaluate alternative materials through LCA methodologies [586, 587]. This not only aids in reducing dependency on scarce resources but also enhances the overall sustainability of electronic products.

LCA enables industries to pinpoint hotspots of environmental impact within the lifecycle of advanced materials, facilitating the implementation of strategies for improvement. For instance, optimizing manufacturing processes, utilizing recycled or alternative feedstocks, and designing products for recyclability are all strategies that can significantly reduce the environmental footprint of materials [588, 589]. Policymakers and researchers rely on LCA data to inform regulations and innovations that promote sustainable material use, thereby driving the transition toward a circular economy [590, 591].

The first stage of the lifecycle involves the extraction and processing of raw materials. Advanced materials often rely on rare or highly processed inputs, such as carbon fibres, specialty polymers, or rare earth elements. These processes can be energy-intensive, result in significant greenhouse gas emissions, and create byproducts that need proper management. For example, carbon fibre production requires high-temperature processes, while mining rare earth metals can lead to habitat disruption and water contamination. LCA in this phase focuses on assessing resource depletion, land use impacts, and the energy and water consumption associated with raw material extraction.

The manufacturing stage involves the production and shaping of advanced materials into usable forms, such as sheets, rods, or complex structures. Advanced manufacturing techniques, including additive manufacturing, autoclaving for composites, and plasma spraying for coatings, often require substantial energy inputs. Additionally, the use of chemicals, solvents, or specialized tooling can introduce environmental and health hazards. LCA evaluates the energy efficiency of these processes, waste generation, and emissions, enabling manufacturers to identify areas for optimization. Advanced manufacturing methods, like 3D printing, are increasingly explored for their potential to reduce material waste and energy consumption.

The usage phase of advanced materials is critical to understanding their lifecycle impacts, as their high-performance characteristics often contribute to improved efficiency and longevity in applications. For instance, lightweight composites reduce fuel consumption in vehicles and aircraft, while high-strength alloys extend the lifespan of critical infrastructure. LCA assesses how these performance benefits offset the environmental costs incurred during earlier lifecycle stages. For example, the reduced operational emissions of an electric vehicle using lightweight composite materials may significantly outweigh the environmental burden of producing those materials.

End-of-life management includes recycling, disposal, or repurposing of advanced materials after their service life. This stage presents unique challenges for advanced materials due to their complex compositions and difficulty in separating components for recycling. For instance, thermoset composites and multi-material assemblies often lack effective recycling methods, leading to landfilling or incineration. LCA evaluates the potential environmental impacts of various end-of-life scenarios, such as emissions from incineration or leaching from landfills. Innovative recycling methods, like chemical recycling for polymers or re-melting for metal alloys, are increasingly being developed to improve the sustainability of advanced materials.

LCA for advanced materials examines not only environmental impacts but also economic and social aspects. Environmental impacts include carbon emissions, resource depletion, and waste generation, while economic considerations focus on production costs, material efficiency, and recycling viability. Social dimensions assess worker safety, community effects of mining or manufacturing, and compliance with ethical sourcing standards. Balancing these factors helps industries adopt more sustainable practices while maintaining the benefits of advanced materials.

Conducting an accurate LCA for advanced materials involves several challenges. Data availability and quality can vary widely, particularly for new materials or emerging manufacturing processes. The complexity of advanced materials, such as composites with multi-layered structures or nanomaterials with unique behaviours, makes modelling their lifecycle impacts difficult. Additionally, the trade-offs between environmental benefits during the usage phase and impacts during production or disposal require careful analysis. For instance, while lightweight composites reduce fuel consumption in transportation, their production often has a high carbon footprint.

The Life Cycle Assessment (LCA) framework, as delineated by the ISO 14040 and ISO 14044 standards, comprises four interdependent phases: Goal and Scope Definition, Life Cycle Inventory (LCI), Life Cycle Impact Assessment (LCIA), and Interpretation. Each phase plays a critical role in ensuring that the assessment is comprehensive and accurate, with the results of one phase informing the others, thus emphasizing the iterative nature of LCA [592, 593].

The first phase of an LCA involves defining the goal and scope of the study. This includes articulating the intended application of the LCA, the reasons for conducting it, the target audience, and whether the results will be used for comparative assertions made public [592, 593]. The ISO standards mandate that the goal must be clearly stated and should encompass the product function, functional unit, system boundaries, assumptions, and data quality requirements [592, 593]. This phase is crucial as it sets the foundation for the entire assessment, ensuring that the study is aligned with its intended purpose and that all stakeholders have a clear understanding of the objectives [592, 593].

Following the goal and scope definition, the LCI phase quantifies the inputs and outputs associated with the product system throughout its life cycle. This involves compiling data on raw materials, energy use, emissions, and waste generation [592-594]. The ISO 14044 standard outlines specific steps for conducting the LCI, including data collection, validation, and allocation procedures if multiple products are involved [592, 593]. The quality of data collected is paramount, as it directly influences the reliability of the LCA results. Practitioners are encouraged to utilize primary data whenever possible, but secondary data can also be employed, provided that its limitations are documented [592-594].

The Life Cycle Impact Assessment (LCIA)phase evaluates the potential environmental impacts associated with the inputs and outputs identified in the LCI. This involves selecting impact categories relevant to the study, classifying inventory results, and applying characterization factors to quantify the contributions of each flow to the selected impact categories [592-594]. The ISO standards emphasize the importance of transparency and justification in the selection of impact categories, ensuring that the assessment reflects the environmental issues pertinent to the geographical context of the study [592-594]. Optional steps such as normalization, grouping, and weighting can further enhance the interpretation of results, although they are not mandatory and should be approached with caution [592-594].

The final phase, interpretation, synthesizes the findings from the LCI and LCIA to draw conclusions and make recommendations. This phase involves identifying significant issues, evaluating the completeness and consistency of the study, and ensuring that the results align with the original goals [592-594]. The interpretation is critical for communicating the results effectively to stakeholders, as it provides insights into the environmental performance of the product system and highlights areas for potential improvement [592-594].

Sample Life Cycle Assessment (LCA): Carbon Fiber Reinforced Polymer (CFRP)

Introduction

Carbon Fiber Reinforced Polymer (CFRP) is an advanced composite material widely used in aerospace, automotive, and renewable energy industries due to its high strength-to-weight ratio, corrosion resistance, and durability. While CFRP offers significant performance benefits, its production and end-of-life management pose environmental challenges. This sample LCA evaluates the environmental impacts of CFRP across its lifecycle, including raw material extraction, manufacturing, usage, and end-of-life stages.

Goal and Scope Definition

The goal of this LCA is to assess the environmental footprint of CFRP used in an automotive application, such as a car hood, over a 10-year lifespan. The functional unit is defined as "one CFRP car hood providing structural performance equivalent to a steel hood." The assessment focuses on global warming potential (GWP), energy consumption, and waste generation.

Life Cycle Inventory (LCI)

1. Raw Material Extraction

- **Inputs**:
  - Carbon fibre precursor (polyacrylonitrile, PAN): 2.5 kg.
  - Epoxy resin: 1.2 kg.
  - Energy for precursor production: 120 MJ/kg PAN.
  - Water for precursor processing: 50 litres.
- **Outputs**:
  - Emissions: 30 kg CO_2-eq from PAN production.
  - Waste: 1.5 kg of solid waste (residual PAN and resin).

2. Manufacturing

- **Processes**:
  - Carbonization of PAN to produce carbon fibre: High-temperature ovens (700–1200°C).
  - Layup and curing process: Autoclave curing at 180°C for 4 hours.
- **Inputs**:
  - Electricity: 300 kWh.
  - Thermal energy: 250 MJ (natural gas).
  - Tooling materials: Minor epoxy waste and fibre trims.
- **Outputs**:
  - Emissions: 75 kg CO_2-eq (energy-intensive carbonization).
  - Waste: 0.5 kg of fibre offcuts and resin scraps.

3. Usage Phase

- **Scenario**:
  - A CFRP hood is 50% lighter than a steel hood, reducing vehicle weight by 15 kg.
  - Fuel savings: 0.2 litres per 100 km.
  - Lifetime vehicle distance: 150,000 km.
- **Inputs**:
  - Fuel savings: 300 litres over 10 years.
- **Outputs**:
  - Avoided emissions: 690 kg CO_2-eq (fuel savings offset manufacturing emissions).

4. End-of-Life Management

- **Scenarios**:
  - o Mechanical recycling (30%): Downcycling into lower-grade composites.
  - o Incineration with energy recovery (70%).
- **Inputs**:
  - o Energy for recycling: 50 MJ/kg CFRP.
  - o Combustion efficiency: 80%.
- **Outputs**:
  - o Recovered energy: 200 MJ (incineration).
  - o Emissions: 20 kg CO_2-eq.
  - o Waste: 0.3 kg non-combustible residue.

Life Cycle Impact Assessment (LCIA)

1. Global Warming Potential (GWP):

- Raw material extraction: 30 kg CO_2-eq.
- Manufacturing: 75 kg CO_2-eq.
- Usage phase: -690 kg CO_2-eq (savings).
- End-of-life: 20 kg CO_2-eq.
- **Net GWP**: -565 kg CO_2-eq (significant net reduction due to fuel savings).

2. Energy Consumption:

- Raw material extraction: 300 MJ.
- Manufacturing: 550 MJ.
- Usage phase: -9,000 MJ (fuel savings).
- End-of-life: 200 MJ recovered.
- **Net energy savings**: -8,050 MJ.

3. Waste Generation:

- Raw material extraction: 1.5 kg.
- Manufacturing: 0.5 kg.
- End-of-life: 0.3 kg non-combustible residue.
- **Total waste**: 2.3 kg.

Interpretation

- **Strengths**:

  o Significant GWP reduction due to fuel savings during the usage phase, demonstrating the long-term environmental benefit of lightweight CFRP components.

  o Net energy savings highlight the material's contribution to energy efficiency in transportation.

- **Weaknesses**:

  o High emissions and energy consumption during raw material extraction and manufacturing due to the energy-intensive carbonization process.

  o Limited recycling options and residual waste at the end-of-life stage.

- **Recommendations**:

  o Improve manufacturing efficiency by integrating renewable energy sources.

  o Develop enhanced recycling technologies, such as chemical recycling, to reduce waste and improve material circularity.

  o Further optimize CFRP designs to maximize weight savings and fuel efficiency.

Conclusion

This LCA demonstrates that while CFRP has a high initial environmental footprint due to energy-intensive production, its lightweight properties result in significant emissions and energy savings during the usage phase, particularly in transportation applications. By addressing challenges in manufacturing and end-of-life management, CFRP can become a more sustainable choice for advanced material applications. This LCA framework can guide industries and policymakers in developing strategies to balance performance and sustainability.

Recycling and Upcycling Strategies

Recycling and upcycling strategies for advanced materials are critical for promoting sustainability and minimizing environmental impacts. Advanced materials, such as composites, nanomaterials, high-performance alloys, and polymers, present unique challenges due to their complex structures and specialized properties.

Mechanical recycling is a prevalent method for recovering materials through physical processes without altering their chemical structure. This method is particularly effective for thermoplastics and metals. The process typically involves shredding, crushing, or grinding materials into smaller pieces for reuse. However, a significant challenge arises with advanced materials like carbon-fibre-reinforced polymers (CFRPs), where mechanical recycling often leads to fibre length reduction and degradation of mechanical properties, limiting their reuse in high-performance applications [595, 596].

Chemical recycling involves breaking down materials into their chemical constituents, allowing for the recovery of monomers or polymers. Techniques such as depolymerization, solvolysis, and pyrolysis are employed to restore materials to their original properties, making them suitable for high-value applications. Despite its advantages, chemical recycling is often energy-intensive and requires specialized equipment, which can increase costs [597].

Thermal recycling recovers energy or materials through high-temperature processes, such as incineration or pyrolysis. The heat generated can be used for power generation or to recover specific components like carbon black or fibres. However, managing emissions and residues from thermal recycling is crucial to minimize environmental impacts [598].

Electrochemical recycling is an emerging method particularly useful for batteries and electronic waste, electrochemical recycling employs processes that separate and recover valuable metals like lithium and cobalt. While this method shows promise, scaling up the technology and ensuring cost-effectiveness remain significant hurdles [599].

Functional upcycling is an approach that transforms waste materials into products with enhanced functionalities. For example, reinforcing recycled plastics with nanomaterials can improve their strength. However, ensuring compatibility between recycled materials and additives can be technically complex [600].

Structural upcycling focuses on repurposing waste materials into new forms with greater value. For instance, scrap CFRPs can be repurposed into lightweight structural panels for automotive or construction applications. A critical challenge is maintaining consistent quality across batches of upcycled materials [595].

Advanced Material Conversion includes techniques such as chemical vapor deposition (CVD) can convert waste materials into nanomaterials or coatings. For example, waste carbon fibres can be transformed into carbon nanotubes for high-tech applications. However, high costs and technical barriers limit widespread adoption [595].

Hybrid Material Design involves blending recycled materials with virgin materials to create hybrid composites with superior performance. These composites find applications in aerospace and automotive industries. Balancing the proportion of recycled content while maintaining performance standards is a significant challenge [595, 600].

The recycling and upcycling of advanced materials face several challenges, including material complexity, contamination, economic viability, technological barriers, and environmental impacts. The complexity of advanced materials, which often consist of multiple layers and coatings, complicates separation and recovery processes. Additionally, impurities in waste streams can degrade the quality of recycled or upcycled materials [595, 599]. The high costs associated with sophisticated recycling processes may limit their adoption, and many technologies are still in early development stages, necessitating further research and scaling [595, 599].

Innovations such as closed-loop recycling systems aim to recover and reuse materials indefinitely, significantly reducing waste. Bio-based recycling agents, including enzymes, are being developed to break down materials in an environmentally friendly manner. Advanced imaging and AI technologies are enhancing waste sorting processes, improving recycling efficiency. Furthermore, self-healing materials can extend the lifespan of products, reducing the need for recycling [599, 601].

Industries are increasingly adopting recycling and upcycling strategies. In aerospace, carbon fibres from decommissioned aircraft are recycled into new structural components. The automotive sector is upcycling plastics and composites from end-of-life vehicles into lightweight parts. Electronics manufacturers are recovering rare earth elements from electronic waste for new devices, while the construction industry is repurposing waste materials into building panels and insulation [595, 599].

Determining and selecting recycling and upcycling strategies for advanced materials necessitates a comprehensive and systematic approach due to the unique characteristics of these materials, such as composites, nanomaterials, and high-performance alloys. These materials often require specialized recovery and reuse methods that take into account their properties, lifecycle stage, economic feasibility, and environmental impact. For instance, carbon fibre-reinforced polymers (CFRPs) present specific challenges due to their composite nature, which includes both carbon fibres and resin matrices that must be addressed during recycling processes [602].

The first critical step in this systematic approach is material characterization. This involves a thorough understanding of the material's composition, physical properties, and potential contaminants. Identifying the chemical constituents—such as polymers, fibres, or metals—and their proportions is essential for tailoring appropriate recycling or upcycling methods. For example, the presence of impurities can complicate recycling processes and may necessitate additional preprocessing steps [603]. Furthermore, the physical properties like mechanical strength and thermal stability significantly influence the feasibility of reuse or repurposing, as they determine whether the recycled material can meet performance requirements for its intended application [602].

Lifecycle assessment (LCA) is another crucial component in evaluating the environmental and economic impacts of recycling and upcycling strategies. LCA provides insights into energy consumption, emissions, and waste generation associated with various recovery methods, allowing for a comparison of the economic feasibility of these strategies against the use of virgin materials [604]. The performance requirements of recycled or upcycled materials must also be considered, as they may need to retain their original characteristics or be suitable for lower-grade applications [605].

The selection of suitable recycling strategies is contingent upon the material's properties and recovery goals. Mechanical recycling is often appropriate for materials that can maintain acceptable properties after physical processing, such as thermoplastics or metals [602]. In contrast, chemical recycling methods, including depolymerization and solvolysis, are better suited for restoring high-quality feedstock from polymers or composites [606]. Thermal recycling can recover energy or specific components from complex materials, while electrochemical recycling is particularly effective for recovering valuable metals and rare elements from electronic waste [607].

Upcycling strategies are chosen based on their potential to enhance material value or repurpose waste. Functional upcycling may involve modifying materials with additives, such as nanomaterials, to improve performance, while structural upcycling focuses on repurposing waste materials into higher-value applications, such as lightweight panels for automotive or construction use [608]. Advanced material conversion can transform waste into entirely new materials, exemplified by converting carbon fibre waste into graphene or nanocomposites [609].

Technical challenges, such as the complexity of separating multi-layered or chemically bonded components, must be addressed when selecting recycling and upcycling strategies. Moreover, material degradation over multiple recycling cycles can affect the quality of the recovered materials, necessitating careful assessment [607]. Contamination levels can further complicate processing and increase costs, often requiring additional cleaning or preprocessing steps [603].

Regulatory and market factors also play a significant role in the feasibility and adoption of recycling and upcycling strategies. Compliance with environmental regulations and adherence to international standards for material recovery are essential for successful implementation [604]. Additionally, market demand for recycled or upcycled products, such as lightweight composites in aerospace or recovered metals in electronics, drives the selection of strategies [610]. Incentives or subsidies for sustainable practices can help offset costs and encourage the adoption of these strategies [604].

The availability of appropriate technologies is another critical factor influencing the selection of recycling and upcycling strategies. Emerging methods, such as enzymatic recycling for polymers or plasma treatment for composites, may offer significant advantages but require careful evaluation for scalability and practicality [606]. Established technologies must also be assessed for their capacity to meet industrial requirements [610].

Finally, pilot testing and validation are necessary before full-scale implementation of recycling and upcycling strategies. Small-scale testing allows for the evaluation of the recovery method's efficiency and ensures that the recycled or upcycled material meets quality standards for its intended application [604]. Cost analysis during pilot testing is crucial for predicting large-scale operational expenses and feasibility [610].

Reducing Carbon Footprint in Material Production

Reducing the carbon footprint in material production is essential for mitigating climate change, conserving resources, and enhancing the sustainability of industrial processes. Material production, especially in energy-intensive sectors such as steel, aluminium, and cement, is a significant contributor to global greenhouse gas (GHG) emissions. For instance, the production of stainless steel in China has been shown to have substantial energy-related GHG emissions, primarily from chemical reactions and energy consumption during manufacturing processes [611]. The carbon footprint associated with these materials can be effectively assessed through Life Cycle Assessment (LCA), which evaluates emissions from raw material extraction to end-of-life disposal [612, 613].

To address the high carbon emissions from these industries, several strategies can be implemented. Optimizing production processes is one approach; for example, the use of Electric Arc Furnace (EAF) technology in steel production, which utilizes a higher proportion of recycled scrap, has demonstrated a significant reduction in GHG emissions compared to traditional methods [614].

An electric arc furnace (EAF) is a furnace primarily used for melting and refining metals, particularly steel, by generating heat through an electric arc. This process is widely utilized in the steelmaking industry due to its efficiency and ability to recycle scrap metal as raw material. EAFs play a critical role in producing high-quality steel with reduced environmental impact.

The operation of an EAF begins with the generation of heat through electrical energy, which melts scrap steel, direct reduced iron (DRI), or other metal inputs. The furnace creates an electric arc between electrodes and the metal charge, producing extremely high temperatures capable of melting the material. This process ensures efficient melting and refining of the metal.

Key components of an EAF include electrodes, the furnace shell, the roof, the power source, and the water cooling system. The electrodes, made of large graphite rods, conduct electricity and generate the electric arc. The furnace shell, lined with refractory material, contains the metal charge and withstands extreme heat. The roof, which can be removed or swung open, houses the electrodes and retains heat within the furnace. High-voltage electrical systems power the arc formation, while the water cooling system prevents overheating of the shell, roof, and other components during operation.

The EAF process starts with charging the furnace. Raw materials, including steel scrap, pig iron, or DRI, are loaded into the furnace, along with fluxes such as lime, which help refine impurities. Once the charge is loaded, the roof is closed, and electricity is passed through the electrodes, forming an electric arc that generates temperatures exceeding 3,000°C (5,432°F). This intense heat melts the metal charge into a molten pool.

After melting, the refining stage begins to remove impurities. Fluxes create a molten slag layer that absorbs impurities like sulfur and phosphorus, while oxygen may be blown into the furnace to oxidize and remove additional contaminants. Throughout the process, temperature control is critical to achieving the desired metallurgical properties. Adjustments are made by varying the arc intensity or adding cooling agents as necessary.

Once the metal reaches the required composition and temperature, it is tapped through a spout into a ladle. Slag is either removed during the tapping process or skimmed off separately. After tapping, the furnace is prepared for the next batch by cleaning residual slag and inspecting the refractory lining.

There are two main types of EAFs: alternating current (AC) EAFs and direct current (DC) EAFs. AC EAFs, the most common type, generate an electric arc between three-phase electrodes and the metal charge. DC EAFs, on the other hand, use a single electrode as the anode and the furnace bottom as the cathode, offering more stable arcs and greater energy efficiency.

EAFs offer several advantages. They are highly efficient at recycling scrap metal, reducing reliance on raw materials such as iron ore. They provide flexibility in producing various steel grades by adjusting the charge composition and refining processes. Additionally, EAFs emit significantly less CO_2 and pollutants compared to traditional blast furnaces when scrap is the primary input.

Despite these benefits, EAFs face challenges. They require large amounts of electrical energy, making them dependent on a reliable and cost-effective power supply. Graphite electrodes degrade over time, adding to operational costs. The quality of the final product is influenced by the purity of the input materials and the efficiency of the refining process.

Electric arc furnaces represent a modern and sustainable solution for metal production. By leveraging their efficiency, flexibility, and lower emissions, EAFs have become a cornerstone of the steel industry and a key technology for advancing sustainability in metal manufacturing.

Additionally, transitioning to cleaner energy sources, such as renewable energy for hydrogen production, can further diminish the carbon footprint associated with material production [615]. The integration of circular economy principles, which emphasize recycling and resource efficiency, is also vital. By adopting these principles, industries can minimize waste and reduce the demand for virgin materials, thus lowering overall emissions [616].

Innovative materials and technologies play a pivotal role in reducing the carbon footprint of industrial processes. For instance, advancements in composite materials can lead to lighter and more efficient products, which in turn can reduce energy consumption during their lifecycle [617]. Furthermore, the development of carbon footprint evaluation tools allows companies to assess and compare the environmental impacts of different production methods and materials, facilitating informed decision-making towards more sustainable practices [618, 619].

Energy Efficiency in Production Processes

Improving energy efficiency in material production is critical for reducing emissions, particularly in sectors such as steelmaking and polymer production. Advanced manufacturing technologies, including precision engineering and digital twins, play a pivotal role in optimizing energy usage by identifying inefficiencies and streamlining operations. For instance, the integration of digital twins allows for real-time monitoring and simulation of production processes, which can lead to significant reductions in energy consumption and waste [620, 621]. These technologies facilitate the identification of energy-intensive processes and enable manufacturers to implement corrective measures, thereby enhancing overall operational efficiency [622].

In steel production, specific advancements such as high-temperature furnaces with improved insulation and continuous casting techniques have been shown to minimize energy waste [623, 624]. Continuous casting, for example, optimizes the flow of materials and energy, reducing losses associated with traditional methods [623]. Furthermore, the implementation of energy recovery systems, such as waste heat recovery, is crucial in capturing and reusing heat generated during production processes. This approach not only lowers overall energy demand but also significantly cuts emissions, as demonstrated in various studies focusing on the steel industry [624, 625].

Similarly, modern polymer production facilities are increasingly adopting automated control systems to enhance energy efficiency during polymerization processes. These systems can adjust parameters in real-time to ensure optimal energy use, thereby reducing unnecessary consumption [626]. The integration of intelligent manufacturing techniques with green energy sources is also being explored to further enhance sustainability in the mechanical engineering sector [626].

Upgrading equipment and retrofitting older plants with energy-efficient technologies are essential strategies for improving energy efficiency. The adoption of advanced materials and technologies, such as energy-efficient thermal insulation, has been shown to contribute significantly to reducing greenhouse gas emissions in various industries (Kono et al., 2016). Moreover, the concept of resource efficiency, which combines energy efficiency and material efficiency, is gaining traction as a means to achieve substantial emissions reductions in the steel sector, which is a major contributor to industrial CO_2 emissions [624, 627].

Transition to Renewable Energy Sources

The transition from fossil fuels to renewable energy sources in material production is a critical strategy for reducing carbon footprints across energy-intensive industries. The reliance on fossil fuels such as coal, oil, and natural gas has significant implications for greenhouse gas (GHG) emissions and energy security. By replacing these traditional energy sources with renewables like solar, wind, and hydropower, industries can substantially decrease their carbon emissions while enhancing their energy independence.

Research indicates that the electrification of production processes, particularly through the use of renewable electricity, is gaining momentum. For instance, electric arc furnaces (EAFs) utilized in steel production can operate on renewable energy, resulting in significantly lower CO_2 emissions compared to conventional blast furnaces, which predominantly rely on fossil fuels [628]. This shift not only mitigates environmental impacts but also aligns with global efforts to transition to sustainable energy systems. The potential for renewable energy to serve as a cleaner alternative is further supported by studies highlighting the role of biomass and other renewable sources in reducing GHG emissions [629, 630].

Figure 48: Electric Arc Furnace. Arnoldius, CC BY-SA 3.0, via Wikimedia Commons.

Moreover, the exploration of renewable-powered electrochemical processes for the synthesis of monomers in polymer manufacturing exemplifies the innovative approaches being adopted to minimize carbon footprints [631]. The integration of renewable energy into these processes not only enhances sustainability but also contributes to the overall efficiency of material production. As industries increasingly adopt renewable technologies, the potential for significant reductions in carbon emissions becomes more tangible, reinforcing the need for continued investment in renewable energy infrastructure [632].

The broader implications of this transition extend beyond mere emissions reductions. By decreasing reliance on fossil fuels, industries can enhance their energy security, reducing vulnerability to fluctuations in fossil fuel markets and geopolitical tensions [633]. The ongoing evolution of energy systems towards renewables is not only an environmental imperative but also a strategic economic move that can foster resilience in energy supply chains.

Low-Carbon Feedstocks and Materials

The transition to low-carbon alternatives in industrial processes, particularly in cement and steel production, is critical for reducing greenhouse gas emissions. Substituting traditional feedstocks with low-carbon alternatives can significantly mitigate emissions at the material input stage. In cement production, for instance, the use of alternative materials such as fly ash, slag, or calcined clay as partial substitutes for clinker has been shown to lower emissions effectively. This approach not only reduces the carbon footprint associated with cement production but also enhances the sustainability of the materials used [634, 635].

In the steel industry, the shift from coke to hydrogen as a reducing agent in direct reduction processes represents a promising low-carbon solution. Hydrogen metallurgy fundamentally eliminates CO_2 emissions during iron and steel production, making it a viable alternative to traditional methods that rely heavily on fossil fuels [636, 637]. The integration of hydrogen in steelmaking processes can reduce carbon emissions by more than 90% compared to conventional blast furnace methods [638]. This transition is supported by advancements in hydrogen production technologies and the increasing availability of renewable energy sources, which are essential for producing green hydrogen [639].

Further, the production of carbon fibre from lignin, a renewable byproduct of the paper industry, exemplifies another innovative approach to creating advanced materials with a lower carbon footprint. Utilizing biobased materials not only reduces reliance on fossil fuels but also promotes a circular economy by repurposing waste products from other industries. However, specific references supporting this claim were not identified in the provided references.

The economic implications of these transitions are also noteworthy. The introduction of carbon pricing mechanisms, such as carbon credits and taxes, can enhance the competitiveness of low-carbon technologies in the steel industry [640]. As industries face growing pressure to decarbonize, the adoption of best available technologies (BAT) and innovative materials will be crucial in meeting ambitious climate targets while maintaining economic viability [634, 641].

Circular Economy Practices

Incorporating circular economy principles into material production is essential for reducing reliance on virgin resources and minimizing waste, which in turn cuts emissions across the lifecycle of materials. The circular economy emphasizes recycling and reusing materials, significantly lowering the energy required for production compared to extracting and processing raw materials. For instance, recycling aluminium is particularly energy-efficient, utilizing only about 5% of the energy required to produce primary aluminium, which results in a drastic reduction of CO_2 emissions by approximately 97% compared to primary production methods [642, 643]. This substantial energy saving illustrates the effectiveness of recycling in mitigating environmental impacts associated with material production.

The upcycling of materials—where waste is transformed into higher-value products—further contributes to carbon footprint reduction. Upcycling strategies not only enhance the value of waste materials but also promote sustainability by creating high-quality feedstocks from post-consumer or industrial waste. For example, advanced material recovery techniques, such as chemical recycling and solvolysis, have been shown to enable the conversion of plastics and composites into valuable materials, thus facilitating a more sustainable material lifecycle [644-646]. Chemical upcycling processes can convert plastic waste into functional polymers, thereby recovering material value while addressing the environmental challenges posed by plastic waste [647, 648].

The potential of upcycling is further exemplified by innovative approaches that transform waste into high-value chemicals and materials. For instance, the conversion of polystyrene waste into valuable chemicals through cascade degradation processes demonstrates the feasibility of upcycling in creating products with significantly higher market value than the original waste [649]. Additionally, the development of new technologies for the upcycling of polyethylene terephthalate (PET) waste into high-value materials without the need for extensive processing steps highlights the advancements in this field [650, 651]. These strategies not only reduce waste but also contribute to the circular economy by ensuring that materials are reused and repurposed effectively.

Process Innovations

The development and adoption of innovative processes in material production are crucial for enhancing sustainability, particularly through technologies such as carbon capture, utilization, and storage (CCUS). CCUS technologies are designed to capture CO_2 emissions generated during industrial processes, subsequently storing the captured carbon underground or converting it into usable products, including building materials and fuels [652]. This approach not only mitigates greenhouse gas emissions but also aligns with the principles of eco-innovation, which emphasizes the reduction of environmental impact through innovative practices [653, 654].

In the cement manufacturing sector, innovations such as Solidia cement and carbon-cured concrete exemplify how chemical processes can sequester CO_2 during production. Solidia cement incorporates a process that allows for the capture of CO_2, which is then utilized to cure concrete, thereby reducing net emissions significantly [653]. Similarly, carbon-cured concrete utilizes captured CO_2 to enhance the strength and durability of concrete, demonstrating a dual benefit of reducing emissions while improving material properties [653].

Furthermore, in the polymer manufacturing industry, integrating enzymatic and photocatalytic processes can lead to more energy-efficient production pathways with lower emissions. Enzymatic processes utilize biological catalysts to facilitate chemical reactions at lower temperatures and pressures, which significantly reduces energy consumption and associated emissions [655, 656]. Research indicates that enzymatic polymerization techniques have gained traction, allowing for the production of various polymer classes, including polyamides and polyesters, through more sustainable methods [656]. Photocatalytic processes also offer a promising avenue for sustainable polymer production by harnessing light energy to drive chemical reactions, thus minimizing reliance on fossil fuels [655].

The integration of these innovative processes not only contributes to the reduction of carbon footprints in material production but also aligns with broader sustainability goals. The implementation of sustainable manufacturing practices has become a focal point for industries worldwide, as evidenced by the increasing attention to sustainable practices in sectors such as textiles and polymers [657-659]. By fostering a culture of eco-innovation, firms can enhance their competitive advantage while contributing to environmental sustainability [660, 661].

Digitalization and Smart Manufacturing

Digital technologies such as artificial intelligence (AI), machine learning, and the Internet of Things (IoT) play a significant role in enhancing production processes through real-time monitoring and optimization. These technologies facilitate the development of predictive maintenance systems that significantly reduce the likelihood of equipment failures, thereby preventing inefficiencies in manufacturing operations. For instance, the integration of digital twin technology allows for continuous monitoring and simulation of production processes, which can lead to optimized performance and reduced waste and energy consumption. Digital twins create a virtual representation of physical assets, enabling manufacturers to analyse data and improve operational efficiency in real time [662-664].

AI-driven models are particularly effective in predicting energy usage and identifying emission hotspots within material production facilities. By leveraging machine learning algorithms, these models can analyse historical data to forecast energy demands and pinpoint areas where emissions can be reduced. This capability allows for targeted interventions that not only enhance energy efficiency but also contribute to sustainability goals [665, 666]. Furthermore, the application of AI in conjunction with IoT devices enables the collection of vast amounts of data, which can be utilized for predictive analytics and decision-making processes that drive operational improvements [667, 668].

Blockchain technology complements these advancements by enhancing transparency and traceability within supply chains [669]. This ensures that sustainable practices are adhered to throughout the production lifecycle. By providing a secure and immutable ledger of transactions, blockchain can facilitate the tracking of materials and processes, thereby promoting accountability and sustainability in manufacturing operations [664, 670]. The integration of these digital technologies not only optimizes production processes but also supports the broader objectives of reducing environmental impact and fostering sustainable industrial practices [671, 672].

Lifecycle Assessments and Sustainable Design

Conducting lifecycle assessments (LCAs) is a critical approach in identifying emission-intensive stages in material production. LCAs provide a comprehensive evaluation of the environmental impacts associated with all stages of a product's life, from raw material extraction through production, use, and disposal. This methodology enables stakeholders to pinpoint specific phases where emissions are highest, thereby facilitating targeted reduction efforts. For instance, Meng et al. [673] highlight that while carbon fibre composites can significantly reduce lifecycle energy use and greenhouse gas (GHG) emissions in applications such as aerospace, the energy-intensive production process of these materials can sometimes negate these benefits if not managed properly. This underscores the importance of LCAs in guiding sustainable practices in material production.

Sustainable design principles, particularly those focusing on longevity and recyclability, further contribute to minimizing emissions throughout the product lifecycle. Nurcahyanie and Rohmadiani [674] discuss the concept of "Design for Longevity," which emphasizes the importance of extending product lifespan through thoughtful design choices. This approach not only enhances resource efficiency but also aligns with sustainability goals by reducing the frequency of product replacements and associated emissions. Moreover, the integration of sustainable manufacturing techniques, such as recycling and the use of eco-friendly materials, is essential in achieving these objectives.

Lightweight materials, such as carbon fibre-reinforced composites (CFRPs), exemplify how innovative design can lead to significant reductions in emissions during the use phase of products, particularly in the automotive and aerospace sectors. The inherent properties of CFRPs, including their high strength-to-weight ratio, allow for lighter vehicle designs, which in turn reduce fuel consumption and emissions during operation. Research indicates that the use of CFRPs can lead to a net decrease in lifecycle emissions, as the fuel savings during use can offset the emissions generated during production [144, 673]. Additionally, advancements in the recyclability of CFRPs, as noted by Amantayeva et al. [675], are crucial for ensuring that these materials do not contribute to long-term environmental waste.

Collaboration and Policy Support

Collaboration across industries and governments is essential for achieving significant reductions in material production emissions. This collaboration can take various forms, including public-private partnerships and industry alliances, which facilitate the sharing of knowledge and best practices. For instance, the Mission Possible Partnership focuses on decarbonizing heavy industries by fostering innovation and collaboration among stakeholders, which is critical for driving systemic change in emission reduction practices [676, 677]. Such partnerships not only enhance resource efficiency but also promote cleaner production methods, thereby contributing to the broader sustainability agenda [677].

Policies such as carbon pricing, subsidies for clean energy, and mandates for low-carbon materials serve as vital incentives for companies to adopt sustainable practices. Carbon pricing mechanisms, including carbon taxes and emissions trading systems, are recognized for their effectiveness in internalizing the costs of carbon emissions, thereby encouraging businesses to reduce their greenhouse gas outputs [678-680]. These mechanisms have been shown to influence corporate behaviour positively, prompting firms to invest in cleaner technologies and practices [681]. Furthermore, subsidies for low-carbon technologies can lower the financial barriers to adopting sustainable practices, making it more feasible for companies to transition towards greener operations [681, 682].

The role of industry partnerships in driving innovation cannot be overstated. Collaborative efforts, such as those seen in the Mission Possible Partnership, enable companies to share insights and strategies that lead to enhanced sustainability outcomes [676, 677]. These partnerships also play a major role in addressing the challenges posed by climate change, as they facilitate the exchange of resources and knowledge necessary for implementing effective emission reduction strategies [677, 683]. By working together, industries can leverage their collective expertise to develop and deploy innovative solutions that significantly lower material production emissions [676].

Future Trends in Sustainable Material Design

The future trends in sustainable material design are increasingly influenced by the principles of the circular economy, advancements in material science, and the integration of innovative design methodologies. As the global community grapples with the challenges of resource depletion and environmental degradation, the focus on sustainable materials is becoming paramount. This synthesis explores these trends through various lenses, including material innovation, circular economy principles, and design methodologies.

One of the most significant trends in sustainable material design is the shift towards materials that are not only renewable but also multifunctional. The rational design of materials, particularly those derived from renewable resources, is gaining traction as researchers and designers seek to create products that minimize environmental impact while maximizing utility [684]. This approach is supported by advances in computational tools and materials databases, which facilitate the prediction, synthesis, and characterization of new materials [685]. The integration of bio-composite materials, which valorise waste products, exemplifies this trend, as they offer high recycling potential and contribute to resource conservation [686].

The circular economy (CE) framework is another critical aspect shaping the future of sustainable material design. CE emphasizes the importance of designing products for longevity, reuse, and recyclability, thereby reducing waste and resource consumption [687]. Designers play a crucial role in this paradigm, as their decisions during the design phase significantly influence the end-of-use (EOU) value recovery of products [687]. The construction industry, in particular, is witnessing a shift towards circular practices, where materials are designed for disassembly and reuse, thus minimizing the environmental impact associated with traditional construction methods [688, 689]. This transition is supported by the development of new business models that prioritize sustainability and resource efficiency [690].

Moreover, the role of education and skill development in promoting circular economy principles cannot be overstated. As the demand for sustainable design practices grows, educational institutions are increasingly incorporating circular economy concepts into their curricula [691]. This educational shift aims to equip future designers with the necessary competencies to innovate within the circular economy framework, fostering a culture of sustainability and responsible resource management [692].

In addition to these trends, the application of life cycle thinking in design processes is gaining prominence. By considering the entire life cycle of a product, designers can make informed decisions that enhance the sustainability of materials used [693]. This approach not only addresses the environmental impacts of materials but also encourages the development of products that are easier to recycle and repurpose, thus contributing to a more sustainable economy [694].

Finally, the integration of social innovation and design thinking into sustainable material design is emerging as a vital trend. This perspective emphasizes the need for collaborative approaches that involve multiple stakeholders in the design process, ensuring that materials and products meet both environmental and social needs [695]. By fostering a culture of innovation and collaboration, the design community can effectively address the barriers to implementing circular economy principles and drive the transition towards a more sustainable future.

Chapter 8

Computational Materials Engineering

Computational Modelling in Material Design

Computational modelling plays a transformative role in material design by enabling researchers and engineers to predict and optimize material properties, behaviours, and performance before physical prototypes are created. It integrates principles of physics, chemistry, and engineering with advanced computational techniques to simulate the complex interactions within materials at multiple scales. This approach accelerates innovation, reduces costs, and enhances the development of advanced materials for various industries.

Computational modelling is an interdisciplinary approach that employs mathematical, physical, and computational techniques to create digital representations of real-world systems. This methodology is pivotal for understanding complex phenomena, predicting outcomes, and addressing problems that are difficult to explore through experimental means alone. By simulating physical systems, researchers can gain insights into their behaviour and interactions, which is essential across various domains such as engineering, biology, and social sciences [696, 697].

The foundation of computational modelling is rooted in a mathematical framework that defines system behaviour through equations, algorithms, or rules. These mathematical constructs can take the form of differential equations, statistical methods, or machine learning algorithms, tailored to specific applications. For instance, weather models incorporate parameters like temperature and humidity to replicate real-world conditions [698]. Various computational algorithms, such as finite element methods and Monte Carlo simulations, are utilized to numerically solve these equations, allowing for the simulation of system dynamics [699, 700]. Visualization tools play a crucial role in interpreting the results, translating numerical data into accessible formats like graphs and animations, which enhances communication and understanding of the modelled phenomena [696].

Computational models can be classified into deterministic and stochastic categories. Deterministic models operate under fixed rules, yielding consistent outputs for given inputs, which is particularly useful for systems with predictable behaviours, such as mechanical systems [700]. Conversely, stochastic models incorporate randomness to account for variability and uncertainty, making them suitable for applications like financial forecasting or ecological modelling [701]. Additionally, mechanistic models grounded in physical laws provide a deeper understanding of system behaviour, while empirical models rely on observed data to establish statistical relationships [702, 703]. Multi-scale models are particularly noteworthy as they integrate phenomena across different scales, from atomic interactions to macroscopic behaviours, thus capturing the complexity of real-world systems [704].

The applications of computational modelling are vast and varied. In engineering, it is employed to optimize the design and performance of structures and materials [705]. In the realm of physics, computational modelling aids in understanding complex phenomena such as fluid dynamics and thermodynamics [706]. In medicine and biology, it facilitates the simulation of disease progression and drug interactions, which is vital for advancing healthcare [703]. Climate science heavily relies on computational models to predict weather patterns and assess the impacts of climate change [707]. Furthermore, in social sciences, computational modelling is instrumental in studying economic systems and human behaviour, providing insights that inform policy and decision-making.

The advantages of computational modelling are significant, including cost-effectiveness, predictive power, and scalability. It reduces the need for expensive physical experiments, allows for rapid iterations in research and development, and can handle systems of varying complexity. However, it is essential to acknowledge the limitations of computational modelling. The accuracy of these models is heavily dependent on the quality of input data and the assumptions made during the modelling process. Validation against experimental data is crucial but can be challenging, particularly for complex systems where simplifications may omit critical details.

Computational modelling allows for accurate predictions of a material's properties based on its atomic structure and composition. Quantum mechanical simulations, such as density functional theory (DFT), are often used to predict electronic, mechanical, thermal, and optical properties at the atomic level. For example, researchers can predict the strength of a composite material or the conductivity of a semiconductor without needing to physically produce it. This predictive capability is crucial for tailoring materials to meet specific performance requirements.

In materials science, modelling provides insights into the fundamental mechanisms governing material behaviour, such as phase transitions, defect formation, and chemical reactions. For instance, molecular dynamics (MD) simulations help study how atoms and molecules interact over time under varying conditions, shedding light on phenomena like corrosion, diffusion, or crack propagation. These insights enable a deeper understanding of why materials behave the way they do, guiding improvements and innovations.

Traditional material discovery relies heavily on trial-and-error experimentation, which can be time-consuming and expensive. Computational modelling streamlines this process by allowing virtual experiments to explore vast compositional and structural spaces. Machine learning algorithms, coupled with computational models, can identify promising material candidates by analysing large datasets and predicting properties based on prior knowledge. This approach has been instrumental in discovering materials for applications such as energy storage, lightweight alloys, and high-temperature ceramics.

Modelling tools enable the optimization of material performance for specific applications. Finite element analysis (FEA) and computational fluid dynamics (CFD) are commonly used to simulate the mechanical, thermal, or fluid-flow performance of materials and structures. For example, FEA can be applied to design lightweight yet durable materials for aerospace components, ensuring they withstand mechanical stresses while minimizing weight. Similarly, CFD helps design materials with optimal thermal conductivity or fluid interaction properties for heat exchangers or biomedical devices.

Advanced computational modelling bridges different scales, from atomic to macroscopic levels, to provide a comprehensive understanding of material behaviour. Multi-scale modelling integrates quantum-level calculations, molecular simulations, and continuum-level mechanics, capturing interactions that span orders of magnitude in size and time. For instance, designing materials for additive manufacturing often requires simulating molecular bonding, microstructural evolution, and final component performance.

Computational modelling supports the development of sustainable materials by enabling lifecycle analysis and environmental impact assessments. Simulations help identify materials that minimize energy consumption, waste, and emissions during production and usage. For example, models can optimize catalytic materials for more efficient chemical processes or design biodegradable polymers with the required mechanical and thermal properties.

Computational modelling drives innovation in industries ranging from aerospace and automotive to electronics and healthcare. In aerospace, it helps design high-performance alloys and composites that reduce fuel consumption. In electronics, models guide the development of semiconductors and nanomaterials with enhanced conductivity and miniaturization. In healthcare, computational approaches are used to design biomaterials for implants and drug delivery systems with tailored biocompatibility and functionality.

By replacing many physical experiments with virtual simulations, computational modelling significantly reduces the time and cost associated with material development. Researchers can quickly iterate through design variations and optimize properties without the need for extensive lab work. This efficiency is particularly valuable in developing materials for cutting-edge applications, where rapid innovation is critical to competitiveness.

Examples of Computational Models Used for Advanced Materials

Computational modelling plays a pivotal role in the design, development, and optimization of advanced materials. These models simulate the behaviour, properties, and interactions of materials across multiple scales, from atomic to macroscopic levels. Below are examples of computational models commonly used for advanced materials:

1. Density Functional Theory (DFT) : Quantum-level modelling of electronic structures.

DFT is used to study the electronic, optical, and magnetic properties of materials. It helps predict the behaviour of materials such as graphene, semiconductors, and superconductors. For example, DFT is widely applied in designing catalysts, such as those used in hydrogen fuel cells, by modelling surface reactions at the atomic level.

In computational materials science, ab initio density functional theory (DFT) serves as a cornerstone for predicting and understanding material behaviour based on quantum mechanical principles. Unlike empirical methods, ab initio DFT requires no external parameters like material constants or experimental data, relying instead on the fundamental properties of electrons and atomic nuclei. This is achieved by solving quantum mechanical equations to describe the electronic structure of materials, providing insights into properties such as conductivity, magnetism, and elasticity.

Contemporary DFT calculations determine the electronic structure of a system through a constructed potential acting on the electrons. This potential comprises two main components: the external potential (V_{ext}) and the effective potential (V_{eff}). The external potential is dictated by the arrangement and elemental composition of atoms in the system, while the effective potential accounts for interelectronic interactions, including exchange and correlation effects. Together, these potentials form the basis for solving the Kohn–Sham equations, a set of Schrödinger-like equations for n non-interacting electrons in a representative supercell of the material. These equations transform the many-electron problem into a manageable framework, making DFT a powerful tool for studying systems ranging from simple crystals to complex alloys and nanomaterials.

The theoretical foundation of DFT traces back to the Thomas–Fermi model, an early attempt to describe electronic structure using electron density as the central variable. However, the breakthrough in DFT came with the Hohenberg–Kohn (HK) theorems, formulated by Walter Kohn and Pierre Hohenberg. These theorems established the theoretical rigor of DFT, demonstrating that the ground-state properties of a system are uniquely determined by its electron density. This insight reduced the complexity of the many-body problem from dealing with $3N$ spatial coordinates for N electrons to just three spatial coordinates. The HK theorems also introduced the concept of energy functionals, where the ground-state electron density minimizes the total energy functional, providing a practical approach to solving the quantum mechanical problem.

The extension of the HK theorems to time-dependent systems led to the development of time-dependent density functional theory (TDDFT), enabling the study of excited states and dynamic processes in materials. This has significant implications for understanding optical properties, electronic excitations, and time-resolved phenomena.

Walter Kohn and Lu Jeu Sham advanced DFT further by introducing the Kohn–Sham (KS) framework, which won them the Nobel Prize in Chemistry. KS DFT simplifies the complex many-body problem of interacting electrons in a static external potential into a system of non-interacting electrons moving within an effective potential. This effective potential combines the external potential with the effects of Coulomb interactions, including exchange and correlation effects. These interactions are represented through approximate exchange–correlation functionals, as the exact functional is unknown. The simplest approach, the local-density approximation (LDA), uses data from a uniform electron gas to approximate the exchange and correlation energy. Despite its simplicity, LDA has been foundational in materials science, with more sophisticated approximations like the generalized gradient approximation (GGA) and hybrid functionals emerging for greater accuracy.

Another less common but theoretically significant approach is orbital-free density functional theory (OFDFT), which avoids the use of Kohn–Sham orbitals altogether. In OFDFT, approximations are applied not only to the exchange–correlation energy but also to the kinetic energy of non-interacting electrons. While computationally efficient, OFDFT is less widely used due to its challenges in accurately modelling kinetic energy.

The advancements in DFT, both in the KS and orbital-free frameworks, have transformed computational materials science. These methods allow researchers to study materials at the atomic scale with remarkable precision, enabling the discovery and design of novel materials for applications in energy, electronics, and nanotechnology.

Worked Examples of Using Ab Initio Density Functional Theory (DFT) in Materials Science

Predicting the Bandgap of Silicon

Objective: To calculate the electronic bandgap of crystalline silicon (Si) using DFT.

Step 1: Structure Setup - The atomic structure of silicon is set up in its diamond cubic crystal form. The lattice parameter is initialized based on experimental data (approximately 5.43 Å).

Step 2: Choice of Functional - The Perdew-Burke-Ernzerhof (PBE) functional, a type of generalized gradient approximation (GGA), is chosen for the initial calculation. Hybrid functionals such as HSE06 can be used for more accurate results.

Step 3: K-point Sampling - A dense k-point grid (e.g., 8 × 8 × 8) is selected to sample the Brillouin zone, ensuring convergence of electronic properties.

Step 4: Self-Consistent Field (SCF) Calculation - The SCF procedure solves the Kohn–Sham equations iteratively to find the ground-state electron density.

Step 5 - Band Structure Calculation Using the converged electron density, the electronic band structure is calculated along high-symmetry paths in the Brillouin zone (e.g., Γ-X-W-K-Γ).

Result: The calculated bandgap using the PBE functional is approximately 0.6 eV, which underestimates the experimental value of 1.1 eV. Employing hybrid functionals like HSE06 increases the accuracy, yielding a bandgap closer to the experimental value.

Adsorption of Hydrogen on a Platinum Surface

Objective: To determine the binding energy of a hydrogen atom adsorbed on a platinum (Pt) surface.

Step 1: Surface Model - A slab model of Pt(111) is created using a periodic supercell with a vacuum layer to prevent interactions between periodic images. A single hydrogen atom is placed at a high-symmetry adsorption site (e.g., atop, bridge, or hollow).

Step 2: Relaxation - The positions of the hydrogen atom and the top layers of the Pt slab are optimized using DFT with a PBE functional to minimize the total energy.

Step 3: Energy Calculation - The total energies of three configurations are calculated:

1. E_{system} : Total energy of the Pt slab with hydrogen adsorbed.

2. E_{Pt} : Total energy of the clean Pt slab.

3. E_{H} : Energy of an isolated hydrogen atom.

Step 4: Binding Energy Calculation - The adsorption energy (E_{ads}) is computed as:

$$E_{ads} = E_{system} - (E_{Pt} + E_{H})$$

Result: The adsorption energy provides insights into the stability of hydrogen on the Pt surface, relevant for catalysis studies in fuel cells.

Diffusion of Lithium in a Solid Electrolyte

Advanced Material Engineering Fundamentals

Objective: To calculate the diffusion barrier of lithium ions in a solid-state electrolyte like Li_3PS_4.

Step 1: Crystal Structure - The bulk structure of Li_3PS_4 is modelled. A defect-free supercell is constructed, and a lithium ion vacancy is introduced for diffusion studies.

Step 2: Pathway Selection - A possible migration pathway for lithium ions between neighbouring lattice sites is identified.

Step 3: Nudged Elastic Band (NEB) Calculation - The NEB method is employed to calculate the energy profile along the diffusion path. Intermediate images of the migrating lithium ion are generated between initial and final positions.

Step 4: Energy Barrier Calculation - The energy barrier is determined from the highest energy along the diffusion path.

Result: The calculated diffusion barrier provides a quantitative measure of ionic conductivity in the solid electrolyte, critical for battery performance.

Mechanical Properties of a High-Entropy Alloy

Objective: To predict the elastic constants of a high-entropy alloy (HEA) using DFT.

Step 1: Alloy Composition - The composition of the HEA (e.g., CoCrFeNiMn) is modelled using a supercell with random atomic arrangements to approximate the alloy structure.

Step 2: Strain Application - Small strains are applied to the supercell in various directions to simulate mechanical deformation.

Step 3: Stress Calculation - For each strained configuration, the total energy is calculated, and the stress tensor is derived.

Step 4: Elastic Constants - The elastic constants (C_{ij}) are determined from the stress-strain relationship:

$$\sigma_i = \sum_j C_{ij}\varepsilon_j$$

where σ is the stress, C_{ij} are the elastic constants, and ε is the strain.

Result: The elastic constants help predict mechanical properties like Young's modulus and shear modulus, which are essential for structural applications.

Defect Formation Energy in Silicon Carbide (SiC)

Objective: To calculate the formation energy of a vacancy defect in 4H-SiC.

Step 1: Defect-Free Supercell - A defect-free supercell of 4H-SiC is constructed, and its total energy ($E_{perfect}$) is calculated.

Step 2: Defect Introduction - A silicon atom is removed to create a silicon vacancy (V_{Si}). The total energy of the supercell with the vacancy (E_{defect}) is calculated.

Step 3: Chemical Potential - The chemical potential of silicon (μ_{Si}) is determined from bulk silicon.

Step 4: Formation Energy - The defect formation energy is given by:

$$E_{\text{form}} = E_{\text{defect}} - E_{\text{perfect}} + \mu_{\text{Si}}$$

Result: The defect formation energy provides insights into the likelihood of vacancy formation, which influences the electrical and thermal properties of SiC.

These examples demonstrate how ab initio DFT methods are used to address challenges in materials science, from electronic and mechanical properties to energy applications and catalytic processes. Each case highlights the predictive power of DFT in understanding and designing advanced materials.

2. Molecular Dynamics (MD) Simulations: Atomic-scale modelling of material behaviour.

MD simulations predict the movement and interaction of atoms and molecules over time. These models are essential for studying nanoscale materials, such as carbon nanotubes, and their mechanical, thermal, and chemical properties. For instance, MD simulations have been used to explore how temperature affects the flexibility and conductivity of nanomaterials.

Molecular dynamics (MD) is a computational simulation technique that allows researchers to analyse and predict the physical movements of atoms and molecules over time. By solving Newton's equations of motion for interacting particles, MD provides a dynamic view of molecular systems, enabling the exploration of phenomena such as molecular motion, thermodynamic behaviour, and structural changes at an atomic scale [708, 709]. The technique employs interatomic potentials or molecular mechanical force fields to calculate forces between particles, which is crucial for understanding the evolution of complex systems [708, 709]. This capability makes MD particularly valuable in fields such as chemical physics, materials science, and biophysics [708, 710].

One of the primary advantages of MD is its ability to handle systems composed of a vast number of particles, which are often too complex for analytical methods [708, 711]. The numerical solutions provided by MD simulations capture intricate details of molecular interactions and dynamics that would otherwise remain inaccessible [708, 711]. However, despite its strengths, MD simulations can accumulate numerical errors over time, necessitating careful algorithm selection and parameter tuning to minimize these inaccuracies [711]. Systems that conform to the ergodic hypothesis can leverage MD to derive macroscopic thermodynamic properties, thus linking microscopic dynamics to macroscopic observables, a fundamental aspect of statistical mechanics [708, 711].

The historical development of MD can be traced back to the mid-20th century, with significant contributions from early pioneers such as Enrico Fermi and Berni Alder, who utilized analogue computers for simulating many-body systems [708, 709]. The advent of digital computing in the 1950s marked a turning point, allowing for more complex simulations. A notable milestone was Aneesur Rahman's 1964 simulations of liquid argon using Lennard-Jones potentials, which laid the groundwork for future applications of MD [708, 709]. Over the years, MD has evolved significantly, integrating advanced computational techniques and algorithms that enhance its applicability across various scientific domains [708, 709].

MD simulations are widely employed in diverse research areas. In materials science, they are used to investigate atomic-scale phenomena such as thin film growth and the behaviour of nanotechnological devices [708, 711]. In structural biology, MD aids in refining protein structures, modelling macromolecular interactions, and simulating processes like protein folding [708, 711]. The pharmaceutical industry also benefits from MD, utilizing it for drug design by analysing ligand binding, protein conformations, and pharmacophore development [708, 711]. Furthermore, MD contributes to theoretical physics by studying systems under extreme conditions, such as high pressures or temperatures, thereby enhancing our understanding of fundamental physical processes [708, 711].

The accuracy of MD simulations is heavily dependent on the quality of the force fields used to describe atomic interactions. While modern force fields account for various interactions, including electrostatic and van der Waals forces, approximations can introduce limitations [708, 709]. For instance, treating hydrogen bonds as simple Coulomb interactions or using vacuum dielectric constants can lead to inaccuracies in simulations [708, 709]. Recent advancements, such as polarizable force fields, aim to address these challenges by providing a more accurate representation of complex environments, particularly in aqueous solutions [708, 709].

In the context of drug discovery, MD simulations have proven invaluable. Researchers utilize MD to refine protein-ligand complexes, identify conserved binding regions, and generate pharmacophore models for virtual screening [708, 710]. However, challenges remain in accurately modelling certain interactions, such as hydrogen bonding and van der Waals forces, due to their quantum mechanical nature and dependence on the surrounding environment [708, 710]. Despite these limitations, MD continues to be a cornerstone of computational science, enabling researchers to explore molecular behaviour with unprecedented detail and paving the way for innovative discoveries in materials design, biophysics, and beyond [708, 710].

Design constraints in molecular dynamics (MD) simulations require careful consideration of computational resources and the scientific goals of the study. The simulation parameters—such as system size, timestep, and total duration—must balance computational feasibility with the need for statistical and physical relevance. Simulations should be long enough to capture the kinetics of natural processes, ensuring meaningful insights, yet not so large as to become computationally prohibitive. For example, protein and DNA dynamics studies often simulate nanoseconds to microseconds, requiring substantial computational power over days or years. Parallel processing algorithms, such as spatial or force decomposition, distribute computational tasks across multiple processors to improve efficiency.

One critical factor in MD simulations is the evaluation of interatomic potentials, which is computationally intensive, particularly for non-bonded interactions like electrostatics and van der Waals forces. These calculations can scale poorly with system size, often described as $O(n^2)$, where n is the number of particles. Advanced methods like particle mesh Ewald summation or spherical cutoff techniques can reduce this complexity, making simulations more manageable. The integration timestep, typically in the femtosecond range, is another key parameter. It must be small enough to resolve the fastest vibrational frequencies within the system while minimizing numerical errors. Techniques like the SHAKE algorithm or multiple timescale methods can extend the timestep for slower interactions, optimizing computational efficiency.

When simulating systems in a solvent, a choice must be made between explicit and implicit solvent models. Explicit solvent simulations, while computationally expensive due to the increased number of particles, provide detailed granularity essential for capturing solute properties and chemical kinetics. Implicit solvents use a mean-field approach, significantly reducing computational load but potentially losing fine details of solvent interactions. In all cases, the simulation box must be large enough to avoid artifacts from boundary conditions, often managed with periodic boundary conditions that mimic a bulk phase.

The choice of ensemble—microcanonical (NVE), canonical (NVT), or isothermal-isobaric (NPT)—affects the physical constraints of the simulation. In the microcanonical ensemble, energy is conserved, and the system's trajectory reflects an exchange between potential and kinetic energy. This ensemble provides insights into isolated systems where heat exchange is absent. In the canonical ensemble, temperature is controlled using thermostats, which approximate energy exchange with a heat bath. Popular thermostats, such as Nosé–Hoover or Langevin dynamics, ensure realistic energy adjustments while maintaining the system's equilibrium distribution. Careful selection of thermostat algorithms and parameters is critical to avoid artifacts, such as the "flying ice cube" effect, where unphysical translations dominate the simulation.

The isothermal-isobaric ensemble is particularly relevant for simulating systems under laboratory-like conditions, where both temperature and pressure are controlled. Barostats manage pressure variations, and for anisotropic systems like lipid bilayers, specialized controls maintain membrane area or surface tension.

Generalized ensembles, such as the replica exchange method (REMD), address the challenge of slow dynamics in systems with multiple energy minima. By allowing temperature exchanges between system replicas, REMD facilitates efficient sampling of conformational space, overcoming energy barriers that hinder exploration in traditional MD simulations.

MD simulations offer unparalleled insights into molecular behaviour, but their success hinges on a well-designed approach that considers computational limits, system-specific requirements, and appropriate ensemble choices. Each decision in the simulation design—from the choice of timestep to the thermostat algorithm—has implications for the accuracy, relevance, and feasibility of the results.

Potentials in molecular dynamics (MD) simulations serve as the mathematical frameworks that define how particles within a system interact. These potential functions, often referred to as "force fields" in chemistry and "interatomic potentials" in materials physics, dictate the energy and forces acting on particles. They are foundational to simulating and analysing molecular and atomic behaviour. Potentials range in complexity from empirical models grounded in classical mechanics to quantum mechanical frameworks, depending on the desired accuracy and the type of system under investigation.

Empirical potentials are commonly employed in MD simulations due to their computational efficiency. These potentials approximate quantum mechanical effects through functional forms fitted to experimental or high-level quantum data. In chemistry, these force fields typically include terms representing bonded interactions (such as bond stretching, angle bending, and dihedral torsions) and non-bonded interactions (like van der Waals forces and electrostatics). Non-bonded interactions, which are inherently long-range, often dominate the computational cost, leading to the development of methods like particle mesh Ewald summation or cut-off radii to optimize calculations. Despite their efficiency, traditional force fields are limited in their ability to simulate chemical reactions since they rely on predefined bonding arrangements. However, reactive force fields, such as ReaxFF, overcome this limitation by enabling bond formation and breaking during simulations.

A key distinction among potentials lies in whether they model interactions using pairwise or many-body approaches. Pair potentials, such as the Lennard-Jones potential, calculate interactions based on atom pairs and are computationally simpler. In contrast, many-body potentials account for interactions among three or more particles, allowing for a more accurate representation of phenomena like bond angles and coordination. Examples of many-body potentials include the Tersoff potential, which models covalent materials like silicon and carbon, and the Embedded Atom Method (EAM), which describes metallic bonding.

When greater accuracy is required, semi-empirical and ab initio potentials provide an alternative. Semi-empirical methods incorporate quantum mechanical principles while relying on parameterization for computational efficiency. Ab initio MD simulations, on the other hand, compute interactions directly from quantum mechanics, often using density functional theory (DFT). This approach enables the study of electronic effects, excited states, and chemical reactions but is computationally intensive, limiting its application to small systems and short timescales.

Hybrid QM/MM methods combine the accuracy of quantum mechanics with the speed of classical mechanics. These methods divide the system into regions, treating chemically active zones quantum mechanically while using classical potentials for the remainder. This approach is particularly useful for studying enzymatic reactions, where the active site is modelled quantum mechanically to capture electronic effects, and the surrounding environment is treated classically.

Coarse-graining and reduced representations simplify simulations of large systems or long timescales by grouping atoms into pseudo-atoms. This reduction sacrifices atomic detail for computational efficiency, enabling studies of phenomena like protein folding, liquid crystal phase transitions, and DNA supercoiling. The parameterization of coarse-grained models is critical to maintaining accuracy, balancing enthalpic and entropic contributions to system behaviour.

In recent years, machine learning force fields (MLFFs) have emerged as a transformative approach. By training on high-level quantum mechanical data, MLFFs can achieve near-ab initio accuracy while being orders of magnitude faster. These models capture complex potential energy surfaces and are particularly useful for systems where traditional force fields struggle to describe intricate interactions.

In MD simulations, selecting the appropriate potential depends on the system's complexity, the desired accuracy, and computational constraints. Each type of potential—whether empirical, many-body, quantum mechanical, or machine learning-based—offers unique advantages and trade-offs, making them indispensable tools in advancing our understanding of molecular and materials behaviour.

Incorporating solvent effects into molecular dynamics (MD) simulations is crucial for accurately representing the behaviour of solutes, as solvents significantly influence solute dynamics through random collisions and frictional drag. These effects are especially relevant in biological and chemical systems where the solvent environment affects the conformation, stability, and interactions of molecules. To balance computational efficiency with accuracy, several strategies are employed to model solvent effects while minimizing the number of solvent molecules explicitly included in the simulation.

One common approach involves using non-rectangular periodic boundary conditions, stochastic boundaries, or solvent shells. These techniques allow the solvent environment closest to the solute to be explicitly simulated while reducing the computational burden associated with distant solvent molecules that contribute less to the solute's behaviour. This selective modelling enables researchers to focus computational resources on the solute, improving simulation efficiency. Additionally, implicit solvent models can incorporate solvent effects without requiring explicit solvent molecules. In such cases, potential mean force (PMF) calculations are used to represent how the free energy changes as a particular coordinate of the solute is varied. The PMF effectively averages out the influence of the solvent, enabling realistic solute behaviour without the direct computational cost of modelling each solvent molecule.

Solvent inclusion is particularly critical when simulating macromolecules like proteins. Without accounting for solvent effects, proteins and other molecules may adopt unrealistic conformations due to the absence of solvent-mediated damping of van der Waals and electrostatic interactions. Even small molecules are affected, potentially collapsing into unnaturally compact states. Explicitly including solvent molecules or using accurate implicit solvent models ensures that the macromolecule or solute interacts realistically with its environment, yielding results that align more closely with experimental observations.

Long-range forces, such as charge-charge or dipole-dipole interactions, introduce additional challenges in solvent modelling. These interactions decay more slowly than short-range forces and can extend beyond the simulation box, making their accurate representation computationally demanding. Solutions like increasing the box size to capture all interactions are computationally expensive and impractical for large systems. Truncating the potential at a fixed distance often results in artifacts near the cutoff. Techniques such as Ewald summation and particle mesh methods address this issue, allowing for the efficient and accurate calculation of long-range forces while preserving the physical realism of the system.

Steered molecular dynamics (SMD) simulations offer another method for exploring solvent effects, particularly in the context of protein and biomolecular studies. SMD applies external forces to manipulate molecular structures, allowing researchers to observe structural changes at the atomic level. By pulling specific atoms or residues at constant velocity or constant force, SMD can simulate mechanical unfolding, stretching, or other processes. These simulations often rely on umbrella sampling to ensure adequate sampling of the system's configurations along a reaction coordinate, enabling the calculation of changes in free energy through PMF analysis. The weighted histogram analysis method (WHAM) is commonly used to analyse umbrella sampling data, providing insights into high- and low-energy states and their contributions to the system's behaviour.

SMD has proven particularly valuable in drug discovery and biomolecular research. For instance, it has been employed to study the stability of Alzheimer's protofibrils, investigate protein-ligand interactions in cyclin-dependent kinase 5, and analyse the effects of electric fields on thrombin-aptamer complexes. These studies highlight the importance of solvent effects and SMD techniques in elucidating molecular mechanisms, guiding drug development, and understanding biomolecular function under varying environmental conditions. By combining solvent modelling with advanced simulation techniques, researchers can achieve a more comprehensive understanding of molecular behaviour in realistic environments.

3. Finite Element Analysis (FEA): Structural and mechanical performance modelling.

FEA divides a material or structure into smaller elements and solves equations to predict stresses, strains, and deformations under load. This method is often used to design lightweight materials for aerospace or automotive applications, such as high-performance composites for aircraft wings or crash-resistant car frames.

Finite Element Analysis (FEA) is a computational technique widely used to model, analyse, and solve complex physical problems in engineering and materials science. By dividing a physical structure into smaller, manageable pieces called finite elements and applying mathematical equations, FEA allows researchers to simulate how materials behave under various conditions. It is particularly useful for predicting material properties, optimizing designs, and studying the interactions between materials and external forces. This makes FEA indispensable for understanding the mechanical, thermal, and electromagnetic properties of materials, facilitating their application in diverse fields.

The FEA process begins with creating a digital representation of the object or material to be analysed using CAD models or other computational geometry tools. This geometry is then divided into finite elements connected at nodes through a process called meshing. The density of the mesh depends on the complexity of the analysis; finer meshes offer more accurate results but require greater computational resources. Material properties, such as Young's modulus, Poisson's ratio, and thermal conductivity, are assigned to each element. Boundary conditions and loads, including constraints like fixed supports or applied forces such as pressure or heat flux, are then defined to simulate the environmental and operational conditions of the material or structure.

Using numerical methods, such as the Galerkin method or direct matrix solvers, the governing equations for each element are solved iteratively to calculate parameters like stress, strain, or displacement. Results are visualized through stress contours, deformation plots, or temperature distributions, enabling researchers to interpret the outcomes and optimize the material or structure.

In materials science, FEA is applied in various ways. For instance, it helps simulate the mechanical properties of composite materials like carbon fibre-reinforced polymers, enabling predictions of stress distribution, deformation, and potential failure modes under load. It is also used in thermal analysis to study heat transfer within materials, optimizing thermal insulation or conductivity. FEA models the stress and strain associated with temperature-dependent phase transitions, such as those in shape memory alloys, and predicts crack initiation and propagation in metals, ceramics, or polymers. Additionally, FEA is crucial in additive manufacturing, analysing residual stresses, thermal gradients, and material distortions to improve processes like 3D printing.

To use FEA, the problem must first be clearly defined, including the objective and the collection of relevant data such as material properties, geometry, and load conditions. A digital model of the geometry is created using CAD or FEA-specific tools, followed by generating an appropriate mesh. Material properties are input, and boundary conditions are applied to simulate the operating environment. The simulation is then run using specialized software like ANSYS, Abaqus, COMSOL, or SolidWorks Simulation. Results are analysed to assess material performance and identify critical regions, and the process is iterated to refine the model, mesh, or conditions based on the findings, optimizing the material or design.

FEA offers several advantages, including detailed insights into material behaviour under various conditions, reducing the need for physical experiments, and enabling the optimization of materials and designs before manufacturing. It also supports the analysis of complex geometries and heterogeneous materials. However, FEA can be computationally intensive, especially for large or fine-mesh models, and results depend on the accuracy of input data and boundary conditions. Simplifications, such as assuming linear material behaviour, may also limit accuracy.

Despite these limitations, FEA is a cornerstone of modern engineering and research. It facilitates the design and analysis of advanced materials for applications ranging from aerospace and automotive industries to electronics and biomedical devices, making it an essential tool in materials science.

4. Phase-Field Modelling: Predicting microstructure evolution.

Phase-field models are used to simulate the formation and evolution of microstructures during processes such as solidification, grain growth, or phase transformations. This method is critical for optimizing materials like superalloys used in jet engines, where the microstructure determines strength and creep resistance.

A phase-field model serves as a robust mathematical framework to tackle interfacial problems, particularly within the realm of materials science. This model is adept at addressing various physical phenomena, including solidification dynamics, fracture mechanics, and even biological processes such as collective cell migration. The core advantage of the phase-field approach lies in its ability to replace sharp boundary conditions at the interface with a continuous description, utilizing partial differential equations that govern the evolution of an auxiliary field known as the phase field. This phase field acts as an order parameter, transitioning smoothly between distinct values in regions corresponding to different phases, thus creating a diffuse interface with a finite width. This characteristic allows researchers to circumvent the computational complexities associated with explicitly modelling sharp boundaries, enabling the analysis of intricate interfacial dynamics [712, 713].

The construction of a phase-field model is fundamentally rooted in its capacity to approximate interfacial dynamics in the sharp interface limit, where the diffuse interface width approaches zero. This alignment with physical realities is critical for the model's applicability. The mathematical foundation of the phase-field model involves partial differential equations that describe the system's behaviour holistically, eliminating the necessity for explicit interface boundary conditions. Early theoretical contributions by Langer and others laid the groundwork for phase-field modelling, which has since evolved into a powerful computational tool, particularly with advancements in computational power and theoretical developments [712, 713]. For instance, the phase-field model has been successfully employed to simulate solidification processes, capturing the transformation from liquid to solid while incorporating terms for interfacial energy and bulk phase stability [714, 715].

One notable advancement in phase-field modelling is the introduction of the thin interface limit by Karma and Rappel, which has significantly enhanced the practicality of simulations by relaxing the stringent requirements for very narrow interfaces. This breakthrough has made phase-field models more computationally feasible, allowing for quantitative simulations of realistic interfacial problems, such as predicting microstructural evolution during solidification and modelling crack propagation in fracture mechanics [716, 717]. The versatility of phase-field models extends to multiphase systems, where multiple order parameters can represent different phases or grain orientations within a material. This capability is particularly advantageous for studying polycrystalline materials and phase transformations in alloys, such as the transformation from austenite to ferrite in ferrous alloys [718, 719].

In the context of fracture mechanics, phase-field models provide a diffuse approach to representing cracks, inherently capturing crack initiation and propagation through a variational formulation. This contrasts with traditional methods that require discrete crack representations and additional criteria for determining crack paths. By coupling elastic and fracture energies, phase-field models offer a natural framework for simulating crack behaviour without the need for additional assumptions or remeshing [720, 721]. Furthermore, in biological systems, phase-field models have been applied to study collective cell migration, where the interactions between cells and their environment lead to coordinated movement. These models can simulate complex phenomena such as wound healing and cancer cell invasion by incorporating phase fields for individual cells and additional variables like chemical gradients [722, 723].

The ongoing advancements in phase-field modelling are marked by the development of alternative energy-density functions and refined numerical techniques, which address challenges such as spontaneous drop shrinkage and enhance the model's capability to handle long-term simulations or systems with multiple interfaces. The phase-field approach has thus become a cornerstone in the study of materials science and beyond, demonstrating its ability to manage complex interfacial dynamics and integrate seamlessly with modern computational tools [724-726].

5. Computational Fluid Dynamics (CFD): Fluid-material interactions and thermal management.

CFD is used to model the behaviour of materials in contact with fluids, such as heat exchangers or porous membranes. Advanced materials like metal foams and thermally conductive polymers are evaluated for their efficiency in dissipating heat or facilitating fluid flow.

Computational Fluid Dynamics (CFD) is a powerful branch of fluid mechanics that utilizes numerical methods and data structures to analyse and solve problems involving fluid flow. In the realm of materials science, CFD plays a critical role in understanding and optimizing processes where fluid interactions significantly impact material properties or production methods. These include heat treatment of metals, casting, additive manufacturing, and coatings. By simulating fluid flow, heat transfer, and related phenomena, CFD enables scientists and engineers to predict outcomes, optimize designs, and improve efficiency in material processing and application.

The fundamental basis of CFD lies in solving equations that describe the conservation of mass, momentum, and energy within a fluid. These equations, often referred to as the Navier-Stokes equations, are derived from fundamental principles of physics, such as Newton's laws of motion and the first law of thermodynamics. Depending on the complexity of the problem, simplifications or modifications of these equations are made. For example, in scenarios involving incompressible fluids or low Mach number flows, the incompressible Navier-Stokes equations are employed. In cases where heat transfer significantly impacts fluid behaviour, energy conservation equations are coupled with the flow equations to model temperature distributions.

In materials science, CFD is particularly relevant in simulating processes like casting, where molten metal flows into molds. By modelling the flow and cooling of the liquid metal, CFD helps predict defects such as air entrapment or uneven solidification, enabling optimization of mold designs and casting parameters. Similarly, in additive manufacturing, CFD simulates the behaviour of molten metal or polymer under the influence of lasers or electron beams, allowing engineers to optimize layer-by-layer deposition processes to minimize defects and residual stresses.

Another critical application of CFD in materials science is in the study of chemical vapor deposition (CVD) and physical vapor deposition (PVD) processes. These methods are used to create thin films or coatings on materials, where the behaviour of gaseous precursors and their interaction with substrates determine the coating's quality and uniformity. CFD models the flow of gases, heat transfer, and chemical reactions in the deposition chamber, providing insights into optimizing deposition rates and achieving uniform coatings.

CFD also addresses challenges associated with multi-phase flows, such as the interaction between gas and liquid phases in metal foaming processes or gas-solid interactions in powder metallurgy. By solving equations for each phase and modelling the interfaces between them, CFD provides detailed predictions of flow dynamics, aiding in the development of high-quality, lightweight materials.

The accuracy and reliability of CFD simulations depend on several factors, including the quality of the computational grid (mesh), the choice of turbulence models, and the boundary conditions applied. In material processes involving turbulent flows, such as in melt stirring or mixing, turbulence models like Reynolds-Averaged Navier-Stokes (RANS) or Large Eddy Simulation (LES) are employed to capture the chaotic behaviour of the fluid. Similarly, for heat transfer simulations, incorporating radiative and conductive heat transfer models enhances the fidelity of the results, especially in high-temperature environments.

To use CFD effectively, the first step is defining the physical problem, including the fluid properties, material boundaries, and process conditions. A computational domain representing the physical geometry is created, followed by discretising the domain into finite elements or volumes to form a mesh. Governing equations are then applied to each mesh element, and numerical solvers iterate to compute flow velocities, pressures, temperatures, and other variables. Post-processing tools visualize the results, enabling detailed analysis of flow patterns, heat distribution, or chemical reactions.

CFD's importance in materials science continues to grow as computational resources and software capabilities advance. It provides a cost-effective, non-invasive way to analyse and optimize processes, reducing the need for physical experimentation while offering insights into complex fluid-material interactions. By integrating CFD with experimental data and machine learning techniques, researchers can further enhance process efficiencies and material performance in diverse applications.

The methodology of CFD involves a stepwise approach to model fluid dynamics, starting with preprocessing, moving through simulation, and culminating in post-processing. This structured process ensures that complex interactions between fluids and materials can be understood and optimized for a variety of applications, such as casting, additive manufacturing, and heat treatment.

The preprocessing stage in CFD is critical for defining the physical and geometric parameters of the problem. Using Computer-Aided Design (CAD) tools, the geometry of the material system is outlined, and the fluid domain—the space occupied by the fluid—is extracted. The domain is then discretised into smaller cells or a mesh. This mesh can be structured, unstructured, or hybrid, and is composed of various element shapes like hexahedra, tetrahedra, or polyhedra. The mesh resolution, or density, is tailored to capture critical flow phenomena; finer meshes are used in areas of high gradients, such as near material interfaces or regions of turbulence. Next, physical models are selected based on the problem's requirements, such as the Navier-Stokes equations for fluid flow, heat transfer equations, or species conservation equations in reactive flows. Boundary conditions and initial conditions are then specified to simulate real-world constraints and behaviours.

Once the preprocessing is complete, the simulation phase begins. Numerical solvers iteratively calculate the fluid's behaviour by solving the governing equations. Depending on the material process being modelled, this simulation can be steady-state or transient. For instance, transient simulations are essential in processes like additive manufacturing, where time-dependent phenomena such as laser-induced melting and solidification occur. The accuracy and stability of the simulation depend on the discretization methods employed. Finite Volume Method (FVM), Finite Element Method (FEM), and Finite Difference Method (FDM) are among the most common discretization techniques. Each has its strengths and is chosen based on the complexity and computational requirements of the problem. FVM is widely used for its efficiency in handling conservation laws, while FEM excels in managing complex geometries and boundary conditions.

Post-processing is the final step, where the results are analysed and visualized. Outputs such as velocity profiles, temperature distributions, or pressure fields are represented through contour plots, vector fields, or animations. In materials science, this step is particularly valuable for interpreting the effects of fluid dynamics on material behaviour. For instance, in casting, post-processing can reveal areas prone to porosity or incomplete filling, while in additive manufacturing, it can highlight regions of thermal stress or distortion.

CFD relies on robust discretization techniques to translate continuous fluid dynamics equations into solvable numerical forms. The Finite Volume Method is frequently used in CFD for materials science applications due to its conservation-centric approach. This method divides the domain into control volumes and ensures flux conservation across their boundaries. For more complex geometries, the Finite Element Method provides flexibility and accuracy but requires careful formulation to maintain conservation properties. The Finite Difference Method, although historically significant, is less common in modern materials-focused CFD due to its challenges in handling irregular geometries. Advanced methods like the Lattice Boltzmann Method (LBM) and Spectral Element Method are also gaining traction, particularly for microscale and multiphase flows.

Turbulence modelling is another critical aspect of CFD, especially in high-Reynolds-number flows encountered in processes like melt stirring or deposition techniques. Models such as Reynolds-Averaged Navier-Stokes (RANS) and Large Eddy Simulation (LES) allow for the approximation of turbulence while balancing computational expense and accuracy. LES, in particular, is increasingly used in materials science for resolving flow structures that influence heat and mass transfer.

CFD applications in materials science are vast and diverse. In casting, CFD simulates molten metal flow and solidification, identifying defects and optimizing mold designs. In additive manufacturing, it models the interaction between energy sources and material to predict melt pool dynamics and thermal gradients. During chemical vapor deposition, CFD helps optimize gas flow and reaction kinetics to ensure uniform coatings. The flexibility of CFD allows researchers to simulate multiphase flows, analyse heat treatment processes, and even study the aerodynamic properties of materials in applications like wind turbines.

The integration of CFD into materials science offers unparalleled capabilities for designing and optimizing processes, reducing experimental costs, and accelerating innovation. As computational power grows and numerical methods evolve, CFD will continue to play a transformative role in understanding and engineering material-fluid interactions.

6. Machine Learning Models: Accelerating material discovery and optimization.

Machine learning (ML) models analyse large datasets to predict material properties or identify novel materials. These models have been used to discover materials for energy storage, such as lithium-ion battery electrodes, by predicting performance based on chemical compositions.

Machine learning (ML) has emerged as a transformative tool in materials science, significantly accelerating the discovery and optimization of materials by leveraging extensive datasets and advanced computational algorithms. This data-driven approach allows researchers to predict material properties, identify novel compositions, and optimize existing materials for specific applications more efficiently than traditional experimental or purely computational methods. The integration of ML with domain expertise provides unprecedented opportunities to tackle complex challenges in materials science, as highlighted by various studies that demonstrate the effectiveness of ML in extracting hidden patterns from large datasets and guiding material selection processes [727, 728].

One of the most impactful applications of ML in materials science is the discovery of advanced materials for energy storage, particularly in lithium-ion batteries, which are essential for renewable energy systems and electric vehicles. ML models analyse the relationships between chemical compositions, structural features, and electrochemical properties to predict the performance of potential electrode materials. This predictive capability enables the identification of high-capacity, long-life, and safe electrode materials without exhaustive experimental testing, thus reducing development time and costs [729-731]. For instance, studies have shown that ML can effectively model the performance of various materials, leading to the discovery of new compounds that meet specific energy storage requirements [732, 733].

Beyond energy storage, ML is revolutionizing other areas of materials science, including structural materials, catalysts, and electronic materials. In the realm of structural materials, ML models predict mechanical properties such as strength, elasticity, and toughness by analysing data from alloy compositions and processing conditions [734, 735]. In catalysis, ML aids in identifying active sites and optimizing catalyst structures for chemical reactions, which is crucial for applications like fuel cells and carbon capture technologies [736, 737]. For electronic materials, ML predicts properties such as band gaps, conductivity, and dielectric constants, facilitating the design of materials for semiconductors, solar cells, and sensors [738, 739].

The success of ML in materials discovery is largely attributed to its ability to process diverse datasets, including experimental results, computational simulations, and high-throughput screening outputs. By employing algorithms such as neural networks, support vector machines, and decision trees, ML models can identify patterns and correlations that may not be immediately apparent to researchers, thus guiding the selection of promising candidates for further study and testing [740-742]. This capability is particularly valuable in the context of inverse design, where ML models suggest compositions or structures that meet desired properties, thereby enabling the design of materials with specific optical, magnetic, or thermal characteristics [743, 744].

In addition to discovery, ML plays a crucial role in optimizing existing materials and processes. For example, ML models can optimize processing parameters, such as heat treatment temperatures or deposition rates, to enhance material performance. Furthermore, ML can predict degradation mechanisms of materials under various conditions, aiding in the development of more durable and reliable products [745, 746]. The ability to optimize both materials and processes through ML not only improves performance but also contributes to cost savings and efficiency in material production [747, 748].

Despite the transformative potential of ML in materials science, challenges remain in ensuring data quality and interpretability. Datasets must be curated carefully to avoid biases or inconsistencies that could lead to inaccurate predictions. Moreover, ML models must be interpretable to provide insights into the underlying mechanisms driving observed trends, enabling researchers to make informed decisions [728, 735]. As the field continues to evolve, the integration of ML with experimental and computational methods will likely lead to significant breakthroughs in materials performance and functionality, addressing critical global challenges in energy, sustainability, and technology [749, 750].

7. Ab Initio Molecular Dynamics (AIMD): Simulating chemical reactions and thermal properties.

AIMD combines quantum mechanics and molecular dynamics to simulate the behaviour of materials at elevated temperatures. For example, it has been applied to study the stability and conductivity of solid electrolytes in next-generation batteries.

The Car–Parrinello Molecular Dynamics (CPMD) method is a cornerstone technique for performing ab-initio molecular dynamics (AIMD), a simulation approach that integrates quantum mechanical calculations with classical molecular dynamics. AIMD methods are used to simulate the motion of atoms and molecules by computing interactions directly from first principles, without relying on pre-defined empirical force fields. CPMD is distinctive in how it efficiently incorporates electronic structure calculations into the dynamics of atomic motion.

In traditional AIMD approaches, such as Born–Oppenheimer Molecular Dynamics (BOMD), the electronic structure problem is solved at every time step to determine forces acting on the nuclei. This iterative process is computationally demanding because it requires solving the Schrödinger equation repeatedly to ensure that the electronic state remains in its ground state for the given nuclear configuration.

The CPMD method addresses this challenge by introducing a novel approach. Instead of solving the electronic structure problem explicitly at every time step, CPMD treats the electronic wavefunctions as fictitious dynamical variables. These variables evolve in time alongside the atomic nuclei under the guidance of an extended Lagrangian. This allows the electronic wavefunctions to stay close to their ground state automatically as the nuclei move, without requiring full self-consistent optimization at every step. The forces acting on the nuclei are derived directly from the electronic wavefunctions at each step, maintaining consistency between the quantum mechanical and classical components of the simulation.

Car–Parrinello Molecular Dynamics (CPMD) is a pivotal computational methodology in materials science for simulating atomic and electronic dynamics at the quantum mechanical level. This method combines classical molecular dynamics with quantum mechanics by treating the nuclei as classical particles and the electrons as quantum mechanical entities. Unlike traditional molecular dynamics, CPMD directly calculates electronic interactions from first principles, enabling accurate modelling of materials and their behaviours under various conditions.

In materials science, CPMD plays a crucial role in studying systems where electronic effects are significant, such as chemical reactions, phase transitions, and the behaviour of materials under extreme conditions. The method excels in cases where empirical potentials fail to capture the complexity of interactions, such as systems involving bond breaking and formation, or those with significant electronic polarization effects. These capabilities make

CPMD invaluable for exploring novel materials and understanding their structural, dynamic, and electronic properties.

The CPMD methodology is built on an extended Lagrangian framework that couples the classical motion of nuclei with the quantum mechanical behaviour of electrons. This coupling allows the system to remain on the electronic ground state throughout the simulation without the need for repeated self-consistent electronic minimization at each time step. Instead, the electronic wavefunctions are treated as fictitious dynamical variables, evolving alongside the nuclei under the constraint of orthonormality. This approach significantly reduces the computational cost compared to Born–Oppenheimer Molecular Dynamics (BOMD), where the electronic structure must be recalculated at every step.

In CPMD simulations, the electronic interactions are typically described using density functional theory (DFT) with a plane-wave basis set and pseudopotentials to represent the core electrons. The Kohn-Sham orbitals, which approximate the electronic wavefunctions, are expanded in a plane-wave basis, enabling efficient calculation of the ground-state electronic density and forces acting on the nuclei. The use of fictitious dynamics to propagate the electronic degrees of freedom ensures that the system remains adiabatic, maintaining the electrons close to their ground state.

One of the method's strengths is its ability to model systems with strong electron-ion coupling and dynamic changes in bonding environments. For instance, CPMD is used to study the melting of materials, high-pressure phase transitions, and the behaviour of complex oxides and semiconductors. It has also been employed to investigate catalytic reactions at surfaces, where the interplay between the electronic structure and the motion of atoms is critical for understanding reaction mechanisms.

Despite its advantages, CPMD has limitations. The computational cost of the method, though lower than BOMD, remains high, restricting its application to systems with a few hundred atoms and timescales of tens of picoseconds. The accuracy of CPMD simulations depends on the choice of pseudopotentials and the plane-wave cutoff, requiring careful parameterization for reliable results. Additionally, the fictitious mass parameter used for the electrons must be small enough to maintain adiabaticity without introducing numerical instability, further constraining the time step of the simulation.

The theoretical framework of CPMD, including its extended Lagrangian and equations of motion, reflects a delicate balance between accuracy and efficiency. The method achieves this by leveraging the Born–Oppenheimer approximation while introducing fictitious dynamics to simplify the treatment of electronic variables. As computational power continues to grow, the applicability of CPMD in materials science is expected to expand, enabling more complex and larger-scale simulations.

8. Monte Carlo Simulations: Statistical modelling of material behaviour.

Monte Carlo methods use random sampling to model the behaviour of materials under various conditions. These simulations are used in polymer science to study chain configurations, crosslinking, and phase separation in advanced polymers like thermosets and hydrogels.

Monte Carlo methods, with their reliance on repeated random sampling, have become indispensable in materials science for solving complex problems where traditional deterministic approaches are impractical. By leveraging randomness to approximate deterministic phenomena, these methods are particularly useful in predicting material behaviour, optimizing processes, and exploring the vast design space of novel materials.

In materials science, Monte Carlo methods are commonly applied to optimize material properties, simulate phase transformations, and study atomic-scale interactions. These techniques allow researchers to account for uncertainties in material inputs and external conditions, enabling more accurate predictions of real-world behaviour. For instance, Monte Carlo simulations are used to model disordered systems, such as amorphous materials or liquids, where coupled degrees of freedom make analytical solutions infeasible. By generating a large number of random configurations, researchers can statistically evaluate the material's properties, such as its density, thermal conductivity, or diffusivity.

One prominent application of Monte Carlo methods in materials science is in phase-field modelling, where the evolution of microstructures during phase transitions is studied. For example, when modelling the solidification of a metal, Monte Carlo algorithms help simulate the growth of grains by randomizing the movement of atoms within a lattice. This randomness replicates the thermal fluctuations inherent in real-world processes, providing insight into grain boundary formation, dendritic growth, and segregation phenomena.

Monte Carlo methods are also employed in the evaluation of thermodynamic properties through statistical mechanics. By sampling configurations of atoms or molecules in a material, the method can estimate macroscopic properties such as free energy, entropy, or specific heat. These calculations are critical for designing materials with tailored properties, especially in fields like energy storage, where optimizing the thermodynamics of lithium-ion batteries is vital.

Another key area of application is the simulation of defects and diffusion in materials. For instance, Monte Carlo algorithms are used to predict how vacancies, dislocations, or impurities move through a crystal lattice. These insights are essential for understanding phenomena such as creep, stress relaxation, and ionic conductivity, which directly impact the performance of structural and functional materials.

Monte Carlo methods are also employed in multiscale modelling, where phenomena at different length and time scales must be integrated. For example, in the simulation of thin-film deposition processes, Monte Carlo methods are used to model atomic-scale events such as adsorption and surface diffusion, which influence the macroscopic properties of the film, like thickness uniformity or crystallinity.

Despite their versatility, Monte Carlo methods face challenges in materials science. The accuracy of these simulations depends heavily on the quality of the random number generators and the number of samples used. For systems with high dimensionality, such as those involving many interacting particles, the computational cost can become prohibitive. Additionally, ensuring convergence to a meaningful solution requires careful design of the algorithm and thorough validation against experimental or theoretical data.

Monte Carlo methods have also been enhanced through hybrid approaches, such as integrating them with molecular dynamics or density functional theory. These combined techniques enable the study of electronic structure changes in materials during dynamic processes, bridging the gap between quantum and classical scales.

Worked Example: Monte Carlo Simulation for Grain Growth in Polycrystalline Materials

Objective: To simulate grain growth in a polycrystalline material using the Monte Carlo method and predict the evolution of the grain structure over time.

Problem Description:

Grain growth in polycrystalline materials is a process where larger grains grow at the expense of smaller ones to reduce the overall grain boundary energy. Understanding this process is critical for controlling the mechanical and thermal properties of materials, such as strength and conductivity.

Monte Carlo methods can simulate grain growth by modelling the evolution of a lattice system, where each lattice site represents a grain. The energy reduction drives the system toward equilibrium.

Steps in the Monte Carlo Simulation:

Define the Initial System:

> o   A 2D square lattice (e.g., 100×100) is used to represent the material.

> o   Each lattice site is assigned a grain orientation label, randomly initialized with integer values (e.g., 1 to 20), representing different grain orientations.

Set the Interaction Energy:

> o   The system's energy is defined by the grain boundary energy between neighbouring lattice sites. The energy for a pair of adjacent sites is higher if their orientations differ, simulating the physical reality of grain boundaries.

Monte Carlo Update Rules:

- o   A random lattice site is selected.

- o   A new orientation (label) is proposed for the site, chosen randomly from its neighbours.

- o   The change in energy (ΔE) is calculated:

$$\Delta E = E_{\text{new}} - E_{\text{old}}$$

where E_{new} and E_{old} are the system's energies before and after the proposed change.

The Metropolis criterion is applied:

- If ΔE≤0, the change is accepted.

- If ΔE>0, the change is accepted with a probability:

$$P = e^{-\Delta E / k_B T}$$

where k_B is Boltzmann's constant and TTT is the system temperature.

Iterate:

- Repeat the process for a sufficient number of Monte Carlo steps (MCS), where one MCS involves attempting updates for all lattice sites.

Output:

- After each MCS, visualize the lattice to observe grain growth.

- Quantify the grain size distribution and average grain size over time.

Example Calculation:

- **Initial Configuration:**

  - o   100×100 lattice initialized with 20 random grain orientations.

- **Parameters:**

  - o   Interaction energy between different grains: J=1 unit.

  - o   Temperature: T=1 unit.

  - o   Simulation duration: 10,000 MCS.

- **Simulation:**

- At each MCS, grain boundaries smooth out as smaller grains shrink and disappear.

  - Larger grains consume smaller ones due to their lower boundary energy.

- **Results:**

  - Initial grain size distribution: Random.

  - Grain size distribution after 10,000 MCS: Fewer grains with larger average size.

  - Visualization: Grain boundaries become less jagged and more uniform, reflecting physical grain growth.

Monte Carlo methods have several applications in material design, validation, and optimization. In material design, they are used to predict the evolution of grain sizes under various thermal treatments, enabling researchers to design processes that achieve specific grain sizes tailored to desired material properties. By understanding the dynamics of grain growth, Monte Carlo simulations help in creating materials with optimized mechanical and thermal characteristics.

These methods also play a crucial role in validation, where simulation results are compared with experimental data obtained through optical or electron microscopy. This comparison ensures that the computational models accurately reflect the physical behaviour of materials, bridging the gap between theoretical predictions and experimental observations.

Optimization is another critical application of Monte Carlo methods. They allow researchers to explore how varying parameters, such as temperature or initial grain distributions, affect the grain growth process. This capability helps in fine-tuning processing conditions to enhance the performance and reliability of materials in specific applications.

Monte Carlo methods effectively simulate grain growth in polycrystalline materials, providing valuable insights into microstructural evolution. By guiding experimental procedures and optimizing material properties, these simulations have become indispensable tools in the development of advanced materials for applications in metals, ceramics, and thin films.

9. Peridynamics: Modelling damage and fracture in materials.

Peridynamics is a non-local continuum mechanics model used to simulate crack propagation and failure in materials. It has been applied to predict the fracture behaviour of ceramics, composites, and other brittle materials used in aerospace and defence industries.

Peridynamics is a modern framework within continuum mechanics that focuses on modelling deformations involving discontinuities, particularly fractures. Unlike traditional continuum mechanics, which relies on differential equations and stress tensors that break down near singularities such as cracks, peridynamics uses integral equations to model material behaviour. This approach allows the same mathematical equations to apply seamlessly across the entire material domain, including at discontinuities.

The fundamental concept of peridynamics is based on the idea of non-local interactions, where material points interact with all other points within a specified finite distance, known as the peridynamic horizon. These interactions are mediated through "bonds," which can be visualized as connections between points within the horizon. These bonds transmit forces, which are functions of the relative displacements and positions of interacting points. This non-local feature aligns peridynamics more closely with molecular dynamics than traditional macroscopic continuum mechanics.

The original formulation, bond-based peridynamics, models interactions as central forces. However, this approach imposes limitations on material properties, such as Poisson's ratio, which cannot exceed certain values. To address this, the state-based peridynamic framework was developed, allowing the force exchanged between two points to depend on the deformation state of all surrounding bonds. This enhancement provides greater flexibility in describing complex material behaviour and accommodates realistic material responses.

One of the primary motivations for peridynamics is its ability to model fractures without the need for additional criteria or specialized methods like stress intensity factors or extended finite element methods (xFEM). Fractures emerge naturally in peridynamic simulations as bonds break when a critical strain is exceeded. This intrinsic handling of discontinuities eliminates the need for separate fracture growth laws and simplifies the modeling of crack propagation and coalescence.

In the peridynamic framework, material bodies are described as continuous point meshes, with each point capable of interacting with others within its peridynamic horizon. The equations of motion are derived from the balance of forces, including internal bond forces and external forces. These equations are integral in nature and consider the cumulative effect of all interactions within the horizon. The displacement field of each point evolves over time based on these forces, governed by the following fundamental equation of peridynamics:

$$\rho \frac{\partial^2 \mathbf{u}(\mathbf{x}, t)}{\partial t^2} = \mathbf{F}(\mathbf{x}, t),$$

where $\mathbf{F}(\mathbf{x}, t)$ represents the total force per unit volume acting on a material point $\mathbf{x}$, including contributions from internal bond forces and external forces.

The equation

$$\rho\frac{\partial^2 \mathbf{u}(\mathbf{x}, t)}{\partial t^2} = \mathbf{F}(\mathbf{x}, t)$$

is a fundamental expression in peridynamics and represents the equation of motion for a material point in the peridynamic framework.

Here is a detailed explanation of the terms:

ρ: This denotes the **mass density** of the material at the point x. It is the mass per unit volume of the material and is a key parameter in determining the inertia of the system.

$\frac{\partial^2 \mathbf{u}(\mathbf{x},t)}{\partial t^2}$: his represents the **second time derivative** of the displacement field $\mathbf{u}(\mathbf{x}, t)$, which is the **acceleration** of the material point located at **x** at time *t*. In peridynamics, $\mathbf{u}(\mathbf{x}, t)$ describes the displacement of a material point from its reference (undeformed) position over time.

$\mathbf{F}(\mathbf{x}, t)$: This is the total **force density** acting on the material point **x** at time *t*. It includes contributions from:

- **Internal forces**: Due to interactions with neighbouring points within the peridynamic horizon.

- **External forces**: Such as applied loads or environmental influences like gravity.

Physical Interpretation:

- The equation is essentially Newton's Second Law of Motion (*F=ma*) expressed in the context of peridynamics. It states that the product of the mass density (ρ) and the acceleration of a material point equals the net force per unit volume acting on that point.

Significance in Peridynamics:

- Unlike classical mechanics, where internal forces are represented using a stress tensor and partial derivatives, peridynamics uses integral equations to calculate $\mathbf{F}(\mathbf{x}, t)$. This force accounts for long-range interactions with neighboring points within the material's **peridynamic horizon**.

- The term $\mathbf{F}(\mathbf{x}, t)$ is computed as the sum of all pairwise interaction forces between the material point **x** and other points within its interaction neighbourhood, along with any external forces.

This equation governs the dynamics of deformation and fracture in the material, enabling the modelling of phenomena such as crack propagation, coalescence, and material failure without requiring special treatments for singularities like cracks.

The non-local nature of peridynamics also means that stress tensors, which are inherently local concepts, are not used. Instead, the force density function, often referred to as the peridynamic kernel, characterizes how internal forces depend on deformation. The kernel captures all constitutive properties of the material and determines how bonds respond to strain. The displacement, strain, and bond forces are calculated iteratively to simulate material deformation and fracture behaviour.

In material science, peridynamics has significant applications in modelling brittle and ductile fractures, delamination in composites, and the behaviour of materials under dynamic loading. Its ability to handle discontinuities and large deformations makes it a powerful tool for understanding complex phenomena, such as crack branching, coalescence, and material failure in polycrystalline structures. By providing a unified framework to address both smooth and fractured regions of a material, peridynamics offers insights into the evolution of microstructures and their influence on macroscopic material properties.

Bond-Based Peridynamics

Bond-based peridynamics is a formulation of continuum mechanics that models the interactions between pairs of material points as bonds, enabling the simulation of deformations and fractures. This approach is based on integral equations rather than the partial differential equations of classical continuum mechanics. It is particularly suited to modelling discontinuities, such as cracks, without requiring additional fracture mechanics laws.

In bond-based peridynamics, the interactions between material points are represented by a force kernel $f(\xi, \eta)$, where $\xi = x' - x$ is the relative position vector between two points in the undeformed configuration, and $\eta = u(x') - u(x)$ is the relative displacement vector between the same points in the deformed configuration. This pairwise interaction force determines how the internal forces within a material respond to deformation. The force kernel is central to defining the material's constitutive properties and dictates how the material behaves under stress or strain.

The bond-based formulation strictly adheres to Newton's Third Law, ensuring that the forces between two interacting points are equal in magnitude and opposite in direction. This principle is mathematically represented as $f(-\eta, -\xi) = -f(\eta, \xi)$. This guarantees the conservation of linear momentum throughout the system, a fundamental requirement for physically accurate simulations.

Bond-based peridynamics also enforces angular momentum conservation by ensuring that the forces between points align with the relative deformed ray vector connecting them. This implies that the force $f(\xi, \eta)$ must be parallel to $\xi + \eta$, the vector describing the bond in the deformed configuration. This condition ensures that the material behaves isotropically and that the forces do not introduce spurious torques.

In the context of hyperelastic materials, the bond forces are derived from a potential energy function $\Phi(\xi, \eta)$, ensuring that the material's behaviour is elastic and reversible. This means the force kernel $f(\xi, \eta)$ can be expressed as the gradient of the potential with respect to the relative displacement: $f(\xi, \eta) = \nabla_\eta \Phi(\xi, \eta)$. This formulation satisfies both the balance of forces and the conservation of angular momentum.

For small deformations where $|\eta| \ll 1$, the force kernel can be linearized around the undeformed state. In this case, the bond forces are expressed as a linear function of the relative displacement, allowing the material's stiffness properties to be represented by a micromodulus tensor $C(\xi)$. This tensor defines the stiffness of the bonds based on their undeformed length |ξ||\xi||ξ| and material properties.

The peridynamic kernel, which governs the bond forces, is often tailored to specific material behaviours. For example, in isotropic elastic materials, the force kernel is proportional to the bond stretch s, defined as the relative elongation of the bond:

$$s = \frac{|\xi + \eta| - |\xi|}{|\xi|}.$$

The corresponding force can then be expressed as:

$$f(\xi, \eta) = c \, s \, \mu(s, t) \, n,$$

where c is the micromodulus constant, $\mu(s, t)$ is a function that accounts for bond failure, and $n = (\xi + \eta)/|\xi + \eta|$ is the unit vector along the bond in the deformed configuration.

In bond-based peridynamics, bond failure is incorporated by introducing a critical stretch s0s_0s0. Bonds are considered broken if their stretch exceeds this critical value, and no force is transmitted through the broken bond. This mechanism naturally models crack initiation and growth, as the breaking of bonds redistributes stress to neighbouring bonds, potentially leading to further failure.

Bond-based peridynamics is widely used to simulate fracture mechanics, including crack propagation and coalescence. However, its assumption of pairwise independent interactions imposes limitations on the material's Poisson ratio, restricting its range to specific values. This limitation is addressed in state-based peridynamics, which considers collective interactions among bonds to model more complex material behaviours.

Bond-based peridynamics provides a powerful framework for simulating the mechanical behaviour of materials, particularly in cases involving fractures and discontinuities. By modelling material interactions as bonds, it captures the essential physics of deformation and failure, offering insights into the behaviour of materials under complex loading conditions.

10. Kinetic Monte Carlo (KMC): Time-dependent processes in materials.

KMC models are used to study slow processes like diffusion, grain growth, or chemical reactions. For instance, KMC is applied to predict the diffusion of lithium ions in battery materials and to optimize their charge-discharge cycles.

The Kinetic Monte Carlo (KMC) method is a computational technique designed to simulate the time evolution of systems where transitions between discrete states occur at known rates. These rates represent the likelihood of specific transitions or processes over time and must be determined through external experimental or theoretical methods, such as molecular dynamics or density functional theory. The KMC method is particularly suited for processes that follow Poisson statistics, such as chemical reactions, diffusion, or defect formation in materials. It is closely related to the Gillespie algorithm, frequently used in stochastic chemical kinetics, and is sometimes referred to as dynamic Monte Carlo.

The fundamental principle of KMC lies in its ability to provide a temporal description of the simulated system. Unlike classical Monte Carlo methods, which focus on equilibrium properties or static configurations, KMC tracks the sequence of events and their timing, making it invaluable for simulating time-dependent processes. At each step, the algorithm selects a specific event from a list of possible transitions, executes it, and updates the system's state. The elapsed time for each event is stochastically determined based on the associated transition rates, ensuring an accurate temporal representation of the system's evolution.

In KMC simulations, two main algorithmic approaches exist: rejection-free KMC (rfKMC) and rejection-based KMC (rKMC). In rejection-free KMC, all possible transitions and their rates are precomputed and stored. The algorithm selects an event probabilistically based on the transition rates, ensuring that every step leads to a valid transition. The time increment for each step is calculated using the cumulative rate of all transitions, and this approach is computationally efficient in terms of time accuracy since no steps are wasted. However, it requires knowledge of all possible transitions and their rates, which can be a computational bottleneck for complex systems.

In contrast, rejection-based KMC simplifies the event selection process by proposing transitions at random and accepting them based on their likelihood relative to a predefined upper bound. While this approach can be more straightforward in terms of data handling, it may involve wasted steps due to rejections, leading to slower progress in terms of simulated time. Despite these inefficiencies, rejection-based KMC is often preferred for systems with complex or dynamically changing transition landscapes, where precomputing all possible rates is impractical.

A significant advantage of KMC is its ability to accurately represent the time evolution of non-equilibrium processes. For example, it can simulate phenomena such as defect mobility in materials, surface growth during deposition, or diffusion processes in alloys. In such cases, the algorithm uses transition rates that reflect the underlying physics or chemistry, such as diffusion coefficients, binding energies, or reaction rate constants. The time intervals between events are determined stochastically to ensure the simulation adheres to the correct statistical distribution of events.

One common application of KMC is in surface growth simulations, where atoms are deposited onto a substrate, diffuse across the surface, and aggregate to form stable structures. In this context, the KMC method tracks the positions of individual atoms and simulates their deposition, diffusion, and aggregation over time. Each transition, such as an atom jumping to a neighbouring site or binding with another atom, has an associated rate based on factors like surface energy or temperature. The method provides insights into the dynamics of thin-film growth, such as layer-by-layer deposition or the formation of island-like structures.

Despite its strengths, KMC has limitations. Its reliance on predefined transition rates means it cannot predict new or unexpected transitions that might occur on longer timescales. Additionally, the assumption of independent Poisson processes may not always hold for real systems, where transitions might be correlated or influenced by external factors. These limitations necessitate careful consideration of the physical assumptions underlying the method and validation of results against experimental data.

Overall, the KMC method is a powerful tool for simulating the temporal evolution of complex systems in physics, chemistry, and materials science. Its ability to model non-equilibrium dynamics and capture stochastic processes makes it invaluable for understanding phenomena ranging from surface diffusion and defect dynamics to chemical kinetics and viscoelastic behaviour in materials. By bridging the gap between microscopic transition rates and macroscopic time evolution, KMC provides critical insights into processes that are otherwise challenging to study experimentally or analytically.

Worked Example: Kinetic Monte Carlo (KMC) Simulation of Grain Boundary Migration in Polycrystalline Materials

Problem Statement:

Grain boundary migration is a critical phenomenon in materials science, influencing properties such as strength, ductility, and conductivity. During processes like annealing, grain boundaries move to reduce the overall energy of the system by minimizing grain boundary area. The migration of grain boundaries occurs via atomic diffusion and the motion of individual atoms along the boundary. This example demonstrates how the Kinetic Monte Carlo (KMC) method can simulate grain boundary migration in polycrystalline materials.

Physical Basis

Grain boundaries are regions of higher energy due to atomic misalignment between adjacent grains. Atoms at the boundary can jump to new positions, driven by local energy gradients. These jumps reduce the total system energy and lead to boundary migration. The transition rates for atomic jumps depend on factors like temperature, activation energy, and local atomic configurations.

The KMC method models this process by considering each possible atomic jump as an event with a probability determined by an Arrhenius-like rate equation:

$$r_{ij} = \nu \exp\left(-\frac{E_a}{k_B T}\right)$$

- r_{ij}: Transition rate for an atom jumping from site i to site j.

- ν : Attempt frequency (pre-exponential factor).

- E_a: Activation energy for the atomic jump.

- k_B: Boltzmann constant.

- T: Temperature.

KMC Simulation Setup

1. Define the Simulation Domain

A polycrystalline material is modelled as a 2D grid of atoms, where each site represents an atom belonging to a specific grain. Grain boundaries are represented as regions where neighbouring sites belong to different grains.

2. Initial State

Assign each atom to a specific grain, creating a polycrystalline structure. Identify grain boundary sites as those where at least one neighbouring atom belongs to a different grain.

3. Transition Events

Possible events are atomic jumps from one site to a neighbouring site. The activation energy E_a depends on the local atomic configuration, being lower for grain boundary atoms compared to bulk atoms.

4. Transition Rates

Calculate the transition rates r_{ij} for all possible jumps using the Arrhenius equation. These rates reflect the probability of each event occurring over time.

5. Event Selection and Execution

Using the rejection-free KMC algorithm:

- Calculate the cumulative transition rate $Q_k = \sum r_{ij}$ for all possible events.

- Generate a random number u to select an event, such that

  $$R_{k,i-1} < uQ_k \leq R_{k,i},$$ where R_k is the cumulative distribution of rates.

- Execute the selected event (move the atom) and update the system state.

6. Time Evolution

Update the simulation time using:

$$\Delta t = \frac{-\ln(1/u')}{Q_k}$$

where u' is another random number, and Q_k is the total rate of all transitions.

Results and Analysis

1. Grain Boundary Motion

As the simulation progresses, atoms at grain boundaries migrate, reducing the total boundary area. Grain growth is observed, with larger grains consuming smaller ones to minimize system energy.

2. Grain Growth Kinetics

The time evolution of the average grain size follows the empirical growth law:

$$\langle R \rangle^n - \langle R_0 \rangle^n = Kt$$

where $\langle R \rangle$ is the average grain radius, n is a growth exponent, K is a temperature-dependent rate constant, and t is time. The KMC simulation can predict the growth exponent and rate constant for different materials and temperatures.

3. Effect of Temperature

Higher temperatures increase transition rates, leading to faster grain boundary migration. This is reflected in the simulation by faster increases in average grain size and reductions in grain boundary area.

4. Microstructure Evolution

The KMC simulation generates snapshots of the evolving microstructure, showing grain growth and boundary motion. These results can be compared to experimental micrographs from techniques like electron backscatter diffraction (EBSD) or optical microscopy.

The application of KMC simulations in grain boundary migration holds significant value in materials design and processing. One critical application lies in optimizing heat treatment processes. By simulating grain growth during annealing, KMC helps predict how microstructures evolve under varying thermal conditions, enabling the design of treatments that achieve desired mechanical properties such as strength, ductility, or hardness.

Another important area is in thin film stability. Grain boundary migration plays a key role in the coarsening of grains in thin films, which impacts their structural and electronic properties. KMC simulations provide insights into the dynamics of grain growth, helping to enhance the performance and reliability of thin films used in electronic devices and integrated circuits.

KMC also aids in alloy design by offering a detailed understanding of how solute atoms influence grain boundary mobility. The interaction between solute atoms and grain boundaries can inhibit or accelerate migration, affecting material strength and other properties. Through KMC simulations, researchers can evaluate the role of different alloying elements, enabling the development of stronger and more resilient materials tailored for specific applications.

11. Crystal Plasticity Finite Element Modelling (CPFEM): Deformation behaviour of crystalline materials.

CPFEM predicts how individual grains in a polycrystalline material respond to mechanical loads. This is crucial for designing high-strength alloys and understanding deformation mechanisms in metals like titanium and nickel superalloys.

Crystal plasticity is a mesoscale computational modelling technique designed to capture the crystallographic anisotropy inherent in the mechanical behaviour of polycrystalline materials. This method goes beyond traditional continuum mechanics by incorporating the physics of deformation mechanisms at the crystal level. Initially, crystal plasticity was primarily used to study deformation through slip, the primary mechanism by which dislocations move on crystallographic planes. However, advanced variations of the technique now also account for other deformation mechanisms such as twinning, where specific portions of the crystal lattice reorient, and phase transformations, which involve structural changes in the crystal lattice. By considering these microscopic phenomena, crystal plasticity models provide a detailed understanding of the relationship between stress and strain at both the macroscopic and microscopic levels. This capability allows for the prediction of not just the overall stress-strain response of materials but also the evolution of crystal texture, localized strain fields, and areas prone to strain localization, which are critical for predicting material failure.

The two most widely used formulations of crystal plasticity are the Crystal Plasticity Finite Element Method (CPFEM) and the spectral formulation. CPFEM relies on finite strain mechanics and the finite element method, making it well-suited for complex geometries and large deformation simulations. In contrast, the spectral formulation leverages the computational efficiency of fast Fourier transforms and operates under the small strain assumption, making it faster for periodic microstructures or cases where small deformations are acceptable.

At the core of crystal plasticity lies the assumption that material deformation is accommodated by slip, which occurs when dislocations move along specific crystallographic planes and directions known as slip systems. The activation of a slip system is governed by Schmid's law, which states that a slip system becomes active when the resolved shear stress on that system exceeds a critical resolved shear stress (CRSS). To model this behaviour, crystal plasticity requires a map of the crystallographic orientation of each grain in the polycrystalline material. These orientations, typically represented using tools like Bunge Euler angles, allow for the transformation of stress and strain tensors between the macroscopic sample reference frame and the crystal's local reference frame.

The mechanics of slip are described using the Schmid tensor, which is constructed as the tensor product of the slip plane normal and the Burgers vector. This tensor is used to calculate the resolved shear stress on each slip system. Each active slip system undergoes a specific amount of shearing, and determining these shear rates is central to crystal plasticity modelling. As deformation progresses, the accumulated strain in the crystal affects the CRSS, which is updated using hardening models such as the Voce hardening law. These updates capture the effects of work hardening, where dislocations interact with one another, and the material becomes stronger as it deforms.

In addition to stress-strain predictions, crystal plasticity models also track the evolution of texture, which is the statistical distribution of grain orientations within the polycrystal. Texture evolution is calculated by updating the crystallographic orientation of each grain as it deforms, which influences the anisotropic mechanical properties of the material. This feature makes crystal plasticity particularly valuable for understanding material behaviour in processes like metal forming, where grain orientation and texture significantly impact the final properties of the material. Overall, crystal plasticity serves as a powerful tool for linking microstructural mechanisms to macroscopic material behaviour, enabling the design and optimization of advanced materials.

Worked Example: Application of CPFEM in Simulating Texture Evolution in a Rolled Aluminium Alloy

Problem Statement

An aluminium alloy sheet undergoes uniaxial rolling, which introduces significant plastic deformation. The aim is to simulate the resulting texture evolution and predict how the crystallographic anisotropy affects mechanical behaviour. CPFEM is employed to link the deformation process with the microscopic slip mechanisms within the polycrystalline material.

Steps in the CPFEM Simulation

1. Material and Microstructure Initialization: The aluminium alloy is modelled as a polycrystalline material. A representative volume element (RVE) consisting of 500 grains is created, where each grain has a unique crystallographic orientation. These orientations are mapped using experimental data (e.g., electron backscatter diffraction (EBSD)) or generated using a statistical texture model.

2. Slip Systems and Constitutive Laws: The FCC crystal structure of aluminium is defined with its $12\{111\}\langle110\rangle$ slip systems. Schmid's law is applied to determine slip activation based on resolved shear stress. The Voce hardening law is used to account for strain hardening, expressed as:

$$\tau_c = \tau_0 + \theta_0\gamma + (\tau_s - \tau_0)(1 - e^{-\gamma/\gamma_s}),$$

where:

- τ_c : Critical resolved shear stress (CRSS)

- τ_0 : Initial CRSS

- τ_s: Saturation CRSS

- θ_0 : Initial hardening rate

- γ : Shear strain

- γ_s : Saturation strain

3. Finite Element Model Setup: The RVE is discretised into finite elements, with each element representing a single grain. The crystallographic orientation of each grain is assigned using the orientation data. Boundary conditions are applied to mimic uniaxial rolling, with one face of the RVE constrained and another subjected to compression.

4. Simulation Execution: During the simulation, the deformation gradient tensor is computed at each time step, which is then decomposed into elastic and plastic components. The slip rates for each slip system are calculated iteratively based on the resolved shear stress and CRSS. The crystallographic orientations of the grains are updated as deformation progresses.

5. Post-Processing: The simulation outputs include:

- Stress-strain curves for the entire RVE.

- Evolution of crystallographic texture, represented by pole figures or orientation distribution functions (ODFs).

- Maps of localized strain and stress within individual grains.

The simulation reveals that the rolling process induces a preferred crystallographic orientation, leading to the formation of a rolling texture dominated by cube and Goss orientations. These preferred orientations are identified through pole figures, which show increased intensity along specific crystallographic directions. This texture evolution highlights how the applied mechanical deformation influences the material's internal structure, contributing to its anisotropic behaviour.

Localized strain distribution is another key outcome of the CPFEM simulation. It demonstrates that strain is not uniformly distributed across all grains. Instead, certain grains experience higher levels of localized strain due to anisotropy in slip activity. Grains with crystallographic orientations that align more favourably with the applied stress exhibit increased slip activity, resulting in greater plastic deformation. This behaviour underscores the importance of grain orientation in determining the mechanical response of polycrystalline materials.

The macroscopic stress-strain response predicted by the simulation closely matches experimental data, validating the accuracy of the CPFEM model. The stress-strain curve captures the characteristic work hardening of the aluminium alloy, which arises from the evolving critical resolved shear stress (CRSS) in the active slip systems. This agreement between simulation and experiment highlights the effectiveness of CPFEM in linking microscopic deformation mechanisms with macroscopic mechanical behaviour.

The practical implications of these results are significant. The simulation provides insights that can optimize rolling processes to achieve desired textures, enhancing the mechanical properties of aluminium sheets, such as increased strength and ductility. By understanding how texture evolves during deformation, engineers can predict and control anisotropic behaviours, such as variations in yield strength and elongation in different directions. These predictions are particularly critical for applications like automotive panels and aerospace components, where mechanical performance along specific orientations is essential.

Moreover, CPFEM serves as a powerful tool for bridging the gap between microstructural mechanisms and macroscopic properties. This capability enables the design of aluminium alloys with tailored performance characteristics for specific applications, showcasing the value of CPFEM in materials design and process optimization.

12. Thermodynamic and Kinetic Models (CALPHAD): Phase diagram predictions and alloy design.

CALPHAD (Calculation of Phase Diagrams) models predict phase stability and transformations in multi-component systems. It is widely used in designing new alloys for extreme environments, such as those encountered in nuclear reactors or deep-space exploration.

CALPHAD, which stands for Computer Coupling of Phase Diagrams and Thermochemistry, is a computational methodology developed in the 1970s by Larry Kaufman. Initially known as the CALculation of PHAse Diagrams, CALPHAD is used to model the thermodynamic properties of materials and simulate their phase behaviour in complex systems. This approach allows researchers to model complex multicomponent systems effectively. For instance, Kattner highlights the historical context of CALPHAD, noting its application in analysing phase equilibria in systems such as Cr-Cu-Ni and Fe-Ni, which underscores its significance in materials science [751]. Furthermore, Lukas et al. emphasize that CALPHAD is essential for understanding the interrelationship between composition, microstructure, and processing conditions in materials research [752]. This capability is particularly valuable in fields like metallurgy and materials engineering, where the precise control of phase behaviour is crucial for optimizing material properties.

A phase diagram, which typically involves the temperature and composition axes, depicts the stable regions of substances or solutions, known as phases, and the regions where two or more phases coexist. The CALPHAD method uses computational tools to assess the equilibrium thermodynamic properties of each phase in a multi-component system, helping to predict the phase behaviour under different conditions. While phase diagrams were originally created as graphical methods to rationalize experimental equilibrium data, the CALPHAD approach allows for a more comprehensive and computationally efficient means to simulate phase behaviour, even in systems with many components and complex interactions.

The CALPHAD methodology begins by collecting all available experimental data on phase equilibria and the thermodynamic properties of materials from various thermochemical and thermophysical studies. Each phase in the system is then described using a mathematical model that incorporates adjustable parameters. These parameters are optimized to fit the available experimental data, including data on coexisting phases, ensuring the consistency and accuracy of the model. Once the thermodynamic models are in place, the phase diagram can be recalculated and refined, offering insights into phase stability, phase transitions, and thermodynamic properties in regions where experimental data may be lacking. The main goal of CALPHAD is to predict the set of stable phases and their properties for conditions that have not been experimentally studied, making it invaluable for predicting behaviour in metastable states or during phase transformations.

A key aspect of CALPHAD's success lies in the modelling of the Gibbs energy for each phase. Since most experimental data are determined under known temperature and pressure conditions, the Gibbs energy is used as a reference for calculating other thermodynamic quantities. It is difficult to exactly capture the Gibbs energy of a multi-component system using analytical expressions, so CALPHAD models rely on approximations and expansions of Gibbs energy in terms of temperature, pressure, and composition. These models are based on identifying key features of phase behaviour and using power series to describe them. By refining the adjustable parameters in the models to best fit the experimental data, CALPHAD provides a reliable method to combine the properties of different sub-systems to model a multi-component system as a whole.

Another crucial factor in the success of CALPHAD is the development of computer software that can perform equilibrium calculations and create phase diagrams based on the mathematical models. Various software tools are available, such as FactSage, MTDATA, PANDAT, and Thermo-Calc, among others. These software packages allow for the calculation of equilibrium under various conditions, including not just temperature and pressure but also factors like constant volume or specific chemical potentials. They are widely used in research and industrial development to reduce the need for extensive experimental work, saving time and resources while providing thermodynamic predictions that would be otherwise unattainable. With the growth of thermodynamic databases, CALPHAD has become an essential tool in materials science, offering accurate predictions for complex multi-component systems and facilitating the design and optimization of materials.

CALPHAD provides a powerful computational framework for modelling phase behaviour and thermodynamic properties in complex materials systems. It allows researchers and engineers to simulate phase diagrams, predict phase stability, and understand material behaviour without relying solely on experimental data. The method is crucial in advancing material design, saving resources, and enabling the optimization of materials for various industrial applications. Through its use in software tools and databases, CALPHAD has become indispensable in materials science research and development.

Example Application of CALPHAD: Designing a High-Strength Alloy for Aerospace Applications

Problem Statement

In aerospace engineering, materials must exhibit high strength, low density, and excellent thermal resistance. Nickel-based superalloys are commonly used due to their exceptional performance under high temperatures and mechanical loads. However, designing a new alloy with optimized properties, such as higher creep resistance and improved oxidation resistance, requires an understanding of complex multi-component phase equilibria. Experimentally testing all potential compositions is time-consuming and expensive. The CALPHAD methodology provides an efficient way to model phase equilibria and predict thermodynamic behaviour in multi-component systems, guiding alloy design with minimal experimental work.

Objective

To design a nickel-based superalloy with enhanced mechanical properties and stability at elevated temperatures by identifying the optimal composition of alloying elements (e.g., Cr, Al, Ti, W, Mo, Co) that results in desirable phases, such as γ (matrix phase) and γ' (strengthening phase), while avoiding deleterious phases like σ or Laves phases.

Steps Using CALPHAD

1. Database Selection and Input Data

Thermodynamic databases specific to nickel-based superalloys are selected (e.g., Thermo-Calc's TCNI database). The database includes assessed thermodynamic descriptions of phases such as γ, γ', σ, and Laves phases, as well as intermetallic compounds.

2. Thermodynamic Modelling

Using CALPHAD software (e.g., Thermo-Calc), the thermodynamic properties of each phase are modelled. Gibbs energy expressions are used for phases based on experimental data and first-principles calculations. Adjustable parameters in these models are fine-tuned to fit experimental phase equilibria data.

3. Phase Diagram Calculation

A phase diagram is calculated for the Ni-Al-Cr system to map out the regions where the γ and γ' phases are stable. The calculation shows how temperature and composition affect phase stability and the formation of secondary phases. For example, it is determined that γ' forms at specific Al and Ti concentrations within the temperature range of interest.

4. Multi-Component Simulation

The model is expanded to include other alloying elements like Co, Mo, and W. Using the CALPHAD approach, multi-component phase diagrams are generated, showing the stability of γ' and potential formation of undesirable phases (e.g., σ phase). This simulation helps narrow down the composition range for further analysis.

5. Optimization

The composition is optimized by maximizing the volume fraction of γ', which improves strength, while avoiding the precipitation of brittle σ or Laves phases. The CALPHAD model predicts the effect of adding elements like Co and W on γ' stability and σ-phase suppression.

6. Property Prediction

Thermodynamic models are combined with kinetic databases to simulate diffusion-controlled processes like coarsening of γ' particles during long-term service. This ensures that the alloy maintains its mechanical properties over time.

Results

Phase Stability: The CALPHAD simulations identify an optimal alloy composition, such as Ni-15Cr-6Al-2Ti-3W (wt%). This composition stabilizes a high volume fraction of γ' phase while avoiding deleterious phases.

Thermal Stability: The model predicts that the γ' phase remains stable up to 900°C, ensuring the alloy's strength at high temperatures.

Kinetic Behaviour: Diffusion simulations indicate that the optimized alloy exhibits slower coarsening of γ' particles, leading to better creep resistance over prolonged exposure to high temperatures.

Validation and Practical Implications

Experimental work is performed to validate the CALPHAD predictions. The alloy is synthesized, and its microstructure is analysed using scanning electron microscopy (SEM) and energy-dispersive X-ray spectroscopy (EDS). Differential scanning calorimetry (DSC) is used to confirm phase transition temperatures. The mechanical properties, including yield strength and creep resistance, are tested, showing good agreement with CALPHAD predictions.

The optimized alloy is then implemented in the manufacture of turbine blades for jet engines, where it demonstrates enhanced performance under operational conditions. The CALPHAD approach significantly reduces development time and cost by minimizing trial-and-error experimentation.

This example demonstrates how CALPHAD can be used to design high-performance materials by predicting phase equilibria, optimizing compositions, and simulating thermodynamic and kinetic behaviour. Its ability to model complex multi-component systems makes it an invaluable tool in the development of advanced materials for demanding applications like aerospace engineering.

13. Reactive Force Field (ReaxFF): Modelling chemical reactions at the atomic scale.

ReaxFF allows for the simulation of chemical bonding and reaction dynamics in materials. This is particularly useful for studying combustion processes in energetic materials or the formation of defects in ceramics.

The Reactive Force Field (ReaxFF) represents a significant advancement in computational chemistry, particularly for modelling chemical reactions and molecular dynamics at the atomic level. It serves as a bridge between quantum mechanical (QM) simulations, which are accurate but computationally intensive, and classical molecular dynamics (MD), which are efficient but limited in their ability to model reactive processes. ReaxFF allows for the simulation of molecular behaviour, capturing the complexities of bond formation and breaking, charge transfer, and dynamic changes in chemical environments, thus providing a comprehensive tool for studying a variety of chemical systems [753-755].

ReaxFF was developed in the early 2000s by Adri van Duin and William Goddard III to address the limitations of classical force fields that assume fixed bond connectivity, making them unsuitable for systems undergoing significant structural changes or chemical reactions. Initially focused on hydrocarbon reactions, ReaxFF has since been adapted for a wide array of chemical systems, including metals, oxides, ceramics, and biomolecules [756-758]. The core of ReaxFF is its bond order formalism, which allows the strength of a bond to be dynamically adjusted based on interatomic distances. This capability enables the force field to accurately model processes such as bond formation, breaking, and charge transfer, which are critical in reactive systems [759, 760].

Key Features and Components:

1. Bond Order Formalism: The bond order (BO) in ReaxFF is calculated as a continuous function of distance, allowing bonds to weaken and break as atoms move apart or form as they come closer together. This dynamic adjustment is essential for accurately simulating chemical reactions [761, 762].

2. Energy Contributions: ReaxFF computes the total potential energy of a system by summing various components that reflect the physics and chemistry of reactive systems. These include bond energy, angle energy, torsional energy, van der Waals interactions, Coulomb interactions, and corrections for over-coordination and under-coordination, as well as charge transfer and polarization effects [763-765].

3. Transferability: One of the strengths of ReaxFF is its parameterization for specific chemical systems, which can be combined and re-parameterized for other systems, enhancing its versatility for modelling diverse materials and reactions [766, 767].

4. Reactive Dynamics: ReaxFF facilitates the simulation of chemical reactions in dynamic environments, making it particularly useful in fields such as combustion, catalysis, and materials science. Its ability to handle bond formation and breaking in real-time is a significant advantage over traditional force fields [768, 769].

Applications of ReaxFF:

1. Combustion and Hydrocarbon Chemistry: Initially developed for hydrocarbon combustion, ReaxFF effectively models processes such as bond breaking in alkanes and radical formation, extending to soot formation and flame propagation [770, 771].

2. Materials Science: ReaxFF is instrumental in studying material behaviours under extreme conditions, such as high pressure and temperature, and is used to model the growth and failure of protective oxide layers on metals, as well as the degradation of polymers [772, 773].

3. Catalysis: The force field aids in understanding catalytic processes at the atomic level, including surface reactions during heterogeneous catalysis and reaction mechanisms in electrocatalysis [774, 775].

4. Biomolecular Reactions: ReaxFF has been successfully applied to simulate complex biochemical reactions, such as enzyme catalysis and drug interactions, showcasing its adaptability to biological systems [776, 777].

5. Environmental Chemistry: The method is also employed to model environmental processes, including pollutant breakdown and mineral dissolution, providing insights into geochemical phenomena.

ReaxFF offers several advantages, including computational efficiency and the ability to model chemical reactions dynamically. However, it also faces challenges, such as parameter sensitivity, where the accuracy of simulations heavily relies on the quality of parameterization. Additionally, while ReaxFF approximates QM results, it may not capture certain electronic effects with the precision of ab initio methods. Despite these limitations, ongoing developments continue to enhance its reliability and expand its applications, solidifying ReaxFF's role as a cornerstone in modern computational materials science and chemistry.

Worked Example: Application of ReaxFF to Simulate Combustion and Oxidation in Hydrocarbon Fuels

Problem Context

The goal is to study the combustion and oxidation mechanisms of a hydrocarbon fuel, such as methane (CH_4), at elevated temperatures. Understanding the reactive processes is critical for improving combustion efficiency, reducing emissions, and designing better fuel systems.

System Description

- **Fuel**: Methane (CH_4), a simple hydrocarbon widely used as a fuel.

- **Oxidizing Agent**: Oxygen (O_2) at high temperatures.

- **Environment**: High-temperature conditions (1000–1500 K) to induce combustion reactions.

- **Objective**: Investigate the detailed reaction pathways, intermediate species, and byproducts of methane combustion.

Methodology

1. System Setup

- **Simulation Software**: LAMMPS (Large-scale Atomic/Molecular Massively Parallel Simulator) with the ReaxFF reactive force field.

- **ReaxFF Parameterization**: Use a parameter set optimized for hydrocarbon reactions and combustion chemistry, such as the one developed for C-H-O systems.

- **Initial Configuration**:

    ○ A simulation box containing methane molecules (CH_4) and oxygen molecules (O_2) at a defined stoichiometric ratio (e.g., 2:1 for complete combustion).

    ○ Periodic boundary conditions to mimic bulk behaviour.

    ○ Initial velocities assigned using a Maxwell-Boltzmann distribution corresponding to the target temperature (e.g., 1200 K).

2. Simulation Procedure

- **Equilibration**: Perform energy minimization and a short equilibration run at the target temperature using the Nose-Hoover thermostat to stabilize the system.

- **Reactive Dynamics**: Run the reactive molecular dynamics simulation using ReaxFF for a time scale of several picoseconds (e.g., 10 ps), capturing the formation and breaking of bonds during combustion.

- **Data Collection**: Monitor bond distances, angles, and the formation of reaction products, as well as energy changes over time.

3. Analysis

- **Reaction Pathways**:

    ○ Identify the initiation of the reaction, such as the formation of a methyl radical ($CH_3\bullet$) through the breaking of a C-H bond in methane.

    ○ Track subsequent reactions, such as $CH_3\bullet + O_2 \rightarrow CH_2O$ (formaldehyde) + $OH\bullet$, and further breakdown of intermediates.

- **Intermediate Species**:

    ○ Capture the formation of reactive species like radicals ($OH\bullet$, $H\bullet$) and molecules (CO, CO_2, H_2O).

- **Energy Profile**:

    o   Examine the energy changes associated with bond formation and breaking.

    o   Evaluate heat release during combustion by tracking changes in the potential energy of the system.

Results

1. Reaction Pathway Analysis

- The simulation reveals that methane combustion begins with the abstraction of a hydrogen atom by oxygen, forming a methyl radical and a hydroxyl radical ($OH\bullet$).

- Subsequent reactions lead to the formation of intermediates like formaldehyde (CH_2O), carbon monoxide (CO), and finally carbon dioxide (CO_2).

2. Intermediate and Product Species

- Key intermediates observed include $CH_3\bullet$, CH_2O, and HCO.

- Final products are CO_2 and H_2O, consistent with complete combustion.

3. Energy Evolution

- A significant decrease in potential energy is observed, corresponding to the release of heat during combustion.

- The simulation provides insight into the energetic barriers of intermediate reactions, aiding in understanding reaction kinetics.

4. Time-Dependent Concentrations

- The simulation tracks the concentration of various species over time, showing the rapid consumption of CH_4 and O_2 and the subsequent formation of products.

The results of the ReaxFF simulation provide detailed insights into the reaction kinetics of combustion processes. These insights enable the optimization of combustion conditions, allowing for maximum efficiency while minimizing the occurrence of unburned hydrocarbons. This optimization can lead to improved energy utilization and cost savings in industrial and automotive applications.

The study also plays a critical role in emission control by identifying key intermediate species formed during combustion. This information helps in designing targeted strategies to reduce harmful emissions such as carbon monoxide (CO) and nitrogen oxides (NO_x), which are major contributors to air pollution and environmental degradation.

In the realm of fuel design, the simulation offers valuable guidance for developing alternative fuels or additives. By understanding the detailed reaction pathways, engineers and scientists can design fuels that not only enhance combustion efficiency but also produce fewer pollutants, paving the way for cleaner and more sustainable energy sources.

Finally, the comparison of simulation results with experimental data serves to validate the accuracy of ReaxFF parameterizations. This validation enhances confidence in predictive models used for reactive systems, ensuring that the simulations provide reliable and reproducible insights. Such confidence is essential for the widespread adoption of ReaxFF in both research and industrial applications, as it bridges the gap between theoretical predictions and practical implementation.

This worked example demonstrates the application of ReaxFF to simulate the complex reactive processes in methane combustion. By providing atomistic insights into reaction mechanisms and energy changes, ReaxFF serves as a powerful tool for advancing combustion science, optimizing industrial processes, and designing cleaner fuels. The methodology and findings are transferable to other reactive systems, such as more complex hydrocarbons, catalytic reactions, or material degradation studies.

14. Topology Optimization Models: Optimizing material distribution for performance.

Topology optimization uses computational algorithms to design materials with specific structural or functional properties, such as lightweight materials for 3D printing or metamaterials with unique mechanical or thermal properties.

Topology optimization is a sophisticated mathematical methodology that focuses on optimizing the material layout within a specified design space to achieve maximum system performance under given loads, boundary conditions, and constraints. Unlike shape or sizing optimization, which are constrained to predefined configurations, topology optimization allows for entirely new shapes to emerge within the design space. This flexibility enables the creation of innovative designs that would be otherwise unattainable.

The conventional approach to topology optimization involves using the finite element method (FEM) to evaluate the performance of a design iteratively. Gradient-based algorithms such as the optimality criteria algorithm and the method of moving asymptotes are often employed to optimize the design. Alternatively, non-gradient-based algorithms like genetic algorithms are also used, depending on the complexity of the problem and computational resources. These methods aim to minimize an objective function, such as compliance, which indirectly maximizes structural stiffness by optimizing material placement.

Topology optimization has found widespread applications across industries such as aerospace, mechanical, biomedical, and civil engineering. It is primarily utilized during the conceptual design phase, where its ability to produce unconventional, free-form structures proves invaluable. However, the outputs of topology optimization are often challenging to manufacture using traditional methods. As a result, designs are frequently refined post-optimization for manufacturability. This limitation is actively being addressed through research on incorporating manufacturing constraints directly into the optimization process. Additive manufacturing, with its capability to produce complex geometries, has greatly enhanced the utility of topology optimization, enabling direct manufacturing of optimized designs and fostering innovation in design for additive manufacturing.

The fundamental problem of topology optimization involves minimizing an objective function that quantifies the performance of the system while satisfying a set of constraints. The objective function often represents compliance, linking the material's stiffness to its structural performance. Material distribution is described by the density field, where the density at a given point indicates the presence or absence of material. The design space defines the allowable region for material placement, accounting for factors such as accessibility and packaging requirements. Constraints, such as volume limitations or stress thresholds, ensure the feasibility and practicality of the optimized design.

The finite element method is commonly employed to solve the governing differential equations that arise from evaluating the state field. This numerical approach allows for accurate modelling of complex geometries and boundary conditions, making it indispensable in topology optimization. Solving the optimization problem can involve either discrete or continuous variables. Discrete formulations, where material density is binary (presence or absence), offer high topological complexity but are computationally expensive and sensitive to parameter variations. In contrast, continuous formulations use interpolative methods, such as the Solid Isotropic Material with Penalization (SIMP) method, to model material properties continuously. SIMP interpolates Young's modulus based on material density, penalizing intermediate densities to favour binary outcomes. While effective, this approach introduces non-convexities that can complicate optimization.

Commercial software solutions for topology optimization vary in their capabilities. Many offer designs that serve as conceptual hints, requiring manual reconstruction for manufacturability. However, advanced solutions now integrate topology optimization with additive manufacturing workflows, enabling direct production of optimized structures. This synergy between computational optimization and manufacturing technology continues to expand the possibilities for innovative and efficient design across multiple industries.

Worked Example: Application of Topology Optimization in Material Science

Objective

Design a lightweight, high-stiffness bracket for aerospace applications to support a specified load, ensuring structural integrity while minimizing material usage. The bracket will be optimized for additive manufacturing to eliminate the need for extensive post-processing and assembly.

Problem Statement

The goal is to minimize the compliance (maximize stiffness) of the bracket under a specific set of loads and constraints, while ensuring that the volume of the material used does not exceed 30% of the design space. The bracket must support a 500 N load applied at a specified point, and it will be connected to a fixed wall through a bolted flange.

Design Space and Constraints

- **Design Space**: A rectangular 3D volume representing the allowable dimensions for the bracket, with specified regions for mounting holes and a load application point excluded from material modification (non-design regions).

- **Constraints**: The total material volume must be ≤ 30% of the design space. Maximum stress values should remain below the yield strength of the material (AlSi10Mg, commonly used in additive manufacturing).

- **Boundary Conditions**: The mounting flange is fixed, and a concentrated load is applied at the designated location.

Material Properties

- **Material**: AlSi10Mg

- **Young's Modulus**: 70 GPa

- **Yield Strength**: 200 MPa

- **Density**: 2.68 g/cm^3

Methodology

1. **Finite Element Method (FEM) Setup**: The design space is discretised into a fine mesh using tetrahedral elements. Boundary conditions and loading are applied to replicate real-world scenarios.

2. **Optimization Algorithm**: The Solid Isotropic Material with Penalization (SIMP) method is employed. The penalization factor $p=3$ is used to favour binary density solutions. The objective function minimizes compliance, and constraints enforce volume and stress limits.

3. **Iterative Solution**: A gradient-based optimization algorithm, such as the Method of Moving Asymptotes (MMA), iteratively adjusts the material distribution. The algorithm evaluates performance using FEM at each step.

4. **Post-Processing**: The optimized topology is smoothed and converted into a manufacturable design suitable for additive manufacturing. Stress analysis is performed on the final design to validate performance.

The results of the topology optimization reveal a truss-like structure in the optimized design, where material is strategically concentrated along load paths. Non-critical regions of the bracket are voided, leading to a significant reduction in weight. This approach ensures that the material is used efficiently, meeting the goal of minimizing the volume while maintaining structural performance.

The final design utilizes only 28% of the initial design space volume, successfully satisfying the imposed volume constraint. This reduction demonstrates the capability of topology optimization to achieve substantial weight savings while adhering to design limitations.

Stress analysis of the optimized bracket shows a maximum stress of 175 MPa, which is safely below the yield strength of 200 MPa for the material used. This confirms the structural integrity of the design under the specified loading conditions, providing confidence in its practical application. Moreover, the compliance of the optimized design is 20% lower than that of a conventional solid bracket, indicating enhanced stiffness and improved mechanical performance.

The design is also highly feasible for additive manufacturing, as it avoids overhangs that would require support structures. This makes the bracket suitable for direct production using selective laser melting (SLM), a widely used additive manufacturing technique.

The weight reduction achieved by this optimization contributes significantly to improved fuel efficiency in aerospace applications, where every gram saved translates to cost and performance benefits. By optimizing the material layout, the bracket exhibits a superior stiffness-to-weight ratio, which is particularly advantageous for components subjected to dynamic loads. Additionally, the direct manufacturability of the design reduces lead times and eliminates the need for assembly, streamlining the production process.

The scalability of this methodology is another noteworthy implication. It can be applied to other aerospace components, enabling lightweight and efficient designs across various applications. This underscores the versatility and value of topology optimization in advancing engineering practices and achieving innovative, high-performance designs.

This example demonstrates how topology optimization models enable the creation of lightweight, high-performance components tailored for additive manufacturing. By integrating computational optimization with advanced manufacturing techniques, material scientists and engineers can achieve innovative designs that meet stringent performance and weight criteria.

15. Continuum Damage Mechanics (CDM): Predicting material degradation and failure.

CDM models simulate the progressive damage in materials due to fatigue, creep, or corrosion. They are widely used in evaluating the long-term reliability of advanced materials in infrastructure and energy applications.

Continuum Damage Mechanics (CDM) is a theoretical framework that facilitates the prediction of material degradation and failure under various loading and environmental conditions. The fundamental principle of CDM is the representation of damage as a continuous variable, which evolves over time and space, rather than as discrete entities like cracks or voids. This continuous representation allows for a nuanced understanding of how materials deteriorate due to mechanisms such as fatigue, creep, and corrosion, thereby providing critical insights into the long-term reliability of materials in advanced engineering applications [778-780].

In the context of fatigue, CDM models are particularly effective in describing damage accumulation under cyclic loading conditions. Fatigue damage occurs when materials are subjected to repeated stress or strain cycles, leading to the gradual initiation and growth of microscopic defects. CDM captures this process by incorporating damage variables into the constitutive equations that relate stress and strain. These damage variables evolve according to predefined laws based on experimental observations, allowing for the simulation of fatigue life and the prediction of failure under specific loading conditions [779, 781, 782]. For instance, Yang et al. [779] proposed a CDM-based model that evaluates the synergistic effects of corrosion and fatigue, demonstrating the model's capability to predict corrosion fatigue crack initiation life in metallic materials.

Creep scenarios are another area where CDM excels, particularly in accounting for time-dependent deformation and degradation of materials at elevated temperatures. Creep damage typically results from sustained stress and high temperatures, which can induce microstructural changes such as grain boundary sliding or void formation. CDM integrates these mechanisms into its formulations, enabling predictions of creep life and the identification of critical regions prone to failure. This capability is invaluable in the design and assessment of components used in high-temperature environments, such as turbines and reactors, where material integrity is paramount [783, 784].

Corrosion damage, resulting from chemical interactions between materials and their environments, is another critical application of CDM. By incorporating corrosion kinetics and their effects on mechanical properties, CDM models can simulate the progressive weakening of materials exposed to aggressive environments, such as saltwater or industrial pollutants. This modelling is essential for evaluating the durability of infrastructure and critical systems, as it allows for the assessment of how corrosion impacts material performance over time [778, 780, 785]. For example, Gastaldi et al. demonstrated the application of CDM in modelling the loss of mechanical strength in biodegradable magnesium alloys due to corrosion, highlighting the framework's versatility [778, 780].

The application of CDM extends to various sectors, including infrastructure and energy, where materials must endure long-term service conditions while maintaining structural integrity. In civil engineering, CDM assists engineers in assessing how fatigue, creep, and corrosion will affect material performance over decades, ensuring the safety and reliability of structures like bridges and dams [786, 787]. Similarly, in the energy sector, CDM models are employed to evaluate the reliability of components in power plants and renewable energy systems, ensuring they meet safety and performance standards throughout their operational lives [785, 788].

The predictive capabilities of CDM provide significant advantages in material design, maintenance planning, and failure prevention. By simulating the progression of damage and identifying potential failure modes, engineers can optimize material selection, enhance design strategies, and implement proactive maintenance schedules. This proactive approach reduces the risk of unexpected failures, enhances safety, and minimizes costs associated with downtime and repairs [781, 782, 789].

Worked Example: Predicting Fatigue Failure in a Structural Steel Beam Using Continuum Damage Mechanics (CDM)

Problem Context

A steel beam is subjected to cyclic loading in a bridge structure. The beam experiences repeated tensile and compressive stresses due to vehicular traffic. Over time, this cyclic loading can lead to fatigue damage, ultimately causing the beam to fail. The objective is to use a CDM model to predict the beam's fatigue life, identify regions of high damage accumulation, and provide insights for maintenance and design improvements.

Material Properties and Loading Conditions

- **Material:** Structural steel (e.g., A36)

- **Young's Modulus (E):** 200 GPa

- **Yield Strength:** 250 MPa

- **Fatigue Strength Coefficient (σ'f):** 500 MPa

- **Fatigue Exponent (b):** −0.1

- **Damage Evolution Law:** Based on Miner's rule, incorporating the cyclic damage evolution equation:

$$\dot{D} = \frac{\Delta\sigma^m}{\sigma_f^m} \cdot N^{-1}$$

Where:

- $\dot{D}$: Damage rate

- $\Delta\sigma$: Stress amplitude

- σ_f: Fatigue strength

- N: Number of cycles

- m: Material-dependent exponent (e.g., 3 for steel)

Cyclic Loading Parameters:

- Stress amplitude: $\Delta\sigma = 150\,\mathrm{MPa}$

- Mean stress: $\sigma_m = 50\,\mathrm{MPa}$

- Load frequency: 10 Hz

Simulation Setup

1. **Finite Element Modelling:** The beam is modelled using the finite element method (FEM), with CDM variables integrated into the constitutive equations to account for progressive fatigue damage.

2. **Damage Initialization:** Damage D is initialized at zero across the beam, representing an undamaged state at the start of the simulation.

3. **Damage Evolution:** The fatigue damage is allowed to accumulate at each time step, with the damage evolution law applied based on the local stress amplitude and the number of cycles.

4. **Failure Criterion:** Failure is defined when the damage variable D reaches a critical threshold, $D_c=0.8$, indicating that the material can no longer sustain the applied loads.

The CDM simulation predicts that the steel beam will fail after approximately 10^6 cycles, corresponding to about 28 days of continuous operation at the specified loading frequency. This provides a clear timeline for when the material is likely to degrade to a critical point under the given conditions. Damage accumulation is observed to be most rapid near the midspan of the beam, where the bending moment and stress amplitude are highest. These critical zones, identified through the simulation, indicate areas requiring focused inspection and potential reinforcement.

As damage progresses, the effective stiffness of the material decreases, leading to greater deformation and localized strain concentrations. This stress-strain response highlights how the material's mechanical properties evolve with fatigue, providing a detailed understanding of the degradation process. When compared with experimental data from fatigue tests on similar steel beams, the simulated fatigue life and damage patterns show strong agreement. This validation enhances confidence in the CDM model's accuracy and reliability for predictive analysis.

The results have significant practical implications. The predicted fatigue life allows for the establishment of targeted maintenance schedules, ensuring that inspections are conducted in critical areas before failure occurs. Insights from the damage distribution suggest potential design improvements, such as adding stiffeners or altering the geometry of the beam, to reduce stress concentrations and enhance fatigue resistance. Understanding the progression of fatigue damage also enables engineers to implement real-time monitoring systems to detect early signs of failure, thereby improving the safety of the structure. Additionally, the study guides the selection of alternative steel grades or surface treatments that can enhance fatigue strength and extend the service life of the beam. These findings illustrate the value of CDM in optimizing structural performance and ensuring long-term reliability.

This example demonstrates how Continuum Damage Mechanics provides a robust framework for predicting material degradation in critical infrastructure, enabling proactive measures to enhance safety, reliability, and performance.

These computational models, often used in combination, provide a powerful toolkit for designing, optimizing, and understanding advanced materials.•They enable researchers to explore material behaviour across scales and accelerate the development of innovative solutions for modern engineering challenges.

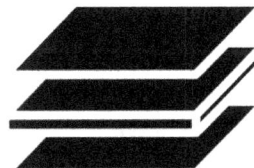

Chapter 9

Challenges in Advanced Materials Engineering

Economic Viability and Scalability

The economic viability and scalability of advanced materials are critical factors that influence their transition from research and development to widespread commercial application. These aspects are essential for industries aiming to integrate advanced materials into their manufacturing processes or consumer products.

Economic viability pertains to the cost-effectiveness of producing and utilizing advanced materials. A significant aspect of this viability is the cost of raw materials. Advanced materials often depend on rare or specialized raw materials, such as rare-earth elements used in high-performance magnets or graphene derived from graphite. The availability, sourcing, and pricing of these materials directly impact their economic viability [790]. For instance, the fluctuating prices of rare-earth elements can hinder the production of certain advanced materials, making them less attractive for commercial applications [790].

Manufacturing costs also play a pivotal role in determining economic viability. The production processes for advanced materials are frequently complex and require specialized equipment and expertise. Techniques such as chemical vapor deposition (CVD) for graphene or additive manufacturing for metal alloys can be costly, and scaling these processes while maintaining cost control is crucial for commercial adoption [48]. Moreover, the performance versus cost trade-off is a vital consideration; materials that offer superior properties, such as high-strength composites in aerospace applications, may justify higher costs due to their benefits in weight savings and fuel efficiency [790].

Market demand and applications are closely linked to economic viability. A robust market for advanced materials in sectors such as automotive, electronics, or renewable energy can enhance their economic viability by improving economies of scale and driving costs down over time [790]. The interplay between market demand and the unique properties of advanced materials can create a favourable environment for their adoption, thereby enhancing their economic viability.

Scalability refers to the ability to produce advanced materials in sufficient quantities to meet industrial or market demand without compromising quality or affordability. One of the primary challenges in scalability is the transition from small-scale lab production to mass manufacturing. Processes optimized for small-scale production may not directly translate to large-scale manufacturing, necessitating significant adjustments in process control and reactor design [48]. For example, synthesizing nanomaterials like carbon nanotubes requires careful consideration of production methods to ensure consistent quality at a larger scale [48].

Infrastructure and equipment are also critical for scaling production. Significant investments in specialized manufacturing facilities or high-throughput equipment are often required, which can be a barrier for small or medium-sized enterprises [790]. Furthermore, ensuring a stable and efficient supply chain is essential for scalability. This includes securing reliable sources of raw materials and developing logistics for material distribution [790].

Process efficiency and standardization are vital for effective scaling. Manufacturing processes must be efficient and standardized to ensure consistent quality, particularly in safety-critical industries like aerospace and healthcare [790]. Innovations in manufacturing processes, such as additive manufacturing and automation, can significantly enhance scalability by reducing production costs and improving efficiency [48].

To enhance economic viability and scalability, several strategies can be employed. Innovations in manufacturing processes, such as additive manufacturing and roll-to-roll printing, can reduce production costs and improve scalability [790]. Additionally, material substitution—developing alternative materials with similar properties but lower production costs—can enhance economic viability [790]. Collaborative research and development initiatives between academic institutions, governments, and industry can also drive down development costs and accelerate the path to scalability [790].

Moreover, integrating circular economy principles, such as recycling and reusing advanced materials, can improve economic viability by reducing raw material dependency and mitigating environmental impacts [790]. Focusing on high-value, niche markets initially can help recoup development costs and refine production processes before broader market entry [790].

Overcoming Technical Limitations in Material Properties

Advanced materials have emerged as pivotal components in various industries due to their enhanced properties, such as increased strength, improved conductivity, and superior thermal resistance. However, these materials also face significant technical limitations that can hinder their practical applications. Addressing these challenges requires innovative strategies in material design, processing, and performance optimization.

Understanding Technical Limitations

Mechanical Properties: Advanced materials, including composites and nanomaterials, often exhibit high strength but may also possess brittleness or low fracture toughness. This brittleness can severely restrict their use in dynamic or high-impact environments, where materials are subjected to sudden loads or stresses [48, 791]. For instance, the mechanical performance of natural fibre composites can be compromised due to poor interfacial adhesion between the fibres and the matrix, leading to inadequate load transfer and failure under stress [791].

Thermal and Electrical Properties: While materials like graphene and carbon nanotubes are known for their exceptional electrical conductivity, they often suffer from low thermal stability, which limits their application in high-temperature environments [792]. The integration of these materials into composites can also be challenging due to their differing thermal expansion coefficients, which can lead to thermal stress and failure [793].

Corrosion and Environmental Degradation: Advanced materials can be susceptible to environmental degradation when exposed to harsh conditions, such as high humidity or corrosive chemicals. This degradation can significantly affect their long-term performance and reliability [794]. For example, the durability of biocompatible materials used in medical applications can be compromised by environmental factors, necessitating the development of more resilient materials [795].

Processability and Manufacturability: The complex structures of advanced materials, such as those found in nanoscale architectures, often complicate their processing and manufacturing at scale. Techniques like additive manufacturing can help overcome some of these challenges by allowing for the creation of complex geometries and tailored material properties [796, 797]. However, the rheological properties of these materials must be well understood to ensure successful processing [797].

Material Compatibility: In composite systems, advanced materials must often interface with conventional materials. Mismatches in thermal expansion, chemical stability, or bonding can lead to performance issues, such as delamination or failure at the interface [798]. This compatibility is crucial for ensuring the structural integrity and functionality of composite materials in practical applications [793].

Strategies to Overcome Technical Limitations

Advanced Material Engineering Fundamentals

Material Engineering and Design Innovations: Tailoring the atomic or molecular structure of advanced materials can effectively address their limitations. For instance, alloying techniques can be employed to develop multi-element alloys that balance strength and ductility, while hybrid structures can combine materials with complementary properties to enhance overall performance [794, 799]. The use of compatibilizers in polymer blends has also been shown to improve interfacial adhesion and mechanical properties [800].

Surface Modifications: Surface engineering techniques, such as applying protective coatings or surface functionalization, can significantly enhance the performance of advanced materials without altering their bulk properties. For example, corrosion-resistant coatings can extend the lifespan of metals in harsh environments [801]. Similarly, surface treatments can improve adhesion and reduce friction in composite materials [794].

Nanostructuring: The incorporation of nanoscale features can dramatically enhance the properties of materials. Techniques such as grain refinement in metals can improve strength through the Hall-Petch effect, while nanocomposites can leverage nanoparticles to enhance strength, conductivity, or thermal resistance [792, 802]. This approach has been particularly effective in biomedical applications, where enhanced material properties are critical [794].

Additive Manufacturing: Advanced manufacturing techniques, including 3D printing, allow for the creation of complex geometries and functionally graded materials (FGMs) that optimize performance across components. This capability is particularly valuable in applications requiring precise material properties [796, 797].

Phase and Microstructure Control: Controlling the phases and microstructures during processing can address issues such as brittleness and thermal stability. Techniques like heat treatment and directional solidification can be employed to achieve desired mechanical properties and improve performance in specific applications [48, 798].

Multi-Scale Modelling and Simulation: Computational tools, including density functional theory (DFT) and molecular dynamics simulations, play a crucial role in predicting material behaviour and guiding experimental designs. These simulations help identify potential weaknesses and optimize properties before manufacturing [803, 804].

Environmental Adaptations: Engineering advanced materials to withstand environmental challenges is essential for their long-term viability. Innovations such as self-healing materials and thermally stable alloys are being developed to enhance durability and performance in extreme conditions [794].

Material Integration and Interface Engineering: Ensuring compatibility in composite systems is vital for performance. The use of interfacial agents and gradient interfaces can improve bonding between disparate materials, reducing thermal or mechanical mismatches and enhancing overall material performance [793, 798].

Graphene and carbon nanotubes represent groundbreaking materials with exceptional strength and conductivity. However, their application is limited by graphene's lack of a bandgap and the challenges associated with mass production. Researchers address these issues through chemical doping and the creation of hybrid composites, which enhance graphene's versatility for use in electronics and structural composites. These advancements have broadened graphene's potential for integrating into high-performance systems.

Nickel-based superalloys, widely employed in jet engines, face the challenge of thermal creep when exposed to extreme temperatures. To mitigate this issue, these superalloys are often coated with thermal barrier ceramics, which provide insulation against high heat. The incorporation of rare-earth elements into their composition further enhances their high-temperature performance, ensuring reliability and efficiency in aerospace applications.

Biodegradable polymers used in medical implants encounter limitations related to mechanical strength and controlled degradation rates. Enhancements in these polymers are achieved by blending them with reinforcing agents and applying surface treatments. These modifications improve their reliability, making them better suited for medical applications where both strength and bio-compatibility are critical.

Machine learning is driving a new era in material design, enabling AI-powered material discovery that identifies advanced materials with tailored properties. Predictive modelling accelerates the process, addressing technical limitations through computational precision. Sustainability and recycling are also gaining prominence, as materials are now being designed with recyclability and environmental impact in mind. These efforts ensure that advanced materials remain viable and environmentally friendly for long-term use.

Emerging fabrication techniques, such as atomic layer deposition (ALD) and two-photon polymerization, offer unparalleled control over material properties at microscopic scales. These advanced methods allow researchers to create materials with precise features and functionalities, pushing the boundaries of what is achievable in manufacturing.

Overcoming technical limitations in advanced materials requires a multidisciplinary approach that integrates innovations in material design, processing, and application. Through cutting-edge technologies and collaborative research, industries can fully harness the potential of advanced materials. This ensures that materials meet the demanding requirements of modern applications while maintaining efficiency, durability, and scalability. These continuous advancements pave the way for transformative solutions across sectors such as aerospace, electronics, energy, and healthcare, shaping a future of enhanced performance and sustainability.

Ensuring Regulatory and Safety Compliance

Ensuring regulatory and safety compliance for advanced materials is a critical aspect of their development and application across various industries. Advanced materials, such as nanomaterials, composites, and functional polymers, are subject to a range of regulatory frameworks that dictate their safe use and environmental impact. These frameworks vary by industry; for instance, materials used in healthcare must adhere to stringent standards set by regulatory bodies like the FDA in the United States and the EMA in Europe, while aerospace materials are regulated by authorities such as the FAA and EASA. Compliance with regulations such as REACH and RoHS is also essential for managing the environmental impact of materials and ensuring their safe use throughout their lifecycle [805, 806].

The unique properties of advanced materials, particularly nanomaterials, necessitate comprehensive risk assessments to identify potential health and environmental risks. Studies have shown that the novel structures and enhanced reactivity of these materials can lead to unforeseen toxicological challenges, including cellular damage and bioaccumulation [807, 808]. Therefore, rigorous evaluations of toxicity, biodegradability, and long-term environmental effects are essential. Safety data sheets (SDS) and compliance documentation play a crucial role in communicating safety protocols to users and ensuring that manufacturers address these concerns effectively [809].

Testing and certification processes are vital for ensuring that advanced materials meet the required safety standards. These processes involve extensive testing to certify properties such as mechanical strength, thermal stability, chemical resistance, and biocompatibility. Organizations like ASTM International and ISO set the standards for these tests, which are conducted under controlled conditions to ensure reliability [810]. For example, biocompatibility testing is particularly critical for materials intended for medical devices, while flame retardancy tests are essential for construction and automotive applications [811]. Certification from accredited testing laboratories further reinforces the credibility of these materials in the market [812].

Lifecycle management and sustainability considerations are increasingly important in regulatory compliance. Advanced materials must not only perform well but also adhere to environmental regulations that promote sustainability and responsible waste management. Compliance with guidelines for recyclability and safe disposal is crucial to prevent pollution and reduce the environmental footprint of these materials [813]. Furthermore, adopting green certifications and circular economy principles can enhance compliance and foster a more sustainable approach to material usage [814].

Worker and consumer safety is another critical aspect of regulatory compliance. Regulations often mandate protective measures for workers handling advanced materials, including the use of personal protective equipment (PPE) and robust handling protocols to minimize exposure to hazardous substances [815]. Consumer safety is equally important, requiring that materials are free from harmful substances and accompanied by clear instructions for safe use [816]. The implementation of effective safety management systems is essential to protect both workers and consumers from potential risks associated with advanced materials [817].

As the field of advanced materials evolves, so too do the regulatory standards governing their use. It is imperative for manufacturers and researchers to stay informed about emerging regulations and adapt their practices accordingly. Global harmonization efforts, led by organizations such as ISO, aim to reduce discrepancies between regional regulations, facilitating international trade and enabling companies to innovate with confidence [818]. This harmonization is crucial for ensuring that advanced materials can be safely marketed and utilized across borders, ultimately fostering public trust in these technologies [819].

Balancing Innovation with Sustainability

Balancing innovation with sustainability in advanced materials is a pressing challenge that requires an integrated approach to material design, production, and lifecycle management. Advanced materials, characterized by their exceptional properties, have transformed various industries; however, their development often leads to significant environmental impacts. This dual challenge necessitates innovative strategies that align material performance with sustainability goals.

The rapid advancement of materials science has led to the creation of high-performance materials such as carbon fibre-reinforced polymers (CFRPs) and lightweight alloys, which are crucial in sectors like automotive and aerospace for enhancing fuel efficiency and reducing emissions [27, 820]. However, the production processes for these advanced materials can be energy-intensive and generate toxic byproducts, raising concerns about their overall environmental footprint [821]. For instance, the lifecycle assessment (LCA) of lightweight materials indicates that while they improve fuel efficiency, their manufacturing processes can contribute to greenhouse gas emissions and resource depletion [822, 823]. Thus, it is essential to evaluate not just the performance benefits but also the environmental costs associated with these innovations.

One promising approach to reconcile innovation with sustainability is the adoption of circular economy principles in material design. This involves creating materials that are not only high-performing but also recyclable, reusable, or biodegradable [824]. For example, bio-based polymers such as polylactic acid (PLA) offer a sustainable alternative to conventional plastics, as they can decompose naturally and reduce reliance on fossil fuels [820, 825]. Additionally, designing metals and composites for easy recovery and recycling can significantly minimize the environmental impact associated with raw material extraction and processing [826]. The integration of circular economy principles into material science not only enhances sustainability but also fosters innovation by encouraging the development of new materials and recycling technologies.

Optimizing production processes for energy efficiency is another critical aspect of balancing innovation with sustainability. Techniques such as additive manufacturing (3D printing) allow for precise material usage, which can reduce waste and energy consumption [827]. Furthermore, advancements in green chemistry are paving the way for more sustainable production methods that minimize or eliminate hazardous substances [828]. For instance, using water-based solvents in the synthesis of nanomaterials can significantly lower environmental toxicity while maintaining material performance [821, 829]. These innovations not only contribute to sustainability but also enhance the economic viability of advanced materials by reducing operational costs.

The incorporation of renewable resources into advanced materials development is vital for achieving sustainability goals. For example, cellulose nanofibers derived from plants are being utilized to create lightweight composites that are both strong and biodegradable [820, 830]. These materials reduce dependency on petroleum-based products and align with environmental objectives. Moreover, integrating advanced materials into renewable energy technologies, such as lightweight components for wind turbines and high-efficiency photovoltaics, supports broader sustainability initiatives [824, 825]. This synergy between material innovation and renewable resource utilization is essential for fostering a sustainable industrial ecosystem.

Lifecycle assessment serves as a crucial tool for evaluating the environmental impacts of materials throughout their lifecycle. By analysing energy use, carbon emissions, and waste generation from production to disposal, LCA helps identify sustainability trade-offs and informs decision-making [822, 831]. This analytical approach is particularly valuable for pinpointing areas in material production that can be improved through innovative practices, ultimately guiding researchers and manufacturers toward more sustainable material choices [823, 826].

Achieving sustainability in advanced materials necessitates collaboration among various stakeholders, including material scientists, engineers, policymakers, and industry leaders. Government initiatives can play a significant role by implementing regulations and incentives that promote sustainable material innovation, such as funding research into green technologies or providing tax credits for eco-friendly materials [827, 832]. Collaborative efforts between academia and industry can accelerate the development of sustainable materials by sharing knowledge and resources, leading to innovative solutions that address both performance and environmental concerns [824, 828].

Several examples illustrate the successful integration of innovation and sustainability in advanced materials. Biodegradable polymers, such as PLA, have emerged as viable alternatives to traditional plastics in packaging, significantly reducing environmental impact [820, 825]. In the automotive sector, lightweight composites not only enhance fuel efficiency but also contribute to lower greenhouse gas emissions during vehicle operation [27]. Additionally, the development of efficient recycling processes for rare earth metals from electronic waste exemplifies how innovation can mitigate resource depletion and environmental harm [821, 829].

The future of material science lies in the ability to harmonize innovation with sustainability. Advances in computational tools, machine learning, and data-driven material discovery enable researchers to predict material performance alongside sustainability metrics [824, 826]. This integrated approach can guide the development of materials that meet high-performance standards while adhering to sustainable practices, ultimately positioning advanced materials as a cornerstone for a sustainable future.

Chapter 10

Emerging Trends in Advanced Materials

Self-Healing Materials

Self-healing materials represent a significant advancement in materials science, offering innovative solutions to enhance the longevity and performance of various materials. These materials are designed to autonomously repair damage, thereby mitigating degradation caused by environmental factors, operational stress, and fatigue. The ability to self-repair is crucial, as micro-damage often goes undetected until it leads to catastrophic failure, compromising the material's thermal, electrical, and mechanical properties [833, 834]. By incorporating internal repair mechanisms, self-healing materials can restore their integrity, significantly extending their lifespan and reducing maintenance costs across multiple industries [835].

The scope of self-healing materials is broad, encompassing polymers, metals, ceramics, and cementitious materials. The mechanisms facilitating self-healing can be classified into intrinsic and extrinsic systems. Intrinsic self-healing materials possess molecular or crystalline structures that realign to bridge cracks autonomously, while extrinsic systems utilize embedded microcapsules or vascular networks that release healing agents upon damage [836, 837]. For instance, Liu et al. [835] discuss the challenges in achieving spontaneous self-healing properties in materials at room temperature, highlighting the complexity of developing effective self-healing mechanisms. Furthermore, advancements in biomimetic materials have inspired the development of self-healing systems that mimic natural processes, enhancing their functionality and application potential [838].

Historically, the concept of self-healing materials can be traced back to ancient practices, such as the Roman use of lime-based mortar, which exhibited self-healing properties due to the incorporation of volcanic ash [839]. This ancient technique demonstrated the potential for materials to autonomously repair themselves over time, a principle that modern science has sought to replicate and enhance. The formal recognition of self-healing materials as a distinct field of research emerged in the 21st century, coinciding with advancements in biomimetic materials and the first international conference on self-healing materials in 2007 [840]. Since then, the field has expanded to include functionalities such as self-lubrication and self-cleaning, further broadening the application of these materials in various sectors, including construction, electronics, and medicine [841].

The impact of self-healing materials on modern materials science is profound. In electronics, self-healing polymers can repair micro-cracks caused by thermal cycling, ensuring consistent performance and reducing electronic waste [833, 834]. In construction, self-healing concrete can autonomously repair micro-cracks, significantly extending the lifespan of infrastructure and reducing maintenance costs [839, 842]. Additionally, self-healing metal alloys are being explored for critical applications in aerospace and automotive industries, where structural integrity is paramount [843]. The potential for biodegradable self-healing polymers in medical applications also highlights the versatility of these materials, offering long-term reliability and reducing the need for secondary surgeries [844].

Despite the promising advancements in self-healing materials, several challenges remain. The scalability of production while maintaining cost-effectiveness is a significant hurdle [835]. Moreover, optimizing the efficiency and speed of the healing process is crucial, particularly in extreme environments [845]. Future research is directed toward developing multifunctional self-healing materials that integrate various properties, such as conductivity and thermal stability, alongside autonomous repair capabilities [846]. The incorporation of smart sensing technologies to monitor damage and activate healing mechanisms in real time represents another promising direction for future developments in this field [840].

Biomimetics: Nature-Inspired Self-Healing Mechanisms in Materials

Biomimetics refers to the science of studying and emulating natural processes and structures to develop innovative materials and systems. In the context of self-healing materials, plants and animals serve as exemplary models due to their intrinsic ability to seal and heal wounds. This natural process typically occurs in two phases: a rapid self-sealing phase to prevent further damage or infection, and a slower self-healing phase that restores structural and mechanical integrity. In plants, self-sealing is particularly crucial to prevent desiccation and protect against pathogenic infections, buying time for the more complex self-healing processes to occur. These biological phenomena have inspired the development of bio-inspired materials that mimic the healing and sealing mechanisms observed in nature.

To create bio-inspired self-repairing materials, scientists abstract functional principles from biological models and translate them into technical applications. This abstraction can take the form of analytical or numerical models that capture the underlying processes observed in nature. Biological systems that rely heavily on physical and chemical processes for healing are especially promising candidates for material design. For instance, academic research has shown how such biomimetic principles can be applied to the development of self-healing systems for polymer composites. These approaches often involve embedding healing agents into materials or designing self-repairing structures that mimic biological configurations.

One notable example involves the replication of skin-like healing processes using a grid of microchannels embedded in an epoxy substrate. This substrate contains dicyclopentadiene (DCPD), a healing agent, alongside Grubbs' catalyst to initiate the healing reaction. When a fracture occurs, the healing agent is released into the crack, where it reacts with the catalyst to form a polymer, effectively repairing the material. This system demonstrates partial recovery of toughness after fractures and can be replenished for multiple healing cycles. However, the process is not indefinitely repeatable due to the accumulation of polymer material in the crack plane over successive cycles.

Several plant systems have served as models for developing biomimetic self-healing materials. For example, the rapid self-sealing process of the twining liana *Aristolochia macrophylla* inspired the creation of a polyurethane (PU) foam coating for pneumatic structures. This biomimetic coating provides high repair efficiencies exceeding 99.9%, while maintaining a low coating weight and thickness. Similarly, latex-bearing plants like the weeping fig (*Ficus benjamina*), the rubber tree (*Hevea brasiliensis*), and spurges (*Euphorbia* spp.) have inspired strategies for sealing lesions using coagulation mechanisms. These natural processes have been adapted to elastomeric materials, demonstrating significant mechanical restoration after macroscopic damage.

The adaptation of biomimetic principles extends across various fields, enabling the design of innovative materials that integrate self-repair capabilities. For instance, microchannel systems inspired by vascular networks in plants and animals enable localized delivery of healing agents to damaged regions. Such systems are particularly effective in polymer composites and structural materials, where they can restore mechanical properties after damage. Another example includes PU foam coatings that emulate plant-based self-sealing mechanisms, offering efficient and lightweight solutions for applications in pneumatic structures.

By incorporating the coagulation-inspired healing mechanisms of latex-bearing plants, researchers have developed elastomeric materials capable of restoring their mechanical properties after substantial damage. These advancements highlight the potential of biomimetic materials to address critical challenges in material longevity and reliability, particularly in industries such as aerospace, automotive, and construction.

While biomimetic materials have made significant strides, challenges remain in optimizing their long-term performance and scalability. For example, systems that rely on replenishable healing agents face limitations in repeated use due to the accumulation of material from previous healing cycles. Moreover, ensuring that these materials function efficiently across diverse environmental conditions requires further research and development. Future efforts aim to refine the integration of biomimetic designs with advanced manufacturing techniques, such as 3D printing, to create customizable and scalable self-healing systems.

Biomimetics represents a transformative approach in materials science, leveraging the principles of natural systems to develop advanced self-healing materials. By studying and mimicking the self-repair mechanisms of plants and animals, researchers have created materials capable of autonomously restoring their functionality after damage. These innovations not only enhance material performance and longevity but also offer sustainable solutions for reducing maintenance and replacement costs. As the field continues to evolve, biomimetic materials are poised to play a pivotal role in shaping the future of engineering and design, bridging the gap between natural processes and technological advancements.

Self-Healing Polymers and Elastomers

Polymers have become an indispensable part of modern life, serving as the foundation for materials like plastics, rubbers, films, fibres, and paints. The immense demand for these materials has driven the development of advanced polymers capable of extending their reliability and lifespan. Self-healing polymers represent a groundbreaking class of materials designed to restore functionality autonomously after damage or fatigue. These materials are categorized into intrinsic and extrinsic types based on their healing mechanisms. Intrinsic systems inherently possess self-healing capabilities, often requiring external triggers like heat or light, while extrinsic systems rely on external healing agents encapsulated in microcapsules or vascular networks.

The self-healing process of autonomous polymers mimics biological responses, involving three critical phases. The first phase, triggering or actuation, occurs immediately upon sustaining damage. Next, material transport to the damaged area enables the healing process, followed by a chemical repair phase involving mechanisms like polymerization, reversible cross-linking, or molecular entanglement. Over time, advancements in polymer science have introduced capsule-based, vascular-based, and intrinsic self-healing systems, each offering unique advantages in addressing material degradation.

Traditional polymers typically degrade through molecular-level processes like sigma bond cleavage, which can occur either homolytically or heterolytically. Homolytic bond cleavage generates radical species that may recombine to repair damage or propagate further damage, while heterolytic cleavage produces ionic species that may recombine or react destructively. Stress at the molecular level can propagate into larger-scale damage, forming microcracks that compromise the material's structural integrity.

Some advanced polymers, however, exhibit reversible behaviour under stress. For instance, Diels-Alder-based polymers undergo reversible cycloaddition reactions, enabling damage repair through stress-induced changes in chemical bonding. Supramolecular polymers, on the other hand, rely on non-covalent interactions like hydrogen bonding, van der Waals forces, and metal coordination for self-healing. Damage in these polymers disrupts these interactions, allowing the material to reform its structural integrity.

Intrinsic self-healing systems are designed to restore material integrity without the need for external healing agents. These systems leverage several strategies, such as reversible chemical reactions, dynamic supramolecular bonds, and molecular diffusion. Diels-Alder and retro-Diels-Alder reactions are widely employed for their thermal reversibility, allowing polymers to revert to their monomeric or oligomeric states under controlled conditions. Similarly, supramolecular interactions and ionomeric clusters act as reversible cross-links, enabling polymers to recover their properties after deformation.

In thermoset matrices, meltable thermoplastic additives are incorporated to facilitate self-healing. Upon heating, these additives diffuse into cracks and form interlocking bonds, mechanically restoring the material. Other strategies involve thiol-based polymers with reversible disulfide bonds and urea-urethane networks that self-heal via room-temperature metathesis reactions.

Extrinsic self-healing polymers rely on external agents embedded within the material, often stored in microcapsules or vascular networks. Capsule-based systems encapsulate healing agents within microstructures that rupture upon damage, releasing the agents to react and restore the material. For instance, systems using dicyclopentadiene (DCPD) and Grubbs' catalyst demonstrate efficient healing through ring-opening metathesis polymerization (ROMP). However, challenges like catalyst cost and optimal capsule design remain critical considerations for commercial applications.

Vascular self-healing systems enhance the efficiency of extrinsic healing by incorporating capillary networks into the material. These networks enable the continuous delivery of healing agents to damaged regions, allowing for repeated and large-scale repairs. Techniques like direct ink writing (DIW) are used to fabricate three-dimensional interconnected networks, improving recovery efficiency and durability.

Emerging technologies have introduced innovative methods to enhance the self-healing capabilities of polymers. For example, carbon nanotube networks embedded in polymer matrices can sense and repair damage by heating and diffusing healing agents to the affected areas. Vitrimers, a new class of dynamic covalent adaptable networks, combine the properties of thermoplastics and thermosets. These polymers exhibit self-healing behaviour through bond exchange mechanisms, making them highly reprocessable and robust.

SLIPS (Slippery Liquid-Infused Porous Surfaces), inspired by the carnivorous pitcher plant, offer self-healing and self-lubricating properties. By embedding lubricating liquids within porous materials, SLIPS create surfaces that can recover from damage while resisting water, oil, and ice. Additionally, sacrificial thread stitching introduces self-healing channels into composite materials by creating hollow networks that can be filled with healing agents.

Self-healing polymers and elastomers hold immense potential across various industries, including aerospace, automotive, construction, and biomedical engineering. They enhance material longevity, reduce maintenance costs, and improve safety by preventing catastrophic failures. Advanced formulations like vitrimer-based bioepoxies and self-healing electronic screens highlight the versatility of these materials in addressing contemporary challenges.

Future research aims to optimize these systems for scalability, sustainability, and commercial relevance. Efforts to integrate self-healing mechanisms into recyclable and environmentally friendly materials align with global sustainability goals. By leveraging advancements in material science, self-healing polymers are poised to revolutionize the design and performance of next-generation materials.

Self-Healing Fibre-Reinforced Polymer Composites

Fibre-reinforced polymer composites (FRPs) are widely used in high-performance applications due to their exceptional strength-to-weight ratio, durability, and versatility. However, they are prone to microcracks and delamination, which compromise their structural integrity over time. Implementing self-healing functionalities into FRPs has been explored to enhance their longevity and reliability. The approaches are predominantly extrinsic, relying on discrete capsule-based systems or continuous vascular systems. Unlike non-filled polymers, intrinsic self-healing mechanisms based on reversible chemical bonds have not yet been successfully applied to FRPs. Research in this field has focused on applying self-healing methods to both simple flat panels and more complex structures, such as T-joints and aircraft fuselages, where traditional repair techniques are less feasible.

The concept of a capsule-based self-healing system was first introduced by White et al. in 2001 and has since been adapted for FRPs. This method involves embedding microcapsules containing a healing agent within the polymer matrix. When a crack propagates through the material, it ruptures the microcapsules, releasing the healing agent into the damage zone. The healing agent reacts with a catalyst present in the matrix or with environmental factors, restoring the material's integrity. While this approach is generally a one-time process because the microcapsules cannot regenerate, it has demonstrated significant effectiveness. Capsule-based systems can restore the material's mechanical properties to nearly 100% and maintain stability over the material's lifetime. This method has gained popularity due to its simplicity and ease of integration into existing composite manufacturing processes.

Vascular systems represent a more sophisticated and versatile approach to self-healing in FRPs. Inspired by biological systems like blood vessels, this method involves embedding a network of hollow channels, or vascules, within the composite structure. When damage occurs, cracks intersect the vascules, causing them to rupture and release a liquid healing agent into the damage plane. This agent flows into the cracks and polymerizes, effectively sealing them and restoring the material's structural properties.

Vascular systems offer several advantages over capsule-based systems. They can deliver larger volumes of healing agent, enabling the repair of extensive damage, and support repeated healing cycles by refilling the vascules. Additionally, these channels can provide multifunctionality, such as thermal management and structural health monitoring, further enhancing the composite's utility.

Several innovative techniques have been developed to incorporate vascular networks into FRPs:

1. **Hollow Glass Fibres (HGFs):** This method integrates hollow glass fibres into the composite, which act as reservoirs for the healing agent. These fibres are lightweight and can maintain the composite's mechanical properties while enabling self-healing functionality.

2. **3D Printing:** Advanced additive manufacturing techniques allow precise placement of vascular networks within the composite structure, offering flexibility in design and customization.

3. **Lost Wax Process:** In this technique, wax rods are embedded in the composite during fabrication. After curing, the wax is melted and removed, leaving behind hollow channels that can be filled with a healing agent.

4. **Solid Preform Route:** This method uses preformed solid structures that are later dissolved or extracted to create a vascular network within the composite.

Self-healing FRPs have transformative potential in industries where durability and reliability are critical, such as aerospace, automotive, and civil engineering. For instance, incorporating vascular systems into aircraft fuselages can enhance safety by enabling real-time damage repair during operation. In structural applications, self-healing FRPs can reduce maintenance costs and extend service life, especially in hard-to-access areas.

Ongoing research focuses on refining these self-healing methods to improve efficiency, scalability, and cost-effectiveness. Efforts are also directed toward integrating intrinsic and extrinsic systems to create hybrid solutions that combine the benefits of both approaches. As advancements continue, self-healing FRPs are poised to revolutionize material design by offering robust, adaptive, and sustainable solutions for modern engineering challenges.

Self-Healing Coatings

Coatings play a critical role in preserving and enhancing the properties of underlying materials by protecting them from environmental factors. They act as barriers against water, oxygen, and other corrosive elements, which can degrade the material or cause it to fail. However, when coatings develop microcracks, these protective properties are compromised, allowing environmental elements to penetrate and cause mechanical degradation, delamination, or electrical failure. Repairing such small-scale damage is challenging, often costly, and sometimes impractical. Self-healing coatings, designed to automatically recover their properties after damage, present a transformative solution by extending the functional lifetime of coatings and maintaining their protective, mechanical, electrical, and aesthetic properties.

Several approaches to self-healing materials have been adapted for coatings, including microencapsulation and the incorporation of reversible chemical and physical bonds. Microencapsulation is one of the most widely employed techniques. In this method, healing agents such as dicyclopentadiene (DCPD) are encapsulated within microcapsules embedded in the coating. When a microcrack occurs, the capsules rupture, releasing the healing agent, which reacts with a catalyst or environmental conditions to repair the crack. This approach, originally developed for epoxy polymers, has been successfully adapted to epoxy adhesive films commonly used in the aerospace and automotive industries.

Beyond microencapsulation, reversible bonding strategies, such as hydrogen bonding, ionic interactions, and Diels-Alder chemical reactions, have been utilized to create self-healing coatings. These systems rely on reversible cross-linking mechanisms, enabling the material to reassemble and restore its integrity under the right conditions, such as heat or light.

One of the most impactful applications of self-healing coatings is in corrosion protection for metallic surfaces. Corrosion not only leads to significant economic costs but also poses environmental challenges. Researchers have encapsulated materials such as isocyanates, DCPD monomers, glycidyl methacrylate (GMA), epoxy resins, linseed oil, and tung oil to demonstrate the effectiveness of self-healing coatings in mitigating corrosion. These coatings extend the lifespan of metallic components by automatically repairing damage, thereby maintaining a continuous protective barrier.

Self-healing coatings have also been explored in electrical applications. For instance, microencapsulated liquid metals or carbon black suspensions have been used to restore electrical conductivity in multilayer microelectronics and battery electrodes. Liquid metal microdroplets embedded in silicone elastomers have been particularly effective for creating stretchable conductors that maintain electrical continuity even when damaged, mimicking the resilience of biological tissues.

In high-temperature environments, self-healing coatings often rely on the formation of glassy layers to repair damage. For instance, silicate-based glass materials have been studied for their ability to self-heal coatings used in thermal barrier systems and space applications, such as heat shields. The self-healing capability in such cases depends on the viscosity of the glass, which determines how effectively it can flow to fill cracks while competing with oxidation or ablation processes. Composite materials based on molybdenum disilicide are of particular interest for improving the self-healing performance of coatings in extreme thermal conditions.

Self-healing coatings have significant economic and ecological implications, reducing maintenance costs and the need for frequent replacements. Their ability to extend the lifetime of materials makes them invaluable in industries such as aerospace, automotive, electronics, and infrastructure. Future research aims to improve the efficiency and scalability of self-healing mechanisms, particularly for advanced applications in high-temperature and extreme environments. Advances in nanotechnology and material science will further enable the development of multifunctional self-healing coatings that not only repair themselves but also provide additional properties such as thermal management, self-cleaning, and enhanced durability.

Self-Healing Cementitious Materials

Cementitious materials, such as concrete, have been integral to construction since the Roman era, showcasing a remarkable natural ability to self-heal. This property was first documented in 1836 by the French Academy of Science and has since been a focus of scientific research to improve its efficiency through chemical and biochemical strategies.

Autogenous healing is the innate capability of cementitious materials to repair cracks without external intervention. This phenomenon primarily occurs due to two processes: the continued hydration of unhydrated cement particles and the carbonation of dissolved calcium hydroxide. In fresh-water environments, this natural healing can effectively close cracks up to 0.2 mm within approximately seven weeks. However, the effectiveness of autogenous healing is often limited in terms of the crack width it can address.

To enhance this natural process and enable the closure of larger cracks, researchers have incorporated superabsorbent polymers (SAPs) into cementitious mixtures. These polymers absorb water, swell, and stimulate further hydration by retaining moisture in the material. Studies show that adding 1% by mass of SAPs relative to cement content can increase hydration activity by nearly 40%, especially if the material is exposed to periodic water contact. This advancement significantly boosts the crack-sealing capability of autogenous healing.

Another approach to self-healing in cementitious materials involves the integration of chemical agents that react upon crack formation to restore the material's integrity. These agents are typically housed in microcapsules or vascular tubes embedded within the concrete. When cracks occur, these capsules or tubes rupture, releasing the healing agents into the damaged area. The agents then react chemically to seal the cracks. Research in this domain has concentrated on improving the durability and efficiency of the housing mechanisms and the encapsulated chemicals to ensure reliable performance over the lifespan of the material.

The incorporation of biological elements, particularly bacteria, into cementitious materials has emerged as an innovative strategy for self-healing. This method leverages the metabolic activity of specific bacteria to induce calcium carbonate precipitation, which seals cracks and prevents water ingress. This concept gained prominence with a 1996 study by H. L. Erlich, which demonstrated the potential of bacterial-induced mineralization in concrete.

At the First International Conference on Self-Healing Materials in 2007, researchers Henk M. Jonkers and Erik Schlangen presented a groundbreaking study where they used alkaliphilic spore-forming bacteria as a self-healing agent in concrete. The bacteria, incorporated directly into cement paste, could survive for up to four months. However, subsequent research addressed this limitation by encasing the bacteria in protective carriers like expanded clay particles or glass tubes. These carriers shield the bacteria from the harsh alkaline environment of concrete, significantly extending their viability and effectiveness. Other protective strategies for bacterial encapsulation have also been explored, paving the way for bio-based self-healing concrete.

The development of self-healing cementitious materials, leveraging autogenous, chemical, and bio-based mechanisms, represents a significant leap in material science. These advancements not only enhance the durability and longevity of concrete structures but also reduce maintenance costs and environmental impact. With ongoing research into improving the efficiency and scalability of these methods, self-healing concrete holds promise for revolutionizing construction practices worldwide.

Self-Healing Ceramics

Ceramics are highly valued for their superior strength at high temperatures, making them ideal for applications where metals might fail. However, their brittleness and sensitivity to flaws challenge their reliability as structural materials. Among ceramics, MAX phase ceramics (denoted as $M_{n+1}AX_n$) have shown remarkable potential for self-healing through an intrinsic mechanism. These ceramics, such as Ti_3AlC_2, Ti_2AlC, and Cr_2AlC, can autonomously repair microcracks caused by thermal stress or wear. When exposed to high temperatures in the presence of air, the A-element within the MAX phase reacts with oxygen to form oxides that fill the crack gaps. This process, first demonstrated with Ti_3AlC_2 at 1200°C in air, can repeat until the healing element is depleted.

The self-healing ability of MAX phases provides a unique advantage over extrinsic systems requiring external agents for repair. The process can even improve the material's properties, such as local strength, depending on the type of oxide formed. On the other hand, conventional ceramics like mullite, alumina, and zirconia lack intrinsic self-healing capabilities but can be engineered for self-repair by embedding second-phase components. For example, embedding silicon carbide (SiC) particles into an alumina matrix allows cracks to be healed through a chemical reaction when the SiC is exposed to oxygen and heat, forming a material that expands to fill the crack gap. Studies on such systems have explored their high-temperature strength and fatigue performance, highlighting the critical role of bonding between the matrix and the healing agent in ensuring effective self-healing.

Self-Healing Metals

Metals exposed to prolonged high temperatures and moderate stresses often experience premature failure due to creep fracture. This failure arises from the formation and growth of cavities, which eventually coalesce into cracks. Developing self-healing mechanisms for these early-stage defects offers a promising approach to extend the lifespan of metallic components. However, achieving self-healing in metals is intrinsically challenging due to their high melting points and low atomic mobility.

The self-healing of metals typically relies on precipitates forming at defect sites to arrest crack growth. In aluminium alloys, underaged treatments show improved creep and fatigue properties compared to peak-hardened alloys, as heterogeneous precipitates form at crack tips and plastic zones. Initial attempts to heal creep damage in steels involved the dynamic precipitation of copper (Cu) or boron nitride (BN) at cavity surfaces. While Cu precipitation can fill defects, its preference for deformation-induced sites is limited, as many spherical Cu precipitates form simultaneously in the matrix.

More recent research has identified gold (Au) atoms as highly efficient healing agents in Fe-based alloys. Au solutes remain dissolved until defects form, at which point they selectively precipitate at cavity surfaces, effectively filling creep cavities. For lower stress levels, up to 80% of creep cavities can be filled with Au precipitates, significantly enhancing the material's creep lifetime. This concept has been demonstrated in binary and ternary alloys, with ongoing efforts to translate it into multicomponent creep steels for practical applications.

The concept of self-healing materials has been explored in various contexts, including polymers and hydrogels, but the application to metals represents a novel frontier. For instance, the self-healing mechanisms observed in hydrogels, as discussed by Tamesue et al. [847] and Zeng et al. [848], highlight the potential for materials to autonomously repair damage. These studies illustrate that self-healing can be achieved through various chemical interactions and structural designs, which could inspire similar approaches in metallic systems. The foundational principles of self-healing, such as the use of dynamic covalent bonds, are critical for understanding how these mechanisms might be adapted for metals [849].

Moreover, the study of fatigue in materials has been a critical area of research, as evidenced by the work of Vila-Cortavitarte et al. [850], which discusses the fatigue response in self-healing asphalt mixtures. This research underscores the importance of understanding fatigue mechanisms in materials, which is directly relevant to the recent findings at Sandia. By drawing parallels between the self-healing mechanisms in asphalt and those in metals, researchers can better understand how to implement these technologies in various applications.

Self-Healing Hydrogels

Hydrogels are unique soft solids composed of three-dimensional polymer networks capable of holding a substantial amount of water. These materials can be derived from both natural and synthetic polymers and are notable for their versatility and biocompatibility. Self-healing hydrogels, a specialized subset of these materials, are engineered to recover their structure and function after being cut or broken. The self-healing mechanism in hydrogels is primarily attributed to non-covalent interactions, such as hydrogen bonding, ionic interactions, and hydrophobic effects, or dynamic covalent chemistry, which allows reversible bond formation and breakage.

One particularly exciting application of self-healing hydrogels is in the biomedical field. Hydrogels that can transition into a fluid-like state and subsequently self-heal are gaining traction for developing injectable hydrogels, which are crucial for minimally invasive delivery systems in tissue regeneration. Additionally, these properties make them highly suitable as inks for 3D bioprinting, enabling the creation of complex biological structures for regenerative medicine. The ability to self-heal ensures structural integrity during and after printing, making these materials essential for advancing bioprinting technologies.

Self-Healing Organic Dyes

The self-healing capability has also been discovered in certain classes of organic dyes, particularly when doped into polymer matrices like PMMA (polymethyl methacrylate). This phenomenon, termed reversible photo-degradation, allows the dyes to recover their structural and optical properties after exposure to light-induced degradation. Unlike traditional recovery processes, which often involve molecular diffusion, the self-healing mechanism here is driven by specific interactions between the dye molecules and the polymer matrix. This interaction stabilizes the dye molecules and facilitates their recovery, making these systems highly valuable for applications requiring prolonged optical performance, such as in light-emitting devices, solar cells, and displays.

Self-Healing of Ice

Ice, a seemingly simple material, has demonstrated a remarkable self-healing capability for repairing micrometre-sized defects on its surface. This process occurs spontaneously over several hours and is governed by unique thermodynamic and molecular mechanisms. Defects on the ice surface create local curvature, leading to increased vapor pressure in these regions. This heightened vapor pressure enhances the volatility of surface molecules, increasing their mobility. As a result, molecules sublimate from the defect edges and condense back onto the surface, effectively repairing the damage.

This self-healing mechanism contrasts with earlier theories suggesting that ice sintering occurred through surface diffusion. The new understanding underscores the importance of sublimation and condensation processes in driving the healing of ice. These insights hold potential for understanding natural phenomena in polar climates and may inspire innovations in cryopreservation, ice-resistant materials, and cold-environment engineering.

Smart and Responsive Materials

Smart materials, often referred to as intelligent or responsive materials, represent a remarkable category of substances engineered to undergo controlled changes in their properties when subjected to external stimuli such as stress, moisture, temperature, pH, light, or electric and magnetic fields. These materials have garnered significant attention due to their versatile applications across various fields, including sensors, actuators, and artificial muscles, particularly in the form of electroactive polymers (EAPs) [851, 852]. Their inherent ability to respond and adapt to environmental changes makes them crucial for the advancement of innovative technologies aimed at enhancing functionality and efficiency across multiple industries [853].

A quintessential application of smart materials can be observed in sportswear equipped with ventilation valves that dynamically respond to temperature and humidity. These valves open during periods of high perspiration and close as the body cools, thereby optimizing comfort and performance [854]. In architectural applications, smart materials facilitate buildings that can adjust to varying atmospheric conditions, such as wind or rain, thereby improving energy efficiency and occupant comfort [855]. In the medical field, smart materials enable the design of drug delivery systems that release medication only upon detecting specific conditions, such as the presence of a viral infection, highlighting their transformative potential in healthcare [852, 856].

Smart materials encompass a diverse array of substances, each exhibiting unique responsive characteristics. For instance, piezoelectric materials generate electrical voltage when stressed and deform under an electric field, making them ideal for sensor and actuator applications [857]. Shape-memory alloys and polymers can deform and revert to their original shape in response to temperature or stress changes, while photovoltaic materials convert light into electricity, forming the backbone of solar energy technologies [853]. Additionally, magnetostrictive materials alter their shape in a magnetic field, and electroactive polymers change volume when exposed to electric fields, further expanding the range of applications for smart materials [855]. Chromogenic systems, including thermochromic, photochromic, and electrochromic materials, change colour in response to environmental stimuli, finding use in products like smart sunglasses and displays [854].

Smart polymers, a subset of smart materials, are particularly noteworthy for their adaptability and versatility. These stimuli-responsive or functional polymers can alter their colour, transparency, conductivity, or shape based on environmental changes such as temperature, pH, or light [852, 858]. Their nonlinear response characteristics allow for significant property alterations even with minor environmental changes, making them highly effective in specialized applications [853]. In biomedical engineering, for instance, pH-responsive polymers are utilized for targeted drug delivery, while humidity-sensitive polymers are employed in self-adaptive wound dressings that maintain optimal healing conditions [852, 856]. The consistent performance of these polymers is vital for critical applications, ensuring reliability in medical treatments and interventions [853].

The pharmaceutical industry has also greatly benefited from the advancements in smart polymers, particularly in drug delivery systems that protect active ingredients from degradation and enable controlled release [852, 856]. For example, hydrogels made from smart polymers can be engineered to release insulin in response to glucose levels, presenting significant potential for diabetes management [852]. Furthermore, smart materials have transcended traditional boundaries, impacting various domains such as biotechnology, where they assist in bioseparations and protein purification, leveraging their reversible structural changes to capture and release target molecules [853].

Despite their immense potential, the implementation of smart materials, especially in biomedicine, faces challenges related to toxicity, compatibility, and the effects of degradation products [856]. Addressing these concerns is crucial for ensuring the safe and effective use of these materials. Nevertheless, ongoing research and advancements are likely to propel smart materials into new applications, including smart toilets for health monitoring and smart irrigation systems that optimize resource use based on soil conditions [852, 856]. The future of smart materials is promising, with the potential for these materials to learn and adapt over time, underscoring their transformative capabilities and the necessity for continued research to overcome existing limitations [853].

Quantum Materials and Topological Insulators

Quantum materials represent a transformative frontier in materials science, characterized by their unique properties arising from quantum mechanical effects. Unlike traditional materials, which can often be described by classical physics, quantum materials exhibit phenomena such as superconductivity, magnetoresistance, and topologically protected states due to the intricate interplay of quantum mechanics, symmetry, and strong interactions among particles [859, 860]. This class of materials includes high-temperature superconductors, quantum spin liquids, Weyl semimetals, and topological insulators, all of which are pivotal in advancing condensed matter physics and materials science [861, 862].

Topological insulators, a significant subset of quantum materials, are defined by their insulating bulk properties while hosting conducting states on their surfaces. This unique behaviour is a consequence of their topological order, which is fundamentally linked to the mathematical concept of topology. The surface states of topological insulators are protected by symmetries, such as time-reversal symmetry, making them robust against perturbations like impurities or defects [863, 864]. This robustness is further enhanced by phenomena such as spin-momentum locking, where the spin of an electron is correlated with its momentum direction, a feature that is particularly promising for applications in spintronics [859, 860].

The physical mechanisms underlying topological insulators are primarily governed by strong spin-orbit coupling, which leads to a band structure where bulk states are insulating while surface states remain gapless. This results in the formation of a "Dirac cone" in the electronic structure, similar to that observed in graphene, which is responsible for the unique conductive properties of these materials [861, 862, 864]. The exploration of these materials has not only expanded our understanding of quantum mechanics but also opened avenues for practical applications in quantum computing, where topologically protected states can be utilized for stable qubits, and in energy-efficient technologies, such as thermoelectric devices [860, 865].

Research into quantum materials and topological insulators is rapidly evolving, with ongoing efforts to discover new materials that exhibit stronger topological effects and higher operational temperatures. The integration of these materials into functional devices is a key focus, as it brings us closer to realizing their potential in transformative technologies like quantum computing and advanced electronics [859, 860, 865]. As such, quantum materials and topological insulators are at the forefront of modern materials science, bridging fundamental physics with practical applications and promising significant advancements in various technological domains.

Topological Insulators

A topological insulator is a unique class of material with dual electrical behaviour: its interior acts as an insulator, preventing the flow of electrons, while its surface supports conductive states, allowing electrons to move freely. This distinction arises from the material's electronic structure and the unique properties of its band topology, which sets it apart from conventional ("trivial") insulators.

In both trivial insulators and topological insulators, an energy gap exists between the valence band (filled with electrons) and the conduction band (where electrons can move freely to conduct electricity). This gap ensures that the material's interior does not conduct electricity under normal conditions. However, in topological insulators, the electronic bands are "twisted" in a way that is topologically distinct from trivial insulators. This "twisting" means that the band structure cannot be continuously transformed into that of a trivial insulator without closing the band gap and passing through a conducting state.

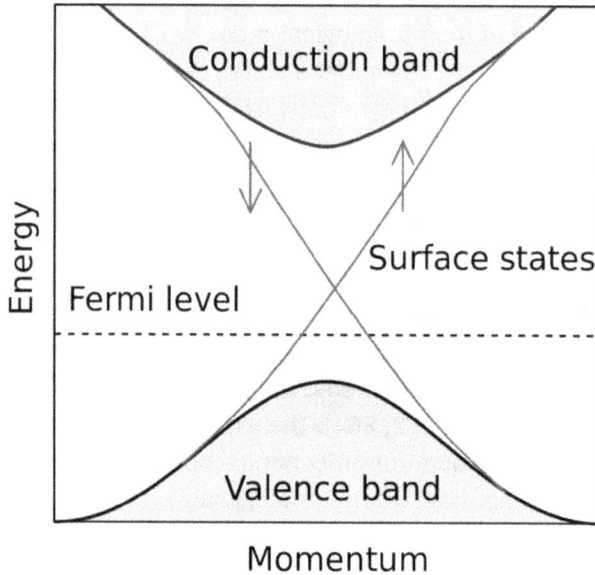

Figure 49: Topological insulator band structure. A13ean, CC BY-SA 4.0, via Wikimedia Commons.

This topological distinction is fundamental. When a topological insulator interfaces with a trivial insulator or vacuum (which is considered trivial), its surface must host conducting states to maintain continuity of the material's electronic properties. This feature is a global property of the material's band structure and is not easily disrupted by local perturbations, as long as the underlying symmetries, such as time-reversal symmetry, are preserved. This robustness is unique to topological insulators and differentiates them from conventional insulators, whose surface states are more susceptible to local disruptions.

A more formal definition of a topological insulator is that it cannot be adiabatically (smoothly and continuously) transformed into a trivial insulator without going through a conducting phase. This means that topological insulators and trivial insulators occupy separate regions in a phase diagram, connected only by intermediate conducting states. This concept marks a departure from the traditional Landau symmetry-breaking theory, which defines most other states of matter. Instead, topological insulators are described by symmetry-protected topological order, where specific symmetries ensure their unique properties.

The properties of a topological insulator, including its surface states, depend heavily on the material's dimensionality and its symmetries. For example, all topological insulators exhibit U(1) symmetry, linked to particle number conservation, and often have time-reversal symmetry in the absence of magnetic fields. These symmetries enable the classification of topological insulators using mathematical invariants, such as Z or Z_2 indices, which determine whether a material is topologically trivial or non-trivial.

One of the most striking features of topological insulators is their exotic surface states. For instance, in three-dimensional (3D) topological insulators with time-reversal symmetry, the surface states exhibit spin-momentum locking. In this phenomenon, the spin of an electron is oriented perpendicularly to its momentum, creating a helical spin texture. This property suppresses backscattering (where electrons reverse direction), as it requires a simultaneous flip in both momentum and spin. This suppression enhances the metallic behaviour of the surface, making it an excellent conductor, even when the bulk remains insulating.

Although topological insulators were initially discovered in quantum mechanical systems, analogous behaviours have been identified in classical systems. For example:

- Photonic topological insulators manipulate the flow of light, ensuring robust and defect-immune light propagation.

- Magnetic topological insulators exhibit similar properties for spin waves or magnons.

- Acoustic topological insulators guide sound waves along specific pathways without scattering.

These analogues showcase the versatility of topological principles beyond electronic systems, extending their potential to a wide range of applications.

The robust surface states of topological insulators have significant implications for technology. Their spin-momentum locking makes them promising candidates for spintronics, where electron spins are used for data storage and processing. Additionally, their ability to suppress scattering could lead to more efficient electronic devices. In quantum computing, the topologically protected states of these materials are being explored for creating stable qubits, resistant to environmental noise.

Topological insulators represent a paradigm shift in our understanding of materials, introducing the interplay of topology and quantum mechanics as a powerful framework. With ongoing research, these materials are poised to revolutionize fields ranging from electronics to photonics and acoustics, offering insights into fundamental physics while driving innovative applications.

One of the defining features of TIs is spin-momentum locking, where the spin of surface electrons is oriented perpendicular to their momentum. This property is a direct consequence of the helical Dirac fermion states present on the surface of 3D TIs. These surface states are symmetry-protected, meaning they are robust against perturbations that preserve time-reversal symmetry. This robustness creates opportunities for hosting exotic quasiparticles like Majorana particles when superconductivity is induced on the surface via proximity effects.

Richard Skiba

Majorana particles, which are theorized to be their own antiparticles, have significant implications for fault-tolerant quantum computing due to their non-abelian statistics.

The non-trivial topology of TIs is characterized by Z_2 topological invariants, mathematical quantities that distinguish TIs from trivial insulators. Unlike traditional quantized transport methods such as spin Hall conductance, the Z_2 invariants require specialized experimental approaches for measurement. These invariants ensure that the conducting surface states persist as long as the material's symmetry is maintained, even though they are not classified as "topological" in the same sense as the robust states in the quantum Hall effect. The periodic table of topological invariants, derived from Altland-Zirnbauer symmetry classes, provides a comprehensive framework for classifying these materials based on their symmetries and spatial dimensions.

The spin-momentum locking in TIs has made them highly promising for spintronic devices, which leverage the spin of electrons rather than their charge for information processing. This property could enable the development of low-power, high-efficiency devices. Additionally, the robustness of TI surface states is being explored for use in quantum computers, particularly in the construction of dissipationless transistors and fault-tolerant qubits based on the quantum Hall and quantum anomalous Hall effects. These advancements have the potential to revolutionize the field of quantum computing by offering energy-efficient and scalable solutions.

Several well-known TIs, such as Bi_2Te_3, Sb_2Te_3, and their alloys, are also excellent thermoelectric materials. Thermoelectric devices convert temperature gradients into electrical energy, and their efficiency is defined by the interplay of thermal conductivity, electrical conductivity, and the Seebeck coefficient. The heavy atomic composition of TIs reduces thermal conductivity, which is beneficial for thermoelectric efficiency. Furthermore, the warping of the bulk band structure, driven by band inversion in TIs, optimizes the Seebeck coefficient and electrical conductivity simultaneously. This unique interplay makes TIs highly effective for thermoelectric applications, offering enhanced power conversion efficiency.

TIs are also being explored in magnetoelectronic and optoelectronic devices. Their surface states can interact with magnetic fields in novel ways, enabling advanced magnetoelectronic components. In optoelectronics, TIs hold promise for creating highly efficient photodetectors and modulators due to their unique electronic structure and spin-polarized surface states. These applications could lead to breakthroughs in technologies ranging from communication systems to advanced imaging.

Topological insulators, with their unique blend of insulating bulk properties and conducting surface states, represent a frontier in materials science. Their spin-momentum locking, robust surface states, and compatibility with superconductivity open avenues for transformative applications in quantum computing and spintronics. Simultaneously, their potential for thermoelectric energy conversion and advanced electronic devices underscores their

368

versatility. As experimental techniques and theoretical understanding continue to evolve, TIs are poised to play a critical role in next-generation technologies, combining fundamental scientific intrigue with practical utility.

The synthesis of topological insulators (TIs) is a critical process that influences their structural and electronic properties, determining their utility in applications ranging from spintronics to quantum computing. Various techniques have been developed for the fabrication of TIs, each with unique advantages suited to specific requirements. Among these methods are metal-organic chemical vapor deposition (MOCVD), physical vapor deposition (PVD), solvothermal synthesis, sonochemical techniques, and molecular beam epitaxy (MBE). Of these, MBE has emerged as the most widely used approach due to its precision and ability to produce high-quality, single-crystal films.

Metal-Organic Chemical Vapor Deposition (MOCVD) and Physical Vapor Deposition (PVD) are popular for large-scale material growth. MOCVD relies on chemical reactions between metal-organic precursors and reactive gases, depositing the material on a substrate. PVD, on the other hand, employs physical processes like thermal evaporation or sputtering to deposit thin films. While these techniques are valuable for various applications, they are less suited for producing the ultra-clean, defect-free surfaces often required for advanced studies of TIs.

Solvothermal synthesis and sonochemical techniques are alternative methods that enable the creation of TIs in solution-based environments. Solvothermal synthesis uses high-temperature and high-pressure solvents to crystallize materials, while sonochemical methods utilize ultrasonic waves to promote chemical reactions. These approaches are particularly effective for producing bulk crystals or nanoparticles of TIs but may face limitations in controlling thin-film uniformity and surface quality.

A key feature of TIs is their layered structure, governed by weak van der Waals interactions. This property enables the exfoliation of thin films from bulk crystals, yielding surfaces that are clean and structurally perfect. Van der Waals epitaxy (VDWE) leverages these interactions to grow layered TIs on various substrates. The flexibility of this method allows for the integration of TIs into heterostructures and complex devices, as the lattice-matching conditions are relaxed due to the weak bonding. This adaptability is crucial for integrating TIs with other materials for use in heterostructures or integrated circuits.

Molecular Beam Epitaxy (MBE) is a vacuum-based technique for growing high-quality crystalline materials layer by layer. In MBE, elements are heated in separate sources until they sublime, and the resulting gaseous atoms condense on a substrate, reacting to form a single crystal. The growth takes place in high or ultra-high vacuum, minimizing contamination and ensuring pristine material properties.

MBE offers significant advantages for synthesizing TIs. Its ability to precisely control growth rates and source material ratios reduces lattice defects and ensures smooth interfaces between layers. This control is particularly important for producing engineered heterostructures with atomically flat surfaces, essential for studying and leveraging the unique properties of TIs. The in-situ compatibility of MBE with characterization techniques like angle-resolved photoemission spectroscopy (ARPES) and scanning tunnelling microscopy (STM) further enhances its utility, allowing real-time analysis of electronic properties and surface structures.

One of the strengths of MBE lies in its ability to grow TIs on a wide variety of substrates. Due to the weak van der Waals bonding, TIs are not strictly constrained by lattice-matching conditions, enabling their growth on substrates such as silicon (Si(111)), sapphire (Al_2O_3), gallium arsenide (GaAs(111)), indium phosphide (InP(111)), cadmium sulphide (CdS(0001)), and yttrium iron garnet ($Y_3Fe_5O_{12}$). This flexibility allows researchers to tailor material properties to specific applications by selecting appropriate substrate materials.

The synthesis of topological insulators is a sophisticated process that underpins their potential in cutting-edge applications. Among the various methods, Molecular Beam Epitaxy stands out for its precision, cleanliness, and compatibility with advanced characterization tools. The ability to grow TIs with smooth, defect-free surfaces and integrate them into diverse substrates underscores the versatility of MBE. As research continues to explore the properties and applications of TIs, advancements in synthesis techniques will play a crucial role in unlocking their full potential in quantum technologies, spintronics, and beyond.

PVD Growth of Topological Insulators

Physical Vapor Deposition (PVD) has emerged as a promising technique for the growth of topological insulators (TIs) due to its simplicity, cost-effectiveness, and reproducibility. Unlike the exfoliation method, which can suffer from inconsistencies and limited scalability, PVD offers a reliable approach for synthesizing single crystals with well-defined crystallographic orientations. Moreover, while Molecular Beam Epitaxy (MBE) remains a gold standard for controlled growth, PVD provides a more accessible alternative without the need for the complex infrastructure and costs associated with MBE.

The PVD technique is particularly suited for growing layered quasi-two-dimensional materials such as bismuth-based topological insulators (Bi_2Se_3, Bi_2Te_3, and their alloys). It enables precise control over several critical parameters, including composition, thickness, size, and surface density on the desired substrate. This control is especially crucial in three-dimensional TIs, where reducing the film thickness minimizes the contribution of bulk electronic channels, allowing the unique surface states to dominate electrical conduction. This enhancement of surface mode contributions is essential for studying and leveraging the quantum properties of TIs.

The majority of research in TIs has focused on bismuth and antimony chalcogenides, such as Bi_2Se_3, Bi_2Te_3, Sb_2Te_3, and their alloys ($Bi_{1-x}Sb_x$, $Bi_{1.1}Sb_{0.9}Te_2S$). These materials exhibit unique van der Waals interactions that relax lattice-matching constraints, enabling their growth on a variety of substrates. Bismuth chalcogenides are particularly valuable not only for their TI properties but also for their applications in thermoelectrics due to their favourable electronic and thermal properties.

One notable feature of bismuth chalcogenides is their low surface energy, which often results in surfaces terminated by tellurium atoms (e.g., Bi_2Te_3). These properties allow for consistent growth on substrates like silicon (Si) and others, although the choice of substrate plays a significant role in determining the overall quality of the material. For instance, while silicon has been successfully used as a substrate for growing Bi_2Te_3, materials like sapphire (Al_2O_3) face challenges due to large lattice mismatches (approximately 15%), which can degrade material properties.

Buffer layers are often introduced to mitigate lattice mismatches and improve the electrical and structural quality of the TIs. For example, Bi_2Se_3 can be grown on $Bi_{2-x}In_xSe_3$ buffer layers, reducing mismatch-induced defects and enhancing material performance.

The substrate choice is pivotal in PVD growth as it influences lattice matching, film nucleation, and overall quality. Lattice mismatch measures the disparity in lattice constants between the substrate and the grown film. A smaller mismatch generally results in fewer defects and better film quality. *Table 5* below summarizes the lattice mismatch of various substrates for Bi_2Se_3, Bi_2Te_3, and Sb_2Te_3:

Table 5: mismatch of various substrates for Bi2Se3, Bi2Te3, and Sb2Te3.

Substrate	Bi_2Se_3 %	Bi_2Te_3 %	Sb_2Te_3 %
Graphene	−40.6	−43.8	−42.3
Si	−7.3	−12.3	−9.7
CaF_2	−6.8	−11.9	−9.2
GaAs	−3.4	−8.7	−5.9
CdS	−0.2	−5.7	−2.8
InP	0.2	−5.3	−2.3
BaF_2	5.9	0.1	2.8
CdTe	10.7	4.6	7.8
Al_2O_3	14.9	8.7	12.0
SiO_2	18.6	12.1	15.5

Despite the challenges posed by lattice mismatches, graphene has emerged as a preferred substrate for TIs due to its weak van der Waals bonding. This allows TIs to grow despite the large lattice mismatch, leveraging the unique properties of graphene to enable high-quality TI films.

Thin films synthesized via PVD typically exhibit textured surfaces characterized by pyramidal single-crystal domains with quintuple-layer steps. The size and distribution of these domains depend on factors like film thickness, lattice mismatch, and interfacial chemistry. However, achieving precise stoichiometry in thin films can be challenging due to the high vapor pressures of the constituent elements. As a result, binary tetradymite TIs are often extrinsically doped as n-type (Bi_2Se_3, Bi2Te3) or p-type (Sb_2Te_3) to compensate for these issues.

PVD offers a versatile and cost-effective method for growing high-quality topological insulators. Its ability to control critical parameters such as thickness, composition, and crystallographic orientation makes it a valuable tool for exploring the unique electronic properties of TIs. By carefully selecting substrates and employing buffer layers, researchers can overcome challenges related to lattice mismatches and stoichiometry, enabling the development of TIs for advanced electronic, spintronic, and quantum applications.

Topological insulators exemplify the potential of quantum materials, bridging fundamental physics with practical applications. Their unique properties—insulating bulk, conducting surface states, and robustness against perturbations—position them as key enablers in fields ranging from quantum computing to energy-efficient technologies. For advanced material developers, leveraging techniques like PVD and MBE to synthesize high-quality TIs while exploring their applications in spintronics, thermoelectrics, and optoelectronics offers a pathway to groundbreaking innovations. Continued research and development in this field will unlock the full potential of TIs, driving transformative advancements across scientific and industrial domains.

Bio-Inspired and Biomimetic Materials

Bio-inspired and biomimetic materials represent a significant intersection between nature and technology, drawing from the vast array of structures and functions that have evolved over billions of years. These materials are designed to address various engineering, medical, and technological challenges by mimicking biological systems. The potential applications of these materials span multiple fields, including aerospace, robotics, healthcare, and sustainable technology, showcasing their versatility and importance in modern science and engineering.

The distinction between bio-inspired and biomimetic materials is important for understanding their applications. Bio-inspired materials are those that take cues from nature but do not replicate its structures or functions directly. For instance, lightweight honeycomb structures in aircraft design are inspired by the efficiency of beehives, providing strength while minimizing weight. On the other hand, biomimetic materials aim to closely replicate specific functions or processes found in nature. A prime example is artificial spider silk, which emulates the remarkable tensile strength and flexibility of natural spider silk, making it suitable for various applications, including medical sutures and lightweight textiles [866].

Advanced Material Engineering Fundamentals

Natural materials often exhibit hierarchical structures that enhance their properties at multiple scales. For example, bone is a composite material that combines collagen, a soft polymer, with hydroxyapatite, a rigid ceramic. This hierarchical arrangement contributes to the bone's exceptional toughness and strength [867]. Similarly, nacre, or mother-of-pearl, consists of alternating layers of soft organic materials and hard aragonite platelets, resulting in extraordinary toughness [868]. The study of such hierarchical structures informs the design of synthetic materials that can mimic these advantageous properties, leading to innovations in fields such as tissue engineering and materials science [869].

One of the most exciting developments in biomimetic materials is their ability to self-heal, a property inspired by biological systems that can repair themselves after damage. Research has shown that materials can be engineered to detect and autonomously heal damage, similar to how biological tissues respond to injury [870]. For instance, self-healing hydrogels have been developed that utilize dynamic covalent bonds to restore their integrity after being cut or damaged [871]. This capability is particularly valuable in applications where durability and longevity are critical, such as in aerospace components and medical devices [872].

The integration of biomimetic principles into material design also holds promise for sustainable development. By emulating nature's efficient processes, researchers can create materials that are not only functional but also environmentally friendly. For example, the development of biodegradable polymers that mimic natural materials can reduce waste and improve sustainability in manufacturing [873]. Furthermore, the concept of "biomimetic promise" suggests that products inspired by nature can contribute significantly to sustainable practices across various industries [874].

Self-healing materials draw inspiration from the natural ability of skin and plants to repair themselves after injury. These materials integrate mechanisms such as microcapsules or vascular networks that release healing agents upon damage. They find applications in automotive coatings, construction materials, and self-repairing electronics, extending the lifespan and reliability of products by enabling automatic restoration of functionality.

Superhydrophobic materials mimic the lotus leaf, which repels water through a combination of micro- and nanoscale roughness and hydrophobic chemistry. These materials are utilized in self-cleaning surfaces, anti-fouling coatings, and water-repellent fabrics, offering enhanced durability and reduced maintenance in various applications.

Spider silk mimics replicate the extraordinary properties of natural spider silk, which is lightweight, flexible, and stronger than steel by weight. These materials are applied in sutures, bulletproof vests, and strong yet lightweight ropes, showcasing their potential for both medical and industrial uses.

Adhesive materials take inspiration from gecko feet, which leverage van der Waals forces for reversible adhesion. These materials have been developed for use in climbing robots, reusable adhesives, and biomedical devices, offering innovative solutions for applications requiring reliable and reversible bonding.

Structural materials, inspired by nacre's brick-and-mortar structure, combine strength and toughness, making them ideal for tough ceramics used in aerospace, automotive, and protective gear. These materials are designed to provide resilience while minimizing weight, enabling efficient performance under demanding conditions.

Optical materials emulate the nanoscale structures of the Morpho butterfly's wings, which manipulate light to create iridescent effects. These materials are applied in anti-counterfeiting measures, advanced displays, and photonic devices, demonstrating their versatility in both security and optical technologies.

Bio-inspired materials have transformative applications across key industries. In healthcare and biomedicine, bio-inspired hydrogels are used for tissue engineering and wound healing, while biomimetic scaffolds support regenerative medicine. Smart drug delivery systems, modelled after cellular processes, offer targeted and efficient treatments. In aerospace and transportation, lightweight yet strong materials inspired by bone and nacre are used in aircraft and automotive components, while anti-icing coatings mimic the water-repellent properties of penguin feathers to enhance safety and performance.

In robotics, soft robotics inspired by the movement of octopus arms and fish fins enable flexible and adaptive motion, while sensors and actuators mimic the properties of human skin or muscles. The energy and sustainability sector benefits from artificial photosynthesis systems for renewable energy production and bio-inspired catalytic surfaces for efficient fuel cells. Consumer products, such as self-cleaning clothes and windows based on lotus leaf-inspired coatings, as well as wearable technology with flexible, bio-inspired designs, bring the advantages of biomimetic materials into everyday life.

Despite their potential, bio-inspired materials face challenges in development. The complexity of replicating intricate hierarchical structures at a large scale poses significant technical and cost-related hurdles. Material scalability remains a challenge, as producing bio-inspired materials in commercially viable quantities is difficult. Effective interdisciplinary collaboration, requiring expertise across biology, chemistry, physics, and engineering, adds to the complexity of development efforts. Additionally, high research, development, and manufacturing costs limit widespread adoption.

Future advancements in bio-inspired materials focus on several areas. Advanced fabrication techniques, such as additive manufacturing and nanoscale fabrication, enable the replication of complex biological architectures. Integration with smart systems, including sensors and actuators, allows the creation of intelligent, responsive materials. Emphasizing sustainability, bio-inspired materials are being developed with biodegradability and environmental impact in mind. Emerging fields, supported by artificial intelligence tools, aim to accelerate the design and optimization of bio-inspired materials.

Bio-inspired and biomimetic materials represent a convergence of nature and technology, offering innovative solutions to challenges in diverse domains. By emulating the efficiency, adaptability, and resilience of biological systems, these materials hold the potential to revolutionize industries while promoting sustainability. As advancements in fabrication techniques, computational modelling, and interdisciplinary collaboration continue, bio-inspired materials are poised to become central to the next generation of material science innovations.

Biomimetics, the practice of emulating nature's strategies to solve human challenges, has garnered significant attention across various fields due to the intricate and diverse solutions provided by biological systems. This interdisciplinary approach leverages the principles of evolution and natural selection, allowing researchers and engineers to create innovative technologies that are efficient, sustainable, and functional. The applications of biomimetics range from architectural designs to advanced robotics, showcasing its versatility and potential for commercial viability [875, 876].

One notable example of a biomimetic application is Murray's law, which was initially formulated to optimize blood vessel diameters to minimize energy costs associated with blood flow. This principle has been adapted for use in engineering systems, such as fluidic networks, where it serves as a design tool to enhance efficiency and reduce material usage [877, 878]. The law's relevance extends beyond biology, influencing the design of microfluidic devices and other engineering applications that require optimal flow characteristics [879, 880].

In the realm of locomotion, nature-inspired innovations have led to significant advancements in transportation and robotics. The Shinkansen 500 Series high-speed train, for instance, adopts the streamlined design of a kingfisher's beak, enhancing aerodynamics and reducing noise [875]. Similarly, biorobots like the BionicKangaroo replicate the energy-efficient hopping of kangaroos, while Kamigami Robots mimic cockroach locomotion, allowing for rapid movement across various surfaces [875]. Shrimp-inspired robots, such as Pleobot, provide insights into metachronal swimming, demonstrating the ecological advantages of this propulsion method [875]. Furthermore, biomimetic flying robots (BFRs) draw inspiration from birds, bats, and insects, utilizing flapping wings to achieve increased manoeuvrability and energy efficiency [875].

Biomimetic architecture exemplifies the integration of biological principles into building design, emphasizing sustainability and energy efficiency. This approach transcends mere aesthetics, as seen in the Eastgate Centre in Zimbabwe, which employs passive cooling systems inspired by termite mounds, significantly reducing energy consumption compared to conventional buildings [881]. Techniques such as double façades and ventilation systems modelled on termite mounds have improved thermal regulation in buildings, showcasing the potential of biomimetic strategies in architectural design [875, 882].

Moreover, the development of structural materials inspired by biological systems highlights the potential for creating lightweight yet strong materials. Natural structures like bone, nacre, and bamboo exhibit exceptional resistance to fracture due to their hierarchical designs [883]. Innovations such as biomorphic mineralization and freeze casting have enabled the replication of these natural architectures, resulting in high-performance composites suitable for various applications, including construction and energy storage [883].

In optics, biomimetic materials have drawn inspiration from the structural coloration found in organisms like butterflies and plants. The vibrant blue of Morpho butterfly wings has been successfully mimicked in optical technologies, influencing the development of sustainable materials for photonics applications [875]. Similarly, plant-based inspirations have led to advancements in biodegradable materials, showcasing the potential for environmentally friendly innovations [875].

Biomimicry has also made significant strides in adhesion technologies, with tree frogs' toe pads inspiring the creation of superior wet traction surfaces and marine mussels' adhesive proteins guiding the development of strong underwater adhesives [875]. Additionally, gecko-inspired dry adhesion systems have facilitated advancements in climbing robots and bio-inspired tapes, demonstrating the practical applications of biomimetic principles in various fields [875].

In agriculture, holistic grazing practices that mimic natural herd movements have shown promise in restoring grasslands, while biomimicry in air conditioning systems and swarm-based space drones illustrates the versatility of nature-inspired designs [875]. Neuromorphic computing, modelled after biological neurons, offers breakthroughs in efficient computation and sensor technologies, further emphasizing the broad impact of biomimetics [875].

Chapter 11

The Advanced Materials Engineer's Toolkit

Essential Equipment for Material Design and Testing

The design and testing of materials require precise and advanced tools to ensure the reliability, efficiency, and suitability of materials for their intended applications. These tools facilitate the development of new materials, evaluate their performance, and ensure compliance with industry standards.

The development and testing of advanced materials demand a combination of synthesis, characterization, and performance evaluation tools. From molecular-level analysis using electron microscopes to large-scale stress testing, each piece of equipment plays a crucial role in ensuring the material meets its intended design specifications. As material science advances, integrating cutting-edge tools with computational modelling will continue to drive innovation, enabling the creation of high-performance materials for diverse applications.

The following provides an overview of the essential equipment used in material design and testing.

1. Material Synthesis Equipment

Material design begins with synthesizing the desired material, whether metallic, polymeric, ceramic, or composite. Specialized synthesis equipment is vital for creating high-quality materials.

- **Furnaces and Kilns**: Used for high-temperature material synthesis, such as sintering ceramics or producing alloys.

- **Chemical Vapor Deposition (CVD) and Physical Vapor Deposition (PVD) Systems**: Essential for thin-film deposition and coating applications in electronics, optics, and surface engineering.

- **Molecular Beam Epitaxy (MBE)**: Enables atomic-level precision in fabricating semiconductor materials and topological insulators.

- **Additive Manufacturing Systems**: Includes 3D printers capable of fabricating complex structures using metals, polymers, or ceramics.

- **Mixers and Homogenizers**: For uniformly mixing components in composite and polymer material development.

Furnaces and kilns are essential tools for high-temperature material synthesis and can be sourced from reputable suppliers like Carbolite Gero, Nabertherm, and MTI Corporation. Specialized manufacturers also offer custom-designed solutions tailored to specific temperature ranges and atmospheric requirements. The cost of these units varies widely depending on their size, temperature range, and advanced features. Benchtop furnaces, capable of reaching up to 1200°C, typically cost between $3,000 and $10,000 (costs given in US Dollars), while high-temperature furnaces with a range of up to 1800°C are priced from $10,000 to $50,000. Large-scale kilns designed for industrial use can cost anywhere from $50,000 to over $200,000. These systems are predominantly used for processes like sintering ceramics, annealing metals, and performing heat treatments, with materials housed in controlled atmosphere chambers to minimize oxidation and undesired reactions. For smaller-scale or lower-temperature operations, tube furnaces or microwave kilns provide cost-effective alternatives. Induction furnaces are also suitable for applications requiring rapid heating of metallic materials.

Chemical Vapor Deposition (CVD) and Physical Vapor Deposition (PVD) systems are critical for thin-film deposition in fields such as semiconductor manufacturing, optical coatings, and surface engineering. These systems are readily available from manufacturers like Veeco, Denton Vacuum, and Angstrom Engineering, and can be purchased or leased depending on project requirements. CVD systems, which employ chemical reactions to deposit materials, range from $50,000 to $500,000, depending on their size, precision, and customization, such as plasma-enhanced configurations. PVD systems, which utilize physical processes like sputtering or evaporation, typically cost between $30,000 and $250,000. These systems are indispensable for high-precision coating applications. For less demanding use cases, spray pyrolysis or spin coating can serve as affordable and effective alternatives, though these methods lack the precision of CVD and PVD systems.

Molecular Beam Epitaxy (MBE) is a highly specialized technique for atomic-level precision in thin-film fabrication, particularly for semiconductors, quantum materials, and topological insulators. MBE systems are typically sourced from companies like Riber, Veeco, or DCA Instruments and are often custom-designed for specific research or industrial applications. The cost of a complete MBE system ranges from $500,000 to $5 million, depending on the complexity of the design, vacuum quality, and desired precision. This method is unparalleled in its ability to produce atomically precise heterostructures, making it critical for cutting-edge material development. However, for less stringent applications, pulsed laser deposition (PLD) or CVD can provide more cost-effective solutions, albeit without the precision offered by MBE.

Additive manufacturing systems, including 3D printers, are widely utilized for fabricating complex geometries in polymers, metals, and ceramics. These systems can be procured from companies like Stratasys, Formlabs, Ultimaker, and EOS for commercial applications, while industrial-grade metal printers are available from Desktop Metal and GE Additive. Desktop polymer 3D printers are relatively affordable, costing between $500 and $10,000, while industrial-grade polymer or ceramic printers range from $10,000 to $100,000. Metal additive manufacturing systems are significantly more expensive, with prices ranging from $50,000 to $1 million depending on the material capabilities and build size. These printers are extensively used for rapid prototyping and producing high-precision parts, particularly in aerospace and biomedical industries. For simpler or mass-production needs, CNC machining or injection molding can serve as viable alternatives, offering cost-effective solutions for machining or molding existing materials.

Mixers and homogenizers play a critical role in ensuring uniform mixing of components, especially in the development of composites, polymers, and ceramic slurries. Lab-scale mixers and homogenizers can be sourced from brands like IKA, Silverson, and Tetra Pak, while large-scale industrial mixers are available from suppliers such as Ross and Netzsch. Lab mixers are relatively affordable, ranging from $500 to $5,000, while industrial homogenizers, depending on their capacity and complexity, cost between $10,000 and $100,000. These devices are essential for maintaining material consistency and ensuring homogeneity in the final product. For smaller-scale operations, manual mixing or basic magnetic stirrers may suffice. In cases where viscous materials are involved, planetary mixers offer an efficient alternative to high-shear homogenizers, ensuring thorough mixing.

2. Characterization and Analysis Tools

To understand a material's properties, characterization tools provide insight into structural, chemical, mechanical, thermal, and electrical behaviours.

- **Scanning Electron Microscopy (SEM)**: Offers high-resolution imaging of material surfaces to analyse microstructures, grain boundaries, and fracture modes.

- **Transmission Electron Microscopy (TEM)**: Provides atomic-scale imaging and diffraction analysis for understanding crystal structures.

- **X-ray Diffraction (XRD)**: Identifies crystallographic structure, phase composition, and lattice parameters of materials.

- **Fourier Transform Infrared Spectroscopy (FTIR)**: Used to determine molecular bonding and chemical composition.

- **Raman Spectroscopy**: Provides vibrational, rotational, and other low-frequency modes in a material, useful for studying molecular structure and crystallinity.

- **Thermal Analysis Equipment**: Includes differential scanning calorimetry (DSC) and thermogravimetric analysis (TGA) to measure phase transitions, thermal stability, and degradation.

- **Spectrophotometers**: Used for optical property characterization, including reflectance, transmittance, and absorbance of materials.

Scanning Electron Microscopy (SEM) systems are sourced primarily from leading manufacturers like JEOL, Hitachi, Thermo Fisher Scientific (FEI), and Zeiss. These systems are usually acquired directly from suppliers or through authorized distributors, with specialized resellers offering used or refurbished units at reduced costs. The price of SEMs varies significantly based on their capabilities. Basic models designed for research and educational purposes range from $50,000 to $100,000, while advanced systems equipped with energy-dispersive X-ray spectroscopy (EDS) are priced between $100,000 and $500,000. High-resolution field-emission SEMs can cost between $500,000 and $1 million. SEMs are used for high-resolution imaging to analyse material surfaces, microstructures, grain boundaries, and fracture modes. They work by directing an electron beam onto a sample, generating signals that produce detailed images and detect compositional variations. This makes them indispensable in materials science, metallurgy, and nanotechnology. Alternatives for lower-resolution imaging include optical or confocal microscopes, while Atomic Force Microscopy (AFM) offers nanoscale imaging, though it lacks SEM's ability to analyse composition.

Transmission Electron Microscopy (TEM) systems are sourced from manufacturers like JEOL, Thermo Fisher Scientific, and Hitachi. These highly specialized tools are typically purchased by large research labs due to the infrastructure and expertise required for their operation. Basic TEM models cost between $500,000 and $1 million, while high-resolution models for atomic-scale imaging can range from $2 million to $10 million. TEMs provide atomic-scale imaging and diffraction analysis to study crystal structures, defects, and interfaces. Samples must be ultra-thin (on the nanometer scale), requiring meticulous preparation techniques. TEMs are vital for advanced materials research, semiconductor analysis, and nanotechnology. For less detailed imaging, SEMs with transmission capabilities (STEM mode) can serve as cost-effective alternatives. X-ray diffraction (XRD) offers crystallographic information without the need for thin sample preparation.

X-ray Diffraction (XRD) equipment can be sourced from Rigaku, Bruker, Malvern Panalytical, and Thermo Fisher Scientific. These systems are used in both academic research and industrial quality control. Benchtop XRD systems cost between $40,000 and $100,000, while advanced research-grade systems range from $150,000 to $500,000. XRD is used to identify crystallographic structures, phase compositions, and lattice parameters by analysing diffraction patterns generated when X-rays are directed at a material. Alternatives for qualitative phase identification include Raman spectroscopy or Fourier Transform Infrared (FTIR) spectroscopy. Neutron diffraction provides similar functionality but is typically limited to large-scale research facilities.

Fourier Transform Infrared Spectroscopy (FTIR) systems are available from companies like Thermo Fisher Scientific, Bruker, PerkinElmer, and Agilent Technologies. They come in portable and benchtop models commonly used in laboratories. The cost of portable FTIR systems ranges from $10,000 to $30,000, research-grade systems range from $30,000 to $100,000, and specialized systems with microscopy capabilities can exceed $100,000. FTIR is used to determine molecular bonding and chemical composition by analysing infrared light absorption. It is extensively applied in polymer research, chemical analysis, and material characterization. Alternatives include Raman spectroscopy, which provides complementary vibrational data, and ultraviolet-visible (UV-Vis) spectroscopy for simpler compositional analyses.

Raman spectrometers can be sourced from manufacturers such as Horiba, Renishaw, Thermo Fisher Scientific, and WITec. Portable models are increasingly popular for field applications. Portable Raman spectrometers range in cost from $15,000 to $50,000, benchtop systems are priced between $50,000 and $250,000, and high-resolution models with confocal capabilities cost $250,000 to $500,000. Raman spectroscopy identifies vibrational, rotational, and other low-frequency modes, making it particularly valuable for studying molecular structures and crystallinity. It is often used to characterize carbon-based materials like graphene. FTIR spectroscopy provides complementary vibrational data, while XRD is better suited for studying crystal structures.

Thermal analysis equipment, including Differential Scanning Calorimetry (DSC) and Thermogravimetric Analysis (TGA), is available from suppliers like TA Instruments, Netzsch, and PerkinElmer. Basic DSC or TGA systems cost between $10,000 and $50,000, while advanced systems with simultaneous thermal analysis (e.g., DSC-TGA) range from $50,000 to $150,000. DSC measures phase transitions such as melting and glass transition temperatures, while TGA evaluates thermal stability and material degradation. These tools are essential in polymer science, metallurgy, and materials development. Alternatives for specific applications include dynamic mechanical analysis (DMA) or rheometry, and simpler setups like calorimeters can provide basic thermal data.

Spectrophotometers are widely available from Agilent Technologies, Shimadzu, Thermo Fisher Scientific, and PerkinElmer. They are offered in models ranging from UV-Vis-only to UV-Vis-NIR capabilities. Basic UV-Vis spectrophotometers cost $5,000 to $15,000, research-grade systems with extended wavelength ranges are priced between $20,000 and $50,000, and specialized spectrophotometers with integrating spheres can cost $50,000 to $100,000. These instruments measure optical properties such as reflectance, transmittance, and absorbance, making them essential for characterizing thin films, coatings, and transparent materials. For simple colorimetric analyses, handheld colorimeters are an alternative, while ellipsometry is a more advanced method for studying optical properties of thin films.

3. Mechanical Testing Equipment

Materials often need to withstand mechanical stresses in real-world applications. Mechanical testing equipment evaluates their strength, elasticity, and toughness.

- **Universal Testing Machines (UTM)**: Measure tensile, compressive, and flexural properties, such as yield strength, modulus of elasticity, and ultimate tensile strength.

- **Hardness Testers**: Includes Vickers, Brinell, and Rockwell hardness testers to evaluate material resistance to deformation.

- **Impact Testers**: Measure a material's ability to absorb energy under sudden forces (e.g., Charpy or Izod impact tests).

- **Fatigue Testers**: Assess how materials respond to repeated cyclic loading over time.

- **Creep Testers**: Evaluate material deformation under sustained load and temperature conditions.

Universal Testing Machines (UTMs) are sourced from manufacturers such as Instron, ZwickRoell, MTS Systems, and Shimadzu. These machines are widely used in research laboratories, educational institutions, and industries for material testing. UTMs are available through direct purchase from manufacturers, authorized distributors, or second-hand equipment resellers for budget-conscious users.

The cost of a UTM depends on its load capacity, features, and level of automation. Basic models for low-capacity testing (up to 10 kN) typically cost between $5,000 and $20,000. High-capacity machines (up to 1,000 kN) with advanced software capabilities can range from $50,000 to $250,000. Fully automated systems designed for high-throughput testing or specialized applications may exceed $500,000.

UTMs measure tensile, compressive, and flexural properties of materials, such as yield strength, modulus of elasticity, and ultimate tensile strength. They are essential in industries like aerospace, construction, and manufacturing to ensure material compliance with standards. Samples are mounted between grips, and loads are applied under controlled conditions while the machine records stress-strain behavior.

For simpler tensile or compressive testing, portable handheld devices may serve as a cost-effective alternative, although with reduced precision. Dedicated tensile or compression testers can also provide targeted solutions for specific testing needs.

Advanced Material Engineering Fundamentals

Hardness testers, including Vickers, Brinell, and Rockwell models, are commonly sourced from manufacturers such as Mitutoyo, Buehler, Wilson, and Struers. These devices are widely available through industrial equipment suppliers, online retailers, and direct sales channels.

Costs for hardness testers vary by type and sophistication. Basic portable hardness testers can cost as little as $1,000 to $5,000. Benchtop models for Vickers or Rockwell testing typically range from $10,000 to $50,000, while advanced systems with automated sample handling and software integration can exceed $100,000.

These testers evaluate a material's resistance to deformation by applying a defined force and measuring the resulting indentation or rebound. Hardness testing is widely used for quality control in metalworking, construction, and polymer manufacturing.

Portable hardness testers, such as rebound testers, are suitable alternatives for field applications or non-destructive testing, though they may lack the precision of traditional benchtop systems.

Impact testers, such as those used for Charpy or Izod tests, are sourced from manufacturers like Tinius Olsen, Instron, and ZwickRoell. These devices are available through specialized material testing equipment suppliers or directly from the manufacturers.

Costs for impact testers depend on their configuration and testing capacity. Basic pendulum-type impact testers cost between $5,000 and $20,000. High-capacity systems designed for advanced testing, such as instrumented impact testers with data acquisition systems, can range from $50,000 to $150,000.

Impact testers measure a material's ability to absorb energy during sudden impact forces, providing critical insights into toughness and fracture resistance. These tests are essential in evaluating materials for applications such as automotive components, pipelines, and structural materials.

Drop-weight impact testers are an alternative for certain industries, offering a different mode of impact energy application. However, these systems may not provide the same data resolution or repeatability as standard pendulum systems.

Fatigue testers are available from manufacturers such as MTS Systems, Instron, and Shimadzu. These systems are typically acquired through direct purchase or specialized distributors, given their complex requirements.

Basic fatigue testers for small-scale research applications cost between $50,000 and $100,000. Industrial-grade systems for large-scale and high-frequency testing range from $200,000 to over $1 million, depending on their capacity and level of customization.

Fatigue testers simulate repeated cyclic loading to evaluate how materials respond to stress over time. They are widely used in aerospace, automotive, and infrastructure industries to predict material longevity under real-world conditions. Fatigue testing is crucial for components like aircraft wings, engine parts, and bridges.

For smaller-scale or less demanding applications, mechanical or electromechanical oscillators can serve as low-cost alternatives, though they may not provide the same level of precision or data acquisition capabilities.

Creep testers are sourced from manufacturers like ZwickRoell, MTS Systems, and ATS (Applied Test Systems). These systems are typically sold directly by manufacturers or through authorized distributors.

Costs for creep testers vary widely based on their temperature and load range. Basic models start around $20,000, while advanced systems for high-temperature and high-stress applications can exceed $100,000. Custom configurations for unique testing requirements may cost significantly more.

Creep testers evaluate material deformation under sustained loads and elevated temperatures over extended periods. These tests are critical in industries like power generation, petrochemical, and aerospace, where materials are subjected to long-term mechanical stresses and high temperatures.

For simpler or shorter-term testing, basic tensile testing machines equipped with elevated-temperature chambers can act as a cost-effective alternative. However, these setups may lack the precision and monitoring capabilities of dedicated creep testing systems.

4. Thermal and Environmental Testing Systems

Materials often face extreme environmental conditions during use. Thermal and environmental testing systems ensure durability and stability.

- **Environmental Chambers**: Simulate conditions such as humidity, temperature extremes, and corrosive environments to test material endurance.

- **Thermal Conductivity Meters**: Measure how well a material conducts heat.

- **Aging Chambers**: Assess the long-term stability and performance of materials under accelerated aging conditions.

- **Cryogenic Testing Systems**: Evaluate material behaviour at extremely low temperatures.

Environmental chambers, used to simulate conditions such as humidity, temperature extremes, and corrosive environments, are widely available from manufacturers such as Thermo Fisher Scientific, ESPEC, Weiss Technik, and Binder. These systems can be purchased directly from manufacturers, through authorized distributors, or via laboratory and industrial equipment suppliers.

The cost of an environmental chamber varies depending on its size, temperature range, and additional features like humidity or corrosive gas control. Small benchtop models typically range from $5,000 to $15,000, while walk-in chambers or systems with advanced capabilities like altitude simulation can cost $50,000 to $200,000 or more. Custom-designed chambers for specific applications may exceed these ranges.

Environmental chambers are used across industries to test material endurance under conditions that mimic real-world scenarios. For example, they evaluate the reliability of electronics in extreme temperatures, the resistance of coatings to humidity, or the durability of construction materials in corrosive environments. Materials are placed in the chamber, and parameters like temperature and humidity are precisely controlled and monitored during testing.

For simpler applications, basic ovens or humidity cabinets can be used, though they lack the precision and versatility of full-scale environmental chambers. Alternatively, smaller climatic chambers can be rented for specific projects to reduce upfront costs.

Thermal conductivity meters, essential for measuring how well materials conduct heat, are available from suppliers like TA Instruments, C-Therm Technologies, NETZSCH, and Decagon Devices. These devices can be purchased through the manufacturers or specialized material testing equipment distributors.

The cost of a thermal conductivity meter depends on the method used (e.g., transient plane source, guarded hot plate) and the level of precision required. Portable units designed for basic measurements start at around $5,000, while research-grade systems with advanced data analysis capabilities can cost between $20,000 and $100,000.

These devices are widely used in materials research, particularly in fields like insulation design, thermal barrier coatings, and energy systems. Thermal conductivity meters work by applying a heat source to a material and measuring the rate of heat transfer. The data is used to optimize material performance in applications such as building insulation or thermal management in electronics.

For basic evaluations, steady-state methods like using a heat flow meter or improvised setups with thermocouples and a heat source can provide a low-cost alternative, although with reduced accuracy and repeatability.

Aging chambers are used to assess the long-term stability and performance of materials under accelerated aging conditions, such as exposure to heat, UV light, or humidity. These chambers are sourced from manufacturers such as Q-Lab, Atlas Material Testing Solutions, and Heraeus. They are available through direct purchase or authorized distributors.

The price of an aging chamber depends on its size, features, and capabilities. Small, benchtop models for UV aging or heat aging tests start at $5,000 to $10,000. Larger, more advanced chambers with programmable controls and multi-factor testing capabilities can range from $20,000 to $100,000 or more.

Aging chambers are critical in industries like automotive, aerospace, and packaging, where materials must withstand long-term environmental exposure. Common tests include evaluating the colourfastness of textiles, the durability of paints, or the lifespan of plastics under UV exposure. Materials are placed in the chamber, where conditions are controlled to simulate years of aging in a condensed timeframe.

For smaller-scale needs, UV lamps or ovens can provide basic aging simulations, but they may lack the precision and multi-factor control of dedicated aging chambers.

Cryogenic testing systems are used to evaluate material behaviour at extremely low temperatures, often required in industries like aerospace, energy, and scientific research. These systems can be sourced from suppliers like Cryogenic Control Systems, Janis Research, and Advanced Research Systems. They are also available through distributors specializing in cryogenic and material testing equipment.

The cost of cryogenic testing systems varies depending on their temperature range, sample size capacity, and integration with other testing systems. Small-scale cryogenic systems start at $10,000 to $30,000. More advanced systems designed for large-scale or highly precise testing can exceed $100,000. Liquid nitrogen or helium supply systems may add additional costs for operation.

Cryogenic testing is essential for understanding how materials behave at temperatures as low as -196°C or lower, such as in cryogenic storage tanks or space exploration components. Materials are subjected to extreme cold to evaluate mechanical properties, fracture toughness, or thermal expansion under such conditions.

As an alternative, basic cryogenic testing can be performed using liquid nitrogen baths or dry ice setups, though these methods lack the controlled environment and precision of dedicated cryogenic systems. Rental options for cryogenic systems can also provide cost-effective solutions for short-term projects.

5. Electrical and Electronic Testing Tools

For materials used in electronics, sensors, or energy storage, electrical testing is essential.

- **Four-Point Probe Systems**: Measure electrical resistivity or conductivity of thin films and bulk materials.

- **Electrochemical Workstations**: Analyse the performance of materials in batteries, fuel cells, or supercapacitors.

- **Dielectric Testers**: Measure a material's insulating properties, including dielectric strength and permittivity.

Four-point probe systems, used to measure the electrical resistivity or conductivity of thin films and bulk materials, can be sourced from specialized manufacturers such as KeithLink Technology, Signatone, Ossila, and Advanced Instrument Technology (AIT). These systems are available through direct purchase or authorized distributors and are commonly found in material research and semiconductor fabrication facilities.

The cost of four-point probe systems depends on their precision, level of automation, and included software. Basic, manual systems for research applications are priced between $1,000 and $5,000, while advanced automated systems for industrial use can range from $10,000 to $50,000 or more.

Four-point probe systems are widely used in material science, electronics, and semiconductor industries to evaluate the resistivity of materials with high precision. The technique involves placing four equally spaced probes on a sample surface, passing a known current through the outer probes, and measuring the voltage drop across the inner probes. This setup minimizes contact resistance, providing accurate resistivity or sheet resistance values for thin films or bulk samples.

For simpler or less precise applications, two-point probe systems or hand-held resistivity meters may be used. However, these alternatives may introduce inaccuracies due to contact resistance effects. For advanced resistivity profiling in thin films, techniques such as van der Pauw measurements can serve as an alternative.

Electrochemical workstations, essential for analysing the performance of materials in batteries, fuel cells, or supercapacitors, can be sourced from manufacturers like Gamry Instruments, Metrohm Autolab, BioLogic Science Instruments, and CH Instruments. These systems are sold through direct sales, academic resellers, or laboratory equipment suppliers.

The price of an electrochemical workstation depends on its capabilities, such as the number of channels, voltage and current ranges, and compatibility with additional techniques like impedance spectroscopy. Entry-level single-channel systems for basic research cost $5,000 to $10,000. Multi-channel or advanced systems designed for industrial applications can cost $20,000 to $100,000 or more.

These workstations are used in electrochemistry to measure properties like charge/discharge cycles, impedance, and reaction kinetics in materials and devices. For instance, they are employed to test battery electrode materials, evaluate catalyst performance in fuel cells, or optimize electrolyte formulations in supercapacitors. The workstations provide precise control over voltage and current, with real-time data collection and analysis.

Alternatives for simpler or preliminary studies include potentiostats or galvanostats with limited features. For some applications, manual cell setups combined with a multimeter and external power source can provide basic insights, though with reduced precision and data analysis capabilities.

Dielectric testers, used to measure a material's insulating properties, including dielectric strength and permittivity, are available from suppliers such as Agilent (now Keysight Technologies), Hioki, AEMC Instruments, and Tonghui Electronics. They can be purchased through specialized distributors or directly from the manufacturers.

The cost of dielectric testers varies depending on the test range and additional capabilities like high-frequency testing or integration with environmental chambers. Basic bench-top dielectric testers start at $5,000 to $15,000, while high-end systems for advanced materials testing can range from $20,000 to $100,000 or more.

Dielectric testers are critical for characterizing materials in the electrical, aerospace, and energy industries. They evaluate a material's ability to resist electrical breakdown, its permittivity, and its dielectric loss, often under varying temperature and humidity conditions. These properties are essential for materials used in capacitors, insulators, and high-voltage components.

Alternatives for basic dielectric testing include LCR meters, which provide limited measurements of permittivity and capacitance. For high-voltage applications, simple breakdown tests using a high-voltage power supply and electrodes can be conducted, though they lack the precision and breadth of data offered by dedicated dielectric testers.

6. Software and Simulation Tools

In modern material design, computational tools complement experimental techniques by predicting material behaviour and optimizing properties.

- **Finite Element Analysis (FEA)**: Simulates mechanical stresses, thermal loads, and fluid dynamics on materials and structures.

- **Molecular Dynamics (MD) Simulations**: Predict atomic-level interactions and material properties based on molecular structures.

- **Computational Fluid Dynamics (CFD)**: Evaluates material performance in fluid flow applications.

- **Materials Databases**: Provide extensive data on existing materials' properties to aid in comparative analysis and design decisions.

Advanced Material Engineering Fundamentals

Finite Element Analysis (FEA) software, used to simulate mechanical stresses, thermal loads, and fluid dynamics in materials and structures, can be sourced from leading providers such as ANSYS, Abaqus (Dassault Systèmes), COMSOL Multiphysics, and Siemens Simcenter. Licenses for these tools are typically purchased directly from the developers or their authorized resellers, and they may also offer academic licenses at discounted rates for educational institutions.

The cost of FEA software varies significantly depending on its capabilities, licensing model, and intended use. Entry-level licenses for academic or small-scale projects may range from $5,000 to $20,000 annually. Advanced licenses for industrial applications, which include features like nonlinear analysis, thermal coupling, and high-performance computing (HPC) support, can cost $30,000 to $100,000 or more per year.

FEA is used extensively in industries such as aerospace, automotive, civil engineering, and biomedical device design to predict material behaviour under various conditions. For example, it can simulate stress distribution in a beam under load, thermal expansion in engine components, or the vibration characteristics of complex assemblies. Users input the material properties, geometry, and boundary conditions, and the software divides the structure into finite elements, solving the equations governing each element to provide detailed insights into performance and failure points.

For smaller-scale or less resource-intensive applications, free or lower-cost alternatives like OpenFOAM, LISA, or CalculiX can be used, though they may lack the advanced features and user-friendly interfaces of premium software.

Molecular Dynamics (MD) simulations, which predict atomic-level interactions and material properties, are conducted using specialized software such as GROMACS, LAMMPS, NAMD, and AMBER. Open-source tools like GROMACS and LAMMPS are freely available and widely used in academic research, while commercial packages like AMBER and Materials Studio offer advanced features and customer support.

The cost of MD simulation tools depends on whether they are open-source or proprietary. Open-source options like GROMACS are free, while commercial packages can range from $5,000 to $20,000 for annual licenses, depending on the number of users and included features.

MD simulations are invaluable for studying molecular interactions, phase transitions, and thermal properties in materials like polymers, metals, and ceramics. By defining the molecular structure and interactions through force fields, users can model and observe atomic behaviour over time. This method is particularly effective for designing new materials, predicting their properties, and optimizing processes such as self-assembly or crystallization.

For simpler molecular studies, quantum chemistry software like Gaussian or open-source tools such as Quantum ESPRESSO may serve as alternatives, focusing on electronic structure calculations rather than dynamic behaviour.

Computational Fluid Dynamics (CFD) software, used to evaluate material performance in fluid flow applications, is available from providers like ANSYS Fluent, OpenFOAM, STAR-CCM+ (Siemens), and COMSOL Multiphysics. Licenses are purchased directly from developers or authorized resellers, with many offering discounted options for academic institutions.

The cost of CFD software varies widely. Open-source tools like OpenFOAM are free, while commercial licenses range from $5,000 to $50,000 annually, depending on the complexity of the simulations and the level of technical support required. High-end systems for industrial applications with HPC capabilities may exceed $100,000 annually.

CFD simulations are critical in industries like aerospace, automotive, energy, and chemical processing. They model how fluids interact with surfaces, enabling the analysis of heat transfer, drag, turbulence, and material erosion. For instance, CFD can be used to optimize the aerodynamics of a car, analyse cooling in electronic devices, or simulate chemical reactions in a reactor.

For simpler fluid dynamics analyses, hand calculations using empirical correlations or basic flow modelling tools like Autodesk CFD may suffice. These alternatives, however, lack the precision and depth of advanced CFD packages.

Materials databases provide extensive data on the properties of existing materials, aiding in comparative analysis and design decisions. These databases are sourced from providers like MatWeb, ASM Materials Information, Granta MI (now Ansys Granta), and SpringerMaterials. Subscriptions can be purchased directly from the database providers, and many universities and research institutions provide access to their students and staff.

The cost of accessing materials databases varies. Basic access to resources like MatWeb is free for limited data, while premium subscriptions for comprehensive access can range from $1,000 to $10,000 per year. Enterprise-level solutions like Granta MI, which integrate material data into product lifecycle management (PLM) systems, may cost $20,000 or more annually.

Materials databases are essential for engineers and scientists to quickly access data on mechanical, thermal, optical, and chemical properties of metals, polymers, ceramics, and composites. They streamline material selection processes by allowing users to filter and compare materials based on specific criteria, such as yield strength, thermal conductivity, or cost.

For those on tighter budgets, open-access resources like the NIST Material Measurement Laboratory or specific academic publications may provide sufficient data for research purposes. However, these sources may not offer the same breadth or ease of use as commercial databases.

7. Specialized Testing Equipment for Niche Applications

Certain materials require specialized testing methods to meet application-specific criteria.

- **Dynamic Mechanical Analysers (DMA)**: Measures viscoelastic properties of polymers and composites.

- **Acoustic Emission Testing**: Monitors crack propagation and structural integrity using sound waves.

- **Magnetic Property Measurement Systems (MPMS)**: Analyse the magnetic properties of materials, crucial for electronic and data storage applications.

- **Nanoindentation Systems**: Provide high-resolution mechanical property measurements at the nanoscale.

Dynamic Mechanical Analysers (DMA) are specialized instruments used to measure the viscoelastic properties of polymers, composites, and other materials by applying oscillatory forces over a range of temperatures and frequencies. These systems are available from manufacturers like TA Instruments, Netzsch, PerkinElmer, and Anton Paar. Research institutions and industrial facilities typically source DMAs directly from these manufacturers or through authorized distributors.

The cost of a DMA system varies depending on its capabilities, temperature range, and frequency options. Basic systems may start around $30,000 to $50,000, while advanced systems with broader temperature ranges and higher precision can cost between $100,000 and $200,000.

DMA is used extensively in industries like aerospace, automotive, and polymers to analyse mechanical properties such as storage modulus, loss modulus, and damping behaviour. These properties are critical for understanding how materials respond under dynamic loading conditions, including elasticity, stiffness, and energy dissipation. For instance, DMAs are used to characterize the glass transition temperature of polymers, which is crucial for selecting materials for specific temperature ranges.

For simpler or lower-cost alternatives, rheometers with oscillatory modes or basic tensile testers may be used, though they may lack the precision and specific insights provided by DMA. Rheometers, for example, are particularly suitable for studying the viscoelasticity of liquid or semi-solid materials.

Acoustic Emission Testing (AET) is a non-destructive evaluation (NDE) technique used to monitor crack propagation and assess structural integrity by detecting sound waves emitted from within a material under stress. This equipment is sourced from NDE equipment providers like Physical Acoustics (a division of Mistras Group), Vallen Systeme, and Olympus Corporation.

AET systems typically range in cost from $20,000 to $100,000, depending on the number of sensors, sensitivity, and additional features such as real-time analysis software. Portable systems for field inspections are available at the lower end of the cost spectrum, while advanced multi-sensor setups for research or industrial applications are more expensive.

AET is widely used in industries such as civil engineering, aerospace, and power generation to detect and monitor defects in materials and structures like pipelines, pressure vessels, and bridges. By analysing the sound waves generated by crack initiation or growth, AET provides early warnings of potential failures, reducing maintenance costs and improving safety.

For smaller-scale or budget-conscious projects, ultrasonic testing (UT) or vibration analysis can provide some insights into structural integrity, although these methods may not detect internal crack propagation as effectively as AET.

Magnetic Property Measurement Systems (MPMS) are highly sensitive instruments designed to measure the magnetic properties of materials, such as magnetization, susceptibility, and coercivity. These systems are commonly sourced from Quantum Design, a leader in MPMS manufacturing, or other specialized providers such as Lake Shore Cryotronics.

MPMS systems are expensive, with costs typically starting at $500,000 and reaching up to $1 million or more, depending on the sensitivity, temperature range (e.g., cryogenic capability), and features like vector magnetometers or alternating current (AC) susceptibility modules.

These systems are critical in materials research for electronics, spintronics, and data storage applications. For example, MPMS can characterize magnetic nanoparticles used in biomedical applications or analyse the performance of magnetic thin films in memory devices. The system operates by measuring the magnetic moment of a sample as it is subjected to varying magnetic fields and temperatures.

For less demanding applications, vibrating sample magnetometers (VSMs) or superconducting quantum interference devices (SQUIDs) can be used as alternatives. These instruments are less expensive and still provide high-quality magnetic measurements, though they may lack the advanced capabilities of MPMS.

Nanoindentation systems are high-resolution instruments used to measure the mechanical properties of materials at the nanoscale, including hardness, elastic modulus, and creep behaviour. These systems are available from manufacturers like Bruker, Hysitron (a part of Bruker), Keysight Technologies, and CSM Instruments.

The cost of nanoindentation systems ranges from $100,000 to $300,000 for basic models and $300,000 to $1 million for systems with advanced features such as in-situ scanning probe microscopy (SPM) or high-temperature testing capabilities.

Nanoindentation is widely used in research and development for advanced materials, coatings, and thin films. For example, it allows precise measurement of the hardness and modulus of microelectronics components, thin films in photovoltaic cells, and wear-resistant coatings in automotive applications. The system works by pressing a sharp indenter (often diamond) into the material and analysing the force-displacement data to derive mechanical properties.

Alternatives for simpler mechanical testing include microhardness testers, such as Vickers or Knoop hardness testers. While these are less expensive (typically $5,000 to $20,000), they cannot achieve the resolution or nanoscale precision of a nanoindenter. Atomic force microscopy (AFM) can also provide mechanical property measurements at the nanoscale but may not be as versatile as dedicated nanoindentation systems.

8. Safety and Quality Assurance Tools

Ensuring safety and consistency in material production is critical.

- **Non-Destructive Testing (NDT) Equipment**: Includes ultrasonic, radiographic, and eddy current methods to detect internal defects without damaging the material.

- **Quality Control Systems**: Employ precision measurement tools such as coordinate measuring machines (CMMs) for dimensional verification.

Non-Destructive Testing (NDT) equipment is essential for evaluating the structural integrity and detecting internal defects in materials without causing damage. Common NDT methods include ultrasonic testing (UT), radiographic testing (RT), eddy current testing (ECT), magnetic particle inspection (MPI), and dye penetrant inspection (DPI). This equipment can be sourced from leading manufacturers like Olympus, GE Inspection Technologies, Magnaflux, and Zetec. Authorized distributors and resellers also provide NDT solutions, often with options for customized packages.

The cost of NDT equipment varies based on the testing method and complexity. Portable ultrasonic flaw detectors range from $5,000 to $50,000, while advanced phased-array ultrasonic systems can exceed $100,000. Radiographic testing systems, particularly those using X-rays, are significantly more expensive, with prices ranging from $50,000 to $500,000. Eddy current testing devices are available for $10,000 to $100,000, depending on their capabilities and features.

NDT equipment is widely used in industries like aerospace, automotive, construction, and oil and gas. Ultrasonic testing is particularly effective for locating internal flaws in welds, pipes, and composite materials. Radiographic testing is employed to examine dense materials, such as metals, for internal voids or cracks. Eddy current testing is preferred for detecting surface and near-surface defects in conductive materials, making it ideal for aircraft component inspection or detecting corrosion under paint.

For simpler or cost-sensitive applications, alternatives like visual inspection, dye penetrant inspection (for surface cracks), or magnetic particle inspection (for ferromagnetic materials) can be used. While these methods are less expensive, they may lack the depth and precision of advanced NDT techniques.

Coordinate Measuring Machines (CMMs) are vital tools for ensuring dimensional accuracy and precision in manufactured components. These systems use probes to measure the geometry of objects and compare them against design specifications. CMMs can be sourced from top manufacturers like Hexagon, Mitutoyo, Zeiss, and Nikon Metrology. Many suppliers offer tailored solutions, including software integration and automation options.

CMM pricing depends on the type, size, and accuracy of the machine. Entry-level benchtop CMMs cost $30,000 to $70,000, while larger, high-precision bridge or gantry CMMs may cost $100,000 to $500,000. Advanced CMMs with multi-sensor capabilities, such as laser scanning or optical measurement, can exceed $1 million.

CMMs are widely used in industries like aerospace, automotive, and precision manufacturing to verify tolerances, alignments, and dimensions of parts. For instance, they ensure the proper alignment of engine components or validate the precision of medical implants. Probes or sensors mounted on the CMM scan the part, generating a detailed 3D representation for comparison with the CAD model.

For smaller-scale or less demanding applications, alternatives include manual measuring tools like micrometers, callipers, and height gauges. Portable coordinate measuring systems, such as articulated arms or laser trackers, provide flexibility and are more affordable, ranging from $10,000 to $100,000. While these alternatives are less precise than CMMs, they offer adequate accuracy for many quality control tasks in production environments.

Key Software Tools for Analysis and Simulation

In the realm of advanced materials development, computational tools are indispensable for simulating material behaviour, optimizing properties, and expediting the innovation process. This synthesis will explore key software tools used in this field, categorized by their specific applications, advantages, and sourcing options.

Finite Element Analysis (FEA) software is crucial for simulating how materials and structures respond to various physical conditions, including mechanical stress and thermal loads. Prominent FEA tools such as ANSYS, COMSOL Multiphysics, and ABAQUS enable engineers to conduct stress analyses, fatigue testing simulations, and predict failure points in structural designs [884]. The cost of these software packages typically ranges from $5,000 to $50,000 annually, depending on the required modules and functionalities [884]. Sourcing can be done directly from developers or through authorized resellers, while open-source alternatives like Code_Aster and CalculiX provide basic FEA capabilities, albeit with fewer advanced features [884].

Molecular dynamics (MD) simulation tools, including LAMMPS, GROMACS, and Materials Studio, are essential for studying atomic-level interactions and predicting material properties. These tools are particularly valuable for designing polymers, nanomaterials, and biomaterials [885]. Applications include understanding molecular phenomena such as diffusion and phase transitions [885]. Open-source tools like LAMMPS are available for free, while commercial platforms like Materials Studio can cost between $10,000 and $50,000 annually [885]. Sourcing options include direct downloads for open-source software and purchases from developers for commercial tools.

CFD software, such as ANSYS Fluent, OpenFOAM, and Star-CCM+, is utilized to analyse fluid flow and heat transfer, which is critical for evaluating materials in fluid-based applications [886, 887]. These tools are instrumental in simulating material performance in fluid flows and optimizing cooling systems [886, 887]. The cost for commercial CFD tools ranges from $10,000 to $60,000 per year, while open-source options like OpenFOAM are freely available but may require specialized expertise to operate effectively [886, 887]. Sourcing is similar to FEA software, with commercial options available from developers and open-source versions downloadable online.

Materials databases, such as MatWeb, Materials Project, and Granta MI, provide extensive property data for thousands of materials, facilitating comparative analysis and early-stage material screening [888]. Subscription fees for these databases can vary widely, from $500 to $50,000 annually, depending on the breadth of data and access levels [888]. Open-access platforms like the Materials Project offer free resources suitable for academic research [888]. These databases are essential for identifying suitable materials based on specific application needs.

For atomic-level electronic structure analysis, quantum mechanical tools such as VASP, Quantum ESPRESSO, and Gaussian are widely used. These platforms utilize DFT to model electronic interactions, making them vital for studying band structures and charge transport in materials [889]. Commercial licenses for software like VASP can cost between $5,000 and $50,000 annually, while free tools like Quantum ESPRESSO are available for non-commercial projects [889]. Sourcing is typically through developers or academic consortia.

Multiphysics tools like COMSOL Multiphysics and ANSYS Multiphysics allow for the simultaneous analysis of multiple physical phenomena, making them ideal for simulating complex material systems [888]. Applications include optimizing multi-functional composites and analysing thermoelectric materials [888]. Licensing costs for these platforms range from

$10,000 to $60,000 annually [888]. Open-source alternatives like Elmer FEM provide basic multiphysics capabilities for limited applications.

Machine learning tools, such as TensorFlow and Scikit-learn, are increasingly utilized in materials discovery to analyse large datasets and predict material properties [890]. These tools can accelerate material discovery and optimize properties, with open-source frameworks available for free and commercial platforms like Citrine Informatics requiring customized pricing [890]. The integration of machine learning into materials science is proving beneficial for predicting various material properties, including band gaps and dielectric constants [890].

The landscape of software tools for materials analysis and simulation is diverse, catering to various needs in design, testing, and optimization. While commercial platforms can be costly, open-source alternatives and academic licenses provide accessible solutions for researchers and developers. The choice of software ultimately depends on the complexity of the material system, the required precision, and the specific application domain.

Skills and Certifications for Advanced Material Engineers

Advanced material engineers play a crucial role in the development and application of innovative materials across various industries, necessitating a robust set of skills and certifications. This following synthesis outlines the essential skills and certifications required for advanced material engineers.

Key Skills

1. Materials Characterization Techniques: Proficiency in materials characterization is fundamental for advanced material engineers. Techniques such as scanning electron microscopy (SEM), transmission electron microscopy (TEM), X-ray diffraction (XRD), and spectroscopic methods (e.g., Raman and FTIR) are essential for analysing microstructures and phase compositions. These skills are critical for the design and improvement of materials, as highlighted by the need for comprehensive characterization in additive manufacturing processes [891-893].

2. Computational and Simulation Skills: The increasing reliance on computational tools necessitates that engineers are adept in software such as ANSYS, COMSOL, MATLAB, and ABAQUS. Skills in molecular dynamics and quantum mechanics-based tools like LAMMPS and VASP are vital for modelling atomic-scale interactions. The literature emphasizes the importance of computational methods in optimizing material properties and processes [894, 895].

3. Design and Process Optimization: Engineers must possess the ability to design materials with specific properties and optimize manufacturing processes. Understanding additive manufacturing, sintering, and other synthesis techniques is essential for creating innovative solutions. The integration of design principles with manufacturing processes is crucial for advancing material applications in industries such as aerospace and biomedical engineering [896, 897].

4. Mechanical Testing and Analysis: A strong foundation in mechanical testing techniques, including tensile, compressive, and fatigue testing, is vital for understanding material behaviour under various conditions. The ability to interpret test results effectively contributes to enhancing material performance, as noted in studies focusing on the mechanical properties of additively manufactured components [892, 895].

5. Knowledge of Emerging Technologies: Staying updated on advancements in nanomaterials, biomimetic materials, and sustainable alternatives is critical. Engineers must understand the principles and applications of these materials to remain competitive in cutting-edge industries [896, 897].

6. Interdisciplinary Collaboration: Material engineers often collaborate with professionals from diverse fields such as chemistry, physics, and biology. Strong communication and teamwork skills are necessary for effective collaboration on multidisciplinary projects, which is increasingly emphasized in engineering education and practice [898, 899].

7. Project Management and Documentation: Efficient project management skills are essential for meeting deadlines and budgets. Documentation skills are also crucial for writing research papers, patents, and compliance reports, as highlighted in the literature on project management practices in engineering [899, 900].

Essential Certifications

1. Fundamentals of Engineering (FE) and Professional Engineer (PE): These certifications validate an engineer's foundational understanding of engineering principles and ethics, serving as critical milestones in an engineering career [899].

2. Six Sigma and Lean Manufacturing Certifications: Certifications such as Six Sigma Green Belt or Black Belt enhance skills in process optimization and quality control, which are vital in material production [899].

3. NACE Corrosion Technician/Technologist Certification: This certification is particularly valuable for engineers working in corrosive environments, ensuring they are equipped to handle materials in challenging conditions [899].

4. Additive Manufacturing Certifications: Organizations like SME and ASTM International offer certifications that equip engineers with essential skills for 3D printing and advanced manufacturing techniques, which are increasingly relevant in various industries [891, 894].

5. AWS Certified Welding Engineer: For engineers involved in welding, this certification ensures expertise in welding processes and metallurgy, which is crucial for material integrity [899].

6. Advanced Degrees and Specializations: Higher academic qualifications, such as a Master's or Ph.D. in material science or related fields, serve as significant credentials, showcasing specialized knowledge [899].

7. ISO and Quality Management Certifications: Certifications like ISO 9001 or AS9100 ensure engineers understand quality management systems, which are essential in industries such as aerospace and automotive [899].

8. Software-Specific Certifications: Proficiency in software tools is validated through certifications from providers like ANSYS and MATLAB, enhancing an engineer's capabilities in computational analysis [899].

9. OSHA and Safety Certifications: For engineers working in hazardous environments, OSHA certifications ensure compliance with safety standards, which is critical for workplace safety [899].

Certifications validate an engineer's expertise and commitment to professional growth. In competitive fields like material science, these credentials can differentiate candidates when pursuing advanced roles or projects. Employers prioritize certifications that align with industry-specific needs, such as additive manufacturing in aerospace or corrosion resistance in the oil and gas sector [899, 900].

Given the dynamic nature of material science, continuous skill and knowledge updates are essential. Participation in workshops, webinars, and conferences hosted by organizations like ASM International and TMS helps engineers stay informed about the latest trends and technologies [899, 900].

Networking and Professional Organizations

Networking and engagement with professional organizations are vital for advanced materials engineers to stay updated on industry trends, foster collaborations, and advance their careers. Professional organizations serve as platforms for knowledge sharing, skill enhancement, and building connections within the global materials science community. Here's an in-depth exploration of the importance of networking and notable professional organizations for advanced materials engineers.

Networking offers numerous benefits for advanced materials engineers. It facilitates the exchange of ideas, exposes professionals to new technologies, and provides opportunities to collaborate on cutting-edge research and projects. Networking also helps engineers gain insights into emerging market trends and technological advancements, ensuring they remain competitive in a rapidly evolving field. Connections made through professional networks often lead to mentorship opportunities, job referrals, and access to exclusive resources like grants, research funding, and industrial partnerships.

Examples of Professional Organizations:

1. ASM International (The Materials Information Society): ASM International is a premier organization for materials engineers, providing a wealth of resources, including technical journals, conferences, and certification programs. Its global network fosters collaboration among professionals from academia, industry, and government. ASM's events and publications cover diverse topics, from metallurgy and additive manufacturing to biomaterials and failure analysis.

2. The Minerals, Metals & Materials Society (TMS): TMS focuses on fostering innovation in minerals, metals, and materials engineering. The society hosts annual conferences, such as the TMS Annual Meeting & Exhibition, which provide a platform for presenting research and discussing advancements in fields like lightweight materials, energy storage, and nanotechnology.

3. Materials Research Society (MRS): The MRS is an interdisciplinary organization that bridges the gap between materials science, engineering, and applied physics. Its biannual meetings are renowned for showcasing cutting-edge research in areas such as quantum materials, biomaterials, and advanced manufacturing. The MRS also offers webinars, publications, and networking opportunities tailored to both early-career and seasoned professionals.

4. American Ceramic Society (ACerS): For engineers specializing in ceramics, glass, and advanced composites, ACerS offers a community dedicated to promoting research and development in these materials. The organization provides access to journals like the *Journal of the American Ceramic Society* and hosts events focused on ceramic applications in energy, aerospace, and electronics.

5. Society for the Advancement of Material and Process Engineering (SAMPE): SAMPE is dedicated to advanced materials and processes, particularly in aerospace, automotive, and defence sectors. Its events and publications emphasize composite materials, additive manufacturing, and material process advancements, offering valuable resources for engineers in these specialized fields.

6. Institute of Materials, Minerals and Mining (IOM3): IOM3 supports professionals working across the entire lifecycle of materials, from extraction and processing to product development and recycling. Based in the UK, it offers networking events, technical resources, and a professional membership framework for engineers worldwide.

7. American Society of Mechanical Engineers (ASME): Although ASME focuses broadly on mechanical engineering, its Materials Engineering Division provides resources specific to materials research and development. ASME's events and journals often explore topics like failure analysis, fatigue testing, and material durability.

8. The European Materials Research Society (E-MRS): E-MRS promotes materials science and engineering research across Europe, offering conferences and publications that address global challenges in energy, sustainability, and health. Its partnerships with organizations worldwide ensure a broad exchange of ideas and collaboration.

9. International Union of Materials Research Societies (IUMRS): IUMRS connects various regional materials societies, fostering international cooperation in research and education. The organization supports global conferences and initiatives to promote innovation and knowledge exchange.

Networking within professional organizations can be achieved through various channels that foster collaboration, knowledge sharing, and career growth. One of the most effective methods is attending conferences and workshops, which provide engineers with opportunities to present their research, engage in technical sessions, and connect with peers who share similar interests. These events serve as dynamic platforms for discussing advancements in materials science, exploring collaborative opportunities, and staying informed about industry trends.

Virtual events, webinars, and online forums offered by many organizations create additional avenues for networking. These platforms enable professionals to participate in discussions, share insights, and seek advice from a global community, all without the constraints of travel. They also facilitate access to expert-led sessions and industry updates in real time, making them invaluable for staying connected and informed.

Joining special interest groups within larger organizations allows engineers to focus their networking efforts on specific areas of interest, such as nanomaterials, additive manufacturing, or sustainable materials. These groups offer targeted resources, discussions, and events tailored to their chosen field, enhancing the relevance and depth of professional interactions.

Mentorship programs offered by professional societies are another powerful networking tool. These programs pair experienced professionals with early-career engineers, promoting knowledge transfer and guidance. By fostering such relationships, mentorship initiatives not only support skill development but also create lasting connections that benefit both mentors and mentees.

Participating in technical competitions or applying for professional awards can further enhance an engineer's visibility within the community. These opportunities provide a platform to showcase expertise, gain recognition, and expand professional circles. Competitions often attract industry leaders and innovators, creating an environment ripe for networking and collaboration. Together, these channels offer comprehensive pathways for building meaningful professional relationships and advancing careers in materials engineering.

Professional organizations often collaborate across regions, facilitating global exchanges of knowledge and fostering multinational partnerships. For example, initiatives like the International Materials Research Congress (IMRC) bring together experts from around the world to address challenges in materials science. Regional events, such as the Asia-Pacific Materials Conference, provide additional opportunities for localized networking and collaboration.

Networking and professional organizations are indispensable for advanced materials engineers seeking to excel in their careers. By engaging with these organizations, engineers gain access to cutting-edge research, professional development opportunities, and a supportive community of peers. As the field of materials science continues to evolve, the connections and resources provided by these networks will play a critical role in shaping the future of materials innovation and application.

Chapter 12

Career Pathways and Opportunities

Educational Background for Advanced Materials Engineers

The educational background for advanced materials engineers is foundational to their ability to innovate and address complex challenges in material science and engineering. Typically, these professionals hold degrees in materials science, materials engineering, or closely related fields such as chemical engineering, mechanical engineering, physics, or chemistry. Their education equips them with the theoretical knowledge, practical skills, and problem-solving capabilities essential for designing, developing, and optimizing materials for diverse applications.

An undergraduate degree in materials science or engineering provides a broad understanding of the fundamental principles governing material behaviour. This includes coursework in thermodynamics, crystallography, mechanics of materials, electronic properties, and phase transformations. Students are also introduced to material characterization techniques, such as microscopy and spectroscopy, as well as processing methods like casting, forging, and additive manufacturing. Practical lab work and collaborative projects form an integral part of undergraduate education, enabling students to apply theoretical concepts to real-world scenarios.

For those aspiring to specialize further, pursuing a master's degree or Ph.D. in materials science or engineering is often essential. Graduate programs offer advanced coursework and opportunities for in-depth research in specialized areas such as nanomaterials, biomaterials, polymer science, energy materials, or quantum materials. During their graduate studies, engineers typically engage in research projects that allow them to tackle complex material challenges, often in collaboration with academic, industrial, or government research institutions. This level of study also emphasizes developing skills in computational modelling, advanced material synthesis, and cutting-edge characterization techniques.

Additionally, interdisciplinary education is becoming increasingly important in advanced materials engineering. Professionals often benefit from knowledge in related fields such as data science, artificial intelligence, environmental science, and systems engineering. For instance, understanding computational modelling and machine learning can enhance material design, while knowledge of sustainability principles is crucial for creating environmentally friendly materials.

To remain competitive and stay abreast of the latest advancements, materials engineers frequently pursue continuing education opportunities. These include certifications, specialized training programs, and short courses offered by universities, professional organizations, and industry associations. Programs like these enable engineers to refine their expertise in areas such as additive manufacturing, failure analysis, or specific software tools like finite element analysis (FEA) and molecular dynamics (MD).

Industries and Roles for Advanced Materials Professionals

Advanced materials professionals are pivotal in various industries, contributing significantly to innovation, sustainability, and the enhancement of product performance. Their expertise spans multiple sectors, including aerospace and defence, automotive, energy and renewables, healthcare and biomedical, electronics and semiconductor, construction and infrastructure, consumer goods, nanotechnology, advanced manufacturing, and academic and government research. Each of these industries presents unique challenges and opportunities where advanced materials play a crucial role.

In the aerospace and defence sectors, the demand for materials that can endure extreme conditions is paramount. Advanced materials engineers focus on developing lightweight, high-strength composites and thermal protection systems. Their roles include materials scientists and structural engineers who specialize in additive manufacturing, crucial for fabricating components for aircraft and space exploration. Research highlights the importance of composite materials in aerospace applications, emphasizing their ability to provide strategic advantages in design and performance under harsh conditions [901].

The automotive industry is undergoing a transformative shift towards electric vehicles (EVs) and sustainability, which has intensified the need for advanced materials. Professionals in this field are tasked with developing lightweight metals and battery materials that enhance vehicle efficiency and safety. Key roles include battery materials researchers and manufacturing engineers who specialize in composites. Studies indicate that advancements in materials such as high-strength steels and metal matrix composites are essential for meeting the industry's demands for weight reduction and improved safety [902-904]. Furthermore, the integration of digital technologies is reshaping manufacturing processes, enabling more sustainable practices [905].

In the energy and renewables sector, advanced materials are critical for enhancing efficiency and environmental performance. Engineers develop materials for renewable energy applications, such as wind turbine blades and solar panels, as well as for energy storage devices like batteries. The oil and gas industry also benefits from advanced materials through the design of corrosion-resistant alloys for pipelines [906]. Roles in this sector include energy materials scientists and process engineers who focus on optimizing material properties for specific applications [907].

The healthcare and biomedical fields rely heavily on advanced materials for creating medical devices, implants, and drug delivery systems. Biomaterials engineers work on developing biocompatible materials that meet stringent safety and performance standards. Research in this area emphasizes the importance of smart materials and tissue scaffolds in regenerative medicine, showcasing the diverse applications of advanced materials in improving patient outcomes [908].

In the electronics and semiconductor industry, the precision of material properties is crucial for the development of microchips and display technologies. Professionals in this sector focus on creating materials with specific electrical, thermal, and optical characteristics. The emergence of fields like quantum computing further underscores the need for innovative materials, with roles including semiconductor materials scientists and R&D specialists [909].

The construction and infrastructure industry is increasingly adopting advanced materials to enhance the sustainability and durability of buildings. Innovations such as self-healing concrete and high-performance insulating materials are being developed to improve energy efficiency and safety. Professionals in this field include civil materials engineers and sustainability consultants who focus on integrating advanced materials into construction practices [910].

In the consumer goods sector, advanced materials are utilized in various applications, from sports equipment to packaging. Materials engineers design lightweight and durable composites that meet consumer demands for performance and sustainability. Roles in this area include product development engineers and materials analysts who drive innovation in consumer products [904].

The field of nanotechnology and advanced manufacturing is rapidly evolving, with professionals designing materials at the atomic and molecular levels. Applications of nanomaterials span electronics, energy storage, and healthcare, highlighting the versatility of advanced materials. Roles include nanomaterials scientists and process engineers who focus on the manufacturing techniques required to produce these innovative materials [17].

Finally, academic and government research institutions play a vital role in advancing the field of materials science. Researchers explore new material properties and develop theoretical models, contributing to fundamental knowledge and practical applications. Positions in this sector include research scientists and laboratory directors who facilitate cross-disciplinary collaboration [911].

Trends in Employment and Salaries

The demand for advanced materials engineers is experiencing significant growth globally, driven by several interrelated factors including technological advancements, sustainability initiatives, and the necessity for innovative solutions across various industries. This demand is particularly pronounced in sectors such as aerospace, automotive, renewable energy, healthcare, and electronics, where the integration of advanced materials is crucial for innovation and efficiency. The transition towards green energy and sustainability has notably accelerated this demand, especially in renewable energy technologies such as solar panels, wind turbines, and energy storage systems, as well as in the development of lightweight materials for electric vehicles [912, 913].

In North America, particularly the United States and Canada, advanced materials engineers are predominantly employed in the aerospace, defence, and advanced manufacturing sectors. The presence of technological hubs like Silicon Valley further emphasizes the demand for materials engineers in the semiconductor and electronics industries [912]. In Europe, countries such as Germany, France, and the U.K. are at the forefront of employing materials engineers, focusing on innovations in automotive, energy, and aerospace sectors. Germany's strong emphasis on advanced manufacturing provides numerous opportunities in material research and development [912, 914]. Meanwhile, in Asia, nations like China, Japan, and South Korea lead in electronics and semiconductor manufacturing, with India emerging as a significant player in renewable energy and infrastructure materials [912, 914].

The global shift towards sustainability is reshaping the landscape for materials engineers. Governments and industries are increasingly prioritizing the development of recyclable, biodegradable, and energy-efficient materials. This trend is particularly evident in the automotive sector, where lightweight materials for electric vehicles are in high demand, as well as in construction, where sustainable composites are gaining traction [912, 913]. The construction industry, for instance, is witnessing a growing emphasis on the use of green materials, which are essential for sustainable building practices. Research indicates that the lack of available sustainable materials can hinder the progress of green construction projects, highlighting the critical need for skilled professionals in this area [915].

Research and development opportunities for advanced materials engineers are expanding, particularly in fields such as nanotechnology, quantum materials, and biomimetic materials. Countries with robust research funding, including the U.S., Germany, Japan, and China, are leading the way in providing numerous opportunities for advanced materials researchers. The integration of advanced materials into various applications, from biomedical devices to energy-efficient technologies, underscores the importance of innovation in this field [912, 914].

The salary trends for advanced materials engineers exhibit significant variability across different regions, industries, and levels of experience. This analysis synthesizes the available data on salary ranges in North America, Europe, Asia, Australia, and the Middle East, highlighting the factors that influence these trends.

In North America, particularly in the United States, entry-level salaries for advanced materials engineers typically range from $65,000 to $85,000 annually. Those with experience, especially in high-demand sectors such as aerospace, defence, and semiconductors, can command salaries between $100,000 and $150,000, with senior roles potentially exceeding $200,000 per year (Das, 2024). In Canada, salaries are generally lower, with entry-level positions offering CAD 60,000 to 80,000, while experienced professionals can earn CAD 100,000 to 130,000 [916].

In Europe, the salary landscape varies significantly by country. In Germany, entry-level engineers earn between €50,000 and €65,000, while senior professionals can earn upwards of €90,000 [916]. The United Kingdom offers entry-level salaries between £30,000 and £40,000, with experienced engineers in high-demand sectors earning between £50,000 and £80,000 [916]. France and the Nordic countries present comparable salary ranges, with early-career professionals earning €40,000 to €60,000 and experienced engineers exceeding €80,000 [916].

In Asia, salary trends also show considerable variation. In China, entry-level salaries for advanced materials engineers range from ¥100,000 to ¥150,000, while senior engineers in sectors like electronics and renewable energy can earn ¥300,000 or more [916]. In Japan and South Korea, entry-level positions typically offer ¥3.5 million to ¥5 million (JPY) or ₩40 million to ₩60 million (KRW), with senior roles exceeding ¥10 million or ₩100 million [916]. In India, salaries are lower, with entry-level engineers earning ₹5 to 10 lakh annually, and experienced professionals earning ₹15 to 25 lakh, particularly in sectors such as electric vehicles and renewable energy [916].

In Australia, salaries for advanced materials engineers range from AUD 70,000 to 100,000 for entry-level positions, while experienced engineers can earn between AUD 120,000 and 150,000, particularly in mining, energy, and infrastructure sectors [916].

Finally, in the Middle East, particularly in oil-rich regions like Saudi Arabia and the UAE, entry-level engineers earn between $40,000 and $60,000 annually. Experienced professionals in high-demand roles can command salaries between $80,000 and $120,000, often tax-free, depending on the employer and location [916].

Overall, the salary trends for advanced materials engineers are influenced by regional economic conditions, industry demand, and the level of specialization required for various roles. These factors collectively shape the compensation landscape for professionals in this field.

The employment landscape for advanced materials engineers is undergoing significant transformation, driven by various emerging trends that are reshaping job roles, salary structures, and industry demands. This synthesis will explore four key trends: the rise of remote and collaborative roles, the increasing importance of certifications, the growth of salaries in green technologies, and the emergence of startups and innovation hubs.

The shift towards remote work has been accelerated by advancements in simulation software and collaborative platforms, particularly in materials engineering. This trend is especially pronounced in research and development (R&D) and academic sectors, where global collaboration is becoming the norm. The ability to work remotely allows engineers to engage with international teams, facilitating knowledge exchange and innovation across borders. This phenomenon is supported by findings that highlight how remote roles are becoming more prevalent in technical fields, enabling professionals to contribute to projects without geographical constraints [917].

In the competitive job market for materials engineers, certifications in specialized areas such as finite element analysis (FEA), additive manufacturing, and nanotechnology are becoming increasingly valuable. These certifications not only enhance employability but also significantly boost earning potential across various regions. Research indicates that professionals with recognized certifications are more likely to secure higher-paying positions, as employers seek candidates who possess verified skills and knowledge in cutting-edge technologies [918]. This trend underscores the necessity for continuous professional development in a rapidly evolving field.

As the global economy shifts towards sustainability, engineers specializing in green technologies, such as battery technology, hydrogen storage, and lightweight materials, are witnessing substantial salary increases. The transition to renewable energy sources is creating a demand for skilled professionals who can contribute to innovative solutions in these areas. Studies have shown that investments in green technologies not only foster job creation but also lead to higher wages for engineers involved in these sectors [919, 920]. The emphasis on environmental sustainability is reshaping the salary landscape, making expertise in green technologies a lucrative career path.

The proliferation of startups, particularly in advanced materials and related fields, is another significant trend affecting employment and salaries. Startups focusing on areas such as quantum computing, nanotechnology, and biomaterials are not only creating new job opportunities but also offering competitive compensation packages, often including equity options for experienced engineers. The role of innovation hubs and incubators is crucial in supporting these startups, providing them with the necessary resources and networks to thrive

[921]. This ecosystem fosters entrepreneurship and innovation, contributing to a dynamic employment landscape for materials engineers.

Entrepreneurial Opportunities in Material Innovation

The field of material innovation is increasingly recognized as a fertile ground for entrepreneurial opportunities, particularly in light of its critical role in addressing pressing global challenges such as sustainability, energy efficiency, and technological advancement. Entrepreneurs can harness advancements in materials science to create innovative products, enhance existing solutions, and develop disruptive technologies across various sectors, thereby contributing to transformative solutions for contemporary issues [922, 923].

A significant area of opportunity lies in the development of sustainable materials. The mounting global pressure to mitigate environmental impacts has led to a heightened demand for materials that are recyclable, biodegradable, or produced with minimal carbon emissions. Innovations such as bioplastics and biodegradable polymers present viable alternatives to conventional plastics, particularly in sectors like packaging and consumer goods [924, 925]. Furthermore, the development of recyclable composites and green concrete alternatives, such as geopolymer concrete, provides substantial opportunities in the construction industry, aligning with the growing emphasis on eco-friendly practices [924].

Energy materials represent another promising domain for innovation, driven by the global transition towards renewable energy sources. Entrepreneurs can focus on the development of next-generation battery materials, including solid-state electrolytes and lithium-sulphur batteries, which offer enhanced performance and sustainability [922]. Advanced photovoltaic materials, such as perovskites and quantum dot solar panels, are also gaining traction, presenting opportunities to improve the efficiency and affordability of solar energy technologies [923]. Additionally, thermoelectric materials that convert waste heat into electricity hold significant potential for enhancing energy efficiency in industrial processes [922].

The advent of additive manufacturing has revolutionized material innovation, creating new markets for advanced materials. Entrepreneurs can explore high-performance printing feedstocks, including metal powders and biocompatible polymers, tailored to meet specific industrial requirements [923]. The emergence of smart materials, such as shape-memory alloys and self-healing substances, further expands the horizons of manufacturing innovation, particularly in sectors like aerospace and automotive, where lightweight materials are critical [923].

In the healthcare and biomedical fields, substantial entrepreneurial potential exists through the development of advanced biomaterials for medical implants, including bioresorbable polymers and titanium alloys that enhance osseointegration (Shin et al., 2019). Innovations in nanomaterials and smart hydrogels are paving the way for targeted drug delivery systems, while flexible and biocompatible materials are driving advancements in wearable sensors and health monitoring devices [923].

Moreover, the intersection of material innovation with electronics and quantum technologies presents additional avenues for entrepreneurship. The development of materials for next-generation semiconductors, photonic devices, and energy-efficient chips is crucial for advancing electronic technologies [923]. The burgeoning market for quantum materials, such as superconductors and topological insulators, is also opening new frontiers in quantum computing, while flexible electronics designed with stretchable materials are set to transform consumer electronics [923].

Addressing environmental challenges such as water scarcity and pollution further underscores the importance of material innovation. The development of nanomaterials and membranes for water purification and desalination is vital for sustainable development [923]. Additionally, materials designed for CO_2 capture and pollutant degradation are essential for environmental remediation efforts [923].

The demand for multifunctional materials has catalysed the development of smart materials across various industries. Self-healing materials are finding applications in construction and automotive sectors, while phase-change materials are ideal for thermal management in energy storage and wearable devices [923]. Materials that respond to environmental stimuli, such as light and temperature, are also gaining traction in sensor and actuator applications [923].

Entrepreneurs can further capitalize on material innovation by focusing on intellectual property. Securing patents for novel materials allows startups to license their innovations, creating sustainable revenue streams [926]. Collaborations with academic institutions for technology transfer can also facilitate commercialization pathways, enhancing the potential for successful ventures in this dynamic field [926, 927].

While the opportunities in material innovation are vast, challenges such as high research and development costs, scalability, and market adoption remain prevalent. Entrepreneurs can navigate these obstacles by collaborating with academic institutions to access cutting-edge technologies and reduce research expenses [926, 927]. Additionally, leveraging funding opportunities, including government grants and venture capital, can provide essential financial support for emerging startups [926, 927]. By focusing on niche markets, entrepreneurs can build expertise and establish a strong presence in specific industries, ultimately contributing to impactful solutions that address global challenges [926, 927].

References

1. Peng, Z., et al., *Ultrasonic Vibration Cutting Of advanced Aerospace Materials: A critical Review of in-Service Functional Performance.* Journal of Intelligent Manufacturing and Special Equipment, 2024. **5**(1): p. 137-169.
2. Umanath, K., et al., *Effect of Hardness on the Wear Behavior of Hybrid Metal Matrix Composites.* Advanced Materials Research, 2014. **984-985**: p. 536-540.
3. Liao, Z., et al., *State-of-the-Art of Surface Integrity in Machining of Metal Matrix Composites.* International Journal of Machine Tools and Manufacture, 2019. **143**: p. 63-91.
4. Abliz, D., et al., *Curing Methods for Advanced Polymer Composites - A Review.* Polymers and Polymer Composites, 2013. **21**(6): p. 341-348.
5. Özden, Ş., et al., *Ballistic Fracturing of Carbon Nanotubes.* Acs Applied Materials & Interfaces, 2016. **8**(37): p. 24819-24825.
6. Shahzad, F., et al., *Electromagnetic Interference Shielding With 2D Transition Metal Carbides (MXenes).* Science, 2016. **353**(6304): p. 1137-1140.
7. El-Atab, N., et al., *Soft Actuators for Soft Robotic Applications: A Review.* Advanced Intelligent Systems, 2020. **2**(10).
8. Kenry, K., J.C. Yeo, and C.T. Lim, *Emerging Flexible and Wearable Physical Sensing Platforms for Healthcare and Biomedical Applications.* Microsystems & Nanoengineering, 2016. **2**(1).
9. Léonard, P.L.Y., et al., *An Introductory Study of the Sustainability Transition for the Aerospace Manufacturing Industry.* 2024.
10. Abdelkader, M., et al., *Ceramics 3D Printing: A Comprehensive Overview and Applications, With Brief Insights Into Industry and Market.* Ceramics, 2024. **7**(1): p. 68-85.
11. Daminabo, S., et al., *Fused Deposition Modeling-Based Additive Manufacturing (3D Printing): Techniques for Polymer Material Systems.* Materials Today Chemistry, 2020. **16**: p. 100248.
12. Huang, Y., et al., *Biosynthesis of Zinc Oxide Nanomaterials From Plant Extracts and Future Green Prospects: A Topical Review.* Advanced Sustainable Systems, 2021. **5**(6).
13. Smith, J.R., *Effects of Machine Learning Algorithms for Predicting and Optimizing the Properties of New Materials in the United States.* European Journal of Physical Sciences, 2023. **6**(1): p. 23-34.
14. Rafin, S., et al., *Power Electronics Revolutionized: A Comprehensive Analysis of Emerging Wide and Ultrawide Bandgap Devices.* Micromachines, 2023. **14**(11): p. 2045.
15. Pujari, R., et al., *Artificial Neural Network Based Wear and Tribological Analysis of Al 7010 Alloy Reinforced With Nanoparticles of SIC for Aerospace Application.* Journal of Machine and Computing, 2023: p. 446-455.

16. Mehta, A., G. Singh, and H. Vasudev, *Processing of Shape Memory Alloys Research, Applications and Opportunities: A Review.* Physica Scripta, 2024. **99**(6): p. 062006.

17. Fairuz, A.M., et al., *Polymer Composite Manufacturing Using a Pultrusion Process: A Review.* American Journal of Applied Sciences, 2014. **11**(10): p. 1798-1810.

18. Muhammad, A., et al., *Investigate the Effect of Different Kinds of Discontinuous Fibers on the Mechanical Properties of Epoxy Matrix Composite Materials.* Engineering and Technology Journal, 2018. **36**(5A): p. 520-522.

19. Henning, F., et al., *Fast Processing and Continuous Simulation of Automotive Structural Composite Components.* Composites Science and Technology, 2019. **171**: p. 261-279.

20. Butenegro, J.A., et al., *Novel Sustainable Composites Incorporating a Biobased Thermoplastic Matrix and Recycled Aerospace Prepreg Waste: Development and Characterization.* Polymers, 2023. **15**(16): p. 3447.

21. Chu, H., et al., *A Hybrid Invisibility Cloak Based on Integration of Transparent Metasurfaces and Zero-Index Materials.* Light Science & Applications, 2018. **7**(1).

22. Jiang, W., Z.L. Mei, and T.J. Cui, *Effective Medium Theory of Metamaterials and Metasurfaces.* 2021.

23. Chen, H., et al., *Ray-Optics Cloaking Devices for Large Objects in Incoherent Natural Light.* Nature Communications, 2013. **4**(1).

24. Zheng, B., et al., *Experimental Realization of an Extreme-Parameter Omnidirectional Cloak.* Research, 2019. **2019**.

25. Prasad, N.V., et al., *Potential Applications of Metamaterials in Antenna Design, Cloaking Devices, Sensors and Solar Cells: a Comprehensive Review.* Journal of Optoelectronic and Biomedical Materials, 2021. **13**(2): p. 23-31.

26. Raza, M.Q., et al., *Transformation Thermodynamics and Heat Cloaking: A Review.* Journal of Optics, 2016. **18**(4): p. 044002.

27. Nguyen-Tran, H.-D., et al., *Effect of Multiwalled Carbon Nanotubes on the Mechanical Properties of Carbon Fiber-Reinforced Polyamide-6/Polypropylene Composites for Lightweight Automotive Parts.* Materials, 2018. **11**(3): p. 429.

28. Mohammadi, H., et al., *Lightweight Glass Fiber-Reinforced Polymer Composite for Automotive Bumper Applications: A Review.* Polymers, 2022. **15**(1): p. 193.

29. Pian, W., Y. Zhou, and T. Xiao, *A Review of the Feasibility of Aluminum Alloys, Carbon Fiber Composites and Glass Fiber Composites for Vehicle Weight Reduction in the Automotive Industry.* Journal of Physics Conference Series, 2023. **2608**(1): p. 012005.

30. Ibhadode, A.O. and R.S. Ebhojiaye, *A New Lightweight Material for Possible Engine Parts Manufacture.* 2019.

31. Qureshi, T. and A. Al-Tabbaa, *Self-Healing Concrete and Cementitious Materials.* 2020.

32. Ihenketu, C., *Reduction in Dilapidation and High Cost of Maintenance on Buildings Using Self-Healing Concrete for Building Construction.* International Research Journal of Modernization in Engineering Technology and Science, 2024.

33. Li, M. and S. Fan, *Designing Repeatable Self-Healing Into Cementitious Materials.* 2016.

34. Wang, X., et al., *Experimental Study on Cementitious Composites Embedded With Organic Microcapsules*. Materials, 2013. **6**(9): p. 4064-4081.

35. Jaafar, M.F.M., et al., *Enhancement of Autogenous Healing on Pre-Cracked PFA Concrete Using Response Surface Methodology (RSM)*. Key Engineering Materials, 2023. **943**: p. 213-223.

36. Adebola, B.A., et al., *Use of Sustainable Materials in Self-Healing Concrete*. 2020.

37. Li, W., et al., *Recent Advances in Intrinsic Self-Healing Cementitious Materials*. Advanced Materials, 2018. **30**(17).

38. Huseien, G.F., et al., *Smart Bio-Agents-Activated Sustainable Self-Healing Cementitious Materials: An All-Inclusive Overview on Progress, Benefits and Challenges*. Sustainability, 2022. **14**(4): p. 1980.

39. Litina, C. and A. Al-Tabbaa, *Development of Sustainable Concrete Repair Materials via Microencapsulated Agents*. Matec Web of Conferences, 2019. **289**: p. 11002.

40. Beglarigale, A., et al., *Sodium Silicate/Polyurethane Microcapsules Synthesized for Enhancing Self-Healing Ability of Cementitious Materials: Optimization of Stirring Speeds and Evaluation of Self-Healing Efficiency*. Journal of Building Engineering, 2021. **39**: p. 102279.

41. Rosario D, R. and M.J. Viado, *Encapsulating Immobilized Ureolytic Bacteria Yields Self-Healing Concrete Apropos Sustainable Transportation Materials: A Review*. E3s Web of Conferences, 2024. **488**: p. 03019.

42. Joachim Osheyor Gidiagba, N., et al., *Economic Impacts and Innovations in Materials Science: A Holistic Exploration of Nanotechnology and Advanced Materials*. Engineering Science & Technology Journal, 2023. **4**(3): p. 84-100.

43. Kennedy, A.J., et al., *A Definition and Categorization System for Advanced Materials: The Foundation for Risk-Informed Environmental Health and Safety Testing*. Risk Analysis, 2019. **39**(8): p. 1783-1795.

44. Patel, P., *Materials Genome Initiative and Energy*. Mrs Bulletin, 2011. **36**(12): p. 964-966.

45. Farcal, L., et al., *Advanced Materials Foresight: Research and Innovation Indicators Related to Advanced and Smart Nanomaterials*. F1000research, 2023. **11**: p. 1532.

46. Bras, R.L., et al., *Challenges in Materials Discovery – Synthetic Generator and Real Datasets*. Proceedings of the Aaai Conference on Artificial Intelligence, 2014. **28**(1).

47. Rambaran, T.F. and R. Schirhagl, *Nanotechnology From Lab to Industry – A Look at Current Trends*. Nanoscale Advances, 2022. **4**(18): p. 3664-3675.

48. Sharma, A.K., et al., *Exploring the Future of Advanced Materials Processing: Innovations and Challenges Ahead: A Review*. E3s Web of Conferences, 2024. **505**: p. 01021.

49. Zamathula Queen Sikhakhane Nwokediegwu, N., et al., *Advanced Materials for Sustainable Construction: A Review of Innovations and Environmental Benefits*. Engineering Science & Technology Journal, 2024. **5**(1): p. 201-218.

50. Lan, G., et al., *Sustainable Carbon Materials Toward Emerging Applications*. Small Methods, 2021. **5**(5).

51. Ikram, M., et al., *Advanced Carbon Functional Materials for Superior Energy Storage*. 2020.

52. Ribeiro, L.N.d.M., et al., *Advances in Hybrid Polymer-Based Materials for Sustained Drug Release*. International Journal of Polymer Science, 2017. **2017**: p. 1-16.

53. Gupta, T.K., et al., *Advances in Carbon Based Nanomaterials for Bio-Medical Applications*. Current Medicinal Chemistry, 2019. **26**(38): p. 6851-6877.

54. Rijkers, B., et al., *Which Firms Create the Most Jobs in Developing Countries? Evidence From Tunisia*. Labour Economics, 2014. **31**: p. 84-102.

55. Esaku, S., *Job Creation, Job Destruction and Reallocation in Sub-Saharan Africa: Firm-Level Evidence From Kenyan Manufacturing Sector*. Cogent Economics & Finance, 2020. **8**(1): p. 1782113.

56. Han, K., *Characterization and Technology of Nanomaterials*. 2016.

57. Sun, J., et al., *Printable Nanomaterials for the Fabrication of High-Performance Supercapacitors*. Nanomaterials, 2018. **8**(7): p. 528.

58. Kim, B., et al., *Recent Application of Nanomaterials to Overcome Technological Challenges of Microbial Electrolysis Cells*. Nanomaterials, 2022. **12**(8): p. 1316.

59. Goryaeva, A.M., et al., *Reinforcing Materials Modelling by Encoding the Structures of Defects in Crystalline Solids Into Distortion Scores*. Nature Communications, 2020. **11**(1).

60. Kostić, M., *The Elusive Nature of Entropy and Its Physical Meaning*. Entropy, 2014. **16**(2): p. 953-967.

61. Baskin, I.I. and Y. Ein-Eli, *Electrochemoinformatics as an Emerging Scientific Field for Designing Materials and Electrochemical Energy Storage and Conversion Devices— An Application in Battery Science and Technology*. Advanced Energy Materials, 2022. **12**(48).

62. Agrawal, A. and A. Choudhary, *Perspective: Materials Informatics and Big Data: Realization of the "Fourth Paradigm" of Science in Materials Science*. Apl Materials, 2016. **4**(5).

63. Song, X., et al., *Modelling of Phase Stability: Integrating Computational Materials Science and Materials Informatics*. 2021.

64. Pablo, J., et al., *New Frontiers for the Materials Genome Initiative*. NPJ Computational Materials, 2019. **5**(1).

65. Ramakrishna, S., et al., *Materials Informatics*. Journal of Intelligent Manufacturing, 2018. **30**(6): p. 2307-2326.

66. Сироткин, О.С., Р.О. Сироткин, and M.Y. Perukhin, *System Analysis and Control of the Influence of a Mixed Type of Chemical Bond in Substances and Materials on Their Structure and Properties*. Key Engineering Materials, 2021. **887**: p. 551-556.

67. Cai, J., et al., *Atomically Precise Bottom-Up Fabrication of Graphene Nanoribbons*. Nature, 2010. **466**(7305): p. 470-473.

68. Kharlamova, M.V., M.I. Paukov, and M.G. Burdanova, *Nanotube Functionalization: Investigation, Methods and Demonstrated Applications*. Materials, 2022. **15**(15): p. 5386.

69. Wang, Y. and W. Zhou, *A Review on Inorganic Nanostructure Self-Assembly*. Journal of Nanoscience and Nanotechnology, 2010. **10**(3): p. 1563-1583.

70. Trixler, F., *Quantum Tunnelling to the Origin and Evolution of Life*. Current Organic Chemistry, 2013. **17**(16): p. 1758-1770.

71. Sloan, P., *Time-Resolved Scanning Tunnelling Microscopy for Molecular Science.* Journal of Physics Condensed Matter, 2010. **22**(26): p. 264001.

72. Miyake, S., M. Wang, and J. Kim, *Silicon Nanofabrication by Atomic Force Microscopy-Based Mechanical Processing.* Journal of Nanotechnology, 2014. **2014**: p. 1-19.

73. Świadkowski, B.M., et al., *Near-Zero Contact Force Atomic Force Microscopy Investigations Using Active Electromagnetic Cantilevers.* Nanotechnology, 2020. **31**(42): p. 425706.

74. Gao, C., *Spin Detection and Manipulation With Scanning Tunneling Microscopy.* Chinese Physics B, 2018. **27**(10): p. 106701.

75. Seyler, H., et al., *Hexa-Peri-Hexabenzocoronene in Organic Electronics.* Pure and Applied Chemistry, 2012. **84**(4): p. 1047-1067.

76. Hosseinian, A., et al., *The Effect of Electric Field and Al Doping on the Sensitivity of Hexa-peri-hexabenzocoronene Nanographene to Chloropicrin.* Applied Organometallic Chemistry, 2018. **32**(10).

77. Yamaguchi, R., et al., *Functionalization of Hexa-peri-hexabenzocoronenes: Investigation of the Substituent Effects on a Superbenzene.* Chemistry - An Asian Journal, 2012. **8**(1): p. 178-190.

78. Beyer, P., et al., *Lattice Matching as the Determining Factor for Molecular Tilt and Multilayer Growth Mode of the Nanographene Hexa-peri-Hexabenzocoronene.* Acs Applied Materials & Interfaces, 2014. **6**(23): p. 21484-21493.

79. Zhang, X., et al., *Synthesis, Self-Assembly, and Charge Transporting Property of Contorted Tetrabenzocoronenes.* The Journal of Organic Chemistry, 2010. **75**(23): p. 8069-8077.

80. Smith, J.N., J.M. Hook, and N.T. Lucas, *Superphenylphosphines: Nanographene-Based Ligands That Control Coordination Geometry and Drive Supramolecular Assembly.* Journal of the American Chemical Society, 2018. **140**(3): p. 1131-1141.

81. Kumar, S., et al., *Polysubstituted Hexa-Cata-Hexabenzocoronenes: Syntheses, Characterization, and Their Potential as Semiconducting Materials in Transistor Applications.* The Journal of Organic Chemistry, 2019. **84**(13): p. 8562-8570.

82. Kang, M.-S., et al., *Synthesis and Characterization of a Contorted Hexabenzocoronene Epoxy Toward High Thermal Stability and Thermal Conductivity.* Bulletin of the Korean Chemical Society, 2023. **44**(7): p. 558-564.

83. Breuer, T., et al., *Self-Assembly of Partially Fluorinated Hexabenzocoronene Derivatives in the Solid State.* Physical Chemistry Chemical Physics, 2016. **18**(48): p. 33344-33350.

84. Rocha-Ortiz, J.S., et al., *Enhancing Planar Inverted Perovskite Solar Cells With Innovative Dumbbell-Shaped HTMs: A Study of Hexabenzocoronene and Pyrene-BODIPY-Triarylamine Derivatives.* Advanced Functional Materials, 2023. **33**(44).

85. Peters, R., et al., *Investigation of the Influence of Hexabenzocoronene in Polyacrylonitrile-Based Precursors for Carbon Fibers.* Fibers, 2023. **11**(2): p. 14.

86. Nolan, D., et al., *Structure–Property Relationships and $^1O_2$ Photosensitisation in Sterically Encumbered Diimine Pt^{II} Acetylide Complexes.* Chemistry - A European Journal, 2013. **19**(46): p. 15615-15626.

87. Zhang, B., et al., *Azepine-Embedded Seco-Hexabenzocoronene-Based Helix Nanographenes: Access to Modification of the Core by N–H Functionalization.* Organic Letters, 2023. **25**(5): p. 732-737.

88. Kang, Z., L. Yang, and S.T. Lee, *Small-Sized Silicon Nanoparticles: New Nanolights and Nanocatalysts.* Nanoscale, 2011. **3**(3): p. 777-791.

89. Zhou, J., Y. Yang, and C. Zhang, *Toward Biocompatible Semiconductor Quantum Dots: From Biosynthesis and Bioconjugation to Biomedical Application.* Chemical Reviews, 2015. **115**(21): p. 11669-11717.

90. Yan, Y., et al., *Recent Advances on Graphene Quantum Dots: From Chemistry and Physics to Applications.* Advanced Materials, 2019. **31**(21).

91. Feng, H. and Z. Qian, *Functional Carbon Quantum Dots: A Versatile Platform for Chemosensing and Biosensing.* The Chemical Record, 2017. **18**(5): p. 491-505.

92. Wang, Y. and A. Hu, *Carbon Quantum Dots: Synthesis, Properties and Applications.* Journal of Materials Chemistry C, 2014. **2**(34): p. 6921.

93. Majood, M., et al., *Carbon Quantum Dots for Stem Cell Imaging and Deciding the Fate of Stem Cell Differentiation.* Acs Omega, 2022. **7**(33): p. 28685-28693.

94. Kausar, A., *Polymer Dots and Derived Hybrid Nanomaterials: A Review.* Journal of Plastic Film & Sheeting, 2021. **37**(4): p. 510-528.

95. Xie, S., et al., *Non-Precious Electrocatalysts for the Hydrogen Evolution Reaction.* 2024. **1**(2): p. 11.

96. Feidenhans'l, A.A., et al., *Precious Metal Free Hydrogen Evolution Catalyst Design and Application.* Chemical Reviews, 2024. **124**(9): p. 5617-5667.

97. Peng, X., et al., *Recent Progress of Transition Metal Nitrides for Efficient Electrocatalytic Water Splitting.* Sustainable Energy & Fuels, 2019. **3**(2): p. 366-381.

98. Shi, L., et al., *Carbon Nanotubes-Promoted Co–B Catalysts for Rapid Hydrogen Generation via NaBH4 Hydrolysis.* International Journal of Hydrogen Energy, 2019. **44**(36): p. 19868-19877.

99. Pak, Y.S., et al., *Defect Formation and Ambivalent Effects on Electrochemical Performance in Layered Sodium Titanate Na₂Ti₃O₇.* Physical Chemistry Chemical Physics, 2023. **25**(4): p. 3420-3431.

100. Rim, C.-H., et al., *Point Defects and Their Impact on Electrochemical Performance in Na_{0.44}MnO₂ for Sodium-Ion Battery Cathode Application.* Physical Chemistry Chemical Physics, 2022. **24**(37): p. 22736-22745.

101. Dedon, L.R., et al., *Nonstoichiometry, Structure, and Properties of BiFeO₃ Films.* Chemistry of Materials, 2016. **28**(16): p. 5952-5961.

102. Hoang, K. and M.D. Johannes, *Defect Physics in Complex Energy Materials.* Journal of Physics Condensed Matter, 2018. **30**(29): p. 293001.

103. Ertekin, E., et al., *Interplay Between Intrinsic Defects, Doping, and Free Carrier Concentration in SrTiO<mml:math xmlns:mml="http://www.w3.org/1998/Math/MathML" Display="inline"><mml:msub><mml:mrow /><mml:mn>3</Mml:mn></Mml:msub></Mml:math>thin Films.* Physical Review B, 2012. **85**(19).

104. Hoang, K. and M.D. Johannes, *Tailoring Native Defects in LiFePO$_4$: Insights From First-Principles Calculations.* Chemistry of Materials, 2011. **23**(11): p. 3003-3013.

105. Bragança, A.M., et al., *How Does Chemisorption Impact Physisorption? Molecular View of Defect Incorporation and Perturbation of Two-Dimensional Self-Assembly.* The Journal of Physical Chemistry C, 2018. **122**(42): p. 24046-24054.

106. Li, X., X. Tang, and Y.-F. Guo, *The Structure Evolution of Titanium–vacancy Complex in a Vanadium-Based Alloy.* Journal of Materials Science, 2020. **56**(6): p. 4433-4445.

107. Kou, L., et al., *Tunable Magnetism in Strained Graphene With Topological Line Defect.* Acs Nano, 2011. **5**(2): p. 1012-1017.

108. Clayton, J.D., *Defects in Nonlinear Elastic Crystals: Differential Geometry, Finite Kinematics, and Second-order Analytical Solutions.* Zamm - Journal of Applied Mathematics and Mechanics / Zeitschrift Für Angewandte Mathematik Und Mechanik, 2013. **95**(5): p. 476-510.

109. Zandiatashbar, A., et al., *Effect of Defects on the Intrinsic Strength and Stiffness of Graphene.* Nature Communications, 2014. **5**(1).

110. Li, M., et al., *Effect of Defects on the Mechanical and Thermal Properties of Graphene.* Nanomaterials, 2019. **9**(3): p. 347.

111. Kähärä, T. and P. Koskinen, *Rippling of Two-Dimensional Materials by Line Defects.* Physical Review B, 2020. **102**(7).

112. Hoogenboom, J.P., et al., *Stacking Faults in Colloidal Crystals Grown by Sedimentation.* The Journal of Chemical Physics, 2002. **117**(24): p. 11320-11328.

113. Spannuth, M., et al., *Stress Transmission Through Three-Dimensional Granular Crystals With Stacking Faults.* Granular Matter, 2004. **6**(4): p. 215-219.

114. Olsson, E., J. Cottom, and Q. Cai, *Defects in Hard Carbon: Where Are They Located and How Does the Location Affect Alkaline Metal Storage?* Small, 2021. **17**(18).

115. Luo, Y.-Y., et al., *Real Time Observation of Partial Dislocations in Thin Colloidal Crystals.* Applied Physics Letters, 2009. **95**(17).

116. Zhao, Q., P. Wei, and Q. Tang, *The Transmission Properties of Elastic Waves Through Multilayers of Spheres With Planar Defects.* Acta Mechanica, 2015. **227**(2): p. 321-331.

117. Waldmann, T., et al., *The Role of Surface Defects in Large Organic Molecule Adsorption: Substrate Configuration Effects.* Physical Chemistry Chemical Physics, 2012. **14**(30): p. 10726.

118. Seuba, J., et al., *Mechanical Properties and Failure Behavior of Unidirectional Porous Ceramics.* Scientific Reports, 2016. **6**(1).

119. Banhart, J. and H.-W. Seeliger, *Recent Trends in Aluminum Foam Sandwich Technology.* Advanced Engineering Materials, 2012. **14**(12): p. 1082-1087.

120. Miyake, K., et al., *The Effect of Particle Shape on Sintering Behavior and Compressive Strength of Porous Alumina.* Materials, 2018. **11**(7): p. 1137.

121. Higaeg, M., et al., *Numerical Modeling of the Porosity Influence on Strength of Structural Materials.* Science of Sintering, 2019. **51**(4): p. 459-467.

122. Moskovic, R., et al., *Understanding Fracture Behaviour of PGA Reactor Core Graphite: Perspective.* Materials Science and Technology, 2014. **30**(2): p. 129-145.

123. Li, D., et al., *Effect of Pore Defects on Mechanical Properties of Graphene Reinforced Aluminum Nanocomposites*. Metals, 2020. **10**(4): p. 468.
124. Feng, R., et al., *Micromechanism of Cold Deformation of Two-Phase Polycrystalline Ti–Al Alloy With Void*. Materials, 2019. **12**(1): p. 184.
125. Dong, Y., C.A. Wang, and J. Zhou, *Effect of YSZ Fiber Addition on Microstructure and Properties of Porous YSZ Ceramics*. Journal of Materials Science, 2012. **47**(17): p. 6326-6332.
126. An, B. and D. Zhang, *Bioinspired Toughening Mechanism: Lesson From Dentin*. Bioinspiration & Biomimetics, 2015. **10**(4): p. 046010.
127. Li, K., Y. Bu, and H. Wang, *Advances on in Situ TEM Mechanical Testing Techniques: A Retrospective and Perspective View*. Frontiers in Materials, 2023. **10**.
128. Meng, X., et al., *In Situ Characterization for Understanding the Degradation in Perovskite Solar Cells*. Solar RRL, 2022. **6**(7).
129. Shen, C., et al., *Development of in Situ Characterization Techniques in Molecular Beam Epitaxy*. Journal of Semiconductors, 2024. **45**(3): p. 031301.
130. Wu, S.P. and Y. Sun, *In Situ Techniques for Probing Kinetics and Mechanism of Hollowing Nanostructures Through Direct Chemical Transformations*. Small Methods, 2018. **2**(11).
131. Agrawal, M., et al., *Innovative Advances and Prospects in in Situ Materials Testing: A Comprehensive Review*. E3s Web of Conferences, 2024. **505**: p. 01031.
132. Petersen, H. and C. Weidenthaler, *A Review of Recent Developments for the <i>in Situ</I>/<i>operando</I> Characterization of Nanoporous Materials*. Inorganic Chemistry Frontiers, 2022. **9**(16): p. 4244-4271.
133. Nejad, A.F., et al., *Hybrid and Synthetic FRP Composites Under Different Strain Rates: A Review*. Polymers, 2021. **13**(19): p. 3400.
134. Hagnell, M.K., et al., *From Aviation to Automotive - A Study on Material Selection and Its Implication on Cost and Weight Efficient Structural Composite and Sandwich Designs*. Heliyon, 2020. **6**(3): p. e03716.
135. Mlýnek, J., et al., *Fabrication of High-Quality Polymer Composite Frame by a New Method of Fiber Winding Process*. Polymers, 2020. **12**(5): p. 1037.
136. Aamir, M., et al., *Recent Advances in Drilling of Carbon Fiber–reinforced Polymers for Aerospace Applications: A Review*. The International Journal of Advanced Manufacturing Technology, 2019. **105**(5-6): p. 2289-2308.
137. Spasenović, J. and I. Blagojević, *Composite Materials in Automotive Industry: A Review*. Industrija, 2021. **49**(2): p. 57-68.
138. Di Trani, N., et al., *Probing Physicochemical Performances of 3D Printed Carbon Fiber Composites During 8-Month Exposure to Space Environment*. Advanced Functional Materials, 2023. **34**(13).
139. Kumar, V., et al., *MXene Reinforced Thermosetting Composite for Lightning Strike Protection of Carbon Fiber Reinforced Polymer*. Advanced Materials Interfaces, 2021. **8**(17).
140. Nagumo, Y., et al., *Fracture Mechanism of Carbon Fiber-Reinforced Thermoplastic Composite Laminates Under Compression After Impact*. Journal of Composite Materials, 2024. **58**(11): p. 1377-1390.

141. Kumar, S. and K.K. Singh, *Tribological Behaviour of Fibre-Reinforced Thermoset Polymer Composites: A Review*. Proceedings of the Institution of Mechanical Engineers Part L Journal of Materials Design and Applications, 2020. **234**(11): p. 1439-1449.

142. Rajak, D.K., et al., *Fiber-Reinforced Polymer Composites: Manufacturing, Properties, and Applications*. Polymers, 2019. **11**(10): p. 1667.

143. Guo, F., et al., *Durability of Fibre Reinforced Polymers in Exposure to Dual Environment of Seawater Sea Sand Concrete and Seawater*. Materials, 2022. **15**(14): p. 4967.

144. Gupta, M.K., V. Singhal, and N.S. Rajput, *Applications and Challenges of Carbon-Fibres Reinforced Composites: A Review*. Evergreen, 2022. **9**(3): p. 682-693.

145. Deng, J., et al., *Fatigue Behaviour of CFRP Bar-Reinforced Seawater Sea Sand Concrete Beams: Deformation Analysis and Prediction*. Buildings, 2023. **13**(9): p. 2273.

146. Singh, M., et al., *Design and Analysis of an Automobile Disc Brake Rotor by Using Hybrid Aluminium Metal Matrix Composite for High Reliability*. Journal of Composites Science, 2023. **7**(6): p. 244.

147. Lajis, M.A., et al., *Mechanical Properties of Recycled Aluminium Chip Reinforced With Alumina (Al_2O_3) Particle*. Materialwissenschaft Und Werkstofftechnik, 2017. **48**(3-4): p. 306-310.

148. Gxowa, Z., et al., *Reinforcement of 2124 Al Alloy With Low Micron SiC and Nano Al_2O_3 Via Solid-State Forming*. Materials Science Forum, 2015. **828-829**: p. 172-178.

149. James, J., et al., *Effect of Wettability and Uniform Distribution of Reinforcement Particle on Mechanical Property (Tensile) in Aluminum Metal Matrix Composite—A Review*. Nanomaterials, 2021. **11**(9): p. 2230.

150. Bodunrin, M.O., K.K. Alaneme, and L.H. Chown, *Aluminium Matrix Hybrid Composites: A Review of Reinforcement Philosophies; Mechanical, Corrosion and Tribological Characteristics*. Journal of Materials Research and Technology, 2015. **4**(4): p. 434-445.

151. Daoud, A., *Microstructure and Tensile Properties of 2014 Al Alloy Reinforced With Continuous Carbon Fibers Manufactured by Gas Pressure Infiltration*. Materials Science and Engineering A, 2005. **391**(1-2): p. 114-120.

152. Wang, C.Y., et al., *The Improvement of Corrosion Resistant for the Cf/Al Composites by Ni-P Coatings*. Key Engineering Materials, 2007. **353-358**: p. 1675-1678.

153. Hajjari, E., M. Divandari, and H. Arabi, *Effect of Applied Pressure and Nickel Coating on Microstructural Development in Continuous Carbon Fiber-Reinforced Aluminum Composites Fabricated by Squeeze Casting*. Materials and Manufacturing Processes, 2011. **26**(4): p. 599-603.

154. Promakhov, V., et al., *High-Temperature Synthesis of Metal–Matrix Composites (Ni-Ti)-TiB2*. Applied Sciences, 2021. **11**(5): p. 2426.

155. Naji, H., S.M. Zebarjad, and S.A. Sajjadi, *The Effects of Volume Percent and Aspect Ratio of Carbon Fiber on Fracture Toughness of Reinforced Aluminum Matrix Composites*. Materials Science and Engineering A, 2008. **486**(1-2): p. 413-420.

156. Xing, C., et al., *Numerical Simulation on Thermal Stresses and Solidification Microstructure for Making Fiber-Reinforced Aluminum Matrix Composites.* Materials, 2022. **15**(12): p. 4166.

157. Chang, K.C., et al., *Influence of Fiber Surface Structure on Interfacial Structure Between Fiber and Matrix in Vapor Grown Carbon Fiber Reinforced Aluminum Matrix Composites.* Materials Transactions, 2009. **50**(6): p. 1510-1518.

158. Galyshev, S., *On the Strength of the CF/Al-Wire Depending on the Fabrication Process Parameters: Melt Temperature, Time, Ultrasonic Power, and Thickness of Carbon Fiber Coating.* Metals, 2021. **11**(7): p. 1006.

159. Li, S.-H. and C.G. Chao, *Effects of Carbon Fiber/Al Interface on Mechanical Properties of Carbon-Fiber-Reinforced Aluminum-Matrix Composites.* Metallurgical and Materials Transactions A, 2004. **35**(7): p. 2153-2160.

160. Veličković, S., et al., *Tribological Characteristics of Al/SiC/Gr Hybrid Composites.* Matec Web of Conferences, 2018. **183**: p. 02001.

161. Liu, X.F., P. Li, and Z.X. Liu, *Modification Process of Carbon Fiber Reinforcement for Aluminum Matrix Composite.* Advanced Materials Research, 2012. **560-561**: p. 899-905.

162. Hui, Q., et al., *Effects of Processing Parameters on Properties of SiCp/Al Composites.* Matec Web of Conferences, 2016. **67**: p. 06016.

163. Padmavathi, K., et al., *Mechanical Characterization of Aluminium-Titania Metal Matrix Composites.* 2020.

164. Zeren, A., *Effect of the Graphite Content on the Tribological Properties of Hybrid Al/SiC/Gr Composites Processed by Powder Metallurgy.* Industrial Lubrication and Tribology, 2015. **67**(3): p. 262-268.

165. Deshmanya, I.B. and G.K. Purohit, *Effect of Forging on Micro-Hardness of Al7075 Based Al2O3 Reinforced Composites Produced by Stir-Casting.* Iosr Journal of Engineering, 2012. **2**(1): p. 20-31.

166. Kim, H.-g., et al., *Effects of Nonuniform Fiber Geometries on the Microstructural Fracture Behavior of Ceramic Matrix Composites.* Mathematical Problems in Engineering, 2019. **2019**(1).

167. Dassios, K.G., *A Geometry-Invariant Fracture Law for Ceramic Matrix Composites.* Journal of Composite Materials, 2013. **49**(1): p. 65-74.

168. Li, H., et al., *Preparation and Characterization of Nextel 720/Alumina Ceramic Matrix Composites via an Improved Prepreg Process.* International Journal of Applied Ceramic Technology, 2022. **19**(4): p. 1970-1980.

169. Li, L., *A Micromechanical Loading/Unloading Constitutive Model of Fiber-reinforced Ceramic-matrix Composites Considering Matrix Crack Closure.* Fatigue & Fracture of Engineering Materials & Structures, 2021. **44**(9): p. 2389-2411.

170. Dassios, K.G., et al., *Nondestructive Damage Evaluation in Ceramic Matrix Composites for Aerospace Applications.* The Scientific World Journal, 2013. **2013**(1).

171. Li, L., P. Reynaud, and G. Fantozzi, *Cyclic-Dependent Damage Evolution in Self-Healing Woven SiC/[Si-B-C] Ceramic-Matrix Composites at Elevated Temperatures.* Materials, 2020. **13**(6): p. 1478.

172. Li, L., et al., *Mechanical Properties and Microstructure Evolution of KD-SA SiC_f/BN/SiC CMCs Oxidized at Different Temperatures.* International Journal of Applied Ceramic Technology, 2024. **21**(3): p. 2183-2195.

173. Jain, N. and D. Koch, *Prediction of Failure in Ceramic Matrix Composites Using Damage-Based Failure Criterion.* Journal of Composites Science, 2020. **4**(4): p. 183.

174. Piotter, V., et al., *Powder Injection Molding of Oxide Ceramic CMC.* Key Engineering Materials, 2019. **809**: p. 148-152.

175. Rashid, A.B., et al., *Breaking Boundaries With Ceramic Matrix Composites: A Comprehensive Overview of Materials, Manufacturing Techniques, Transformative Applications, Recent Advancements, and Future Prospects.* Advances in Materials Science and Engineering, 2024. **2024**: p. 1-33.

176. Hui, M., et al., *First Printing of Continuous Fibers Into Ceramics.* Journal of the American Ceramic Society, 2018. **102**(6): p. 3244-3255.

177. Wrbanek, J.D., G.C. Fralick, and D. Zhu, *Ceramic Thin Film Thermocouples for SiC-based Ceramic Matrix Composites.* Thin Solid Films, 2012. **520**(17): p. 5801-5806.

178. Shrivastava, S., et al., *Ceramic Matrix Composites: Classifications, Manufacturing, Properties, and Applications.* Ceramics, 2024. **7**(2): p. 652-679.

179. Yin, X., et al., *Fibre-Reinforced Multifunctional SiC Matrix Composite Materials.* International Materials Reviews, 2016. **62**(3): p. 117-172.

180. Yuan, S. and Q. Wu, *Experimental Investigation and Optimization in Rotary Ultrasonic Drilling of C/C Composites.* Materials Science Forum, 2016. **874**: p. 313-319.

181. Jing, J.-Y., Q. Fu, and R. Yuan, *Nanowire-Toughened CVD-SiC Coating for C/C Composites With Surface Pre-Oxidation.* Surface Engineering, 2018. **34**(1): p. 47-53.

182. Xu, Y. and T. Gao, *Optimizing Thermal-Elastic Properties of C/C–SiC Composites Using a Hybrid Approach and PSO Algorithm.* Materials, 2016. **9**(4): p. 222.

183. Feng, T., et al., *Anti-Oxidation Property of the ZrB₂–SiC–Si-coated Low-Density C/C Composites.* Surface Engineering, 2018. **34**(1): p. 40-46.

184. Li, G., et al., *Preparation and Properties of C/SiC Bolts via Precursor Infiltration and Pyrolysis Process.* Rare Metals, 2011. **30**(S1): p. 572-575.

185. Ravichandran, D., et al., *3D Printing Carbon-Carbon Composites With Multilayered Architecture for Enhanced Multifunctional Properties.* 2024.

186. Kumar, S., et al., *Synthesis of Polycarbosilane, Polymer Impregnation Pyrolysis Based C SiC Composites and Prototype Development.* Defence Science Journal, 2019. **69**(6): p. 599-606.

187. Wang, Q., W.-L. Wu, and W. Li, *Compression Properties of Interlayer and Intralayer Carbon/Glass Hybrid Composites.* Polymers, 2018. **10**(4): p. 343.

188. Wu, W.-L., et al., *The Effects of Hybridization on the Flexural Performances of Carbon/Glass Interlayer and Intralayer Composites.* Polymers, 2018. **10**(5): p. 549.

189. Feng, N.L., et al., *Mechanical Properties and Water Absorption of Kenaf/Pineapple Leaf Fiber-reinforced Polypropylene Hybrid Composites.* Polymer Composites, 2019. **41**(4): p. 1255-1264.

190. Atmakuri, A., et al., *Analysis of Mechanical and Wettability Properties of Natural Fiber-Reinforced Epoxy Hybrid Composites.* Polymers, 2020. **12**(12): p. 2827.

191. Hu, P., et al., *Quasistatic Compression Properties of Fiber Hybrid Ceramic Matrix Composites With Different Hybrid Ratios and Fiber Dispersions.* Advanced Engineering Materials, 2023. **25**(13).

192. Al-Maadeed, S., et al., *Date Palm Wood Flour/Glass Fibre Reinforced Hybrid Composites of Recycled Polypropylene: Mechanical and Thermal Properties.* Materials & Design (1980-2015), 2012. **42**: p. 289-294.

193. Ebrahimnezhad-Khaljiri, H. and R. Eslami-Farsani, *Thermal and Mechanical Properties of Hybrid Carbon/Oxidized Polyacrylonitrile Fibers-epoxy Composites.* Polymer Composites, 2015. **38**(7): p. 1412-1417.

194. Russo, P., et al., *Bio-Polyamide 11 Hybrid Composites Reinforced With Basalt/Flax Interwoven Fibers: A Tough Green Composite for Semi-Structural Applications.* Fibers, 2019. **7**(5): p. 41.

195. Silva, R.V.d., et al., *Hybrid Composites With Glass Fiber and Natural Fibers of Sisal, Coir, and Luffa Sponge.* Journal of Composite Materials, 2020. **55**(5): p. 717-728.

196. Razzaq, M.E.A., S.E. Moma, and M.S. Rabbi, *Mechanical Properties of Biofiber/Glass Reinforced Hybrid Composites Produced by Hand Lay-Up Method: A Review.* Materials Engineering Research, 2021. **3**(1): p. 144-155.

197. Bahrami, M., et al., *Hybridization Effect on Interlaminar Bond Strength, Flexural Properties, and Hardness of Carbon–Flax Fiber Thermoplastic Bio-Composites.* Polymers, 2023. **15**(24): p. 4619.

198. Neto, J.S.S., et al., *A Review of Recent Advances in Hybrid Natural Fiber Reinforced Polymer Composites.* Journal of Renewable Materials, 2022. **10**(3): p. 561-589.

199. Ng, L.F., M.Y. Yahya, and C. Muthukumar, *Mechanical Characterization and Water Absorption Behaviors of Pineapple Leaf/Glass Fiber-reinforced Polypropylene Hybrid Composites.* Polymer Composites, 2021. **43**(1): p. 203-214.

200. Nunna, S., et al., *A Review on Mechanical Behavior of Natural Fiber Based Hybrid Composites.* Journal of Reinforced Plastics and Composites, 2012. **31**(11): p. 759-769.

201. Gairola, S., et al., *Static and Dynamic Mechanical Behavior of Intra-hybrid Jute/Sisal-reinforced <scp>polypropylene</Scp> Composites: Effect of Stacking Sequence.* Polymer Composites, 2024. **45**(8): p. 7049-7058.

202. Kaushik, D., et al., *Static and Dynamic Mechanical Behavior of Intra-hybrid Jute/Sisal and Flax/Kenaf Reinforced Polypropylene Composites.* Polymer Composites, 2022. **44**(1): p. 515-523.

203. Jariwala, H.R. and P. Jain, *A Review on Mechanical Behavior of Natural Fiber Reinforced Polymer Composites and Its Applications.* Journal of Reinforced Plastics and Composites, 2019. **38**(10): p. 441-453.

204. Ramesh, M., K. Palanikumar, and K.H. Reddy, *Influence of Fiber Orientation and Fiber Content on Properties of Sisal-jute-glass Fiber-reinforced Polyester Composites.* Journal of Applied Polymer Science, 2015. **133**(6).

205. Both, A.K., et al., *Green Chemical Approach to Fabricate Hemp Fiber Composites for Making Sustainable Hydroponic Growth Media.* Acs Agricultural Science & Technology, 2021. **1**(5): p. 499-506.

206.   Momeni, S., et al., *Valorization of Hemp Hurds as Bio-Sourced Additives in PLA-Based Biocomposites.* Polymers, 2021. **13**(21): p. 3786.

207.   Tan, H., et al., *Behavior of Sisal Fiber Concrete Cylinders Externally Wrapped With Jute FRP.* Polymer Composites, 2015. **38**(9): p. 1910-1917.

208.   Maurya, H.O., et al., *Development and Characterization of Microwave -processed <scp>linear Low-density Polyethylene</Scp> Based Sisal/Jute Hybrid Laminates.* Polymer Composites, 2024. **45**(9): p. 8306-8320.

209.   Balaji, A., et al., *Study on Mechanical and Morphological Properties of Sisal/Banana/Coir Fiber-Reinforced Hybrid Polymer Composites.* Journal of the Brazilian Society of Mechanical Sciences and Engineering, 2019. **41**(9).

210.   Liu, Y.-Y., et al., *Biobased High-Performance Epoxy Vitrimer With UV Shielding for Recyclable Carbon Fiber Reinforced Composites.* Acs Sustainable Chemistry & Engineering, 2021. **9**(12): p. 4638-4647.

211.   Prakash, R.A., et al., *High-Speed Edge Trimming of CFRP and Online Monitoring of Performance of Router Tools Using Acoustic Emission.* Materials, 2016. **9**(10): p. 798.

212.   Wahab, M.S., et al., *Laser Cutting Characteristic on the Laminated Carbon Fiber Reinforced Plastics (CFRP) Composite of Aerospace Structure Panel.* Advanced Materials Research, 2012. **576**: p. 503-506.

213.   Fan, W., et al., *Random Process Model of Mechanical Property Degradation in Carbon Fiber-Reinforced Plastics Under Thermo-Oxidative Aging.* Journal of Composite Materials, 2016. **51**(9): p. 1253-1264.

214.   Li, Y., et al., *Nanoscale SiO<sub>2</sub>/ZrO<sub>2</sub> Particulate-Reinforced Titanium Composites for Bone Implant Materials.* Key Engineering Materials, 2012. **520**: p. 242-247.

215.   Sayuti, M., et al., *Effect of Mould Vibration on Mechanical Properties of Particulate Reinforced Aluminium Alloy Matrix Composite.* Advanced Materials Research, 2012. **445**: p. 475-480.

216.   Steiner, S.A., R. Li, and B.L. Wardle, *Circumventing the Mechanochemical Origins of Strength Loss in the Synthesis of Hierarchical Carbon Fibers.* Acs Applied Materials & Interfaces, 2013. **5**(11): p. 4892-4903.

217.   V, N., et al., *Investigation of Microcapsules Based Self-Healing Composites Embedded With Carbon Nanotubes for Improved Healing Efficiency.* 2024.

218.   Rozhbiany, F.A.R. and S.R. Jalal, *Influence of Reinforcement and Processing on Aluminum Matrix Composites Modified by Stir Casting Route.* Advanced Composites Letters, 2019. **28**.

219.   Sahoo, S., B.B. Jha, and A. Mandal, *Powder Metallurgy Processed TiB₂-reinforced Steel Matrix Composites: A Review.* Materials Science and Technology, 2021. **37**(14): p. 1153-1173.

220.   Amin, M., et al., *An Evaluation of Mechanical Properties on Kenaf Natural Fiber/Polyester Composite Structures as Table Tennis Blade.* Journal of Physics Conference Series, 2017. **914**: p. 012015.

221.   Noori, M., et al., *Properties of Aluminium-SiC<sub>P</Sub> Composites (Review).* Advanced Materials Research, 2017. **1143**: p. 72-78.

222. Tavarageri, K.B., *Fiber-Reinforced Polymer Composites: A Study*. Journal of Production and Industrial Engineering, 2021. **2**(1).

223. Pugalethi, P., M. Jayaraman, and A. Natarajan, *Evaluation of Mechanical Properties of Aluminium Alloy 7075 Reinforced With SiC and Al₂O₃ Hybrid Metal Matrix Composites*. Applied Mechanics and Materials, 2015. **766-767**: p. 246-251.

224. Özyürek, D., et al., *Effect of Al2O3 Amount on Microstructure and Wear Properties of Al–Al2O3 Metal Matrix Composites Prepared Using Mechanical Alloying Method*. Powder Metallurgy and Metal Ceramics, 2010. **49**(5-6): p. 289-294.

225. Harish, P., et al., *Effect of Alumina and Graphene on Mechanical and Tribological Behaviour of Al-7075 Hybrid Composite*. Applied Engineering Letters Journal of Engineering and Applied Sciences, 2019. **4**(3): p. 79-87.

226. Kotarska, A., T. Poloczek, and D. Janicki, *Characterization of the Structure, Mechanical Properties and Erosive Resistance of the Laser Cladded Inconel 625-Based Coatings Reinforced by TiC Particles*. Materials, 2021. **14**(9): p. 2225.

227. Vostřák, M., et al., *Comparison of Mechanical Properties of Laser Cladded WC and (TiW)C1-x in Nibased Alloy Coatings*. 2020.

228. Rana, S., et al., *Effects of Aluminum and Silicon Carbide on Morphological and Mechanical Properties of Epoxy Hybrid Composites*. Polymers and Polymer Composites, 2022. **30**.

229. Zimmermann, M.V.G., et al., *Influence of Flexographic Photopolymer-Plate Residue Incorporation on the Mechanical Properties of Glass-Fiber-Reinforced Polyester Composites*. Materials Research, 2023. **26**.

230. Islam, F., et al., *Mechanical and Interfacial Characterization of Jute Fabrics Reinforced Unsaturated Polyester Resin Composites*. Nano Hybrids and Composites, 2019. **25**: p. 22-31.

231. Ram, K. and P.K. Bajpai, *Performance Analysis of Hybrid Bio-Composites Developed Through Different Processing Routes for Automobile Application*. Proceedings of the Institution of Mechanical Engineers Part C Journal of Mechanical Engineering Science, 2023. **237**(20): p. 4769-4780.

232. Landowski, M., M.K. Budzik, and K. Imielińska, *On Degradation of Glass/Polyester Laminate Immersed in Water*. Advances in Materials Science, 2011. **11**(1).

233. Rydarowski, H. and M. Kozioł, *Repeatability of Glass Fiber Reinforced Polymer Laminate Panels Manufactured by Hand Lay-Up and Vacuum-Assisted Resin Infusion*. Journal of Composite Materials, 2014. **49**(5): p. 573-586.

234. Sałasińska, K., et al., *The Effect of Manufacture Process on Mechanical Properties and Burning Behavior of Epoxy-Based Hybrid Composites*. Materials, 2022. **15**(1): p. 301.

235. Kim, S.-Y., et al., *Mechanical Properties and Production Quality of Hand-Layup and Vacuum Infusion Processed Hybrid Composite Materials for GFRP Marine Structures*. International Journal of Naval Architecture and Ocean Engineering, 2014. **6**(3): p. 723-736.

236. Abdellah, M.Y., et al., *A Comparative Study to Evaluate the Essential Work of Fracture to Measure the Fracture Toughness of Quasi-Brittle Material*. Materials, 2022. **15**(13): p. 4514.

237. Song, Y., et al., *A Novel CAE Method for Compression Molding Simulation of Carbon Fiber-Reinforced Thermoplastic Composite Sheet Materials*. Journal of Composites Science, 2018. **2**(2): p. 33.

238. Xie, J., et al., *Process Optimization for Compression Molding of Carbon Fiber–Reinforced Thermosetting Polymer*. Materials, 2019. **12**(15): p. 2430.

239. Manalu, J., et al., *Characterization of Eco-friendly Composites for Automotive Applications Prepared by the Compression Molding Method*. Polymer Composites, 2024. **45**(9): p. 8104-8118.

240. Bhuyan, M., et al., *Decreasing the Cycle Time for In-mold Coating (IMC) of Sheet Molding Compound (SMC) Compression Molding*. Polymer Engineering & Science, 2019. **59**(6): p. 1158-1166.

241. Teuwsen, J., S.K. Hohn, and T.A. Osswald, *Direct Fiber Simulation of a Compression Molded Ribbed Structure Made of a Sheet Molding Compound With Randomly Oriented Carbon/Epoxy Prepreg Strands—A Comparison of Predicted Fiber Orientations With Computed Tomography Analyses*. Journal of Composites Science, 2020. **4**(4): p. 164.

242. Zhang, T., Y. Zhao, and B. Zhang, *A Method Based on the Time–temperature Superposition Principle to Predict Pressurization Time in Compression Molding*. Journal of Applied Polymer Science, 2018. **135**(36).

243. Kühn, C., et al., *Experimental and Numerical Analysis of Fiber Matrix Separation During Compression Molding of Long Fiber Reinforced Thermoplastics*. Journal of Composites Science, 2017. **1**(1): p. 2.

244. Li, W.D., et al., *The Processing Characteristics and Mechanical Properties of Semi-Prepreg RTM Composites*. Advanced Materials Research, 2013. **721**: p. 153-158.

245. Ornaghi, H.L., et al., *Mechanical and Dynamic Mechanical Analysis of Hybrid Composites Molded by Resin Transfer Molding*. Journal of Applied Polymer Science, 2010. **118**(2): p. 887-896.

246. Ding, Y. and Y. Jia, *Sensitivity Analysis on Resin Transfer Molding Processes With Edge Effect and Curing Reaction Characteristics*. Polymer Composites, 2014. **36**(11): p. 2008-2016.

247. Seuffert, J., L. Kärger, and F. Henning, *Simulation of the Influence of Embedded Inserts on the RTM Filling Behavior Considering Local Fiber Structure*. Key Engineering Materials, 2017. **742**: p. 681-688.

248. Wang, K., et al., *Flow Pattern Control in Resin Transfer Molding Using a Model Predictive Control Strategy*. Polymer Engineering & Science, 2017. **58**(9): p. 1659-1665.

249. Robinson, M. and J.B. Kosmatka, *Analysis of the Post-Filling Phase of the Vacuum-Assisted Resin Transfer Molding Process*. Journal of Composite Materials, 2013. **48**(13): p. 1547-1559.

250. Nakanishi, E., S. Maki, and S. Matsumoto, *Molding of C-FRP Plate With Using Induction Heating*. Advanced Materials Research, 2011. **410**: p. 345-348.

251. Suzuki, Y., et al., *Dual-Energy X-Ray Computed Tomography for Void Detection in Fiber-Reinforced Composites.* Journal of Composite Materials, 2019. **53**(17): p. 2349-2359.

252. Guan, Q., et al., *Curing Kinetics and Mechanism of Novel High Performance Hyperbranched Polysiloxane/Bismaleimide/Cyanate Ester Resins for Resin Transfer Molding.* Journal of Applied Polymer Science, 2011. **122**(1): p. 304-312.

253. Grössing, H., et al., *Flow Front Advancement During Composite Processing: Predictions From Numerical Filling Simulation Tools in Comparison With Real-World Experiments.* Polymer Composites, 2015. **37**(9): p. 2782-2793.

254. Goergen, C., D. May, and P. Mitscháng, *Integration of rCF in Resin Transfer Pressing Process.* Journal of Reinforced Plastics and Composites, 2020. **39**(9-10): p. 361-372.

255. Bankov, B., et al., *A Multi-Step Approach for Analyzing the Number of Polymer Injection Molding Gate Spots and Their Position in T-RTM Technology.* Environment Technology Resources Proceedings of the International Scientific and Practical Conference, 2024. **3**: p. 20-23.

256. Chen, C.-H. and P.-H. Chen, *Hybrid Fibre Reinforced Epoxy Composites for Pultrusion: Mechanical and Thermal Properties.* Polymers and Polymer Composites, 2011. **19**(6): p. 459-468.

257. Vedernikov, A., et al., *Pultruded Materials and Structures: A Review.* Journal of Composite Materials, 2020. **54**(26): p. 4081-4117.

258. Silva, F.J.G., et al., *Saving Energy in the GFRP Pultrusion Process.* Journal of Research Updates in Polymer Science, 2013.

259. Tucci, F., et al., *Injection Pultrusion of Glass-Reinforced Epoxy: Cure Kinetics, Rheology, and Force Analysis.* Polymers, 2024. **16**(12): p. 1642.

260. Tipboonsri, P., et al., *Optimization of Thermoplastic Pultrusion Parameters of Jute and Glass Fiber-Reinforced Polypropylene Composite.* Polymers, 2023. **16**(1): p. 83.

261. Minchenkov, K., et al., *Thermoplastic Pultrusion: A Review.* Polymers, 2021. **13**(2): p. 180.

262. Zhu, Y., et al., *The Material Heterogeneity Effect on the Local Resistance of Pultruded GFRP Columns.* Materials, 2023. **17**(1): p. 153.

263. Özkılıç, Y.O., E. Madenci, and L. Gemi, *Tensile and Compressive Behaviors of the Pultruded GFRP Lamina.* Turkish Journal of Engineering, 2020. **4**(4): p. 169-175.

264. Alajarmeh, O., et al., *Fatigue Behavior of Unidirectional Fiber-reinforced Pultruded Composites With High Volume Fiber Fraction.* Fatigue & Fracture of Engineering Materials & Structures, 2023. **46**(6): p. 2034-2048.

265. Huang, S., et al., *A Novel Combining Method for Composite Groove Structure Fabrication.* Crystals, 2023. **13**(12): p. 1644.

266. Srebrenkoska, S., et al., *Effect of Process Parameters on Thermal and Mechanical Properties of Filament Wound Polymer-Based Composite Pipes.* Polymers, 2023. **15**(13): p. 2829.

267. Demirci, M.T. and Ö.S. Şahin, *Low-velocity Impact Response and Inspection of Damage Propagation for Basalt Fiber Reinforced Filament Wound Pipes.* Polymer Composites, 2022. **43**(7): p. 4626-4644.

268. Jois, K.C., et al., *Towpreg Manufacturing and Characterization for Filament Winding Application.* Polymer Composites, 2024. **45**(9): p. 7893-7905.

269. Kadri, K., A. Ben Abdallah, and S. Ballut, *Hydrogen Storage Vessels of Type 4 and Type 5.* 2024.

270. Hou, X., et al., *A Dynamic Modeling Approach for a High-Speed Winding System With Twin-Rotor Coupling.* Textile Research Journal, 2020. **90**(21-22): p. 2533-2551.

271. Liu, Y., et al., *Study on the Resin Infusion Process Based on Automated Fiber Placement Fabricated Dry Fiber Preform.* Scientific Reports, 2019. **9**(1).

272. Yassin, K. and M. Hojjati, *Processing of Thermoplastic Matrix Composites Through Automated Fiber Placement and Tape Laying Methods.* Journal of Thermoplastic Composite Materials, 2017. **31**(12): p. 1676-1725.

273. Lemaire, E., S.H.S. Zein, and M. Bruyneel, *Optimization of Composite Structures With Curved Fiber Trajectories.* Composite Structures, 2015. **131**: p. 895-904.

274. Bahar, M. and M. Sinapius, *Adaptive Feeding Roller With an Integrated Cutting System for Automated Fiber Placement (AFP).* Journal of Composites Science, 2020. **4**(3): p. 92.

275. Rakhshbahar, M. and M. Sinapius, *A Novel Approach: Combination of Automated Fiber Placement (AFP) and Additive Layer Manufacturing (ALM).* Journal of Composites Science, 2018. **2**(3): p. 42.

276. Hirsch, P., et al., *Processing and Analysis of Hybrid Fiber-Reinforced Polyamide Composite Structures Made by Fused Granular Fabrication and Automated Tape Laying.* Journal of Manufacturing and Materials Processing, 2024. **8**(1): p. 25.

277. Miao, L., et al., *Modeling of Contact Mechanics and Defects Investigation in Automated Fiber Placement.* Polymer Composites, 2024. **45**(7): p. 5992-6007.

278. Hoa, S.V., M.D. Hoang, and J. Simpson, *Manufacturing Procedure to Make Flat Thermoplastic Composite Laminates by Automated Fibre Placement and Their Mechanical Properties.* Journal of Thermoplastic Composite Materials, 2016. **30**(12): p. 1693-1712.

279. Oromiehie, E., et al., *In Situ Process Monitoring for Automated Fibre Placement Using Fibre Bragg Grating Sensors.* Structural Health Monitoring, 2016. **15**(6): p. 706-714.

280. Radzi, M.K.F.M., et al., *Optimizing Injection Parameters of Kenaf Filler Polypropylene Composite by Taguchi Method.* Materials Science Forum, 2017. **894**: p. 81-84.

281. Jiang, B., et al., *Effect of Thermal Gradient on Interfacial Behavior of Hybrid Fiber Reinforced Polypropylene Composites Fabricated by Injection Overmolding Technique.* Polymer Composites, 2020. **41**(10): p. 4064-4073.

282. Azenha, J.J.A., M. Gomes, and A.J. Pontes, *High Strength Injection Molded Thermoplastic Composites.* Polymer Engineering & Science, 2018. **58**(4): p. 560-567.

283. Xu, A., et al., *Molding of PBO Fabric Reinforced Thermoplastic Composite to Achieve High Fiber Volume Fraction.* Polymer Composites, 2013. **34**(6): p. 953-958.

284. Liou, G.-Y., et al., *Fabrication and Property Characterization of Long-Glass-Fiber-Reinforced Polypropylene Composites Processed Using a Three-Barrel Injection Molding Machine.* Polymers, 2022. **14**(6): p. 1251.

285. Chen, D., et al., *Warpage of Injection-Molded Automotive B Pillar Trim Fabricated With Ramie Fiber-Reinforced Polypropylene Composites.* Journal of Reinforced Plastics and Composites, 2015. **34**(14): p. 1144-1152.

286. Bex, G., et al., *Effect of Process Parameters on the Adhesion Strength in Two-component Injection Molding of Thermoset Rubbers and Thermoplastics.* Journal of Applied Polymer Science, 2018. **135**(29).

287. Harrigan, W.C., *Processing of Aluminum Metal-Matrix Composites.* 2018: p. 375-386.

288. Bezerra, C.E.A., et al., *Features of the Processing of AA2124 Aluminum Alloy Metal Matrix Composites Reinforced by Silicon Nitride Prepared by Powder Metallurgy Techniques.* Materials Science Forum, 2014. **802**: p. 108-113.

289. Etemadi, R., et al., *Pressure Infiltration Processes to Synthesize Metal Matrix Composites – A Review of Metal Matrix Composites, the Technology and Process Simulation.* Materials and Manufacturing Processes, 2018. **33**(12): p. 1261-1290.

290. Campbell, F.C., *Structural Composite Materials.* 2010.

291. Iqbal, A., Y. Arai, and W. Araki, *Effect of Hybrid Reinforcement on Crack Initiation and Early Propagation Mechanisms in Cast Metal Matrix Composites During Low Cycle Fatigue.* Materials & Design (1980-2015), 2013. **45**: p. 241-252.

292. Zhu, M., et al., *Study on Microstructure and Abrasive Wear Properties of in-Situ TiC Reinforced High Chromium Cast Iron Matrix Composite.* Materials Research Express, 2022. **9**(3): p. 036517.

293. Radhika, C., et al., *A Review on Additive Manufacturing for Aerospace Application.* Materials Research Express, 2024. **11**(2): p. 022001.

294. Kurzynowski, T., A. Pawlak, and I. Smolina, *The Potential of SLM Technology for Processing Magnesium Alloys in Aerospace Industry.* Archives of Civil and Mechanical Engineering, 2020. **20**(1).

295. Parandoush, P., et al., *Additive Manufacturing of Continuous Carbon Fiber Reinforced Epoxy Composite With Graphene Enhanced Interlayer Bond Toward <scp>ultra-high</Scp> Mechanical Properties.* Polymer Composites, 2021. **43**(2): p. 934-945.

296. Hilal, H., et al., *Investigating the Influence of Process Parameters on the Structural Integrity of an Additively Manufactured Nickel-Based Superalloy.* Metals, 2019. **9**(11): p. 1191.

297. Rivera, A.D.P.F., F.d.C. Magalhães, and J. Rubio, *Experimental Characterization of PLA Composites Printed by Fused Deposition Modelling.* Journal of Composite Materials, 2022. **57**(5): p. 941-954.

298. Pant, M., et al., *A Review of Additive Manufacturing in Aerospace Application.* Revue Des Composites Et Des Matériaux Avancés, 2021. **31**(2): p. 109-115.

299. Gupta, N., et al., *Advanced Composite Manufacturing Using Additive Manufacturing and Robotic Techniques.* E3s Web of Conferences, 2023. **430**: p. 01118.

300. Rejeski, D., F. Zhao, and Y. Huang, *Research Needs and Recommendations on Environmental Implications of Additive Manufacturing.* Additive Manufacturing, 2018. **19**: p. 21-28.

301. Ferro, C.G., S. Varetti, and P. Maggiore, *Experimental Evaluation of Mechanical Compression Properties of Aluminum Alloy Lattice Trusses for Anti-Ice System Applications.* Machines, 2024. **12**(6): p. 404.

302.   Karolewska, K. and B. Ligaj, *Mechanical Properties Comparison of Ti6Al4V Produced by Different Technologies Under Static Load Conditions*. Matec Web of Conferences, 2019. **290**: p. 08010.

303.   Dass, A. and A. Moridi, *State of the Art in Directed Energy Deposition: From Additive Manufacturing to Materials Design*. Coatings, 2019. **9**(7): p. 418.

304.   Matúš, M., et al., *Effect of Software for FDM Additive Manufacturing on Geometric Accuracy and Print Quality*. Global Journal of Engineering and Technology Advances, 2022. **13**(3): p. 110-120.

305.   Wong, K.V. and A. Hernandez, *A Review of Additive Manufacturing*. Isrn Mechanical Engineering, 2012. **2012**: p. 1-10.

306.   Khosravani, M.R., et al., *Characterization of 3d-Printed PLA Parts With Different Raster Orientations and Printing Speeds*. Scientific Reports, 2022. **12**(1).

307.   Li, H., *Current Status and Prospects of Three-Dimensional Printing Application*. Applied and Computational Engineering, 2023. **11**(1): p. 192-202.

308.   Binner, J., et al., *Selection, Processing, Properties and Applications of Ultra-High Temperature Ceramic Matrix Composites, UHTCMCs – A Review*. International Materials Reviews, 2020. **65**(7): p. 389-444.

309.   Li, L., *Hysteresis of Ceramic-Matrix Composites*. 2019.

310.   Mansour, R. and G.N. Morscher, *Mode I Interlaminar Fracture Behavior of 2D Woven Ceramic Matrix Composites*. International Journal of Applied Ceramic Technology, 2018. **16**(2): p. 735-745.

311.   Dhanasekar, S., S. Baskar, and S. Vishvanathperumal, *Halloysite nanotubes effect on cure and mechanical properties of EPDM/NBR nanocomposites*. Journal of Inorganic and Organometallic Polymers and Materials, 2023. **33**(10): p. 3208-3220.

312.   Chinnamahammad Bhasha, A. and K. Balamurugan, *Fabrication and property evaluation of Al 6061+ x%(RHA+ TiC) hybrid metal matrix composite*. SN Applied Sciences, 2019. **1**(9): p. 977.

313.   Imal, M. and M. Ermurat, *Design of lightweight electric vehicle and application for efficiency challenge marathon competition*. Int. J. Eng. Sci. Technol, 2022. **6**(6): p. 19-27.

314.   Naik, N., et al., *A review on composite materials for energy harvesting in electric vehicles*. Energies, 2023. **16**(8): p. 3348.

315.   Mansour, R. and G.N. Morscher, *Mode I interlaminar fracture behavior of 2D woven ceramic matrix composites*. International Journal of Applied Ceramic Technology, 2019. **16**(2): p. 735-745.

316.   Srikanth, V. and K.R.N. Reddy, *Mechanical and Thermal Characteristics of CF/Cochlospermum gossypium composites*.

317.   Dhanasekar, S., et al., *A comprehensive study of ceramic matrix composites for space applications*. Advances in Materials Science and Engineering, 2022. **2022**(1): p. 6160591.

318.   Binner, J., et al., *Selection, processing, properties and applications of ultra-high temperature ceramic matrix composites, UHTCMCs–a review*. International Materials Reviews, 2020. **65**(7): p. 389-444.

319. Salas, M.F., et al., *Nanotechnological Applications in Dermocosmetics*. European Journal of Pharmaceutical Research, 2023. **3**(1): p. 1-7.
320. Gatoo, M.A., et al., *Physicochemical Properties of Nanomaterials: Implication in Associated Toxic Manifestations*. Biomed Research International, 2014. **2014**: p. 1-8.
321. Mohamed, A.T., *Emerging nanotechnology applications in electrical engineering*. 2021: IGI Global.
322. Bayda, S., et al., *The History of Nanoscience and Nanotechnology: From Chemical–Physical Applications to Nanomedicine*. Molecules, 2019. **25**(1): p. 112.
323. Hrkach, J.S., et al., *Preclinical Development and Clinical Translation of a PSMA-Targeted Docetaxel Nanoparticle With a Differentiated Pharmacological Profile*. Science Translational Medicine, 2012. **4**(128).
324. Zhang, Y., et al., *Nanotechnology in Cancer Diagnosis: Progress, Challenges and Opportunities*. Journal of Hematology & Oncology, 2019. **12**(1).
325. Diallo, M.S. and N.A. Fromer, *Nanotechnology for Sustainable Development: Retrospective and Outlook*. Journal of Nanoparticle Research, 2013. **15**(11).
326. Vogel, E.M., et al., *Challenges of Nanotechnology in Cosmetic Permeation With Caffeine*. Brazilian Journal of Biology, 2022. **82**.
327. Joni, I.M., et al., *Nanotechnology: Development and Challenges in Indonesia*. 2018.
328. Patra, J.K. and S. Gouda, *Application of Nanotechnology in Textile Engineering: An Overview*. Journal of Engineering and Technology Research, 2013. **5**(5): p. 104-111.
329. Rodrigues, R., *The Implications of High-Rate Nanomanufacturing on Society and Personal Privacy*. Bulletin of Science Technology & Society, 2006. **26**(1): p. 38-45.
330. Johnson, D.G., *Ethics and Technology 'In the Making': An Essay on the Challenge of Nanoethics*. Nanoethics, 2007. **1**(1): p. 21-30.
331. Chen, T., et al., *Aptamer-Conjugated Nanomaterials for Bioanalysis and Biotechnology Applications*. Nanoscale, 2011. **3**(2): p. 546-556.
332. Eisenstat, J., et al., *A Comparative Review of Material Properties for Current and Future Dental Filling Nanomaterials*. International Journal of Engineering Materials and Manufacture, 2021. **6**(4): p. 225-241.
333. Ge, Y., et al., *Two-Dimensional Nanomaterials With Unconventional Phases*. Chem, 2020. **6**(6): p. 1237-1253.
334. Choi, Y.H. and K.H. Kim, *Fabrication of Nanomaterial Devices for Field Emission Applications*. Advanced Materials Research, 2012. **463-464**: p. 739-742.
335. Vance, M.E., et al., *Nanotechnology in the Real World: Redeveloping the Nanomaterial Consumer Products Inventory*. Beilstein Journal of Nanotechnology, 2015. **6**: p. 1769-1780.
336. Khan, A., et al., *Cellulosic Nanomaterials in Food and Nutraceutical Applications: A Review*. Journal of Agricultural and Food Chemistry, 2017. **66**(1): p. 8-19.
337. Daramy, K., et al., *Investigating the Impact of Shear Flow on Nanoparticle-Protein Interactions*. British Journal of Pharmacy, 2022. **7**(2).
338. Li, W., et al., *A Review of Recent Applications of Ion Beam Techniques on Nanomaterial Surface Modification: Design of Nanostructures and Energy Harvesting*. Small, 2019. **15**(31).

339. Landsiedel, R., et al., *Safety Assessment of Nanomaterials Using an Advanced Decision-Making Framework, the DF4nanoGrouping.* Journal of Nanoparticle Research, 2017. **19**(5).

340. Hendren, C.O., et al., *A Functional Assay-Based Strategy for Nanomaterial Risk Forecasting.* The Science of the Total Environment, 2015. **536**: p. 1029-1037.

341. Chellamuthu, P., et al., *Biogenic Control of Manganese Doping in Zinc Sulfide Nanomaterial Using Shewanella Oneidensis MR-1.* Frontiers in Microbiology, 2019. **10**.

342. Wani, M.Y., et al., *Nanotoxicity: Dimensional and Morphological Concerns.* Advances in Physical Chemistry, 2011. **2011**(1).

343. Harrison, D., et al., *A Review of the Aquatic Environmental Transformations of Engineered Nanomaterials.* Nanomaterials, 2023. **13**(14): p. 2098.

344. Umeda, Y., et al., *PEG-Attached PAMAM Dendrimers Encapsulating Gold Nanoparticles: Growing Gold Nanoparticles in the Dendrimers for Improvement of Their Photothermal Properties.* Bioconjugate Chemistry, 2010. **21**(8): p. 1559-1564.

345. Chauhan, A.S., *Dendrimers for Drug Delivery.* Molecules, 2018. **23**(4): p. 938.

346. Yang, K., et al., *Host–Guest Chemistry of Dendrimer–Drug Complexes. 6. Fully Acetylated Dendrimers as Biocompatible Drug Vehicles Using Dexamethasone 21-Phosphate as a Model Drug.* The Journal of Physical Chemistry B, 2011. **115**(10): p. 2185-2195.

347. Wang, W., et al., *Protective Effect of PEGylation Against Poly(amidoamine) Dendrimer-induced Hemolysis of Human Red Blood Cells.* Journal of Biomedical Materials Research Part B Applied Biomaterials, 2010. **93B**(1): p. 59-64.

348. Sharma, A. and J. Das, *Small Molecules Derived Carbon Dots: Synthesis and Applications in Sensing, Catalysis, Imaging, and Biomedicine.* Journal of Nanobiotechnology, 2019. **17**(1).

349. Nkele, A.C. and F.I. Ezema, *Diverse Synthesis and Characterization Techniques of Nanoparticles.* 2021.

350. Low, Z.H., I. Ismail, and K.S. Tan, *Sintering Processing of Complex Magnetic Ceramic Oxides: A Comparison Between Sintering of Bottom-Up Approach Synthesis and Mechanochemical Process of Top-Down Approach Synthesis.* 2018.

351. Olatomiwa, A.L., et al., *Graphene Synthesis, Fabrication, Characterization Based on Bottom-Up and Top-Down Approaches: An Overview.* Journal of Semiconductors, 2022. **43**(6): p. 061101.

352. Singh, J., et al., *'Green' Synthesis of Metals and Their Oxide Nanoparticles: Applications for Environmental Remediation.* Journal of Nanobiotechnology, 2018. **16**(1).

353. Lancaster, C.A., et al., *Uniting Top-Down and Bottom-Up Strategies Using Fabricated Nanostructures as Hosts for Synthesis of Nanomites.* The Journal of Physical Chemistry C, 2020. **124**(12): p. 6822-6829.

354. Radjenović, B. and M. Radmilović-Radjenović, *Top Down Nano Technologies in Surface Modification of Materials.* Open Physics, 2011. **9**(2): p. 265-275.

355. Planillo, J. and F. Alves, *Fabrication and Characterization of Micrometer Scale Graphene Structures for Large-Scale Ultra-Thin Electronics.* Electronics, 2022. **11**(5): p. 752.

356. Wang, P., et al., *Controlled Syntheses and Multifunctional Applications of Two-Dimensional Metallic Transition Metal Dichalcogenides.* Accounts of Materials Research, 2021. **2**(9): p. 751-763.

357. Xu, H., et al., *Macromolecular Self-Assembly and Nanotechnology in China.* Philosophical Transactions of the Royal Society a Mathematical Physical and Engineering Sciences, 2013. **371**(2000): p. 20120305.

358. Manojlović, J., *Introduction to Nanotechnology and Molecular Self-Assembly.* Facta Universitatis Series Automatic Control and Robotics, 2018. **17**(2): p. 105.

359. Palma, M.I.M., et al., *Synthesis and Properties of Platinum Nanoparticles by Pulsed Laser Ablation in Liquid.* Journal of Nanomaterials, 2016. **2016**: p. 1-11.

360. Song, R., C. Wang, and Y. Jiang, *Synthesis of Metal/Poly(diphenylsilylenemethylene) Nanocomposite Thin Films by Pulsed Laser Ablation.* Journal of Wuhan University of Technology-Mater Sci Ed, 2015. **30**(1): p. 6-9.

361. Ye, F., et al., *Defect-Rich MoSe$_2$ 2h/1t Hybrid Nanoparticles Prepared From Femtosecond Laser Ablation in Liquid and Their Enhanced Photothermal Conversion Efficiencies.* Advanced Materials, 2023. **35**(30).

362. Wang, Y., et al., *Au-NP-Decorated Crystalline FeOCl Nanosheet: Facile Synthesis by Laser Ablation in Liquid and Its Exclusive Gas Sensing Response to HCl at Room Temperature.* Advanced Materials Interfaces, 2016. **3**(9).

363. Zhou, L., et al., *Onion-Structured Spherical MoS$_2$ Nanoparticles Induced by Laser Ablation in Water and Liquid Droplets' Radial Solidification/Oriented Growth Mechanism.* The Journal of Physical Chemistry C, 2017. **121**(41): p. 23233-23239.

364. Al-Ogaidi, M. and I. Al-Ogaidi, *Evaluating the Antibacterial Activity of AgGO Nanocomposite Against Clinical Isolate Bacteria.* Journal of Southwest Jiaotong University, 2019. **54**(6).

365. Dell'Aglio, M., et al., *Investigation on the Material in the Plasma Phase by High Temporally and Spectrally Resolved Emission Imaging During Pulsed Laser Ablation in Liquid (PLAL) for NPs Production and Consequent Considerations on NPs Formation.* Plasma Sources Science and Technology, 2019. **28**(8): p. 085017.

366. Godja, N.-C. and F.-D. Munteanu, *Hybrid Nanomaterials: A Brief Overview of Versatile Solutions for Sensor Technology in Healthcare and Environmental Applications.* Biosensors, 2024. **14**(2): p. 67.

367. Maduraiveeran, G., B.-R. Adhikari, and A. Chen, *Nanomaterials-Based Electrochemical Detection of Chemical Contaminants.* RSC Advances, 2014. **4**(109): p. 63741-63760.

368. Sable, H., et al., *Review—Nanosystems-Enhanced Electrochemical Biosensors for Precision in One Health Management.* Journal of the Electrochemical Society, 2024. **171**(3): p. 037527.

Advanced Material Engineering Fundamentals

369. Nie, Y., et al., *Synthesis and Structure-Dependent Optical Properties of ZnO Nanocomb and ZnO Nanoflag*. The Journal of Physical Chemistry C, 2017. **121**(46): p. 26076-26085.

370. Zeng, S., et al., *Nanomaterials Enhanced Surface Plasmon Resonance for Biological and Chemical Sensing Applications*. Chemical Society Reviews, 2014. **43**(10): p. 3426.

371. Gerdan, Z., Y. Saylan, and A. Denizli, *Biosensing Platforms for Cardiac Biomarker Detection*. Acs Omega, 2024. **9**(9): p. 9946-9960.

372. Park, H. and H. Choi, *Synthesis of Nanoparticles and One-Dimensional Nanomaterials*. 2009: p. 14-42.

373. Paramasivam, G., et al., *Nanomaterials: Synthesis and Applications in Theranostics*. Nanomaterials, 2021. **11**(12): p. 3228.

374. Li, H., et al., *Phase Engineering of Nanomaterials for Clean Energy and Catalytic Applications*. Advanced Energy Materials, 2020. **10**(40).

375. Aragay, G., F. Pino, and A. Merkoçi, *Nanomaterials for Sensing and Destroying Pesticides*. Chemical Reviews, 2012. **112**(10): p. 5317-5338.

376. Chen, S., et al., *Fe(III)-Tannic Acid Complex Derived Fe$_3$C Decorated Carbon Nanofibers for Triple-Enzyme Mimetic Activity and Their Biosensing Application*. Acs Biomaterials Science & Engineering, 2019. **5**(3): p. 1238-1246.

377. Yuan, X., et al., *Cellular Toxicity and Immunological Effects of Carbon-Based Nanomaterials*. Particle and Fibre Toxicology, 2019. **16**(1).

378. Al-Abed, S.R., et al., *Environmental Aging Alters Al(OH)$_3$ coating of TiO$_2$ nanoparticles Enhancing Their Photocatalytic and Phototoxic Activities*. Environmental Science Nano, 2016. **3**(3): p. 593-601.

379. Dong, Y., et al., *20.2: Ultra-Bright, Highly Efficient, Low Roll-Off Inverted Quantum-Dot Light Emitting Devices (QLEDs)*. Sid Symposium Digest of Technical Papers, 2015. **46**(1): p. 270-273.

380. Seol, M., et al., *Highly Efficient and Durable Quantum Dot Sensitized ZnO Nanowire Solar Cell Using Noble-Metal-Free Counter Electrode*. The Journal of Physical Chemistry C, 2011. **115**(44): p. 22018-22024.

381. Yun, H., et al., *Nanocrystal Size-Dependent Efficiency of Quantum Dot Sensitized Solar Cells in the Strongly Coupled CdSe Nanocrystals/TiO$_2$ System*. Acs Applied Materials & Interfaces, 2016. **8**(23): p. 14692-14700.

382. Tian, J. and G. Cao, *Design, Fabrication and Modification of Metal Oxide Semiconductor for Improving Conversion Efficiency of Excitonic Solar Cells*. Coordination Chemistry Reviews, 2016. **320-321**: p. 193-215.

383. Kim, H., et al., *Hybrid-Type Quantum-Dot Cosensitized ZnO Nanowire Solar Cell With Enhanced Visible-Light Harvesting*. Acs Applied Materials & Interfaces, 2012. **5**(2): p. 268-275.

384. Tian, J., et al., *Enhanced Performance of CdS/CdSe Quantum Dot Cosensitized Solar Cells via Homogeneous Distribution of Quantum Dots in TiO$_2$ Film*. The Journal of Physical Chemistry C, 2012. **116**(35): p. 18655-18662.

385. Nozik, A.J., *Nanoscience and Nanostructures for Photovoltaics and Solar Fuels*. Nano Letters, 2010. **10**(8): p. 2735-2741.

386. Cardoso, J., et al., *Germanium Quantum Dot Grätzel-Type Solar Cell*. Physica Status Solidi (A), 2018. **215**(24).

387. Huang, F., et al., *High Efficiency CdS/CdSe Quantum Dot Sensitized Solar Cells With Two ZnSe Layers*. Acs Applied Materials & Interfaces, 2016. **8**(50): p. 34482-34489.

388. Thabet, A., S. Abdelhady, and Y. Mobarak, *Design Modern Structure for Heterojunction Quantum Dot Solar Cells*. International Journal of Electrical and Computer Engineering (Ijece), 2020. **10**(3): p. 2918.

389. Kim, D.Y., et al., *Type-Ii InAs/GaAsSb Quantum Dot Solar Cells With GaAs Interlayer*. Ieee Journal of Photovoltaics, 2018. **8**(3): p. 741-745.

390. Li, Q., et al., *Core–shell ZnO@TiO$_2$ Hexagonal Prism Heterogeneous Structures as Photoanodes for Boosting the Efficiency of Quantum Dot Sensitized Solar Cells*. Dalton Transactions, 2024. **53**(6): p. 2867-2875.

391. Wan, X., et al., *Study on Preparation of CdS Quantum Dots for Dye Sensitized Solar Cells*. Advanced Materials Research, 2014. **1070-1072**: p. 608-611.

392. Oshima, R., et al., *High-density Quantum Dot Superlattice for Application to High-efficiency Solar Cells*. Physica Status Solidi (C), 2010. **8**(2): p. 619-621.

393. Kramer, I.J. and E.H. Sargent, *Colloidal Quantum Dot Photovoltaics: A Path Forward*. Acs Nano, 2011. **5**(11): p. 8506-8514.

394. Cappelluti, F., A. Musu, and A. Khalili, *Study of Light-Trapping Enhanced Quantum Dot Solar Cells Based on Electrical and Optical Numerical Simulations*. 2016.

395. Shen, H., et al., *High-Efficiency, Low Turn-on Voltage Blue-Violet Quantum-Dot-Based Light-Emitting Diodes*. Nano Letters, 2015. **15**(2): p. 1211-1216.

396. Roelofs, K.E., et al., *Recombination Barrier Layers in Solid-State Quantum Dot-Sensitized Solar Cells*. 2012: p. 003040-003043.

397. Li, X., *Applications of ZnSe Quantum Dots for Solar Energy Harvesting*. E3s Web of Conferences, 2024. **520**: p. 03024.

398. McDaniel, H., et al., *Engineered CuInSe$_x$S$_{2-x}$ Quantum Dots for Sensitized Solar Cells*. The Journal of Physical Chemistry Letters, 2013. **4**(3): p. 355-361.

399. Semonin, O.E., et al., *Peak External Photocurrent Quantum Efficiency Exceeding 100% via MEG in a Quantum Dot Solar Cell*. Science, 2011. **334**(6062): p. 1530-1533.

400. Roelofs, K.E., T.P. Brennan, and S.F. Bent, *Interface Engineering in Inorganic-Absorber Nanostructured Solar Cells*. The Journal of Physical Chemistry Letters, 2014. **5**(2): p. 348-360.

401. Yang, Z., et al., *Quantum Dot-Sensitized Solar Cells Incorporating Nanomaterials*. Chemical Communications, 2011. **47**(34): p. 9561.

402. Tang, J., et al., *Quantum Junction Solar Cells*. Nano Letters, 2012. **12**(9): p. 4889-4894.

403. Acquavia, M.A., et al., *Natural Polymeric Materials: A Solution to Plastic Pollution From the Agro-Food Sector*. Polymers, 2021. **13**(1): p. 158.

404. Rosenboom, J.G., R. Langer, and G. Traverso, *Bioplastics for a Circular Economy*. Nature Reviews Materials, 2022. **7**(2): p. 117-137.

405. Atiwesh, G., et al., *Environmental Impact of Bioplastic Use: A Review*. Heliyon, 2021. **7**(9): p. e07918.

406. Arikan, E.B. and H. Bilgen, *Production of Bioplastic From Potato Peel Waste and Investigation of Its Biodegradability.* International Advanced Researches and Engineering Journal, 2019. **3**(2): p. 93-97.

407. Santana, I., M. Felix, and C. Bengoechea, *Seaweed as Basis of Eco-Sustainable Plastic Materials: Focus on Alginate.* Polymers, 2024. **16**(12): p. 1662.

408. Vu, D.H., et al., *Production of Polyhydroxyalkanoates (PHAs) By <i>Bacillus Megaterium</I> Using Food Waste Acidogenic Fermentation-Derived Volatile Fatty Acids.* Bioengineered, 2021. **12**(1): p. 2480-2498.

409. Poltronieri, P., *Polyhydroxyalkanoate Production in Biofermentor Monitored Through Biosensor Application.* International Journal of Biosensors & Bioelectronics, 2018. **4**(5).

410. Pilco, C.J., et al., *Use of Organic Fruit Residues to Obtain Bioplastics.* Migration Letters, 2024. **21**(S5): p. 1382-1400.

411. Molina-Besch, K., *Use Phase and End-of-Life Modeling of Biobased Biodegradable Plastics in Life Cycle Assessment: A Review.* Clean Technologies and Environmental Policy, 2022. **24**(10): p. 3253-3272.

412. Sulaeman, B., et al., *Development of Bioplastics From Tawaro's Environmentally Friendly Sago Starch (Metroxylon).* Eastern-European Journal of Enterprise Technologies, 2023. **5**(12 (125)): p. 6-16.

413. Fòlino, A., et al., *Biodegradation of Wasted Bioplastics in Natural and Industrial Environments: A Review.* Sustainability, 2020. **12**(15): p. 6030.

414. Ali, S., I. Isha, and Y.-C. Chang, *Ecotoxicological Impact of Bioplastics Biodegradation: A Comprehensive Review.* 2023.

415. Schmidt, B.V.K.J., *Polymer Chemistry: Fundamentals and Applications.* Beilstein Journal of Organic Chemistry, 2021. **17**: p. 2922-2923.

416. Bettinger, C.J., et al., *Amino Alcohol-Based Degradable Poly(ester Amide) Elastomers.* Biomaterials, 2008. **29**(15): p. 2315-2325.

417. Amsden, B.G., et al., *Synthesis and Characterization of a Photo-Cross-Linked Biodegradable Elastomer.* Biomacromolecules, 2004. **5**(6): p. 2479-2486.

418. Wei, Y., *Nonclassical or Reactivation Chain Polymerization: A General Scheme of Polymerization.* Journal of Chemical Education, 2001. **78**(4): p. 551.

419. Amsden, B.G., *Curable, Biodegradable Elastomers: Emerging Biomaterials for Drug Delivery and Tissue Engineering.* Soft Matter, 2007. **3**(11): p. 1335.

420. Bettinger, C.J., *Synthetic Biodegradable Elastomers for Drug Delivery and Tissue Engineering.* Pure and Applied Chemistry, 2010. **83**(1): p. 9-24.

421. He, X., R. Huang, and L. Tang, *Advanced Elasticity and Biodegradability of Bio-Based Copolyester Elastomer Achieved by the Decrystalization of Isosorbide and Flexibility of 1,6-Hexanediol.* 2024.

422. Momeni, S., et al., *The Effect of Poly (Ethylene Glycol) Emulation on the Degradation of PLA/Starch Composites.* Polymers, 2021. **13**(7): p. 1019.

423. Doppalapudi, S., et al., *Biodegradable Polymers—an Overview.* Polymers for Advanced Technologies, 2014. **25**(5): p. 427-435.

424. Göktürk, E. and H. Erdal, *Poliglikolik Asit' in (PGA) Biyomedikal Uygulamaları.* Sakarya University Journal of Science, 2017: p. 1-1.

425. Tang, D., et al., *Poly(urea Ester): A Family of Biodegradable Polymers With High Melting Temperatures.* Journal of Polymer Science Part a Polymer Chemistry, 2016. **54**(24): p. 3795-3799.

426. Gümüş, S., G. Özkoç, and A. Aytaç, *Plasticized and Unplasticized PLA/organoclay Nanocomposites: Short- and Long-term Thermal Properties, Morphology, and Nonisothermal Crystallization Behavior.* Journal of Applied Polymer Science, 2011. **123**(5): p. 2837-2848.

427. Hsu, S.-T., H. Tan, and Y.L. Yao, *Effect of Laser-Induced Crystallinity Modification on Biodegradation Profile of Poly(L-Lactic Acid).* Journal of Manufacturing Science and Engineering, 2013. **136**(1).

428. Eckel, F., et al., *Influence of Microbial Biomass Content on Biodegradation and Mechanical Properties of Poly(3-Hydroxybutyrate) Composites.* Biodegradation, 2023. **35**(2): p. 209-224.

429. Puchalski, M., et al., *Molecular and Supramolecular Changes in Polybutylene Succinate (PBS) and Polybutylene Succinate Adipate (PBSA) Copolymer During Degradation in Various Environmental Conditions.* Polymers, 2018. **10**(3): p. 251.

430. Xue, Z., *A Review on Biodegradable Polymer: Shortcomings, Developments, and Future Direction.* Applied and Computational Engineering, 2023. **23**(1): p. 86-95.

431. Shamsuddin, I.M., et al., *Biodegradable Polymers for Sustainable Environmental and Economic Development.* Moj Bioorganic & Organic Chemistry, 2018. **2**(4).

432. Silva, T.F.d., et al., *Preparation and Characterization of Antistatic Packaging for Electronic Components Based on Poly(lactic Acid)/Carbon Black Composites.* Journal of Applied Polymer Science, 2018. **136**(13).

433. Rheinberger, T., et al., *RNA-Inspired and Accelerated Degradation of Polylactide in Seawater.* Journal of the American Chemical Society, 2021. **143**(40): p. 16673-16681.

434. Kanzariya, R., et al., *Kinetics of Biomass and Polyhydroxyalkanoates Synthesis Using Sugar Industry Waste as Carbon Substrate by Alcaligenes Sp. NCIM 5085.* Journal of Environmental Biology, 2023. **44**(4): p. 612-622.

435. Li, X., et al., *Evaluating the Biological Impact of Polyhydroxyalkanoates (PHAs) on Developmental and Exploratory Profile of Zebrafish Larvae.* RSC Advances, 2016. **6**(43): p. 37018-37030.

436. Doat, O., et al., *Controlling the Marine Biodegradation Profile and Mechanical Properties of Poly(ε-Caprolactone) With Hydrophobic Water-Responsive Linkages.* Acs Applied Polymer Materials, 2023. **6**(1): p. 244-252.

437. Vastano, M., et al., *Conversion of No/Low Value Waste Frying Oils Into Biodiesel and Polyhydroxyalkanoates.* Scientific Reports, 2019. **9**(1).

438. Nandini, A., D. Nagarajan, and J.-S. Chang, *Production of Lactic Acid from Microalgal Biomass Chlorella Vulgar ESP-31 as a Feedstock Using PVA Immobilized Bacteria L. Plantarum 23.* Nusantara Science and Technology Proceedings, 2020: p. 165-169.

439. Nagarajan, D., et al., *Lactic Acid Production From Renewable Feedstocks Using Poly(vinyl Alcohol)-Immobilized <i>Lactobacillus Plantarum</I> 23.* Industrial & Engineering Chemistry Research, 2020. **59**(39): p. 17156-17164.

440. Hwang, H.J., et al., *Lactic Acid Production From Seaweed Hydrolysate of Enteromorpha Prolifera (Chlorophyta)*. Journal of Applied Phycology, 2011. **24**(4): p. 935-940.

441. Parate, R., et al., *Bioglycerol (<scp>C3</Scp>) Upgrading to 2,3-butanediol (<scp>C4</Scp>) by Cell-free Extracts of <scp><i>Enterobacter Aerogenes</I> NCIM</scp> 2695*. Journal of Chemical Technology & Biotechnology, 2021. **96**(5): p. 1316-1325.

442. Wang, L., et al., *Fermentation of Sweet Sorghum Derived Sugars to Butyric Acid at High Titer and Productivity by a Moderate Thermophile Clostridium Thermobutyricum at 50 °C*. Bioresource Technology, 2015. **198**: p. 533-539.

443. Wang, S. and L. Copeland, *Molecular Disassembly of Starch Granules During Gelatinization and Its Effect on Starch Digestibility: A Review*. Food & Function, 2013. **4**(11): p. 1564.

444. Yamak, H.B., *Thermal, Mechanical and Antibacterial Properties of Ldpe/Starch Bio-Based Polymer Blends for Food Packing Applications*. Journal of the Turkish Chemical Society Section a Chemistry, 2016. **3**(3): p. 637-637.

445. Aleksanyan, K.V., С.З. Роговина, and N.E. Ivanushkina, *Novel Biodegradable Low-density Polyethylene–poly(lactic Acid)–starch Ternary Blends*. Polymer Engineering & Science, 2020. **61**(3): p. 802-809.

446. Tyński, P., et al., *Properties of Biodegradable Films Based on Thermoplastic Starch and Poly(butylene Succinate) With Plant Oil Additives*. 2019: p. 257-261.

447. Liu, X., et al., *New Evidences of Accelerating Degradation of Polyethylene by Starch*. Journal of Applied Polymer Science, 2013. **130**(4): p. 2282-2287.

448. de Souza, F.M. and R.K. Gupta, *Bacteria for Bioplastics: Progress, Applications, and Challenges*. Acs Omega, 2024. **9**(8): p. 8666-8686.

449. Inoue, K., et al., *Catalytic Conversion of Ethanol to Propylene by H-ZSM-11*. Reaction Kinetics Mechanisms and Catalysis, 2010. **101**(1): p. 227-235.

450. Ayala, H., et al., *Valorization of Cocoa's Mucilage Waste to Ethanol and Subsequent Direct Catalytic Conversion Into Ethylene*. Journal of Chemical Technology & Biotechnology, 2022. **97**(8): p. 2171-2178.

451. Xin, H., et al., *Catalytic Dehydration of Ethanol Over Post-Treated ZSM-5 Zeolites*. Journal of Catalysis, 2014. **312**: p. 204-215.

452. Wu, L., et al., *Hybrid Biological–Chemical Approach Offers Flexibility and Reduces the Carbon Footprint of Biobased Plastics, Rubbers, and Fuels*. Acs Sustainable Chemistry & Engineering, 2018. **6**(11): p. 14523-14532.

453. Adkins, J., et al., *Engineering Microbial Chemical Factories to Produce Renewable "Biomonomers"*. Frontiers in Microbiology, 2012. **3**.

454. Coltelli, M.B., et al., *Preparation and Compatibilization of PBS/Whey Protein Isolate Based Blends*. Molecules, 2020. **25**(14): p. 3313.

455. Jiménez-Rosado, M., et al., *Injection Molding Versus Extrusion in the Manufacturing of Soy Protein -based Bioplastics With Zinc Incorporated*. Journal of Applied Polymer Science, 2021. **139**(7).

456. Jariyasakoolroj, P., P. Leelaphiwat, and N. Harnkarnsujarit, *Advances in Research and Development of Bioplastic for Food Packaging.* Journal of the Science of Food and Agriculture, 2019. **100**(14): p. 5032-5045.

457. Aversa, C., et al., *Injection-stretch Blow Molding of Poly (Lactic Acid)/Polybutylene Succinate Blends for the Manufacturing of Bottles.* Journal of Applied Polymer Science, 2021. **139**(4).

458. Gurram, R., et al., *A Solvent-Free Approach for Production of Films From Pectin and Fungal Biomass.* Journal of Polymers and the Environment, 2018. **26**(11): p. 4282-4292.

459. Capezza, A.J., et al., *Biodegradable Fiber-Reinforced Gluten Biocomposites for Replacement of Fossil-Based Plastics.* Acs Omega, 2023. **9**(1): p. 1341-1351.

460. Kobayashi, S., *Lipase-Catalyzed Polyester Synthesis - A Green Polymer Chemistry.* Proceedings of the Japan Academy Series B, 2010. **86**(4): p. 338-365.

461. Todea, A., et al., *Achievements and Trends in Biocatalytic Synthesis of Specialty Polymers From Biomass-Derived Monomers Using Lipases.* Processes, 2021. **9**(4): p. 646.

462. Jiao, Y. and K. Loos, *Enzymatic Synthesis of Biobased Polyesters and Polyamides.* Polymers, 2016. **8**(7): p. 243.

463. Çavdur, T.T. and P. Aniş, *The Effect of Enzymatic Modification on the Dyeability of Polyester Fabric With Reactive Dye.* Aatcc Journal of Research, 2020. **7**(6): p. 41-47.

464. Amaraweera, S.M., et al., *Development of Starch-Based Materials Using Current Modification Techniques and Their Applications: A Review.* Molecules, 2021. **26**(22): p. 6880.

465. Pyser, J.B., et al., *State-of-the-Art Biocatalysis.* Acs Central Science, 2021. **7**(7): p. 1105-1116.

466. Mao, Y., et al., *Flow-through Enzymatic Reactors Using Polymer Monoliths: From Motivation to Application.* Electrophoresis, 2021. **42**(24): p. 2599-2614.

467. Al-Khairy, D., et al., *Closing the Gap Between Bio-Based and Petroleum-Based Plastic Through Bioengineering.* Microorganisms, 2022. **10**(12): p. 2320.

468. Dey, A., et al., *A Performance Study on 3d-Printed Bioplastic Pots From Soybean by-Products.* Sustainability, 2023. **15**(13): p. 10535.

469. Yali, W., *Application of Genetically Modified Organism (GMO) Crop Technology and Its Implications in Modern Agriculture.* International Journal of Applied Agricultural Sciences, 2022. **8**(1): p. 1.

470. Liu, B.R., et al., *Cell-Penetrating Peptides for Use in Development of Transgenic Plants.* Molecules, 2023. **28**(8): p. 3367.

471. Haile, G., M. Adamu, and T. Tekle, *The Effects of Genetically Modified Organisms (GMO) on Environment and Molecular Techniques to Minimize Its Risk.* American Journal of Polymer Science and Technology, 2020. **6**(4): p. 32.

472. Turnbull, C., M. Lillemo, and T. Hvoslef-Eide, *Global Regulation of Genetically Modified Crops Amid the Gene Edited Crop Boom – A Review.* Frontiers in Plant Science, 2021. **12**.

473. Wadood, B., et al., *Genetically Modified Plants and Climate Change.* Journal of Arable Crops and Marketing, 2022. **4**(1): p. 71-79.

474. Gui, M.S.Z., et al., *Comparison of the Yield and Properties of Bio-Oil Produced by Slow and Fast Pyrolysis of Rice Husks and Coconut Shells.* Applied Mechanics and Materials, 2014. **625**: p. 626-629.

475. Prasetiawan, H., et al., *Preliminary Study on the Bio-Oil Production From Multi Feed-Stock Biomass Waste via Fast Pyrolysis Process.* Journal of Advanced Research in Fluid Mechanics and Thermal Sciences, 2023. **103**(2): p. 216-227.

476. Afrah, B.D., et al., *Effect of Pyrolysis Temperature and Biomass Composition on Bio-Oil Characteristics.* Ecological Engineering & Environmental Technology, 2024. **25**(3): p. 264-274.

477. Shi, K., et al., *Review of Catalytic Pyrolysis of Biomass for Bio-Oil.* 2011: p. 317-321.

478. Aqsha, A., M.M. Tijani, and N. Mahinpey, *Catalytic Pyrolysis of Straw Biomasses (Wheat, Flax, Oat and Barley Straw) and the Comparison of Their Product Yields.* 2014.

479. Bandara, J.C., M.S. Eikeland, and B.M.E. Moldestad, *Analysis of the Effect of Steam-to-Biomass Ratio in Fluidized Bed Gasification With Multiphase Particle-in-Cell CFD Simulation.* 2017.

480. Nzihou, A., G. Flamant, and B.R. Stanmore, *Synthetic Fuels From Biomass Using Concentrated Solar Energy – A Review.* Energy, 2012. **42**(1): p. 121-131.

481. Hardian, R., et al., *Artificial Intelligence: The Silver Bullet for Sustainable Materials Development.* Green Chemistry, 2020. **22**(21): p. 7521-7528.

482. Zhang, B.G.X., et al., *Bioactive Coatings for Orthopaedic Implants—Recent Trends in Development of Implant Coatings.* International Journal of Molecular Sciences, 2014. **15**(7): p. 11878-11921.

483. Hu, L.-X., et al., *Biological Evaluation of the Copper/Low-Density Polyethylene Nanocomposite Intrauterine Device.* Plos One, 2013. **8**(9): p. e74128.

484. Ramezani, M. and Z.M. Ripin, *An Overview of Enhancing the Performance of Medical Implants With Nanocomposites.* Journal of Composites Science, 2023. **7**(5): p. 199.

485. Danso, H., *Identification of Key Indicators for Sustainable Construction Materials.* Advances in Materials Science and Engineering, 2018. **2018**(1).

486. Gunansyah, G., *Between the Sustainable Development Narrative and the Environmental Crisis.* Biokultur, 2022. **11**(1): p. 28-44.

487. Levorová, J., et al., *In Vivo Study on Biodegradable Magnesium Alloys: Bone Healing Around WE43 Screws.* Journal of Biomaterials Applications, 2017. **32**(7): p. 886-895.

488. Kim, S.J., et al., *The Potential Role of Polymethyl Methacrylate as a New Packaging Material for the Implantable Medical Device in the Bladder.* Biomed Research International, 2015. **2015**: p. 1-8.

489. Bakar, M.A.A., et al., *Effect of Carbon Fibre Ratio to the Impact Properties of Hybrid Kenaf/Carbon Fibre Reinforced Epoxy Composites.* Applied Mechanics and Materials, 2013. **393**: p. 136-139.

490. Swolfs, Y., et al., *Global Load-Sharing Model for Unidirectional Hybrid Fibre-Reinforced Composites.* Journal of the Mechanics and Physics of Solids, 2015. **84**: p. 380-394.

491. Hu, L., et al., *Hydrogel-Based Flexible Electronics.* Advanced Materials, 2023. **35**(14).

492. Wang, S., et al., *Skin-Inspired Electronics: An Emerging Paradigm.* Accounts of Chemical Research, 2018. **51**(5): p. 1033-1045.

493. Zhao, F., et al., *Multifunctional Nanostructured Conductive Polymer Gels: Synthesis, Properties, and Applications.* Accounts of Chemical Research, 2017. **50**(7): p. 1734-1743.

494. Cheng, H., et al., *Poly(vinyl Alcohol), Tannic Acid, and Silver-Based Hydrogel Strain Sensors With "Fish Scale-Like" Surfaces.* Acs Applied Polymer Materials, 2023. **5**(6): p. 4146-4158.

495. Chen, L., et al., *Highly Thermally Stable, Green Solvent Disintegrable, and Recyclable Polymer Substrates for Flexible Electronics.* Macromolecular Rapid Communications, 2020. **41**(19).

496. Wang, Y., et al., *Stretchable, Biodegradable Dual Cross-Linked Chitin Hydrogels With High Strength and Toughness and Their Potential Applications in Flexible Electronics.* Acs Sustainable Chemistry & Engineering, 2023. **11**(18): p. 7083-7093.

497. Asta, M., *Computational Materials Discovery and Design.* Jom, 2014. **66**(3): p. 364-365.

498. Exl, L., et al., *Magnetic Microstructure Machine Learning Analysis.* Journal of Physics Materials, 2018.

499. Durbacă, I., et al., *Approaches to the Evaluation of the Mechanical Properties of Single-Layer Composite Plates Made of Recyclable Polymeric and Protein Materials.* 2020.

500. El-Azab, A., *Why Materials Theory?* Materials Theory, 2017. **1**(1).

501. Tao, C., et al., *Mechanical Properties of Cemented Particulate Composite: A 3D Micromechanical Model.* Materials, 2021. **14**(14): p. 3875.

502. Kalaiselvi, V., R. Mathammal, and P. Anitha, *Synthesis and Characterization of Hydroxyapatite Nanoparticles Using Wet Chemical Method.* International Journal of Advanced Science and Engineering, 2017. **4**(2): p. 571.

503. Toda, K., et al., *Synthesis of Nano-Sized Materials Using Novel Water Assisted Solid State Reaction Method.* Key Engineering Materials, 2018. **777**: p. 163-167.

504. Kocazorbaz, E.K., *Euphorbia Rigida Yaprak Ekstraktından Gümüş Nanopartiküllerin Sentezi, Optimizasyonu Ve Karakterizayonu Ve Antimikrobiyal Potansiyellerinin Araştırılması.* Bilecik Şeyh Edebali Üniversitesi Fen Bilimleri Dergisi, 2021. **8**(2): p. 512-522.

505. Золотаренко, О.Д., et al., *Creation of 3D-products using carbon nanostructures and 3D-printing technologies (FDM, CJP, SLA, SLS).* Поверхня, 2023(15 (30)): p. 110-134.

506. Gürünlü, B., *Investigation of Alternative Techniques for Graphene Synthesis.* 2021.

507. Chitranshi, M., et al., *Carbon Nanotube Sheet-Synthesis and Applications.* Nanomaterials, 2020. **10**(10): p. 2023.

508. Pai, R.K. and M. Cotlet, *Highly Stable, Water-Soluble, Intrinsic Fluorescent Hybrid Scaffolds for Imaging and Biosensing.* The Journal of Physical Chemistry C, 2011. **115**(5): p. 1674-1681.

509. Wan, Z., et al., *Machine Learning Prediction of the Optimal Carrier Concentration and Band Gap of Quaternary Thermoelectric Materials via Element Feature Descriptors.* International Journal of Quantum Chemistry, 2021. **121**(18).

510. Cha, H.W., et al., *Transmission Electron Microscopy Specimen Preparation of Delicate Materials Using Tripod Polisher.* Applied Microscopy, 2016. **46**(2): p. 110-115.

511. Leer, B.V., et al., *New Workflows Broaden Access to S/Tem Analysis and Increase Productivity.* Microscopy Today, 2018. **26**(1): p. 18-25.

512. Wang, W., X. Wu, and J. Zhang, *Graphene and Other 2D Material Components Dynamic Characterization and Nanofabrication at Atomic Scale.* Journal of Nanomaterials, 2015. **2015**(1).

513. Fultz, B. and J.M. Howe, *Transmission Electron Microscopy and Diffractometry of Materials.* 2013.

514. Velez, N.R., et al., *Development of Quantitative <i>In Situ</I> TEM Nanomechanical Testing for Polymers.* Microscopy and Microanalysis, 2017. **23**(S1): p. 742-743.

515. Anber, E.A., et al., *Thermal Stability of High Entropy Alloys During in Situ TEM Heating.* Microscopy and Microanalysis, 2018. **24**(S1): p. 1928-1929.

516. Sorgente, D. and L. Tricarico, *Characterization of a Superplastic Aluminium Alloy ALNOVI-U Through Free Inflation Tests and Inverse Analysis.* International Journal of Material Forming, 2012. **7**(2): p. 179-187.

517. Dai, X. and Y. Ni, *Testing Extrusion Flow Stress and Friction Factor via Inverse Analysis.* Advanced Materials Research, 2011. **381**: p. 128-134.

518. Iyer, A., et al., *Data-Centric Mixed-Variable Bayesian Optimization for Materials Design.* 2019.

519. Schmaltz, S. and K. Willner, *Identification of Material Behavior via a Finite Element Model Updating Strategy.* Pamm, 2014. **14**(1): p. 439-440.

520. Cao, J., *Inner Sequential Single Solid Method for Layout Optimization of Multi-Materials.* Journal of Physics Conference Series, 2022. **2235**(1): p. 012091.

521. Zhang, L., et al., *Experiment Research on Complex Optimization Algorithm-Based Adaptive Iterative Learning Control for Electro-Hydraulic Shaking Tables.* Electronics, 2023. **12**(8): p. 1797.

522. Alayón, C., K. Säfsten, and G. Johansson, *Conceptual Sustainable Production Principles in Practice: Do They Reflect What Companies Do?* Journal of Cleaner Production, 2017. **141**: p. 693-701.

523. Nordin, N., H. Ashari, and M.G. Hassan, *Drivers and Barriers in Sustainable Manufacturing Implementation in Malaysian Manufacturing Firms.* 2014: p. 687-691.

524. Abdul-Rashid, S.H., et al., *The Impact of Sustainable Manufacturing Practices on Sustainability Performance.* International Journal of Operations & Production Management, 2017. **37**(2): p. 182-204.

525. Hermawan, A.N., et al., *The Effect of Sustainable Manufacturing on Environmental Performance Through Government Regulation and Eco-Innovation.* International Journal of Industrial Engineering and Operations Management, 2023. **6**(4): p. 299-325.

526. Rani, S., *Impact of Sustainable Manufacturing Practices on Financial Performance of MSME in Coimbatore District.* International Journal of Engineering and Advanced Technology, 2019. **9**(1s4): p. 1033-1036.

527. Bouazza, Y., A. Lajjam, and B. Dkhissi, *Green Manufacturing and Environmental Sustainability Case of Morocco.* Advanced Engineering Forum, 2023. **50**: p. 125-134.

528. Yuan, C., Q. Zhai, and D. Dornfeld, *A Three Dimensional System Approach for Environmentally Sustainable Manufacturing*. Cirp Annals, 2012. **61**(1): p. 39-42.

529. Gretsch, O., E.C. Salzmann, and A. Kock, *University-industry Collaboration and Front-end Success: The Moderating Effects of Innovativeness and Parallel Cross-firm Collaboration*. R and D Management, 2019. **49**(5): p. 835-849.

530. Moellers, T., C. Visini, and M. Haldimann, *Complementing Open Innovation in Multi-business Firms: Practices for Promoting Knowledge Flows Across Internal Units*. R and D Management, 2018. **50**(1): p. 96-115.

531. Hydle, K.M., et al., *Digital Transformation Through Collaborative Platformization: A Study of Incumbent-Entrepreneur Relations*. 2021.

532. Deininger, M., et al., *Novice Designers' Use of Prototypes in Engineering Design*. Design Studies, 2017. **51**: p. 25-65.

533. Camburn, B., et al., *Design Prototyping Methods: State of the Art in Strategies, Techniques, and Guidelines*. Design Science, 2017. **3**.

534. Coutts, E., A. Wodehouse, and J. Robertson, *A Comparison of Contemporary Prototyping Methods*. Proceedings of the Design Society International Conference on Engineering Design, 2019. **1**(1): p. 1313-1322.

535. Mahtani, R., K. Umstead, and C. Gill, *Efficacious Prototyping for Early Stage Industrial Design: Understanding What Matters in Prototyping to Make Prototyping Matter More*. 2019.

536. Suteja, J. and M.A. Hadiyat, *Optimization of Material Removal Rate and Dimensional Errors in Subtractive Rapid Prototyping of Polycarbonate Material*. Materials Science Forum, 2020. **975**: p. 235-241.

537. Sharma, A. and G. Luthra, *Significance of ISO 10993 Standards in Ensuring Biocompatibility of Medical Devices: A Review*. Journal of Pharmaceutical Research International, 2023. **35**(8): p. 23-34.

538. Panwar, A.S., et al., *Intermetallic Microstructural and Mechanical Response of Niti (Nitinol) Shape Memory Alloy With Laser Welding to Various Materials*. 2022. **7**(2): p. 127-136.

539. Hartl, D.J., et al., *Standardization of Shape Memory Alloy Test Methods Toward Certification of Aerospace Applications*. Smart Materials and Structures, 2015. **24**(8): p. 082001.

540. Inselman, D.W., C.J. Medberry, and W. Czaja, *Bacterially Derived Medical Devices: How Commercialization of Bacterial Nanocellulose and Other Biofabricated Products Requires Challenging of Standard Industrial Practices*. Journal of Biomedical Materials Research Part B Applied Biomaterials, 2021. **109**(11): p. 1953-1959.

541. Morricone, S., et al., *Commercialization Strategy and IPO Underpricing*. Research Policy, 2017. **46**(6): p. 1133-1141.

542. Stein, A. and B. Ramaseshan, *Towards the Identification of Customer Experience Touch Point Elements*. Journal of Retailing and Consumer Services, 2016. **30**: p. 8-19.

543. Lilja, J., M. Eriksson, and P. Ingelsson, *Commercial Experiences From a Customer Perspective*. The TQM Journal, 2010. **22**(3): p. 285-292.

544. Huang, L. and Y. Xu, *A Approach to Identify the Commercialization Potential of New Technology*. 2010.

545. Llanes, R.P., *E-Commerce as a Tool to Boost the Development of Cuban Agribusiness Companies.* Scientia Et Technica, 2020. **25**(1): p. 120-126.

546. Pererva, P., et al., *Formation of Intellectual Property Commercialization Strategies.* Eastern-European Journal of Enterprise Technologies, 2024. **1**(13 (127)): p. 80-91.

547. Harvey, J.P., et al., *Experimental Methods in Chemical Engineering: Differential Scanning Calorimetry—DSC.* The Canadian Journal of Chemical Engineering, 2018. **96**(12): p. 2518-2525.

548. Schick, C., *Differential Scanning Calorimetry (DSC) of Semicrystalline Polymers.* Analytical and Bioanalytical Chemistry, 2009. **395**(6): p. 1589-1611.

549. Leyva-Porras, C., et al., *Application of Differential Scanning Calorimetry (DSC) and Modulated Differential Scanning Calorimetry (MDSC) in Food and Drug Industries.* Polymers, 2019. **12**(1): p. 5.

550. Righetti, M.C., *Crystallization of Polymers Investigated by Temperature-Modulated DSC.* Materials, 2017. **10**(4): p. 442.

551. Gunaratne, L.M.W.K. and R.A. Shanks, *Thermal Memory of Poly(3-hydroxybutyrate) Using Temperature-modulated Differential Scanning Calorimetry.* Journal of Polymer Science Part B Polymer Physics, 2005. **44**(1): p. 70-78.

552. Rodríguez-Pacheco, L.C., et al., *Low-Cost and Novel Arduino®-Load Cell-Based Prototype to Determine Transition Temperatures.* Polímeros, 2024. **34**(1).

553. Soygun, K., G. Bolayır, and A. Boztuğ, *Mechanical and Thermal Properties of Polyamide Versus Reinforced PMMA Denture Base Materials.* The Journal of Advanced Prosthodontics, 2013. **5**(2): p. 153.

554. Roy, S., et al., *Montmorillonite–multiwalled Carbon Nanotube Nanoarchitecture Reinforced Thermoplastic Polyurethane.* Polymer Composites, 2014. **37**(6): p. 1775-1785.

555. Jager, S., et al., *Dynamic Thermo-Mechanical Properties of Various Flowable Resin Composites.* Journal of Clinical and Experimental Dentistry, 2016: p. 0-0.

556. Jeż, K., B. Jeż, and P. Pietrusiewicz, *Influence of Isothermal Heating on the Curie Temperature of FeCoB Bulk Amorphous Alloy.* Revista De Chimie, 2019. **70**(9): p. 3158-3162.

557. Carlstedt, M., et al., *Application of Lorentz Force Eddy Current Testing and Eddy Current Testing on Moving Nonmagnetic Conductors.* International Journal of Applied Electromagnetics and Mechanics, 2014. **45**(1-4): p. 519-526.

558. Chon, G.-B., et al., *Order-Disorder Transformation in Fe<SUB>50</SUB>Co<SUB>50</SUB> Particles Synthesized by Polyol Process.* Materials Transactions, 2010. **51**(4): p. 707-711.

559. Tan, Y., X. Wang, and R. Moreau, *An Innovative Contactless Method for Detecting Defects in Electrical Conductors by Measuring a Change in Electromagnetic Torque.* Measurement Science and Technology, 2015. **26**(3): p. 035602.

560. Lu 鲁, S.帅., X.学. Ma 马, and S.松. Liu 刘, *Preliminary Electromagnetic Analysis of the COOL Blanket for CFETR.* Plasma Science and Technology, 2024. **26**(1): p. 015601.

561. Brauer, H., et al., *Lorentz Force Eddy Current Testing: A Novel NDE-technique.* Compel the International Journal for Computation and Mathematics in Electrical and Electronic Engineering, 2014. **33**(6): p. 1965-1977.

562.	Uhlig, R.P., et al., *Lorentz Force Sigmometry: A Contactless Method for Electrical Conductivity Measurements.* Journal of Applied Physics, 2012. **111**(9).

563.	Otterbach, J.M., et al., *Comparison of Defect Detection Limits in Lorentz Force Eddy Current Testing and Classical Eddy Current Testing.* Journal of Sensors and Sensor Systems, 2018. **7**(2): p. 453-459.

564.	Cheng, L., et al., *Vector-Based Eddy-Current Testing Method.* Applied Sciences, 2018. **8**(11): p. 2289.

565.	Zec, M., et al., *Finite Element Analysis of Nondestructive Testing Eddy Current Problems With Moving Parts.* Ieee Transactions on Magnetics, 2013. **49**(8): p. 4785-4794.

566.	Ahmed, J., et al., *Enhanced Electrocatalytic Activity of Copper–Cobalt Nanostructures.* The Journal of Physical Chemistry C, 2011. **115**(30): p. 14526-14533.

567.	Belozerov, E.V., et al., *High-Strength Magnetically Hard Fe-Cr-Co-Based Alloys With Reduced Content of Chromium and Cobalt.* The Physics of Metals and Metallography, 2012. **113**(4): p. 319-325.

568.	Mohamed, M.A., et al., *Temperature-Dependent Rigidity and Magnetism of Polyamide 6 Nanocomposites Based on Nanocrystalline Fe-Ni Alloy of Various Geometries.* Express Polymer Letters, 2016. **10**(10): p. 822-834.

569.	Zhang, Z.R., et al., *Magnetoelastic Properties of Epoxy Resin Based $Tb_xHo_{0.9-x}Nd_{0.1}(Fe_{0.8}Co_{0.2})_{1.93}$ Particulate Composites.* Materials Science-Poland, 2017. **35**(1): p. 81-86.

570.	Zhu, Y., et al., *A Review of Optical NDT Technologies.* Sensors, 2011. **11**(8): p. 7773-7798.

571.	Hess, P. and A.M. Lomonosov, *Solitary Surface Acoustic Waves and Bulk Solitons in Nanosecond and Picosecond Laser Ultrasonics.* Ultrasonics, 2010. **50**(2): p. 167-171.

572.	Shetu, M.S.A., *Review of Nondestructive Testing Methods for Aerospace Composite Materials.* 2024. **3**(1): p. 30-41.

573.	Wang, B., et al., *Non-Destructive Testing and Evaluation of Composite Materials/Structures: A State-of-the-Art Review.* Advances in Mechanical Engineering, 2020. **12**(4): p. 168781402091376.

574.	Huke, P., et al., *Novel Trends in Optical Non-Destructive Testing Methods.* Journal of the European Optical Society Rapid Publications, 2013. **8**: p. 13043.

575.	Zhao, Q., et al., *Digital Shearography for NDT: Phase Measurement Technique and Recent Developments.* Applied Sciences, 2018. **8**(12): p. 2662.

576.	Lu, S., et al., *Real-Time Monitoring of Low-Velocity Impact Damage for Composite Structures With the Omnidirection Carbon Nanotubes' Buckypaper Sensors.* Structural Health Monitoring, 2018. **18**(2): p. 454-465.

577.	Asif, M., et al., *Identification of an Effective Nondestructive Technique for Bond Defect Determination in Laminate Composites—A Technical Review.* Journal of Composite Materials, 2018. **52**(26): p. 3589-3599.

578.	Figueroa, A.I., et al., *Structural and Magnetic Properties of Amorphous Co-W Alloyed Nanoparticles.* Physical Review B, 2011. **84**(18).

579. Kiseleva, T.Y., et al., *Magnetodeformational Anisotropy of FeGa/PU Hybrid Nanocomposite via Particle Concentration and Spatial Orientation.* Solid State Phenomena, 2015. **233-234**: p. 607-610.

580. Tian, J., et al., *Bonded Terfenol-D Composites With Low Eddy Current Loss and High Magnetostriction.* Rare Metals, 2010. **29**(6): p. 579-582.

581. Yang, C., et al., *Metal-Based Magnetic Functional Fluids With Amorphous Particles.* RSC Advances, 2014. **4**(103): p. 59541-59547.

582. Antonacci, A., et al., *Predictive Sustainability Analysis Applied to an Automotive Design Case Study.* Iop Conference Series Materials Science and Engineering, 2024. **1306**(1): p. 012041.

583. Delogu, M., et al., *Environmental and Economic Life Cycle Assessment of a Lightweight Solution for an Automotive Component: A Comparison Between Talc-Filled and Hollow Glass Microspheres-Reinforced Polymer Composites.* Journal of Cleaner Production, 2016. **139**: p. 548-560.

584. Ellingsen, L.A.W., et al., *Life Cycle Assessment of a Lithium-Ion Battery Vehicle Pack.* Journal of Industrial Ecology, 2013. **18**(1): p. 113-124.

585. Tarne, P., et al., *Introducing a Product Sustainability Budget at an Automotive Company—one Option to Increase the Use of LCSA Results in Decision-Making Processes.* The International Journal of Life Cycle Assessment, 2018. **24**(8): p. 1461-1479.

586. Álvarez-del-Castillo, M.D., et al., *Environmental Impact of Chicken Feathers Based Polypropylene Composites Developed for Automotive and Stationary Applications and Comparison With Glass-Fibre Analogues.* Waste and Biomass Valorization, 2022. **13**(11): p. 4585-4598.

587. Mair-Bauernfeind, C., et al., *Comparing the Incomparable? A Review of Methodical Aspects in the Sustainability Assessment of Wood in Vehicles.* The International Journal of Life Cycle Assessment, 2020. **25**(11): p. 2217-2240.

588. Popa, N., et al., *Holistic Approach to Sustainability of Bridges.* Steel Construction, 2018. **11**(3): p. 179-183.

589. Akponeware, A., et al., *Exploring the Development of a BIM-Enabled Process Framework for LCA of Rail Tracks.* 2020.

590. Ellingsen, L.A.W., et al., *Environmental Screening of Electrode Materials for a Rechargeable Aluminum Battery With an AlCl3/EMIMCl Electrolyte.* Materials, 2018. **11**(6): p. 936.

591. Allard, S., *DataONE: Facilitating eScience Through Collaboration.* Journal of Escience Librarianship, 2012: p. 4-17.

592. Moretti, C., et al., *Reviewing ISO Compliant Multifunctionality Practices in Environmental Life Cycle Modeling.* Energies, 2020. **13**(14): p. 3579.

593. Finnveden, G., et al., *Recent Developments in Life Cycle Assessment.* Journal of Environmental Management, 2009. **91**(1): p. 1-21.

594. Schaubroeck, T., et al., *Attributional &Amp; Consequential Life Cycle Assessment: Definitions, Conceptual Characteristics and Modelling Restrictions.* Sustainability, 2021. **13**(13): p. 7386.

595. Asmatulu, E., J.M. Twomey, and M. Overcash, *Recycling of Fiber-Reinforced Composites and Direct Structural Composite Recycling Concept.* Journal of Composite Materials, 2013. **48**(5): p. 593-608.

596. Thomason, J., L. Yang, and R. Meier, *The Properties of Glass Fibres After Conditioning at Composite Recycling Temperatures.* Composites Part a Applied Science and Manufacturing, 2014. **61**: p. 201-208.

597. Döhlert, P., et al., *Introducing Students to Feedstock Recycling of End-of-Life Silicones via a Low-Temperature, Iron-Catalyzed Depolymerization Process.* Journal of Chemical Education, 2015. **92**(4): p. 703-707.

598. Bogacka, M., et al., *PV Waste Thermal Treatment According to the Circular Economy Concept.* Sustainability, 2020. **12**(24): p. 10562.

599. Li, L., et al., *Sustainable Recovery of Cathode Materials From Spent Lithium-Ion Batteries Using Lactic Acid Leaching System.* Acs Sustainable Chemistry & Engineering, 2017. **5**(6): p. 5224-5233.

600. Martey, S., et al., *Hybrid Chemomechanical Plastics Recycling: Solvent-free, High-Speed Reactive Extrusion of Low-Density Polyethylene.* Chemsuschem, 2021. **14**(19): p. 4280-4290.

601. Kusano, R. and Y. Kusano, *Applications of Plasma Technologies in Recycling Processes.* Materials, 2024. **17**(7): p. 1687.

602. Tan, T., et al., *Upcycling Plastic Wastes Into Value-Added Products by Heterogeneous Catalysis.* Chemsuschem, 2022. **15**(14).

603. Zhang, R., et al., *Systematic Study of Al Impurity for NCM622 Cathode Materials.* Acs Sustainable Chemistry & Engineering, 2020. **8**(26): p. 9875-9884.

604. Caldera, S., et al., *Evaluating Barriers, Enablers and Opportunities for Closing the Loop Through 'Waste Upcycling': A Systematic Literature Review.* Journal of Sustainable Development of Energy Water and Environment Systems, 2022. **10**(1): p. 1-20.

605. Zhao, S., et al., *Recovery Methods and Regulation Status of Waste Lithium-Ion Batteries in China: A Mini Review.* Waste Management & Research the Journal for a Sustainable Circular Economy, 2019. **37**(11): p. 1142-1152.

606. Chang, L. and Y. Xia, *Excavating the Potential of Photo- and Electroupcycling Platforms Toward a Sustainable Future for Waste Plastics.* Small Science, 2023. **4**(2).

607. Su, X., *Electrochemical Separations for Metal Recycling.* The Electrochemical Society Interface, 2020. **29**(3): p. 55-61.

608. Pavec, A.L., et al., *Friction Magazine: The Upcycling of Manufacture for Structural Design.* International Journal of Space Structures, 2021. **36**(4): p. 281-293.

609. Savas, P.E., et al., *Carbon Dioxide Sorbent From Construction and Textile Plastic Waste.* Advanced Sustainable Systems, 2023. **7**(6).

610. Diaz, L.A., et al., *Comprehensive Process for the Recovery of Value and Critical Materials From Electronic Waste.* Journal of Cleaner Production, 2016. **125**: p. 236-244.

611. Jing, R., et al., *Assessments of Greenhouse Gas (GHG) Emissions From Stainless Steel Production in China Using Two Evaluation Approaches.* Environmental Progress & Sustainable Energy, 2019. **38**(1): p. 47-55.

612. Thøgersen, J. and K.S. Nielsen, *A Better Carbon Footprint Label*. Journal of Cleaner Production, 2016. **125**: p. 86-94.

613. Panagiotopoulou, V.C., P. Stavropoulos, and G. Chryssolouris, *A Critical Review on the Environmental Impact of Manufacturing: A Holistic Perspective*. The International Journal of Advanced Manufacturing Technology, 2021. **118**(1-2): p. 603-625.

614. Liu, Z., et al., *Insights Into the Regional Greenhouse Gas (GHG) Emission of Industrial Processes: A Case Study of Shenyang, China*. Sustainability, 2014. **6**(6): p. 3669-3685.

615. Ekinci, F. and M. Mert, *Turkey's Green Economy Initiative: An Experimental Evaluation of Hydrogen Energy*. Çukurova Üniversitesi Mühendislik Fakültesi Dergisi, 2023. **38**(2): p. 463-471.

616. Yan, W., X. He, and H. Zhang, *A Multisource Data-driven Approach for Carbon Footprint Analysis of Remanufacturing Systems*. Energy Science & Engineering, 2023. **11**(12): p. 4446-4462.

617. Gu, L., et al., *Carbon Footprint Analysis of Bamboo Scrimber Flooring—Implications for Carbon Sequestration of Bamboo Forests and Its Products*. Forests, 2019. **10**(1): p. 51.

618. Patel, N., M. Feofilovs, and F. Romagnoli, *Carbon Footprint Evaluation Tool for Packaging Marketplace*. Environmental and Climate Technologies, 2023. **27**(1): p. 368-378.

619. Zhou, Y., et al., *Carbon Footprint and Water Footprint Assessment of Down Jackets*. Aatcc Journal of Research, 2023. **10**(5): p. 300-310.

620. Wang, G., et al., *Energy Efficiency Optimization Based on Digital Twin Workshops*. Academic Journal of Engineering and Technology Science, 2022. **5**(13): p. 15-25.

621. Pires, F., et al., *Digital Twin Based What-if Simulation for Energy Management*. 2021: p. 309-314.

622. Li, W., *Research on the Impact of Digital Economy on Industrial Energy Efficiency*. Frontiers in Business Economics and Management, 2022. **7**(1): p. 115-119.

623. Zhao, P., et al., *Optimization Method for Multi-Process Energy Consumption in Iron and Steel Enterprises Under the Coordinated Use of Comprehensive Energy*. Journal of Physics Conference Series, 2024. **2729**(1): p. 012017.

624. Hernandez, A.G. and J.M. Cullen, *How Resource-Efficient Is the Global Steel Industry?* Resources Conservation and Recycling, 2018. **133**: p. 132-145.

625. Shinkevich, A.I., S.S. Kudryavtseva, and I. Ershova, *Modelling of Energy Efficiency Factors of Petrochemical Industry*. International Journal of Energy Economics and Policy, 2020. **10**(3): p. 465-470.

626. Chen, S., *Advancing Sustainability in Mechanical Engineering: Integration of Green Energy and Intelligent Manufacturing*. Applied and Computational Engineering, 2024. **66**(1): p. 150-155.

627. Kono, J., et al., *Factors for Eco-Efficiency Improvement of Thermal Insulation Materials*. Key Engineering Materials, 2016. **678**: p. 1-13.

628. Bogdanov, D., et al., *Radical Transformation Pathway Towards Sustainable Electricity via Evolutionary Steps*. Nature Communications, 2019. **10**(1).

629. NA, N., *Potentials of biomass briquetting and utilization: the Nigerian perspective.* Pacific International Journal, 2020. **3**(1): p. 07-12.

630. Yadav, K.K., et al., *Review on Evaluation of Renewable Bioenergy Potential for Sustainable Development: Bright Future in Energy Practice in India.* Acs Sustainable Chemistry & Engineering, 2021. **9**(48): p. 16007-16030.

631. Alonso, D.M., J.Q. Bond, and J.A. Dumesic, *Catalytic Conversion of Biomass to Biofuels.* Green Chemistry, 2010. **12**(9): p. 1493.

632. Karabegović, I., *Tendency of Global Capacity Development of Renewable Energy Sources in the World in the Last Ten Years.* Contemporary Materials, 2021. **12**(2).

633. Ozili, P.K. and E. Özen, *Global Energy Crisis: Impact on the Global Economy.* SSRN Electronic Journal, 2023.

634. Bataille, C., et al., *A Review of Technology and Policy Deep Decarbonization Pathway Options for Making Energy-Intensive Industry Production Consistent With the Paris Agreement.* Journal of Cleaner Production, 2018. **187**: p. 960-973.

635. Pérez–Fortes, M., et al., *CO2 Capture and Utilization in Cement and Iron and Steel Industries.* Energy Procedia, 2014. **63**: p. 6534-6543.

636. Wan, F., et al., *Research of the Impact of Hydrogen Metallurgy Technology on the Reduction of the Chinese Steel Industry's Carbon Dioxide Emissions.* Sustainability, 2024. **16**(5): p. 1814.

637. Bhaskar, A., M. Assadi, and H.N. Somehsaraei, *Decarbonization of the Iron and Steel Industry With Direct Reduction of Iron Ore With Green Hydrogen.* Energies, 2020. **13**(3): p. 758.

638. Kolbe, N., et al., *Carbon Utilization Combined With Carbon Direct Avoidance for Climate Neutrality in Steel Manufacturing.* Chemie Ingenieur Technik, 2022. **94**(10): p. 1548-1552.

639. wang, c., S. Walsh, and A. Feitz, *A Techno-Economic Analysis of Australian Green Steel Production From Hydrogen.* 2022.

640. Toktarova, A., et al., *Pathways for Low-Carbon Transition of the Steel Industry—A Swedish Case Study.* Energies, 2020. **13**(15): p. 3840.

641. Arens, M., et al., *Pathways to a Low-Carbon Iron and Steel Industry in the Medium-Term – The Case of Germany.* Journal of Cleaner Production, 2017. **163**: p. 84-98.

642. Wędrychowicz, M., et al., *Recycling of Electrical Cables—Current Challenges and Future Prospects.* Materials, 2023. **16**(20): p. 6632.

643. Kaya, A.Y., et al., *Archives of Foundry Engineering.* 2020.

644. Roy, P.S., et al., *Strategic Approach Towards Plastic Waste Valorization: Challenges and Promising Chemical Upcycling Possibilities.* Chemsuschem, 2021. **14**(19): p. 4007-4027.

645. Worch, J.C. and A.P. Dove, *100th Anniversary of Macromolecular Science Viewpoint: Toward Catalytic Chemical Recycling of Waste (And Future) Plastics.* Acs Macro Letters, 2020. **9**(11): p. 1494-1506.

646. Korley, L.T.J., et al., *Toward Polymer Upcycling—adding Value and Tackling Circularity.* Science, 2021. **373**(6550): p. 66-69.

647. Li, Z., et al., *Chemical Upcycling of Poly(3-Hydroxybutyrate) (P3HB) Toward Functional Poly(amine-<i>alt</I>-Ester) via Tandem Degradation and Ring-Opening Polymerization*. Macromolecules, 2022. **55**(21): p. 9697-9704.

648. Razzaq, A., et al., *Upcycled Polyvinyl Chloride (PVC) Electrospun Nanofibers From Waste PVC-Based Materials for Water Treatment*. Acs Applied Engineering Materials, 2023. **1**(7): p. 1924-1936.

649. Xu, Z., et al., *Cascade Degradation and Upcycling of Polystyrene Waste to High-Value Chemicals*. Proceedings of the National Academy of Sciences, 2022. **119**(34).

650. Zhang, H., et al., *Catalytic Amounts of an Antibacterial Monomer Enable the Upcycling of Poly(Ethylene Terephthalate) Waste*. Advanced Materials, 2023. **35**(20).

651. Li, J., et al., *Co-Upcycling of Plastic Waste and Biowaste via Tandem Transesterification Reactions*. Jacs Au, 2024.

652. Skiba, R., *Insight into Carbon Capture and Storage (CCS) Technology*. Organic & Medicinal Chem IJ, 2020. **9**(5).

653. Adeyemo, O.K., *A call to strengthen eco-innovation using indigenous resources and waste products*. Proceedings of the Nigerian Academy of Science, 2022. **15**(1).

654. Ceptureanu, S.I., et al., *Eco-Innovation Capability and Sustainability Driven Innovation Practices in Romanian SMEs*. Sustainability, 2020. **12**(17): p. 7106.

655. Silvianti, F., et al., *Green Pathways for the Enzymatic Synthesis of Furan-Based Polyesters and Polyamides*. 2020: p. 3-29.

656. Maniar, D., et al., *Enzymatic Polymerization of Dimethyl 2,5-Furandicarboxylate and Heteroatom Diamines*. Acs Omega, 2018. **3**(6): p. 7077-7085.

657. Harsanto, B., et al., *Sustainability Innovation in the Textile Industry: A Systematic Review*. Sustainability, 2023. **15**(2): p. 1549.

658. Muhardi, M., et al., *The Implementation of Sustainable Manufacturing Practice in Textile Industry: An Indonesian Perspective*. Journal of Asian Finance Economics and Business, 2020. **7**(11): p. 1041-1047.

659. Rauter, R., et al., *Open Innovation and Its Effects on Economic and Sustainability Innovation Performance*. Journal of Innovation & Knowledge, 2019. **4**(4): p. 226-233.

660. Albort-Morant, G., A. Leal-Millán, and G. Cepeda-Carrión, *The Antecedents of Green Innovation Performance: A Model of Learning and Capabilities*. Journal of Business Research, 2016. **69**(11): p. 4912-4917.

661. Abreu, M.C.S.d., et al., *Collaboration in Achieving Sustainable Solutions in the Textile Industry*. Journal of Business and Industrial Marketing, 2020. **36**(9): p. 1614-1626.

662. Deng, Y., *Digital Twin-Based Modeling of Complex Systems for Smart Aging*. Discrete Dynamics in Nature and Society, 2022. **2022**(1).

663. Zhao, W., et al., *Construction Method of Digital Twin System for Thin-Walled Workpiece Machining Error Control Based on Analysis of Machine Tool Dynamic Characteristics*. Machines, 2023. **11**(6): p. 600.

664. Stanke, J., et al., *Development of a Hybrid DLT Cloud Architecture for the Automated Use of Finite Element Simulation as a Service for Fine Blanking*. The International Journal of Advanced Manufacturing Technology, 2020. **108**(11-12): p. 3717-3724.

665. Asif, R., *Deep Neural Networks for Future Low Carbon Energy Technologies: Potential, Challenges and Economic Development*. 2020.

666. Lee, D. and L. Chen, *Sustainable Air-Conditioning Systems Enabled by Artificial Intelligence: Research Status, Enterprise Patent Analysis, and Future Prospects.* Sustainability, 2022. **14**(12): p. 7514.

667. Turgay, S., Ö. Bilgin, and N. Akar, *Digital Twin Based Flexible Manufacturing System Modelling with Fuzzy Approach.* Advances in Computer, Signals and Systems, 2022. **6**(7): p. 10-17.

668. Yildiz, E., C. Møller, and A. Bilberg, *Demonstration and Evaluation of a Digital Twin-Based Virtual Factory.* The International Journal of Advanced Manufacturing Technology, 2021. **114**(1-2): p. 185-203.

669. Skiba, R., *Blockchain technology as a health and safety contributor in the transport and logistics industry–human resource requirements.* International Journal of Innovative Science and Research Technology, 2020. **5**(4): p. 544-560.

670. Arrieta, A.B., et al., *Explainable Artificial Intelligence (XAI): Concepts, Taxonomies, Opportunities and Challenges Toward Responsible AI.* Information Fusion, 2020. **58**: p. 82-115.

671. dong, j., J. Hu, and Z. Luo, *Quality Monitoring of Resistance Spot Welding Based on Digital Twin.* 2023.

672. Yusupbekov, N., et al., *Application of Digital Twin Technologies in Mining Industrial Branch.* E3s Web of Conferences, 2023. **417**: p. 05016.

673. Meng, F., et al., *Comparing Life Cycle Energy and Global Warming Potential of Carbon Fiber Composite Recycling Technologies and Waste Management Options.* Acs Sustainable Chemistry & Engineering, 2018. **6**(8): p. 9854-9865.

674. Nurcahyanie, Y.D. and L.D. Rohmadiani, *Design for Longevity and Design for X: Concepts, Applications, and Perspectives.* Tibuana, 2023. **6**(1): p. 58-64.

675. Amantayeva, A., et al., *Challenges and Opportunities of Implementing Industry 4.0 in Recycling Carbon Fiber Reinforced Composites.* Advances in Science and Technology, 2022. **116**: p. 67-73.

676. Niesten, E. and A. Jolink, *Motivations for Environmental Alliances: Generating and Internalizing Environmental and Knowledge Value.* International Journal of Management Reviews, 2020. **22**(4): p. 356-377.

677. Ramsheva, Y.K. and A. Remmen, *Industrial Symbiosis in the Cement Industry - Exploring the Linkages to Circular Economy.* 2018: p. 35-53.

678. Wags Numoipiri Digitemie, N. and N. Ifeanyi Onyedika Ekemezie, *Assessing the Role of Carbon Pricing in Global Climate Change Mitigation Strategies.* Magna Scientia Advanced Research and Reviews, 2024. **10**(2): p. 022-031.

679. Stella Emeka-Okoli, N., et al., *Review of Carbon Pricing Mechanisms: Effectiveness and Policy Implications.* International Journal of Applied Research in Social Sciences, 2024. **6**(3): p. 337-347.

680. Bhatnagar, M., S. Taneja, and P. Kumar, *The Effectiveness of Carbon Pricing Mechanism in Steering Financial Flows Toward Sustainable Projects.* International Journal of Environmental Impacts Management Mitigation and Recovery, 2023. **6**(4): p. 183-196.

681. Zhao, S., et al., *Impact of Carbon Tax and Subsidy Policies on Original Equipment Manufacturers and Remanufacturing Companies From the Perspective of Carbon*

Emissions. International Journal of Environmental Research and Public Health, 2022. **19**(10): p. 6252.

682. Miškovičová, Z., et al., *An Overview Analysis of Current Research Status in Iron Oxides Reduction by Hydrogen.* Metals, 2024. **14**(5): p. 589.

683. Dzhengiz, T., A. Riandita, and A. Broström, *Sustainability-Oriented Textile/Fashion Partnerships: Mechanisms and Levels of Change.* Academy of Management Proceedings, 2022. **2022**(1).

684. Percec, S. and A.C. Albertsson, *Rational Design of Multifunctional Renewable-Resourced Materials.* Biomacromolecules, 2019. **20**(2): p. 569-572.

685. Alberi, K., et al., *The 2019 Materials by Design Roadmap.* Journal of Physics D Applied Physics, 2018. **52**(1): p. 013001.

686. Jammoukh, M., et al., *Bio-Charge Elastic Characterization for a Qualitative Perspective of Innovative Bio-Composite Materials.* Iraqi Journal of Science, 2021: p. 90-95.

687. Liu, C., F. Zhao, and J.W. Sutherland, *A Design Method to Improve End-of-Use Product Value Recovery for Circular Economy.* Journal of Mechanical Design, 2019. **141**(4).

688. Ghufran, M., et al., *Circular Economy in the Construction Industry: A Step Towards Sustainable Development.* Buildings, 2022. **12**(7): p. 1004.

689. Kanters, J., *Circular Building Design: An Analysis of Barriers and Drivers for a Circular Building Sector.* Buildings, 2020. **10**(4): p. 77.

690. Petrova, V., *Exploring the Opportunities for Sustainable Management of Critical Raw Materials in the Circular Economy.* The Eurasia Proceedings of Science Technology Engineering and Mathematics, 2023. **26**: p. 664-671.

691. Esparragoza, I.E. and J.A. Mesa, *A Case Study Approach to Introduce Circular Economy in Sustainable Design Education.* 2019.

692. Sumter, D., et al., *Circular Economy Competencies for Design.* Sustainability, 2020. **12**(4): p. 1561.

693. Goldsworthy, K. and D. Ellams, *Collaborative Circular Design. Incorporating Life Cycle Thinking Into an Interdisciplinary Design Process.* The Design Journal, 2019. **22**(sup1): p. 1041-1055.

694. Antonini, E., et al., *Reversibility and Durability as Potential Indicators for Circular Building Technologies.* Sustainability, 2020. **12**(18): p. 7659.

695. Deniz, D., *Sustainable Design Thinking and Social Innovation for Beating Barriers to Circular Economy.* 2021.

696. Mann, J., et al., *Mechanism-Based Organization of Neural Networks to Emulate Systems Biology and Pharmacology Models.* Scientific Reports, 2024. **14**(1).

697. Hendriks, C., et al., *Introducing a Mechanistic Model in Digital Soil Mapping to Predict Soil Organic Matter Stocks in the <scp>Cantabrian</Scp> Region (<scp>Spain</Scp>).* European Journal of Soil Science, 2020. **72**(2): p. 704-719.

698. Kraikivski, P., *A Dynamic Mechanistic Model of Perceptual Binding.* Mathematics, 2022. **10**(7): p. 1135.

699. Lagergren, J., et al., *Forecasting and Uncertainty Quantification Using a Hybrid of Mechanistic and Non-Mechanistic Models for an Age-Structured Population Model.* Bulletin of Mathematical Biology, 2018. **80**(6): p. 1578-1595.

700. Tøndel, K. and H. Martens, *Analyzing Complex Mathematical Model Behavior by Partial Least Squares Regression -based Multivariate Metamodeling*. Wiley Interdisciplinary Reviews Computational Statistics, 2014. **6**(6): p. 440-475.

701. Breughe, M., S. Eyerman, and L. Eeckhout, *A Mechanistic Performance Model for Superscalar in-Order Processors*. 2012.

702. Liao, X., et al., *Modified Mechanistic Model Based on Gaussian Process Adjusting Technique for Cutting Force Prediction in Micro -End Milling*. Mathematical Problems in Engineering, 2019. **2019**(1).

703. Cabral, J.S., L.M. Valente, and F. Härtig, *Mechanistic Simulation Models in Macroecology and Biogeography: State -of-art and Prospects*. Ecography, 2016. **40**(2): p. 267-280.

704. Boult, V.L. and L.C. Evans, *Mechanisms Matter: Predicting the Ecological Impacts of Global Change*. Global Change Biology, 2021. **27**(9): p. 1689-1691.

705. Anand, R.S., et al., *Mechanistic Modeling of Micro-Drilling Cutting Forces*. The International Journal of Advanced Manufacturing Technology, 2016. **88**(1-4): p. 241-254.

706. Ashena, R., et al., *Mechanistic Modeling of Annular Two-Phase Flow While Underbalanced Drilling in Iran*. 2010.

707. Jalihal, A.P., et al., *Modeling and Analysis of the Macronutrient Signaling Network in Budding Yeast*. Molecular Biology of the Cell, 2021. **32**(21).

708. Perilla, J.R., et al., *Molecular Dynamics Simulations of Large Macromolecular Complexes*. Current Opinion in Structural Biology, 2015. **31**: p. 64-74.

709. Kühne, T.D., *Second Generation Car–Parrinello Molecular Dynamics*. Wiley Interdisciplinary Reviews Computational Molecular Science, 2014. **4**(4): p. 391-406.

710. Lindorff-Larsen, K., et al., *How Fast-Folding Proteins Fold*. Science, 2011. **334**(6055): p. 517-520.

711. Bond, S.D. and B. Leimkuhler, *Molecular Dynamics and the Accuracy of Numerically Computed Averages*. Acta Numerica, 2007. **16**: p. 1-65.

712. Anderson, D.M., G.B. McFadden, and A.A. Wheeler, *A Phase-Field Model With Convection: Sharp-Interface Asymptotics*. Physica D Nonlinear Phenomena, 2001. **151**(2-4): p. 305-331.

713. Chen, L., *Phase-Field Models for Microstructure Evolution*. Annual Review of Materials Science, 2002. **32**(1): p. 113-140.

714. Li, J., J. Wang, and Y. Gao, *Phase Field Simulation of the Columnar Dendritic Growth and Microsegregation in a Binary Alloy*. Chinese Physics B, 2008. **17**(9): p. 3516-3522.

715. Ohno, M., *Quantitative Phase-Field Modeling of Nonisothermal Solidification in Dilute Multicomponent Alloys With Arbitrary Diffusivities*. Physical Review E, 2012. **86**(5).

716. Shimono, Y., M. Oba, and S. Nomoto, *Solidification Simulation of Direct Energy Deposition Process by Multi-Phase Field Method Coupled With Thermal Analysis*. Modelling and Simulation in Materials Science and Engineering, 2019. **27**(7): p. 074006.

717. Sadamoto, K., Y. Tadano, and S. Morita, *A Three-Dimensional Formulation of Phase Field Model Representing Polycrystalline Solidification*. Key Engineering Materials, 2019. **794**: p. 208-213.

718. Yang, C., et al., *Multi-Phase-Field Simulation of Austenite Peritectic Solidification Based on a Ferrite Grain*. Chinese Physics B, 2021. **30**(1): p. 018201.

719. Shibuta, Y., M. Ohno, and T. Takaki, *Solidification in a Supercomputer: From Crystal Nuclei to Dendrite Assemblages*. Jom, 2015. **67**(8): p. 1793-1804.

720. Badalassi, V., H.D. Ceniceros, and S. Banerjee, *Computation of Multiphase Systems With Phase Field Models*. Journal of Computational Physics, 2003. **190**(2): p. 371-397.

721. Masuda, K., *Phase-Field Modelling for Atoms*. 2023.

722. Berti, A. and C. Giorgi, *A Phase-Field Model for Liquid–Vapor Transitions*. Journal of Non-Equilibrium Thermodynamics, 2009. **34**(3).

723. Xiao, R.Z., et al., *Phase-Field Modeling of Solute Trapping in Single-Phase Alloys During Directional Solidification*. Advanced Materials Research, 2010. **154-155**: p. 401-406.

724. Feng, X., Y. He, and C. Li, *Analysis of Finite Element Approximations of a Phase Field Model for Two-Phase Fluids*. Mathematics of Computation, 2006. **76**(258): p. 539-571.

725. Ofori-Opoku, N., et al., *Complex Order Parameter Phase-Field Models Derived From Structural Phase-Field-Crystal Models*. Physical Review B, 2013. **88**(10).

726. Boussinot, G., et al., *Strongly Out-of-Equilibrium Columnar Solidification During Laser Powder-Bed Fusion in Additive Manufacturing*. Physical Review Applied, 2019. **11**(1).

727. Noguchi, S., H. Wang, and J. Inoue, *Identification of Microstructures Critically Affecting Material Properties Using Machine Learning Framework Based on Metallurgists' Thinking Process*. Scientific Reports, 2022. **12**(1).

728. Morgan, D. and R. Jacobs, *Opportunities and Challenges for Machine Learning in Materials Science*. Annual Review of Materials Science, 2020. **50**(1): p. 71-103.

729. Zeng, S., et al., *Atom Table Convolutional Neural Networks for an Accurate Prediction of Compounds Properties*. NPJ Computational Materials, 2019. **5**(1).

730. Li, W., R. Jacobs, and D. Morgan, *Predicting the Thermodynamic Stability of Perovskite Oxides Using Machine Learning Models*. Computational Materials Science, 2018. **150**: p. 454-463.

731. Lu, S., et al., *Accelerated Discovery of Stable Lead-Free Hybrid Organic-Inorganic Perovskites via Machine Learning*. Nature Communications, 2018. **9**(1).

732. Liang, T., et al., *Design of High Strength and Electrically Conductive Aluminium Alloys by Machine Learning*. Materials Science and Technology, 2022. **38**(2): p. 116-129.

733. Yuyama, S. and H. Kaneko, *Simultaneous Design of Gas Separation Membranes and Schemes Through Combined Process and Materials Informatics*. Industrial & Engineering Chemistry Research, 2023. **62**(44): p. 18541-18551.

734. Zipoli, F., et al., *Prediction of Phase Diagrams and Associated Phase Structural Properties*. Industrial & Engineering Chemistry Research, 2022. **61**(24): p. 8378-8389.

735. Schmidt, J., et al., *Recent Advances and Applications of Machine Learning in Solid-State Materials Science*. NPJ Computational Materials, 2019. **5**(1).

736. Sami, Y., et al., *Selecting Machine Learning Models to Support the Design of Al/CuO Nanothermites*. The Journal of Physical Chemistry A, 2022. **126**(7): p. 1245-1254.

737. Butler, K.T., et al., *Machine Learning for Molecular and Materials Science.* Nature, 2018. **559**(7715): p. 547-555.

738. Zhang, L. and Z. Li, *Machine Learning for Materials Classifications From Images.* Journal of Physics Conference Series, 2022. **2369**(1): p. 012081.

739. Jacobs, R., et al., *The Materials Simulation Toolkit for Machine Learning (MAST-ML): An Automated Open Source Toolkit to Accelerate Data-Driven Materials Research.* Computational Materials Science, 2020. **176**: p. 109544.

740. Tawfik, S.A., et al., *Predicting Thermal Properties of Crystals Using Machine Learning.* Advanced Theory and Simulations, 2019. **3**(2).

741. Wu, T., *High Throughput Screening of Thermal Interface Materials by Machine Learning.* Applied and Computational Engineering, 2024. **61**(1): p. 77-86.

742. Verma, D., *Multiscale Modelling and Characterization of Coupled Damage-Healing for Materials in Concurrent Computational Homogenization Approach Using Machine Learning.* Turkish Journal of Computer and Mathematics Education (Turcomat), 2018. **9**(2): p. 598-609.

743. Negi, P., *Application of Machine Learning in Predicting the Fatigue Behaviour of Materials Using Deep Learning.* Turkish Journal of Computer and Mathematics Education (Turcomat), 2018. **9**(2): p. 541-553.

744. Pan, Z., Y. Zhou, and L. Zhang, *Photoelectrochemical Properties, Machine Learning, and Symbolic Regression for Molecularly Engineered Halide Perovskite Materials in Water.* Acs Applied Materials & Interfaces, 2022. **14**(7): p. 9933-9943.

745. Popova, M., et al., *OpenChem: A Deep Learning Toolkit for Computational Chemistry and Drug Design.* 2020.

746. Xia, D., et al., *Prediction of Material Properties by Neural Network Fusing the Atomic Local Environment and Global Description: Applied to Organic Molecules and Crystals.* E3s Web of Conferences, 2021. **267**: p. 02059.

747. Sungphueng, P. and K. Amnuyswat, *Thermoelectric Prediction From Material Descriptors Using Machine Learning Technique.* Current Applied Science and Technology, 2023. **23**(6).

748. Kauwe, S.K., et al., *Can Machine Learning Find Extraordinary Materials?* 2019.

749. Rekatsinas, C., G. Giannakopoulos, and E. Karkaletsis, *Machine-Learning-Driven Health Monitoring Diagnostics Focused on Composite Structures Utilizing Smart Layerwise Spectral Elements.* 2023: p. 935-946.

750. Stanev, V., et al., *Machine Learning Modeling of Superconducting Critical Temperature.* NPJ Computational Materials, 2018. **4**(1).

751. Kattner, U.R., *The Thermodynamic Modeling of Multicomponent Phase Equilibria.* Jom, 1997. **49**(12): p. 14-19.

752. Lukas, H.L., S.G. Fries, and B. Sundman, *Computational Thermodynamics.* 2007.

753. Rice, B.M., et al., *Parameterizing Complex Reactive Force Fields Using Multiple Objective Evolutionary Strategies (MOES): Part 2: Transferability of ReaxFF Models to C–H–N–O Energetic Materials.* Journal of Chemical Theory and Computation, 2015. **11**(2): p. 392-405.

754. Rimsza, J., L. Deng, and J. Du, *Molecular Dynamics Simulations of Nanoporous Organosilicate Glasses Using Reactive Force Field (ReaxFF)*. Journal of Non-Crystalline Solids, 2016. **431**: p. 103-111.

755. Larentzos, J.P. and B.M. Rice, *Transferable Reactive Force Fields: Extensions of ReaxFF-lg to Nitromethane*. The Journal of Physical Chemistry A, 2017. **121**(9): p. 2001-2013.

756. Raju, M., et al., *Reactive Force Field Study of Li/C Systems for Electrical Energy Storage*. Journal of Chemical Theory and Computation, 2015. **11**(5): p. 2156-2166.

757. Dittner, M., et al., *Efficient Global Optimization of Reactive Force-field Parameters*. Journal of Computational Chemistry, 2015. **36**(20): p. 1550-1561.

758. Larentzos, J.P., et al., *Parameterizing Complex Reactive Force Fields Using Multiple Objective Evolutionary Strategies (MOES). Part 1: ReaxFF Models for Cyclotrimethylene Trinitramine (RDX) and 1,1-Diamino-2,2-Dinitroethene (FOX-7)*. Journal of Chemical Theory and Computation, 2015. **11**(2): p. 381-391.

759. Hjertenæs, E., A.Q. Nguyen, and H. Koch, *A ReaxFF Force Field for Sodium Intrusion in Graphitic Cathodes*. Physical Chemistry Chemical Physics, 2016. **18**(46): p. 31431-31440.

760. Ashraf, C. and A.C.T.v. Duin, *Extension of the ReaxFF Combustion Force Field Toward Syngas Combustion and Initial Oxidation Kinetics*. The Journal of Physical Chemistry A, 2017. **121**(5): p. 1051-1068.

761. Mueller, J.E., A.C.T.v. Duin, and W.A. Goddard, *Development and Validation of ReaxFF Reactive Force Field for Hydrocarbon Chemistry Catalyzed by Nickel*. The Journal of Physical Chemistry C, 2010. **114**(11): p. 4939-4949.

762. Agrawalla, S. and A.v. Duin, *Development and Application of a ReaxFF Reactive Force Field for Hydrogen Combustion*. The Journal of Physical Chemistry A, 2011. **115**(6): p. 960-972.

763. Gomzi, V., I.M. Šapić, and A. Vidak, *ReaxFF Force Field Development and Application for Toluene Adsorption on MnMO$_x$ (M = Cu, Fe, Ni) Catalysts*. The Journal of Physical Chemistry A, 2021. **125**(50): p. 10649-10656.

764. Lele, A., P. Krstić, and A.C.T.v. Duin, *ReaxFF Force Field Development for Gas-Phase hBN Nanostructure Synthesis*. The Journal of Physical Chemistry A, 2022. **126**(4): p. 568-582.

765. Cheung, S., et al., *ReaxFF$_{MgH}$ Reactive Force Field for Magnesium Hydride Systems*. The Journal of Physical Chemistry A, 2005. **109**(5): p. 851-859.

766. Jung, C., L. Braunwarth, and T. Jacob, *Grand Canonical ReaxFF Molecular Dynamics Simulations for Catalytic Reactions*. Journal of Chemical Theory and Computation, 2019. **15**(11): p. 5810-5816.

767. Trnka, T., I. Tvaroška, and J. Koča, *Automated Training of ReaxFF Reactive Force Fields for Energetics of Enzymatic Reactions*. Journal of Chemical Theory and Computation, 2017. **14**(1): p. 291-302.

768. Li, X., et al., *ReaxFF Molecular Dynamics Simulations of Thermal Reactivity of Various Fuels in Pyrolysis and Combustion*. Energy & Fuels, 2021. **35**(15): p. 11707-11739.

769. Qiao, S., et al., *Reactive Molecular Dynamics Simulation on the Disintegration of Kapton and Upilex-S During Atomic Oxygen Impact*. Frontiers in Materials, 2023. **10**.

770.   Chen, H., et al., *Modification of Short-Range Repulsive Interactions in ReaxFF Reactive Force Field for Fe–Ni–Al Alloy*. Chinese Physics B, 2021. **30**(8): p. 086110.

771.   Huang, Q., et al., *Adaptive Force Field Parameter Optimization for Expanding Reaction Simulations Within Wide-Ranged Temperature*. The Journal of Physical Chemistry A, 2024. **128**(12): p. 2487-2497.

772.   Heijmans, K., et al., *Reactive Grand-Canonical Monte Carlo Simulations for Modeling Hydration of $MgCl_2$*. Acs Omega, 2021. **6**(48): p. 32475-32484.

773.   Duin, A.C.T.v., et al., *ReaxFF Reactive Force Field for Solid Oxide Fuel Cell Systems With Application to Oxygen Ion Transport in Yttria-Stabilized Zirconia*. The Journal of Physical Chemistry A, 2008. **112**(14): p. 3133-3140.

774.   Shchygol, G., et al., *ReaxFF Parameter Optimization With Monte-Carlo and Evolutionary Algorithms: Guidelines and Insights*. Journal of Chemical Theory and Computation, 2019. **15**(12): p. 6799-6812.

775.   Kim, S.-Y. and Y. Qi, *Property Evolution of ALD-Al_2O_3 Coated and Uncoated Si Electrodes*. Ecs Meeting Abstracts, 2014. **MA2014-01**(3): p. 342-342.

776.   Rahaman, O., et al., *Development of a ReaxFF Reactive Force Field for Aqueous Chloride and Copper Chloride*. The Journal of Physical Chemistry A, 2010. **114**(10): p. 3556-3568.

777.   Hong, S., et al., *Computational Synthesis of MoS_2 Layers by Reactive Molecular Dynamics Simulations: Initial Sulfidation of MoO_3 Surfaces*. Nano Letters, 2017. **17**(8): p. 4866-4872.

778.   Gastaldi, D., et al., *Continuum Damage Model for Biodegradable Magnesium Alloy Stent*. Advanced Materials Research, 2010. **138**: p. 85-91.

779.   Yang, X., W. Fan, and Z. Li, *A Continuum Damage Model for Prediction of Crack Initiation Life of Pitting Corrosion and Fatigue*. International Journal of Damage Mechanics, 2022. **31**(6): p. 797-814.

780.   Gastaldi, D., et al., *Continuum Damage Model for Bioresorbable Magnesium Alloy Devices — Application to Coronary Stents*. Journal of the Mechanical Behavior of Biomedical Materials, 2011. **4**(3): p. 352-365.

781.   Gao, J., *Life Prediction for LY12CZ Notched Plate Based on the Continuum Damage Mechanics and the Genetic Algorithm and Radial Basis Function Method*. Journal of Theoretical and Applied Mechanics, 2018: p. 1109.

782.   Giancane, S., et al., *Fatigue Life Prediction of Notched Components Based on a New Nonlinear Continuum Damage Mechanics Model*. Procedia Engineering, 2010. **2**(1): p. 1317-1325.

783.   Kang, Z., et al., *Creep-fatigue Deformation Characteristics and Life Prediction Model of Inconel 718 Superalloy Under Hybrid Stress–strain-controlled Mode*. Fatigue & Fracture of Engineering Materials & Structures, 2024. **47**(6): p. 2251-2267.

784.   Mahmoudi, A., B. Mohammadi, and H. Hosseini-Toudeshky, *Damage Behaviour of Laminated Composites During Fatigue Loading*. Fatigue & Fracture of Engineering Materials & Structures, 2019. **43**(4): p. 698-710.

785. Saconi, F., et al., *Experimental Characterization and Numerical Modeling of the Corrosion Effect on the Mechanical Properties of the Biodegradable Magnesium Alloy WE43 for Orthopedic Applications.* Materials, 2022. **15**(20): p. 7164.

786. Wang, Y., W. Zhang, and Y. Zheng, *Experimental Study on Corrosion Fatigue Performance of High-Strength Steel Wire With Initial Defect for Bridge Cable.* Applied Sciences, 2020. **10**(7): p. 2293.

787. Sun, B., et al., *Auto-adaptive Multiblock Cycle Jump Algorithm for Fatigue Damage Simulation of Long-span Steel Bridges.* Fatigue & Fracture of Engineering Materials & Structures, 2018. **42**(4): p. 919-928.

788. Jia-yu, T., et al., *Computational Modeling of the Corrosion Process and Mechanical Performance of Biodegradable Stent.* Journal of Physics Conference Series, 2021. **1888**(1): p. 012019.

789. Berlińska, J., et al., *Mathematical Modelling of the Degradation Behaviour of Biodegradable Metals.* Biomechanics and Modeling in Mechanobiology, 2016. **16**(1): p. 227-238.

790. Nwakamma Ninduwezuor-Ehiobu, N., et al., *Exploring Innovative Material Integration in Modern Manufacturing for Advancing U.S. Competitiveness in Sustainable Global Economy.* Engineering Science & Technology Journal, 2023. **4**(3): p. 140-168.

791. Luna, P., J. Lizarazo-Marriaga, and A. Mariño, *Compatibilization of Natural Fibres as Reinforcement of Polymeric Matrices.* 2019. **2**: p. 239-249.

792. Sarker, F., et al., *Ultrahigh Performance of Nanoengineered Graphene-Based Natural Jute Fiber Composites.* Acs Applied Materials & Interfaces, 2019. **11**(23): p. 21166-21176.

793. Dawod, N., et al., *Metal–Ceramic Compatibility in Dental Restorations According to the Metallic Component Manufacturing Procedure.* Materials, 2023. **16**(16): p. 5556.

794. Rahmani, B., et al., *Manufacturing and Hydrodynamic Assessment of a Novel Aortic Valve Made of a New Nanocomposite Polymer.* Journal of Biomechanics, 2012. **45**(7): p. 1205-1211.

795. Li, C., et al., *Surface Modification of Decellularized Heart Valve by the POSS–PEG Hybrid Hydrogel to Prepare a Composite Scaffold Material With Anticalcification Potential.* Acs Applied Bio Materials, 2022. **5**(8): p. 3923-3935.

796. Srivastva, S., et al., *Emerging Applications of Advanced Materials Processing in Healthcare and Biotechnology.* E3s Web of Conferences, 2024. **505**: p. 01028.

797. Banerjee, R. and S.S. Ray, *Role of Rheology in Morphology Development and Advanced Processing of Thermoplastic Polymer Materials: A Review.* Acs Omega, 2023. **8**(31): p. 27969-28001.

798. Mircea, A.C., et al., *Experimental Study Regarding the Influence of Fibre to Matrix Compatibility on General Performance of Fibre Engineered Cementitious Materials (FECM).* Matec Web of Conferences, 2019. **289**: p. 04005.

799. Nuhiji, B., et al., *Tooling Materials Compatible With Carbon Fibre Composites in a Microwave Environment.* Composites Part B Engineering, 2019. **163**: p. 769-778.

800. Weng, F., et al., *Preparation and Properties of Compatible Starch-PCL Composites: Effects of the NCO Functionality in Compatibilizer.* Starch - Stärke, 2020. **72**(3-4).

801. Ren, M., et al., *Special Interface Structure and Properties for Compatible Packaging Film From Biodegradable Poly (Butylene Adipate-co-terephthalate)/Corn Starch Composite.* Polymer Engineering & Science, 2024. **64**(6): p. 2824-2840.

802. Ogihara, N., et al., *Biocompatibility and Bone Tissue Compatibility of Alumina Ceramics Reinforced With Carbon Nanotubes.* Nanomedicine, 2012. **7**(7): p. 981-993.

803. Chen, Q., et al., *Predicting the Glass Transition Temperature and Solubility Parameter Between Rubber/Silica and Rubber/Resins via All-atom Molecular Dynamics Simulation.* Polymer International, 2024. **73**(9): p. 770-778.

804. Wang, H., et al., *Study of the Epoxy/Amine Equivalent Ratio on Thermal Properties, Cryogenic Mechanical Properties, and Liquid Oxygen Compatibility of the Bisphenol a Epoxy Resin Containing Phosphorus.* High Performance Polymers, 2019. **32**(4): p. 429-443.

805. Evanoff, B., et al., *Results of a Fall Prevention Educational Intervention for Residential Construction.* Safety Science, 2016. **89**: p. 301-307.

806. Tamers, S.L., et al., *Research Methodologies for Total Worker Health ®.* Journal of Occupational and Environmental Medicine, 2018. **60**(11): p. 968-978.

807. Lyu, S., et al., *Relationships Among Safety Climate, Safety Behavior, and Safety Outcomes for Ethnic Minority Construction Workers.* International Journal of Environmental Research and Public Health, 2018. **15**(3): p. 484.

808. Chen, H., et al., *The Impact of Wearable Devices on the Construction Safety of Building Workers: A Systematic Review.* Sustainability, 2023. **15**(14): p. 11165.

809. Duan, P., J. Zhou, and W. Fan, *Safety Tag Generation and Training Material Recommendation for Construction Workers: A Persona-Based Approach.* Engineering Construction & Architectural Management, 2022. **31**(1): p. 115-135.

810. Wameyo, E.S., M. Kiambigi, and J. Okaka, *A Strategy for Effective Safety Management in Construction Sites in Kenya.* East African Journal of Engineering, 2023. **6**(1): p. 228-246.

811. Mahadewi Natalia Wardaniyagung, N., *Implementation of Safety Assurance at PT. X.* World Journal of Advanced Research and Reviews, 2023. **18**(2): p. 356-360.

812. Abduhamidov Abdurahmon Abdulatif o'g'li, N., *A Comprehensive Analysis of the Impact of Globalization on Auditing Standards.* Kokand University Herald, 2023. **9**: p. 61-63.

813. Howard, J.D.J., *Nonstandard Work Arrangements and Worker Health and Safety.* American Journal of Industrial Medicine, 2016. **60**(1): p. 1-10.

814. Chen, Q. and R. Jin, *A Comparison of Subgroup Construction Workers' Perceptions of a Safety Program.* Safety Science, 2015. **74**: p. 15-26.

815. Gaang, L., B. Juhyeon, and S.H. Lee, *Paired Electrodes- And Contraint Independent Components Analysis-Based Denoising to Alleivate Motion Artifacts in Electroencephalogram Collected at Construction Sites.* 2022.

816. Asadi, S., E. Karan, and A. Mohammadpour, *Advancing Safety by in-Depth Assessment of Workers Attention and Perception.* International Journal of Safety Science, 2017. **01**(03): p. 46-60.

817. Damayanti, F., et al., *Analysis of the Effect of Employee Status on Construction Worker's Safety Behavior Using Structural Equation Model.* Eastern-European Journal of Enterprise Technologies, 2022. **6**(10 (120)): p. 54-62.

818. Harvey, E.J., P. Waterson, and A.R.J. Dainty, *Beyond ConCA: Rethinking Causality and Construction Accidents.* Applied Ergonomics, 2018. **73**: p. 108-121.

819. N, Y.D., S. Soeparmi, and Y. Sardjono, *Analysis of Safety and Health of Radiation Officer at Pilot Plant BNCT.* Indonesian Journal of Physics and Nuclear Applications, 2017. **2**(1): p. 42.

820. Shanmugam, K., et al., *Advanced High-Strength Steel and Carbon Fiber Reinforced Polymer Composite Body in White for Passenger Cars: Environmental Performance and Sustainable Return on Investment Under Different Propulsion Modes.* Acs Sustainable Chemistry & Engineering, 2019. **7**(5): p. 4951-4963.

821. Breister, A.M., et al., *Microbial Dark Matter Driven Degradation of Carbon Fiber Polymer Composites.* 2020.

822. Sakib, M.N. and A.A. Iqba, *Epoxy Based Nanocomposite Material for Automotive Application- A Short Review.* International Journal of Automotive and Mechanical Engineering, 2021. **18**(3).

823. Kim, H.C., et al., *Life Cycle Assessment of Vehicle Lightweighting: Novel Mathematical Methods to Estimate Use-Phase Fuel Consumption.* Environmental Science & Technology, 2015. **49**(16): p. 10209-10216.

824. Hou, W., et al., *Bending Behavior of Single Hat-Shaped Composite T-Joints Under Out-of-Plane Loading for Lightweight Automobile Structures.* Journal of Reinforced Plastics and Composites, 2018. **37**(12): p. 808-823.

825. Cecchin, A., et al., *What Is in a Name? The Rising Star of the Circular Economy as a Resource-Related Concept for Sustainable Development.* Circular Economy and Sustainability, 2021. **1**(1): p. 83-97.

826. Kasner, R., et al., *Sustainable Wind Power Plant Modernization.* Energies, 2020. **13**(6): p. 1461.

827. Ramakrishna, S., W. Hu, and R. Jose, *Sustainability in Numbers by Data Analytics.* Circular Economy and Sustainability, 2022. **3**(2): p. 643-655.

828. Abdelhafeez, I.A. and S. Ramakrishna, *Promising Sustainable Models Toward Water, Air, and Solid Sustainable Management in the View of SDGs.* Materials Circular Economy, 2021. **3**(1).

829. Bui, T.D., et al., *Sustainable Supply Chain Management Towards Disruption and Organizational Ambidexterity: A Data Driven Analysis.* Sustainable Production and Consumption, 2021. **26**: p. 373-410.

830. Wang, K. and F. Taheri, *Comparisons of the Performance of Novel Lightweight Three-Dimensional Hybrid Composites Against GLARE Fiber–Metal Laminate.* Processes, 2023. **11**(10): p. 2875.

831. Karoonsit, B., et al., *Performance Evaluation for Ultra-Lightweight Epoxy-Based Bipolar Plate Production With Cycle Time Reduction of Reactive Molding Process.* Polymers, 2022. **14**(23): p. 5226.

832. Zhao, Y., et al., *Characterizing the Conductivity and Enhancing the Piezoresistivity of Carbon Nanotube-Polymeric Thin Films.* Materials, 2017. **10**(7): p. 724.

833. Hager, M.D., et al., *Self-Healing Materials*. Advanced Materials, 2010. **22**(47): p. 5424-5430.

834. Ferguson, J.B., B.F. Schultz, and P.K. Rohatgi, *Self-Healing Metals and Metal Matrix Composites*. Jom, 2014. **66**(6): p. 866-871.

835. Liu, Z., et al., *Biomimetic Materials With Multiple Protective Functionalities*. Advanced Functional Materials, 2019. **29**(28).

836. Xie, Z., et al., *Hydrogen Bonding in Self-Healing Elastomers*. Acs Omega, 2021. **6**(14): p. 9319-9333.

837. Song, Y., et al., *Towards Dynamic but Supertough Healable Polymers Through Biomimetic Hierarchical Hydrogen -Bonding Interactions*. Angewandte Chemie, 2018. **130**(42): p. 14034-14038.

838. Diesendruck, C.E., et al., *Biomimetic Self-Healing*. Angewandte Chemie, 2015. **54**(36): p. 10428-10447.

839. Müller, W.E.G., et al., *Transformation of Construction Cement to a Self-Healing Hybrid Binder*. International Journal of Molecular Sciences, 2019. **20**(12): p. 2948.

840. Sayadi, S., I. Mihai, and A. Jefferson, *Biomimetic Materials in Construction Industry: The Necessity of Simulation*. 2024.

841. Chen, J., et al., *Repetitive Biomimetic Self-Healing of Ca2+-Induced Nanocomposite Protein Hydrogels*. Scientific Reports, 2016. **6**(1).

842. Huang, H., et al., *Self-Healing in Cementitious Materials: Materials, Methods and Service Conditions*. Materials & Design, 2016. **92**: p. 499-511.

843. Sharma, S., G. Nandan, and R.K. Tyagi, *Self-Healing Metal Matrix Composite of Nitinol Wire-Reinforced A356 Alloy Matrix by Stir Casting Technology*. Engineering Research Express, 2023. **5**(3): p. 035038.

844. Liu, Y., et al., *Biomimetic Strain-Stiffening in Chitosan Self-Healing Hydrogels*. Acs Applied Materials & Interfaces, 2022. **14**(14): p. 16032-16046.

845. Zhao, D., et al., *UV Light Curable Self-Healing Superamphiphobic Coatings by Photopromoted Disulfide Exchange Reaction*. Acs Applied Polymer Materials, 2019. **1**(11): p. 2951-2960.

846. Blaiszik, B.J., et al., *Autonomic Restoration of Electrical Conductivity*. Advanced Materials, 2011. **24**(3): p. 398-401.

847. Tamesue, S., et al., *Linear Versus Dendritic Molecular Binders for Hydrogel Network Formation With Clay Nanosheets: Studies With ABA Triblock Copolyethers Carrying Guanidinium Ion Pendants*. Journal of the American Chemical Society, 2013. **135**(41): p. 15650-15655.

848. Zeng, L., et al., *A Highly Stretchable, Tough, Fast Self-Healing Hydrogel Based on Peptide–Metal Ion Coordination*. Biomimetics, 2019. **4**(2): p. 36.

849. Dahlke, J., et al., *How to Design a Self-Healing Polymer: General Concepts of Dynamic Covalent Bonds and Their Application for Intrinsic Healable Materials*. Advanced Materials Interfaces, 2018. **5**(17).

850. Vila-Cortavitarte, M., et al., *Laboratory and Statistical Analysis of the Fatigue Response of Self-Healing Asphalt Mixtures Containing Metal by-Products*. Coatings, 2021. **11**(4): p. 385.

851. Wei, M., et al., *Stimuli-Responsive Polymers and Their Applications*. Polymer Chemistry, 2017. **8**(1): p. 127-143.

852. Zhao, D., R. Rajan, and K. Matsumura, *Dual Thermo- And pH-Responsive Behavior of Double Zwitterionic Graft Copolymers for Suppression of Protein Aggregation and Protein Release*. Acs Applied Materials & Interfaces, 2019. **11**(43): p. 39459-39469.

853. Schattling, P., F.D. Jochum, and P. Théato, *Multi-Stimuli Responsive Polymers – The All-in-One Talents*. Polymer Chemistry, 2014. **5**(1): p. 25-36.

854. Maji, T., et al., *Dual-Stimuli-Responsive <scp>l</Scp>-Serine-Based Zwitterionic UCST-Type Polymer With Tunable Thermosensitivity*. Macromolecules, 2015. **48**(14): p. 4957-4966.

855. Samanta, S., et al., *Polyacetals: Water-Soluble, pH-Degradable Polymers With Extraordinary Temperature Response*. Macromolecules, 2016. **49**(5): p. 1858-1864.

856. Lu, X., et al., *Biodegradable Temperature- And pH Dually-Responsive Poly(β-Amino Ester)*. Destech Transactions on Engineering and Technology Research, 2017(apetc).

857. Zhang, Z., et al., *A Synthetic, Transiently Thermoresponsive Homopolymer With UCST Behaviour Within a Physiologically Relevant Window*. Angewandte Chemie, 2019. **58**(23): p. 7866-7872.

858. Lerch, A., et al., *Structural Insights Into Polymethacrylamide-Based LCST Polymers in Solution: A Small-Angle Neutron Scattering Study*. Macromolecules, 2021. **54**(16): p. 7632-7641.

859. Keimer, B. and J.E. Moore, *The Physics of Quantum Materials*. Nature Physics, 2017. **13**(11): p. 1045-1055.

860. Basov, D.N., R.D. Averitt, and D. Hsieh, *Towards Properties on Demand in Quantum Materials*. Nature Materials, 2017. **16**(11): p. 1077-1088.

861. Hu, C., et al., *Realization of an Intrinsic Ferromagnetic Topological State in MnBi ₈ Te ₁₃*. Science Advances, 2020. **6**(30).

862. Wang, J. and S.-C. Zhang, *Topological States of Condensed Matter*. Nature Materials, 2017. **16**(11): p. 1062-1067.

863. Qi, X.L. and S.C. Zhang, *Topological Insulators and Superconductors*. Reviews of Modern Physics, 2011. **83**(4): p. 1057-1110.

864. Hasan, M.Z. and C.L. Kane, *<i>Colloquium</I>: Topological Insulators*. Reviews of Modern Physics, 2010. **82**(4): p. 3045-3067.

865. Soluyanov, A.A., et al., *Type-Ii Weyl Semimetals*. Nature, 2015. **527**(7579): p. 495-498.

866. Zafar, M.S., et al., *Biomimetic Aspects of Restorative Dentistry Biomaterials*. Biomimetics, 2020. **5**(3): p. 34.

867. Zhou, C., et al., *Biomimetic Fabrication of a Three-Level Hierarchical Calcium Phosphate/Collagen/Hydroxyapatite Scaffold for Bone Tissue Engineering*. Biofabrication, 2014. **6**(3): p. 035013.

868. Barthelat, F., *Architectured Materials in Engineering and Biology: Fabrication, Structure, Mechanics and Performance*. International Materials Reviews, 2015. **60**(8): p. 413-430.

869. Kong, D., et al., *A Biomimetic Structural Material With Adjustable Mechanical Property for Bone Tissue Engineering*. Advanced Functional Materials, 2023. **34**(8).

870. Ma, W., et al., *Preparation and Characterization of Microcapsule Filled With Ethyl Phenyl Acetate and Its Self-Healing Application*. 2015.

871. Deshpande, S.R., et al., *DNA -Responsive Polyisocyanopeptide Hydrogels With Stress -Stiffening Capacity*. Advanced Functional Materials, 2016. **26**(48): p. 9075-9082.

872. Tan, Y., et al., *Progress and Roadmap for Intelligent Self-Healing Materials in Autonomous Robotics*. Advanced Materials, 2020. **33**(19).

873. Hu, D., et al., *Chitosan-Based Biomimetically Mineralized Composite Materials in Human Hard Tissue Repair*. Molecules, 2020. **25**(20): p. 4785.

874. Speck, O. and T. Speck, *An Overview of Bioinspired and Biomimetic Self-Repairing Materials*. Biomimetics, 2019. **4**(1): p. 26.

875. Uchiyama, Y., E. Blanco, and R. Kohsaka, *Application of Biomimetics to Architectural and Urban Design: A Review Across Scales*. Sustainability, 2020. **12**(23): p. 9813.

876. Pohl, G. and W. Nachtigall, *Biomimetics for Architecture &Amp; Design*. 2015.

877. Haškovec, J., P.A. Markowich, and G. Pilli, *Murray's Law for Discrete and Continuum Models of Biological Networks*. Mathematical Models and Methods in Applied Sciences, 2019. **29**(12): p. 2359-2376.

878. Stephenson, D.B., et al., *Generalizing Murray's Law: An Optimization Principle for Fluidic Networks of Arbitrary Shape and Scale*. Journal of Applied Physics, 2015. **118**(17).

879. Chen, Y., et al., *Blood Physiological and Flow Characteristics Within Coronary Artery Circulatory Network for Human Heart Based on Vascular Fractal Theory*. Advances in Mechanical Engineering, 2014. **12**(7).

880. Zheng, X., et al., *Bio-Inspired Murray Materials for Mass Transfer and Activity*. Nature Communications, 2017. **8**(1).

881. Garcia-Holguera, M., et al., *Ecosystem Biomimetics for Resource Use Optimization in Buildings*. Building Research & Information, 2015. **44**(3): p. 263-278.

882. Chayaamor-Heil, N. and N. Hannachi-Belkadi, *Towards a Platform of Investigative Tools for Biomimicry as a New Approach for Energy-Efficient Building Design*. Buildings, 2017. **7**(1): p. 19.

883. Jin, J., et al., *Hierarchy Design in Metal Oxides as Anodes for Advanced Lithium -Ion Batteries*. Small Methods, 2018. **2**(11).

884. Ateshian, G.A., et al., *Finite Element Framework for Computational Fluid Dynamics in FEBio*. Journal of Biomechanical Engineering, 2018. **140**(2).

885. Mannodi-Kanakkithodi, A., et al., *Comprehensive Computational Study of Partial Lead Substitution in Methylammonium Lead Bromide*. Chemistry of Materials, 2019. **31**(10): p. 3599-3612.

886. Xiao, Q. and J. Wang, *CFD–DEM Simulations of Seepage-Induced Erosion*. Water, 2020. **12**(3): p. 678.

887. Battista, N.A., *Suite-Cfd: An Array of Fluid Solvers Written in MATLAB and Python*. Fluids, 2020. **5**(1): p. 28.

888. Li, C. and K. Zheng, *Methods, Progresses, and Opportunities of Materials Informatics*. Infomat, 2023. **5**(8).

889. Shibiao, H., et al., *VASP Porting and Parallel Optimization on GPU Like Accelerator.* 2023: p. 63.

890. Oliynyk, A.O. and A. Mar, *Discovery of Intermetallic Compounds From Traditional to Machine-Learning Approaches.* Accounts of Chemical Research, 2017. **51**(1): p. 59-68.

891. Huning, A., et al., *Advancement of Certification Methods and Applications for Industrial Deployments of Components Derived From Advanced Manufacturing Technologies.* 2022.

892. Gray, G.T., et al., *Structure/Property (Constitutive and Dynamic Strength/Damage) Characterization of Additively Manufactured 316L SS.* Epj Web of Conferences, 2015. **94**: p. 02006.

893. Steuben, J.C., A. Iliopoulos, and J.G. Michopoulos, *Discrete Element Modeling of Particle-Based Additive Manufacturing Processes.* Computer Methods in Applied Mechanics and Engineering, 2016. **305**: p. 537-561.

894. Saraçyakupoğlu, T., *Certification Steps for the Additively Manufactured Aviation-Grade Parts.* The European Journal of Research and Development, 2022. **2**(4): p. 33-42.

895. Chen, Z., et al., *A Review on Qualification and Certification for Metal Additive Manufacturing.* Virtual and Physical Prototyping, 2021. **17**(2): p. 382-405.

896. Jones, R., et al., *Modelling the Variability and the Anisotropic Behaviour of Crack Growth in SLM Ti-6Al-4v.* Materials, 2021. **14**(6): p. 1400.

897. Saraçyakupoğlu, T., *The Qualification of the Additively Manufactured Parts in the Aviation Industry.* American Journal of Aerospace Engineering, 2019. **6**(1): p. 1.

898. Tamayo, J.A., et al., *Additive Manufacturing of Ti6Al4V Alloy via Electron Beam Melting for the Development of Implants for the Biomedical Industry.* Heliyon, 2021. **7**(5): p. e06892.

899. Surono, S., *Development of Micro Credential Design for Project Management to Improve the Quality of Engineering Practices.* International Journal of Social Service and Research, 2023. **3**(11): p. 2910-2920.

900. Wang, X., L. Xue-wen, and J. Luo, *Course of Engineering Project Management for Automatic Engineering.* International Journal for Innovation Education and Research, 2020. **8**(9): p. 304-309.

901. Georgantzinos, S.K., et al., *Composites in Aerospace and Mechanical Engineering.* Materials, 2023. **16**(22): p. 7230.

902. Ma, Y., *Medium-Manganese Steels Processed by Austenite-Reverted-Transformation Annealing for Automotive Applications.* Materials Science and Technology, 2017. **33**(15): p. 1713-1727.

903. Veličković, S., et al., *Application of Nanocomposites in the Automotive Industry.* Mobility and Vehicle Mechanics, 2019. **45**(3): p. 51-64.

904. Abedsoltan, H., *Applications of Plastics in the Automotive Industry: Current Trends and Future Perspectives.* Polymer Engineering & Science, 2023. **64**(3): p. 929-950.

905. Llopis-Albert, C. and F. Rubio, *Impact of Digital Transformation on the Automotive Industry.* Technological Forecasting and Social Change, 2021. **162**: p. 120343.

906. Jahromi, F.T., et al., *Additive Manufacturing of Polypropylene Micro and Nano Composites Through Fused Filament Fabrication for Automotive Repair Applications.* Polymers for Advanced Technologies, 2022. **34**(3): p. 1059-1074.

907. Nunes, D.M., R.D.S.G. Campilho, and F. Silva, *Design of a Transfer System for the Automotive Industry.* Proceedings of the Institution of Mechanical Engineers Part E Journal of Process Mechanical Engineering, 2022. **236**(5): p. 2044-2055.

908. Adebisi, A.A., M.A. Maleque, and M.M. Rahman, *Metal Matrix Composite Brake Rotor: Historical Development and Product Life Cycle Analysis.* International Journal of Automotive and Mechanical Engineering, 2011. **4**: p. 471-480.

909. Schäper, T., et al., *Determinants of Idea Sharing in Crowdsourcing: Evidence From the Automotive Industry.* R and D Management, 2020. **51**(1): p. 101-113.

910. Krauklis, A.E., et al., *Composite Material Recycling Technology—State-of-the-Art and Sustainable Development for the 2020s.* Journal of Composites Science, 2021. **5**(1): p. 28.

911. Kušar, J., et al., *Concurrent Realisation and Quality Assurance of Products in the Automotive Industry.* Concurrent Engineering, 2014. **22**(2): p. 162-171.

912. Adeniyi Kehinde Adeleke, N., et al., *Process Development in Mechanical Engineering: Innovations, Challenges, and Opportunities.* Engineering Science & Technology Journal, 2024. **5**(3): p. 901-912.

913. Orenuga, O.S., O. Adebisi, and I. Adediran, *Emerging Trends in Sustainable Materials for Green Building Constructions.* Key Engineering Materials, 2024. **974**: p. 13-22.

914. Suchek, N., et al., *Innovation and the Circular Economy: A Systematic Literature Review.* Business Strategy and the Environment, 2021. **30**(8): p. 3686-3702.

915. Yousif, Y., M.S. Misnan, and M.Z. Ismail, *The Influence of Labelled Green Building Materials on the Performance of Green Construction Projects.* Iop Conference Series Earth and Environmental Science, 2023. **1274**(1): p. 012028.

916. Suman Das, N., N. Joyeshree Biswas, and N. Iqtiar Md Siddique, *Mechanical Characterization of Materials Using Advanced Microscopy Techniques.* World Journal of Advanced Research and Reviews, 2024. **21**(3): p. 274-283.

917. Karimov, I., *Impact of Alternative Energy Investments on Employment in Azerbaijan.* Agora International Journal of Juridical Sciences, 2024. **18**(1): p. 191-203.

918. Haryono, C.B.C. and N.T.P. Sari, *Adoption of Green Innovation in SMEs: A Literature Review.* 2023: p. 41-48.

919. Li, Z. and Y. Wu, *Does Managerial Myopia Hinder Green Technological Innovations? An Examination Based on Chinese-Listed Heavy Polluters.* Frontiers in Environmental Science, 2023. **11**.

920. Finkelstein-Shapiro, A. and V. Nuguer, *Climate Policies, Labor Markets, and Macroeconomic Outcomes in Emerging Economies.* 2023.

921. Ştefan, I., et al., *The Role of Romanian Startup Hubs – A Bridge Between a Business Idea and the Reality of the Economic Sector.* Proceedings of the International Conference on Business Excellence, 2023. **17**(1): p. 800-811.

922. Šūmakaris, P., R. Korsakienė, and D. Ščeulovs, *Determinants of Energy Efficient Innovation: A Systematic Literature Review.* Energies, 2021. **14**(22): p. 7777.

923. Shin, J., C. Kim, and H. Yang, *Does Reduction of Material and Energy Consumption Affect to Innovation Efficiency? The Case of Manufacturing Industry in South Korea.* Energies, 2019. **12**(6): p. 1178.

924. Arranz, N., et al., *Incentives and Inhibiting Factors of Eco-Innovation in the Spanish Firms.* Journal of Cleaner Production, 2019. **220**: p. 167-176.

925. Turner, J., E.I. Daniel, and E. Chinyio, *The Application for Innovative Methods and Materials for Greater Sustainability in Residential Buildings in the UK: "A Bibliometric Review".* Discover Sustainability, 2024. **5**(1).

926. Nambisan, S. and R.A. Baron, *Entrepreneurship in Innovation Ecosystems: Entrepreneurs' Self–Regulatory Processes and Their Implications for New Venture Success.* Entrepreneurship Theory and Practice, 2013. **37**(5): p. 1071-1097.

927. Schiuma, G. and D. Carlucci, *Managing Strategic Partnerships With Universities in Innovation Ecosystems: A Research Agenda.* Journal of Open Innovation Technology Market and Complexity, 2018. **4**(3): p. 25.

Index

3

3D Printing, 2, 11, 16, 17, 21, 162, 163, 164, 165, 167, 191, 194, 219, 220, 228, 284, 320, 355, 366, 370, 376, 379, 423, 436, 447

A

Abrasion Resistance, 139
Additive Manufacturing, 11, 16, 17, 20, 21, 56, 162, 163, 164, 165, 166, 219, 220, 228, 284, 308, 320, 323, 324, 325, 356, 357, 358, 359, 363, 364, 365, 366, 370, 379, 398, 402, 403, 421, 422, 423, 424, 425, 426, 427, 428, 432, 434, 436, 455, 481, 491, 492
Advanced Ceramics, 30, 49, 50
Advanced Coatings, 53, 79, 176, 184, 282
Advanced Polymers, 10, 56, 330, 376, 377
Advanced Testing, 408
Aerospace Applications, 48, 49, 56, 72, 140, 161, 162, 215, 222, 225, 237, 239, 257, 348, 356, 358, 364, 367, 429, 444, 446, 470
Aerospace Materials, 2, 368, 436
Aging Chambers, 409, 410
Alternative Energy, 322, 493
Amorphous Materials, 93, 330
Anisotropy, 282, 343, 344, 345, 472
Atomic Force Microscopy (AFM), 90, 92, 405
Atomic Structure, 8, 22, 23, 24, 35, 36, 246, 307, 311

B

Bio-Based Polymers, 370
Biocompatibility, 4, 54, 56, 177, 183, 197, 216, 225, 309, 369, 384, 470, 486
Biocomposites, 449, 465
Biodegradable Polymers, 4, 10, 44, 54, 56, 191, 195, 196, 197, 198, 206, 308, 396, 433, 463
Bio-Inspired Materials, 375, 398

Biomaterials, 3, 4, 8, 9, 10, 12, 14, 16, 17, 19, 20, 24, 28, 39, 40, 44, 54, 56, 100, 309, 420, 424, 428, 429, 433, 434, 458, 460, 462, 467, 470, 490
Biomimicry, 399, 400, 491
Bioplastics, 3, 184, 189, 190, 191, 192, 193, 194, 204, 205, 206, 207, 208, 209, 210, 211, 213, 214, 433, 462, 464, 465
Biosensors, 72, 188, 459, 462

C

Carbon Nanotubes, 2, 4, 23, 26, 40, 62, 65, 103, 123, 141, 172, 173, 175, 176, 180, 183, 186, 187, 188, 228, 291, 314, 364, 365, 367, 436, 437, 442, 449, 472, 486
Catalyst Design, 442
Catalytic Materials, 308
Ceramic Composites, 169
Chemical Vapor Deposition (CVD), 131, 161, 180, 402
Circular Economy, 11, 189, 205, 284, 296, 299, 300, 304, 305, 364, 369, 370, 462, 473, 474, 478, 479, 487, 488, 493
Coatings, 6, 13, 28, 33, 47, 49, 53, 54, 55, 69, 78, 79, 81, 87, 92, 98, 100, 101, 107, 127, 139, 141, 165, 173, 175, 180, 181, 184, 190, 191, 192, 216, 228, 255, 266, 281, 284, 291, 292, 323, 326, 366, 375, 380, 381, 397, 398, 406, 409, 418, 445, 450, 455, 466, 489
Composite Materials, 3, 26, 62, 70, 92, 113, 117, 120, 122, 139, 140, 141, 142, 143, 148, 149, 152, 155, 156, 157, 169, 176, 217, 218, 219, 246, 256, 257, 258, 279, 281, 285, 296, 304, 320, 366, 378, 419, 425, 428, 437, 444, 446, 447, 448, 449, 451, 452, 453, 454, 455, 456, 472, 473, 479, 490
Compression Testing, 229, 243, 249, 258
Computational Materials Science, 17, 309, 310, 352, 439, 481, 482
Conductive Polymers, 7, 10, 187, 217, 323

Corrosion Resistance, 22, 23, 33, 34, 55, 65, 122, 125, 127, 149, 153, 162, 163, 169, 173, 286, 423
Crystal Growth, 67
Crystallography, 3, 23, 56, 427

D

Defect Engineering, 38, 56
Deformation Mechanisms, 343, 346
Dielectric Properties, 62, 68
Dielectric Strength, 22, 411, 412
Diffraction Analysis, 404, 405
Dimensional Stability, 50, 73, 259
Dislocations, 23, 24, 37, 38, 63, 64, 65, 66, 80, 81, 128, 235, 236, 240, 260, 279, 330, 343, 344, 443
Dispersions, 447
Ductility, 21, 22, 23, 31, 32, 35, 38, 49, 64, 67, 69, 125, 128, 129, 229, 239, 240, 241, 246, 248, 249, 257, 259, 340, 342, 346, 366
Dynamic Mechanical Analysis (DMA), 220, 221, 250, 254, 255, 256

E

Elasticity, 22, 26, 39, 45, 46, 49, 93, 108, 139, 195, 229, 242, 243, 247, 248, 259, 309, 326, 406, 407, 416, 463
Electrochemical Workstations, 411
Electrode Materials, 45, 326, 412, 473
Electron Beam Lithography, 179
Electron Microscopy, 16, 23, 55, 64, 81, 86, 87, 89, 90, 96, 220, 221, 258, 333, 350, 404, 405, 421, 468
Emerging Technologies, 17, 422
Energy Efficiency, 4, 7, 11, 187, 214, 223, 273, 284, 290, 295, 296, 297, 302, 370, 386, 399, 429, 433, 434, 475
Energy Harvesting, 456, 457, 461
Energy Storage, 3, 4, 7, 10, 15, 17, 24, 26, 27, 28, 30, 31, 36, 43, 62, 68, 69, 70, 72, 74, 86, 117, 136, 178, 187, 188, 256, 280, 308, 326, 330, 399, 411, 424, 429, 430, 435, 439, 483
Environmental Degradation, 12, 28, 52, 121, 149, 169, 190, 213, 304, 354, 365
Environmental Impact, 3, 2, 10, 11, 12, 15, 16, 171, 182, 199, 211, 213, 223, 224, 283, 284, 285, 286, 290, 291, 292, 294, 296, 297, 300, 302, 304, 305, 308, 365, 368, 370, 371, 383, 398, 433, 462, 473, 474, 478
Environmental Testing, 409
Epitaxy, 392, 393, 402, 403, 443

F

Fatigue Testing, 108, 229, 247, 249, 258, 260, 420, 422, 425
Fiber Reinforcement, 446
Finite Element Analysis (FEA), 319, 413, 420
Flexible Electronics, 139, 183, 217, 434, 467
Fracture Mechanics, 321, 322, 336, 338
Fracture Toughness, 128, 129, 140, 164, 247, 365, 411, 445, 451
Functionally Graded Materials, 366

G

Gallium Arsenide, 11, 61, 274, 393
Gallium Nitride, 65, 69
Grain Boundaries, 21, 24, 38, 66, 67, 68, 69, 70, 81, 86, 236, 259, 332, 333, 340, 342, 404
Graphene, 4, 25, 26, 27, 28, 40, 44, 47, 50, 51, 61, 62, 64, 65, 68, 73, 81, 86, 103, 123, 124, 141, 175, 177, 180, 183, 186, 187, 188, 220, 221, 228, 293, 309, 363, 365, 367, 388, 394, 395, 406, 440, 441, 442, 443, 450, 455, 458, 468, 485
Green Chemistry, 183, 370, 466, 475
Green Manufacturing, 469

H

Heat Treatments, 133, 402
High-Entropy Alloys, 65, 73
High-Performance Materials, 21, 113, 117, 120, 135, 165, 350, 370, 401
High-Performance Polymers, 3, 6
Hybrid Materials, 220
Hydrogen Storage, 93, 153, 154, 188, 433, 453
Hydrothermal Synthesis, 183

I

Impact Resistance, 73, 122, 136, 138, 139, 140, 141, 143

Industrial Applications, 3, 6, 17, 27, 53, 55, 70, 117, 120, 152, 182, 183, 193, 199, 200, 208, 213, 254, 263, 273, 277, 282, 348, 355, 403, 412, 414, 415, 417

Injection Molding, 145, 158, 193, 207, 208, 403, 447, 452, 454, 465

Innovation Ecosystems, 493

Insulating Materials, 51, 429

Ion Beam Techniques, 457

Isotropic Materials, 242

L

Lattice Dynamics, 94

Life Cycle Analysis, 492

Lightweight Alloys, 308, 370

Liquid Crystals, 252

M

Machine Learning in Materials, 481

Magnetic Materials, 267

Magnetostrictive Materials, 386

Manufacturing Processes, 11, 17, 71, 117, 137, 153, 164, 177, 223, 224, 228, 264, 284, 285, 294, 363, 364, 370, 379, 422, 429, 445, 454

Material Characterization, 3, 75, 95, 97, 221, 250, 260, 292, 405, 427

Material Innovation, 5, 7, 304, 371, 433, 434, 435

Material Interfaces, 262, 324

Material Properties, 4, 8, 19, 38, 59, 81, 93, 95, 110, 126, 137, 144, 148, 159, 162, 167, 181, 182, 216, 218, 219, 220, 222, 227, 235, 246, 251, 255, 256, 258, 259, 261, 263, 266, 267, 271, 281, 301, 306, 319, 320, 323, 326, 330, 333, 334, 336, 337, 347, 356, 357, 360, 365,366, 368, 393, 394, 413, 414, 420, 421, 422, 429, 430, 457, 481, 482

Material Recycling, 492

Material Selection, 216, 230, 233, 235, 242, 249, 326, 360, 415, 444

Material Stability, 110

Materials Chemistry, 441

Materials Databases, 218, 304, 413, 415

Materials Informatics, 439, 440, 481, 491

Materials Testing, 413, 444

Mechanical Design, 479

Mechanical Properties, 2, 3, 4, 6, 7, 21, 22, 37, 40, 45, 52, 57, 60, 61, 64, 67, 71, 72, 81, 90, 92, 120, 121, 122, 123, 125, 126, 127, 128, 132, 133, 136, 137, 138, 139, 140, 141, 143, 144, 145, 148, 149, 151, 152, 153, 155, 156, 158, 159, 160, 162, 163, 165, 166, 167, 173, 175, 177, 180, 186, 191, 193, 197, 202, 206, 208, 217, 218, 219, 220, 221, 225, 233, 237, 242, 247, 254, 255, 256, 257, 291, 312, 313, 320, 326, 342, 344, 346, 348, 349, 350, 360, 362, 365, 366, 367, 373, 375, 376, 379, 411, 416, 417, 418, 422, 437, 443, 445, 446, 447, 448, 449, 450, 451, 453, 454, 455, 456, 463, 464, 467, 471, 485, 486

Metal Matrix Composites, 2, 69, 125, 126, 138, 142, 177, 429, 436, 446, 450, 454, 488

Metamaterials, 3, 5, 355, 437

Microfabrication, 179, 180

Microstructural Properties, 86

Molecular Dynamics, 227, 308, 313, 315, 316, 318, 319, 328, 329, 331, 334, 338, 350, 353, 367, 413, 414, 422, 428, 480, 483, 484, 486

N

Nanocomposites, 114, 141, 173, 176, 177, 193, 228, 293, 366, 443, 456, 463, 466, 472, 492

Nanocrystals, 44, 176, 184, 460

Nanoelectronics, 65, 171, 183, 187

Nanomaterials, 3, 4, 6, 7, 8, 10, 12, 13, 14, 16, 19, 20, 23, 24, 44, 45, 94, 103, 117, 140, 141, 171, 172, 173, 174, 175, 176, 177, 178, 180, 181, 182, 183, 184, 185, 186, 187, 188, 217, 219, 220, 283, 285, 290, 291, 292, 293, 309, 310, 314, 364, 365, 368, 370, 420, 422, 426, 427, 430, 434, 436, 438, 439, 441, 442, 445, 456, 457, 458, 459, 460, 461, 468

Nanoparticles, 21, 23, 44, 53, 54, 55, 82, 86, 172, 173, 174, 175, 176, 180, 181, 182, 184, 185, 186, 187, 188, 219, 228, 366, 392, 417, 437, 441, 458, 459, 460, 468, 472

Nanoscale Properties, 178

Nanostructures, 23, 28, 40, 41, 44, 81, 86, 92, 100, 103, 116, 173, 180, 181, 185, 220, 444, 457, 458, 461, 468, 471

Nanotechnology, 3, 2, 17, 26, 28, 40, 42, 43, 81, 83, 86, 92, 171, 172, 176, 178, 179, 180, 181,

183, 228, 310, 381, 404, 405, 424, 428, 430, 431, 432, 433, 438, 440, 456, 457, 459
Non-Destructive Testing (NDT), 261, 418

O

Optical Coatings, 402
Optical Properties, 16, 22, 42, 44, 61, 70, 72, 108, 185, 228, 307, 310, 385, 406, 459
Optoelectronic Materials, 114
Organic Electronics, 40, 43, 440

P

Permeability, 191, 193, 266, 267, 270
Perovskite Materials, 482
Phase Diagrams, 252, 346, 347, 348, 349, 482
Phase Transition, 20, 108, 110, 250, 251, 252, 254, 255, 280, 308, 317, 320, 328, 329, 330, 347, 350, 404, 406, 414, 420
Physical Properties, 2, 23, 42, 66, 123, 195, 292
Physical Vapor Deposition (PVD), 392, 393, 402
Piezoelectric Materials, 5, 386
Plasticity, 46, 49, 343, 344
Polymer Composites, 2, 51, 52, 162, 375, 378, 436, 444, 447, 448, 449, 450, 451, 452, 453, 454, 455, 471, 473, 487
Polymers, 2, 3, 6, 8, 10, 11, 23, 26, 27, 39, 40, 47, 49, 56, 74, 90, 94, 100, 101, 108, 109, 110, 120, 122, 123, 124, 131, 140, 142, 159, 164, 165, 166, 167, 168, 169, 177, 190, 192, 193, 194, 195, 196, 197, 198, 199, 200, 201, 202, 203, 204, 206, 207, 208, 217, 221, 222, 229, 234, 240, 241, 242, 252, 253, 254, 255, 256, 257, 284, 285, 291, 292, 293, 294, 300, 301, 320, 351, 367, 368, 370, 371, 373, 374, 376, 377, 378, 379, 380, 382, 384, 386, 387, 402, 403, 414, 415, 416, 420, 434, 436, 437, 444, 447, 448, 449, 450, 452, 453, 454, 456, 461, 462, 463, 465, 468, 470, 486, 488, 489, 492
Porosity, 72, 73, 74, 119, 131, 133, 135, 145, 160, 162, 264, 325, 443
Powder Metallurgy, 126, 159, 323, 446, 449, 450, 454
Process Engineering, 425
Process Optimization, 346, 422, 423, 451
Processing Techniques, 3, 6, 19, 22, 56, 142, 168, 194, 208, 219, 220, 280

Prototype Development, 447

Q

Quantum Dots, 4, 44, 81, 114, 173, 175, 176, 178, 184, 185, 187, 188, 441, 460, 461
Quantum Materials, 4, 387, 388, 395, 403, 424, 428, 431, 434, 489

R

Recycling Technologies, 290, 370, 478
Reflectivity, 32, 95, 278
Refractive Index, 5, 22
Renewable Materials, 213, 448
Rheological Properties, 366
Rheology, 452, 486

S

Safety Standards, 225, 369, 423
Self-Healing Materials, 4, 217, 292, 367, 373, 374, 375, 376, 380, 382, 384
Semiconductor Materials, 86, 402, 429
Shape Memory Alloys, 225, 320, 437
Simulation Techniques, 319
Smart Hydrogels, 434
Smart Materials, 3, 6, 9, 16, 165, 386, 387, 429, 434, 470
Solar Cell Materials, 117
Spectroscopy, 16, 23, 40, 54, 55, 86, 89, 92, 95, 96, 97, 100, 102, 103, 114, 116, 350, 393, 404, 405, 412, 427
Spintronics, 66, 187, 388, 390, 392, 393, 395, 417
Strain Rate Sensitivity, 248
Structural Integrity, 6, 10, 26, 30, 34, 35, 50, 53, 55, 64, 71, 86, 103, 115, 122, 125, 126, 129, 135, 136, 137, 138, 142, 143, 148, 153, 201, 217, 254, 257, 261, 264, 270, 280, 281, 356, 358, 360, 366, 374, 377, 378, 385, 416, 417, 418, 455
Structural Materials, 66, 68, 87, 202, 241, 326, 375, 383, 399, 408, 443
Superalloys, 2, 33, 51, 52, 64, 68, 163, 249, 321, 343, 348, 349, 367
Supercapacitors, 43, 185, 188, 411, 412, 439
Surface Chemistry, 98
Surface Energy, 186, 339, 394

Surface Functionalization, 186, 366
Surface Modification, 79, 123, 366, 457, 458, 485
Surface Roughness, 90, 92
Sustainability, 4, 1, 2, 6, 10, 11, 12, 13, 14, 15, 16, 17, 18, 31, 54, 119, 124, 137, 138, 169, 172, 182, 183, 185, 190, 195, 205, 207, 209, 213, 214, 215, 216, 223, 283, 284, 285, 290, 294, 296, 297, 298, 299, 300, 301, 302, 303, 304, 327, 368, 369, 370, 371, 378, 396, 397, 398, 399, 425, 428, 429, 430, 431, 433, 434, 436, 438, 462, 465, 469, 472, 473, 474, 475, 476, 477, 479, 486, 487, 488, 490, 493
Sustainability Metrics, 371
Synthetic Materials, 396

T

Temperature Resistance, 122, 162
Tensile Strength, 22, 40, 45, 48, 122, 125, 126, 132, 133, 136, 138, 140, 141, 143, 153, 169, 176, 186, 202, 229, 230, 231, 232, 233, 235, 243, 246, 247, 249, 257, 259, 260, 396, 406, 407
Thermal Analysis, 4, 52, 108, 110, 112, 250, 251, 256, 257, 258, 259, 260, 320, 404, 406, 480
Thermal Barrier Coatings, 30, 75, 410
Thermal Conductivity, 21, 22, 25, 26, 27, 32, 35, 50, 51, 52, 68, 70, 125, 131, 132, 133, 134, 138, 140, 186, 215, 216, 308, 320, 330, 391, 409, 410, 415, 441
Thermal Expansion, 33, 50, 52, 128, 144, 258, 281, 365, 366, 411, 414
Thermal Management, 11, 70, 126, 127, 187, 322, 379, 381, 410, 435
Thermal Stresses, 160, 445
Thermodynamic Properties, 8, 314, 330, 346, 347, 348, 349

Thermosetting Polymers, 150, 195, 255
Thin Films, 40, 76, 78, 79, 81, 82, 86, 90, 93, 94, 100, 105, 107, 114, 180, 181, 281, 323, 333, 342, 392, 395, 406, 411, 412, 417, 418, 459, 488
Topological Insulators, 4, 66, 387, 388, 389, 390, 392, 393, 395, 402, 403, 434, 490
Toughness, 2, 22, 23, 33, 45, 49, 51, 64, 65, 66, 67, 68, 69, 71, 72, 73, 74, 108, 121, 122, 123, 128, 129, 136, 140, 158, 186, 229, 326, 375, 396, 397, 406, 408, 467
Transparent Materials, 406
Tribological Properties, 446
Tribology, 446

V

Vapor Deposition, 21, 127, 176, 182, 220, 291, 323, 325, 363, 392
Vibration Analysis, 417
Viscosity, 121, 123, 148, 193, 381

W

Wear Resistance, 47, 69, 70, 125, 126, 127, 128, 129, 140, 141, 142, 162, 222

X

X-Ray Diffraction (XRD), 75, 76, 80, 220, 221, 258

Y

Yield Strength, 66, 128, 229, 231, 232, 233, 234, 235, 237, 238, 239, 248, 346, 350, 357, 358, 360, 406, 407, 415

www.ingramcontent.com/pod-product-compliance
Lightning Source LLC
Chambersburg PA
CBHW080409270326
41929CB00018B/2957